cpo science

A member of
School Specialty
Science

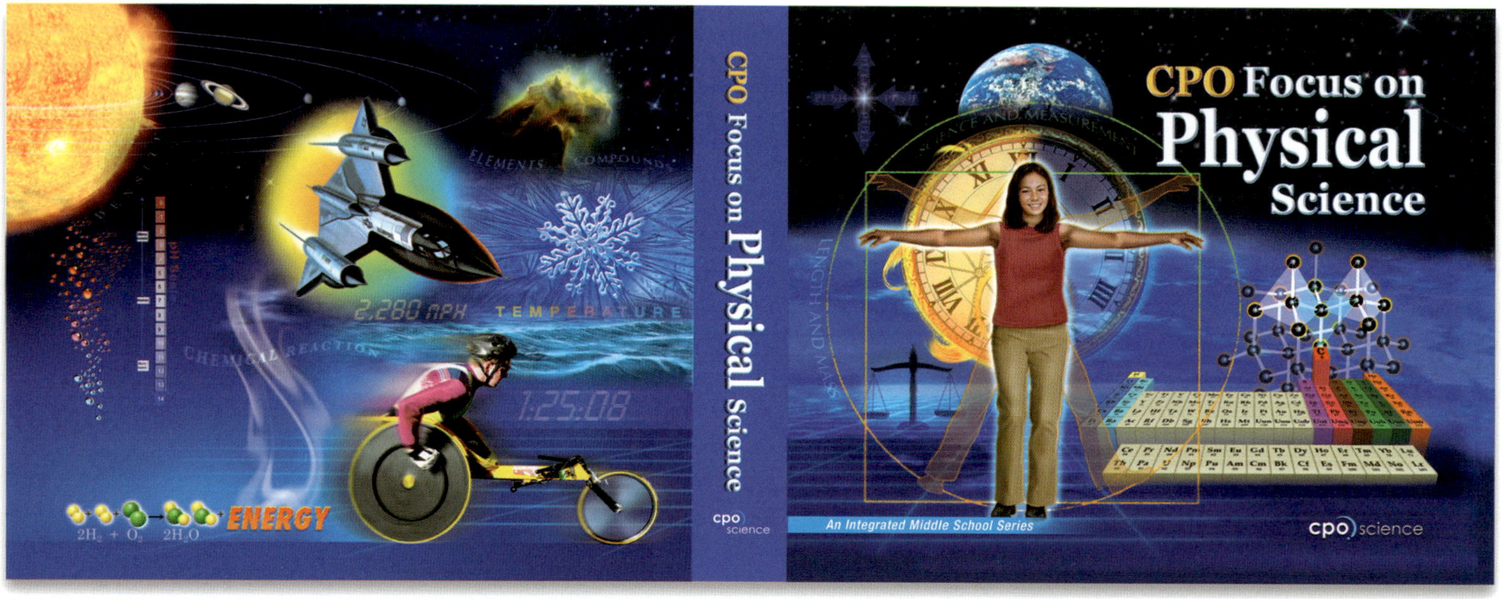

CPO Focus on Physical Science
First Edition

ISBN-10: 1-58892-259-6
ISBN-13: 978-1-58892-259-5
6 7 8 9 10 11 - QWE- 12 11 10 09 08

CPO Science
80 Northwest Boulevard
Nashua, New Hampshire 03063
(866)588-6951
http://www.cposcience.com
Printed and Bound in the United States of America

Credits

Writers

Thomas C. Hsu, Ph.D. – Author
Nationally recognized innovator in science and math education and the founder of CPO Science. He holds a doctorate in applied plasma physics from the Massachusetts Institute of Technology, and has taught students across the nation at the elementary, secondary, and college level. Tom has worked with more than 12,000 teachers and administrators at grades K-12, and he is well known as a consultant, workshop leader, and developer of curriculum and equipment for inquiry-based learning in science and math.

Stacy Kissel
B.S., Civil and Environmental Engineering, Carnegie Mellon University; M.Ed., Physics Education, Boston College.

Has nine years' experience teaching physics, math, and integrated science at Brookline (Mass.) High School. Stacy has been a curriculum developer with CPO Science for the five years.

Patsy Eldridge
B.S., Biology, Grove City College; M.Ed., Tufts University.

Experienced high school physical science teacher and national hands-on science trainer and presenter. As an adjunct professor for Endicott College in Beverly, MA, and the College of Charleston, developed content-intensive physical science courses for pre- and in-service educators. Technical background in the medical device field. Patsy has developed curriculum and training materials with CPO Science for six years.

Sylvia Gutman – Content review
B.S., Science Education, University of New Orleans; M.S., Curriculum and Instruction, University of San Diego.

Teacher of the Year Award from the Lake Elsinore (Calif.) Unified School District and the 2001 California Science Teachers Association Distinguished Science Award for Middle School.

Jodye Selco, Ph.D. – Content review

Professor, Center for Education and Equity in Math, Science, and Technology, California State Polytechnic University, Pomona, Calif.

Kelly McAlister - Content review
B.S., Biology, M.Ed., Science Education.

Eighth-grade team leader and mentor teacher at Gage Middle School in Riverside, California. Kelly has taught life and physical science and served as department chair.

Michael Vela – Content review

Ph.D., Inorganic chemistry, Brandeis University.

Teaches 10th- and 11th-grade chemistry at Concord-Carlisle High School in Concord, Massachusetts. Taught 11th-grade chemistry at Lexington High School, Massachusetts.

Melissa Vela – Content review
B.A., Earth and Environmental Science, Lehigh University; M.S., Agricultural and Biological Engineering, Cornell University; M.Ed., Curriculum/Instruction, Boston College.

Melissa has taught six years of 9th grade earth and space science at Lexington High School (Lexington, MA). She also taught 6th grade physical science and 8th grade algebra at Weston Middle School (Weston, MA).

Senior Editor

Lynda Pennell – Executive Vice President
B.A., English; M.Ed., Administration, Reading Disabilities, Northeastern University; CAGS Media, University of Massachusetts, Boston.

Nationally known for high school restructuring and integrating academic and career education. Served as director at an urban school for five years and has 17 years of teaching/administrative experience in Boston public schools. Lynda has led development at CPO Science for six years.

Editorial Consultants

Julie Dalton
B.A., English; M.S. English education; taught English in middle and high school.

Has worked with the CPO Science staff as our primary copy editor. Reviews all text material and informs us of proper conventions and helps in maintaining the consistency of our text. Julie is a copy editor for *The Boston Globe* and has been a journalist for over 25 years.

Art and Illustration

Polly Crisman – Designer, illustrator
B.F.A., University of New Hampshire.

Graphic artist with expertise in advertising and marketing design, freelance illustrating, and caricature art. Polly is the CPO Science primary book illustrator and manages all files.

Bruce Holloway – Cover designs
Pratt Institute, N.Y.; Boston Museum School of Fine Arts.

Created all CPO Science book covers and many of the CPO specific designs. Expertise in product design, advertising, and three-dimensional exhibit design. Commissioned for the New Hampshire Duck Stamp for 1999 and 2003.

Jesse Van Valkenburgh - Designer, illustrator
B.F.A., Illustration, Rochester Institute of Technology.

Has worked in prepress film production and design. Jesse has worked at PC Connection in Merrimack, N.H., as a graphic designer doing catalog and direct mailing design, logo design, and illustration.

James Travers
A.A.S., Business and Commercial Art, Akron University.

Has held positions as a graphic designer and art development manager. He is a commissioned artist.

Connections

Catherine C. Reed
B.S., Secondary Education, Akron University; M.Ed., Administration and Curriculum, Baldwin-Wallace College.

Taught middle school science and worked as a middle school specialist for 20 years. Catherine's work has included curriculum and professional development in the area of inquiry learning.

Sharon O. Faulkner is an educational freelance writer who lives in Andover, Mass.

John K. Manos is an educational freelance writer who lives in Evanston, Ill.

Laura J. Tierney is a freelance writer living in Chelmsford, Mass.

Laine Ives
B.A., Gordon College; graduate course work at Cornell University's Shoals Marine Laboratory and Wheelock College.

Taught high school English overseas and environmental education at a middle school in New England.

Assessment

Mary Ann Erickson
B.S, Naval Architecture and Marine Engineering, Massachusetts Institute of Technology.

Ran a technical writing consulting business, writing process control manuals for water treatment plants, software design documentation for simulation software, and operator manuals for mining equipment.

David H. Bliss
B.S., Science, Cornell University; M.Ed., Zoology

Has taught for 39 years in the science field: biology, chemistry, earth science, and physics. David has been science department chairman of Mohawk Central School District in Mohawk, N.Y.

Kelly A. Story
B.S., Chemistry, Gordon College; M.S., Chemistry, University of Massachusetts, Lowell.

Taught chemistry and maintains a position as lab instructor at Gordon College, Wenham, Mass.

Erik Benton - Principal investigations editor
B.F.A., University of Massachusetts, physics minor.

Taught for eight years in public and private schools, focusing on inquiry and experiential learning.

Equipment Design

Thomas Narro – Senior Vice President
B.S., Mechanical Engineering, Rensselaer Polytechnic Institute.

Accomplished design and manufacturing engineer; experienced consultant in corporate reengineering and industrial-environmental acoustics.

Danielle Dzurik
Bachelor of Industrial Design, Auburn University.

Focused on product development, new product design, and improving older designs.

Material Support

Michael Grady – Tech services coordinator
A.A., Arranging and Performing, Berklee College of Music; Medical Technician Diploma, Middlesex Community College.

Professional musician for more than 20 years, with 10 years experience in customer service fields.

Kathryn Gavin – Quality Control and Purchasing Manager

Responsible for quality control and purchasing and works with product engineering on all new development. Kerry has been assuring total quality of CPO Science equipment for ten years.

Lisa LaChance – Senior materials specialist
A.S., Accounting.

Lisa evaluates samples to ensure that materials meet project requirements, developing and managing the release of specifications. She also evaluates changes to current products to enhance them and/or reduce manufacturing costs.

Reviewers

Nancy Joplin
English-language arts department chair
Ray Wiltsey Middle School, Ontario, Calif.

Jodye Selco, Ph.D.
Professor, Center for Education and Equity in Math, Science, and Technology
California State Polytechnic University, Pomona, Calif.

Brian E. Goodrow
Physical science teacher
Apple Valley Middle School, Apple Valley, Calif.

Philip L. Hunter
Science department chair
Johnson Middle School, Westminster, Calif.

Bianca N. McRae
Science teacher, department chair
Menifee Valley Middle School, Menifee, Calif.

Tia L. Shields
Life science/health and English language learning teacher
Nicolas Junior High School, Fullerton, Calif.

Kelly McAllister
Science teacher
Gage Middle School, Riverside, Calif.

Brad Joplin
Science teacher
Ray Wiltsey Middle School, Ontario, Calif.

Tony Heinzman
Science teacher
Apple Valley Middle School, Apple Valley, Calif.

Sylvia Gutman
Science teacher, department chair
David A. Brown Middle School
Wildomar, Calif.
Lake Elsinore Unified School District

Technical Support

Tracy Morrow – Framemaker specialist, technical editor and trainer
B.A., English, Texas A&M University; M.A., English, Sam Houston State University.

Taught middle school in Klein, Texas, a suburban region outside Houston, for nine years; at Tomball College in Tomball, Texas, for five years; and worked as a technical writer in the oil and gas, airlines, and real estate industries. Tracy offers consulting services and technical training; her expertise is in the editing program Framemaker.

Special Thanks for Photos

Dr. Geerat Vermeij
Professor of Marine Ecology and Paleoecology, University of California, Davis

Mary Doval Graziose
University of California, Davis

Curtis Ebbesmeyer
Beachcombers' and Oceanographers' International Association

Jim White
Beachcombers' and Oceanographers' International Association

Dr. Adam Dziewonski
Harvard University

Sarah Herve
John Day Fossil Beds National Monument

Joan Buhrman
American Society of Civil Engineers

John M. Watson
USGS

David A. Abel
Freelance photographer

On each page of the student text, there are aids to help you find information, understand concepts, and answer questions. The following introduction includes sample pages with indicators that point out the page contents and reading aids.

Unit Pages and Chapter Pages

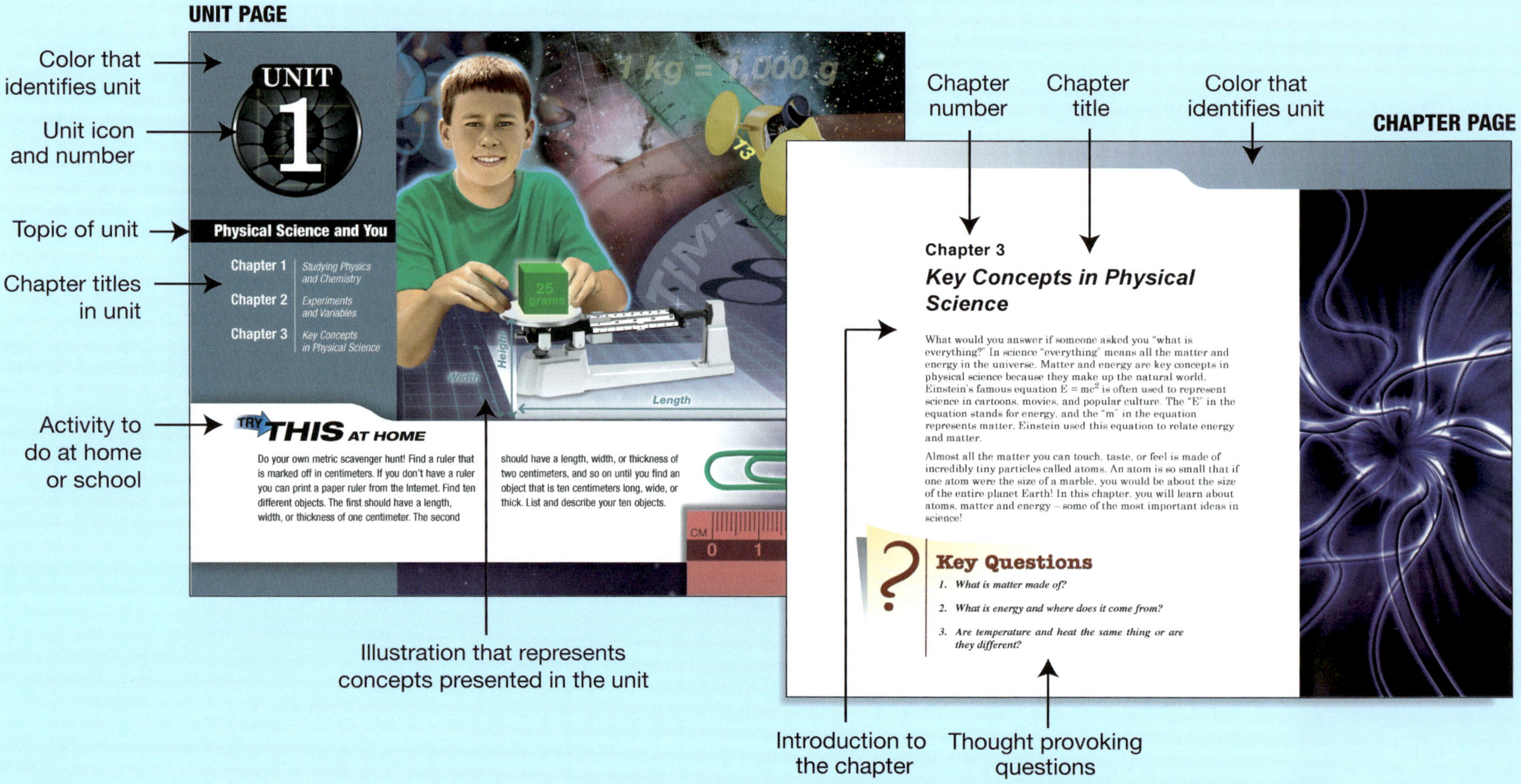

Student Text Pages

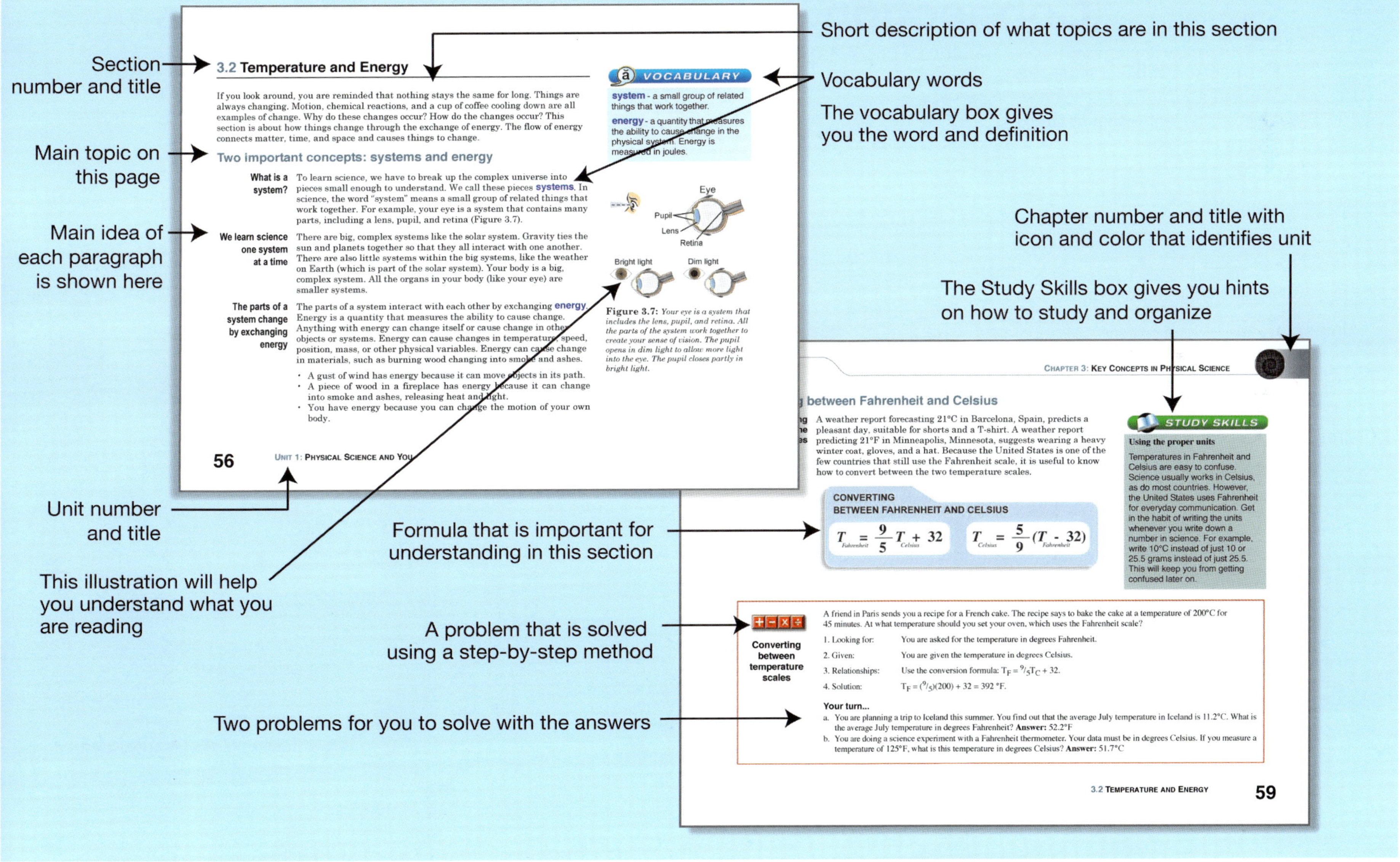

3.2 Temperature and Energy

If you look around, you are reminded that nothing stays the same for long. Things are always changing. Motion, chemical reactions, and a cup of coffee cooling down are all examples of change. Why do these changes occur? How do the changes occur? This section is about how things change through the exchange of energy. The flow of energy connects matter, time, and space and causes things to change.

Two important concepts: systems and energy

What is a system? To learn science, we have to break up the complex universe into pieces small enough to understand. We call these pieces **systems**. In science, the word "system" means a small group of related things that work together. For example, your eye is a system that contains many parts, including a lens, pupil, and retina (Figure 3.7).

We learn science one system at a time There are big, complex systems like the solar system. Gravity ties the sun and planets together so that they all interact with one another. There are also little systems within the big systems, like the weather on Earth (which is part of the solar system). Your body is a big, complex system. All the organs in your body (like your eye) are smaller systems.

The parts of a system change by exchanging energy The parts of a system interact with each other by exchanging **energy**. Energy is a quantity that measures the ability to cause change. Anything with energy can change itself or cause change in other objects or systems. Energy can cause changes in temperature, speed, position, mass, or other physical variables. Energy can cause change in materials, such as burning wood changing into smoke and ashes.

- A gust of wind has energy because it can move objects in its path.
- A piece of wood in a fireplace has energy because it can change into smoke and ashes, releasing heat and light.
- You have energy because you can change the motion of your own body.

VOCABULARY

system - a small group of related things that work together.

energy - a quantity that measures the ability to cause change in the physical system. Energy is measured in joules.

Figure 3.7: *Your eye is a system that includes the lens, pupil, and retina. All the parts of the system work together to create your sense of vision. The pupil opens in dim light to allow more light into the eye. The pupil closes partly in bright light.*

56 UNIT 1: PHYSICAL SCIENCE AND YOU

CHAPTER 3: KEY CONCEPTS IN PHYSICAL SCIENCE

…g between Fahrenheit and Celsius

A weather report forecasting 21°C in Barcelona, Spain, predicts a pleasant day, suitable for shorts and a T-shirt. A weather report predicting 21°F in Minneapolis, Minnesota, suggests wearing a heavy winter coat, gloves, and a hat. Because the United States is one of the few countries that still use the Fahrenheit scale, it is useful to know how to convert between the two temperature scales.

CONVERTING BETWEEN FAHRENHEIT AND CELSIUS

$$T_{Fahrenheit} = \frac{9}{5}T_{Celsius} + 32 \qquad T_{Celsius} = \frac{5}{9}(T_{Fahrenheit} - 32)$$

STUDY SKILLS

Using the proper units

Temperatures in Fahrenheit and Celsius are easy to confuse. Science usually works in Celsius, as do most countries. However, the United States uses Fahrenheit for everyday communication. Get in the habit of writing the units whenever you write down a number in science. For example, write 10°C instead of just 10 or 25.5 grams instead of just 25.5. This will keep you from getting confused later on.

Converting between temperature scales

A friend in Paris sends you a recipe for a French cake. The recipe says to bake the cake at a temperature of 200°C for 45 minutes. At what temperature should you set your oven, which uses the Fahrenheit scale?

1. Looking for: You are asked for the temperature in degrees Fahrenheit.
2. Given: You are given the temperature in degrees Celsius.
3. Relationships: Use the conversion formula: $T_F = {}^9/_5 T_C + 32$.
4. Solution: $T_F = ({}^9/_5)(200) + 32 = 392$ °F.

Your turn...

a. You are planning a trip to Iceland this summer. You find out that the average July temperature in Iceland is 11.2°C. What is the average July temperature in degrees Fahrenheit? **Answer:** 52.2°F

b. You are doing a science experiment with a Fahrenheit thermometer. Your data must be in degrees Celsius. If you measure a temperature of 125°F, what is this temperature in degrees Celsius? **Answer:** 51.7°C

3.2 TEMPERATURE AND ENERGY 59

Connection Pages and Activity Pages

CONNECTION

The **Connection** is like a magazine article about an interesting science fact. There is a Connection at the end of each chapter.

Title of the Connection

Main idea heading

Photos and illustrations to support your understanding

Unit color and Chapter number

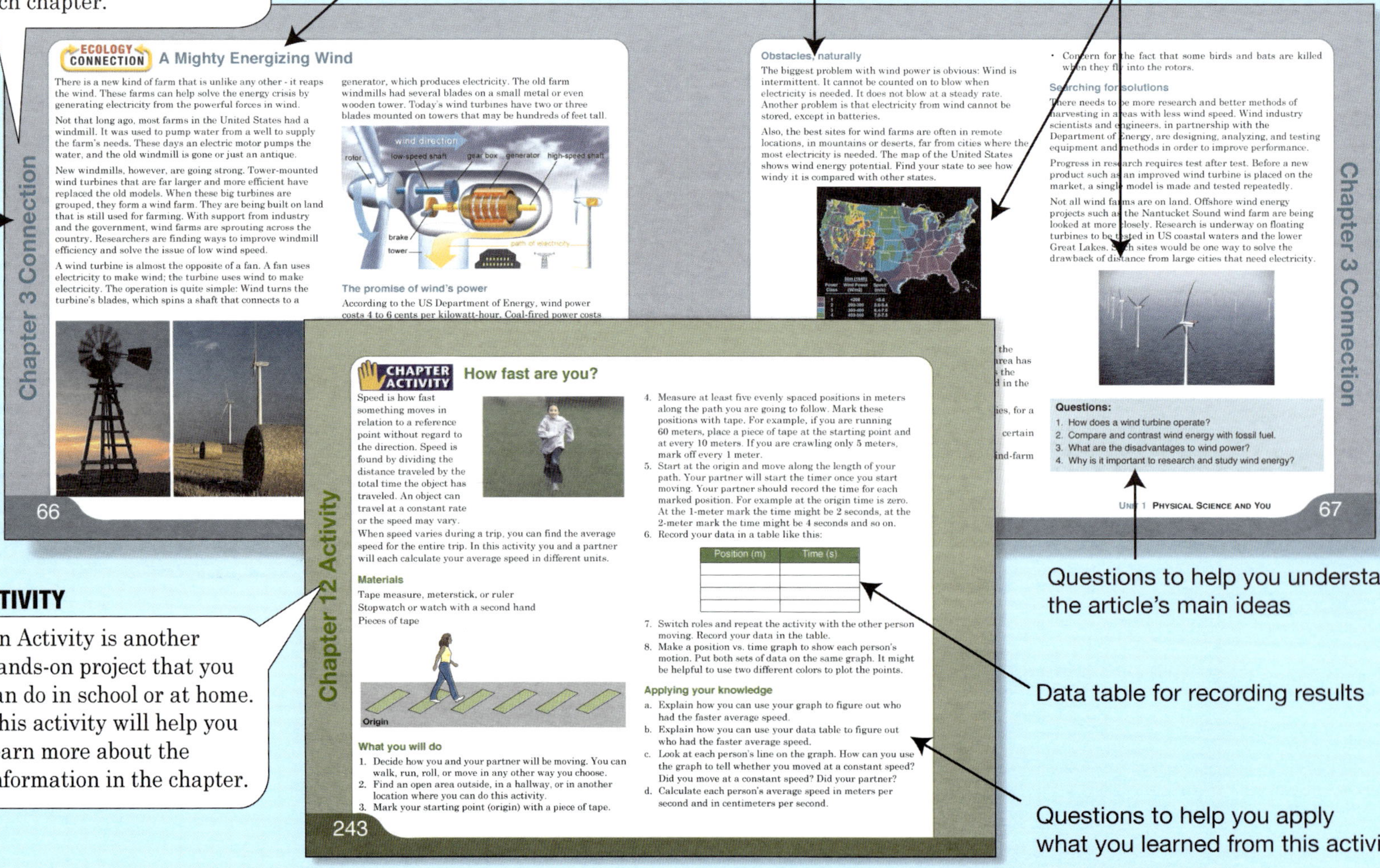

Questions to help you understand the article's main ideas

ACTIVITY

An Activity is another hands-on project that you can do in school or at home. This activity will help you learn more about the information in the chapter.

Data table for recording results

Questions to help you apply what you learned from this activity

Assessment Pages

SECTION REVIEW

By answering these questions, you will have a quick check on what you remembered from the section.

CHAPTER 3: KEY CONCEPTS IN PHYSICAL SCIENCE

3.2 Section Review

1. Write a paragraph about a system inside your home or school building. Describe what the system does as a whole. Describe at least three parts of the system. For each part, describe how it contributes to the function of the whole system.
2. Scientists would like to understand many things that are large and complex, like the ecology of Earth. Scientists divide complex things into smaller groups called systems because:
   a. It is easier to understand a small group than a large complex thing.
   b. There is not enough money to study the entire complex thing.
3. Which is the higher temperature: 30°C or 60°F?
4. A cook warms up 1 bowl of soup from 20°C to 60°C (Figure 3.15). Another cook warms up 10 identical bowls of soup from 20°C to 60°C. Which of the following statements is FALSE?
   a. Both cooks use the same amount of heat (energy) because the temperatures are the same.
   b. Both cooks use different amounts of heat (energy) even though the temperatures are the same.
5. Describe the flow of energy as a cup of hot coffee cools down as it sits on a table.
6. True or false: If the same amount of heat is applied to water, steel, and wood, the temperature of each one will rise by the same amount.
7. Imagine you are the teacher of a science class. A student brings in a newspaper article that claims the world will run out of energy by the year 2050 because all the oil will be pumped out of the planet. The student is confused because she has learned in science that energy can never be created or destroyed. How would you explain to her what "running out of energy" means in the article.

CHALLENGE

Research what is going on in your community around energy conservation. Write about a project that is anticipated to save energy. How much energy might be saved?

CHALLENGE

Every month your family pays an electric bill for energy you have used. Research the cost of electricity in your area. How much does it cost for 1 million joules? This is the amount of energy used by a single electric light bulb in three hours.

1 bowl of soup

Do they take th… to reach the…

10 b…

Figure 3.15:

3.2 TEMPERATURE AND E…

This gives you an interesting way to learn more about information in the section

Graphs, diagrams, or charts will help you in answering questions

CHAPTER TEST

These questions are answered after reading the chapter.

This part of the review asks you to fill in sentences with vocabulary words

This tells you where to find the information

Some questions ask you to write out short answers or solve a problem

Chapter 3 Assessment

Vocabulary

Select the correct term to complete the sentences.

kilogram	gram	atom
element	compound	system
energy	Calorie	newton
heat	thermometer	Fahrenheit
Celsius	force	conservation of energy

Section 3.1

1. The basic SI unit of mass, approximately equal to 1000 grains of rice, is the ____.
2. A pure substance that cannot be broken down into other substances is known as a(n) ____.
3. A combination of elements that most often has properties different than the elements from which it is made is called a(n) ____.
4. The smallest particle of matter which retains the identity of the element it comes from is called a(n) ____.
5. An amount of mass equal to one-thousandth of a kilogram is a(n) ____.

Section 3.2

6. A small group of related things that work together may be called a ____.
7. A quantity that measures the ability to cause change in a physical system in units of joules is known as ____.
8. A unit of energy larger than a joule used to measure the energy available in food is the ____.
9. An SI unit used to measure force, equal to less than one-quarter of a pound, is the ____.
10. A natural law that says energy cannot be created or destroyed is the law of ____.
11. An action, measured in newtons, that has the ability to transer energy or change motion is called a(n) ____.
12. A(n) ____ is an instrument used to measure temperature.
13. The temperature scale on which the freezing point of water is 32 degrees and boiling point of water is 212 degrees is the ____ scale.
14. The temperature scale on which the freezing point of water is 0 degrees and boiling point of water is 100 degrees is the ____ scale.
15. When thermal energy flows from hot to cold it is called ____.

Concepts

Section 3.1

1. Scientists find it useful to use both grams and kilograms for measuring mass. Why is it necessary to have two different SI units of mass?
2. Mass describes:
   a. an object's size.
   b. the amount of matter in an object.
   c. the type of elements in an object.
3. What laboratory instrument is most often used to measure mass?
4. Identify the following substances as an *element* or a *compound*:
   a. ____ table salt (sodium chloride)
   b. ____ oxygen gas
   c. ____ rust
   d. ____ iron

CHAPTER 3 PHYSICAL SCIENCE AND YOU 69

Table of Contents

UNIT 1 Physical Science and You

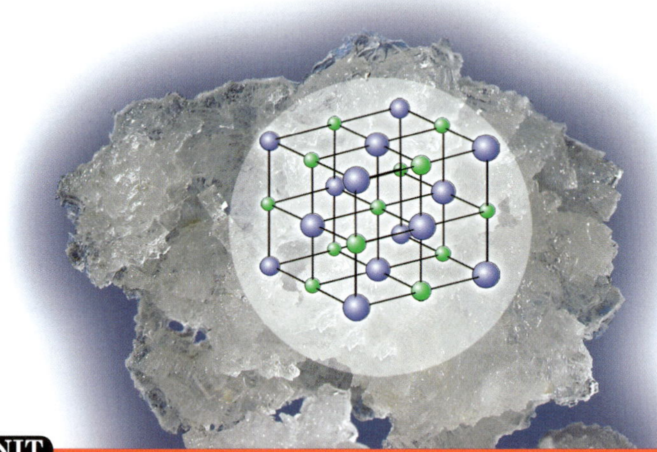

UNIT 2 Properties of Matter

UNIT 3 Atoms and the Periodic Table

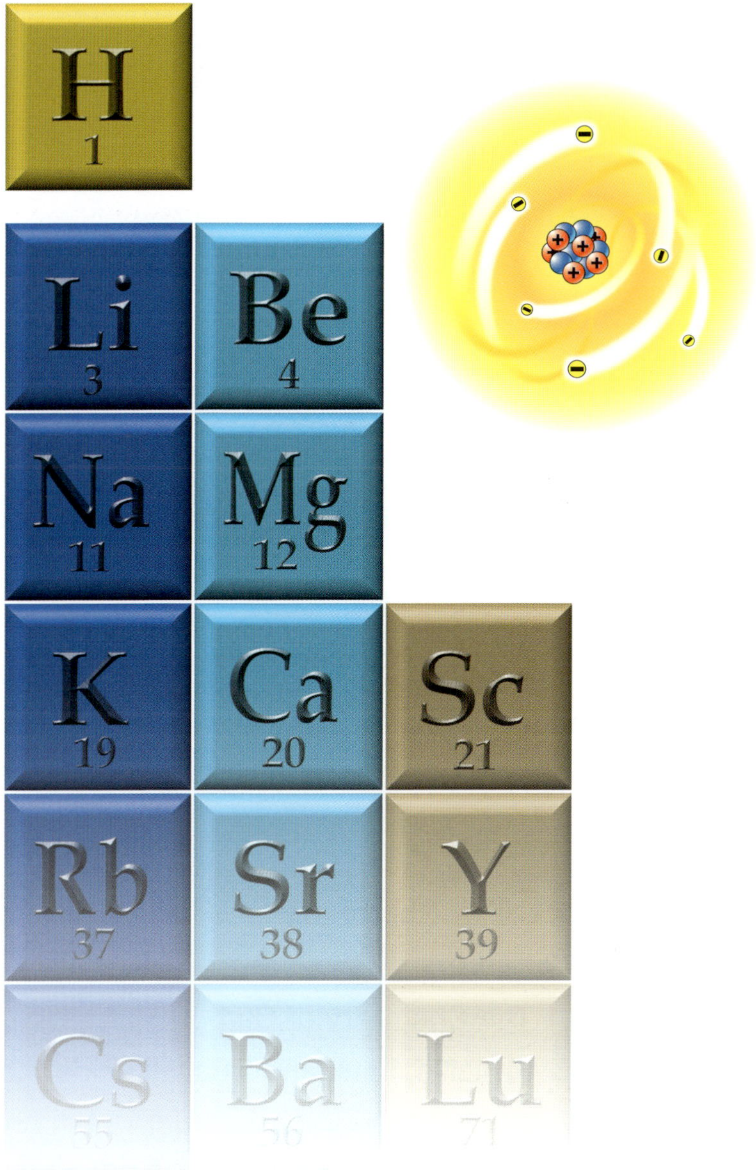

Table of Contents

UNIT 4 Matter and Change

UNIT 5 Motion and Force

UNIT 6

Astronomy

UNIT 1

Physical Science and You

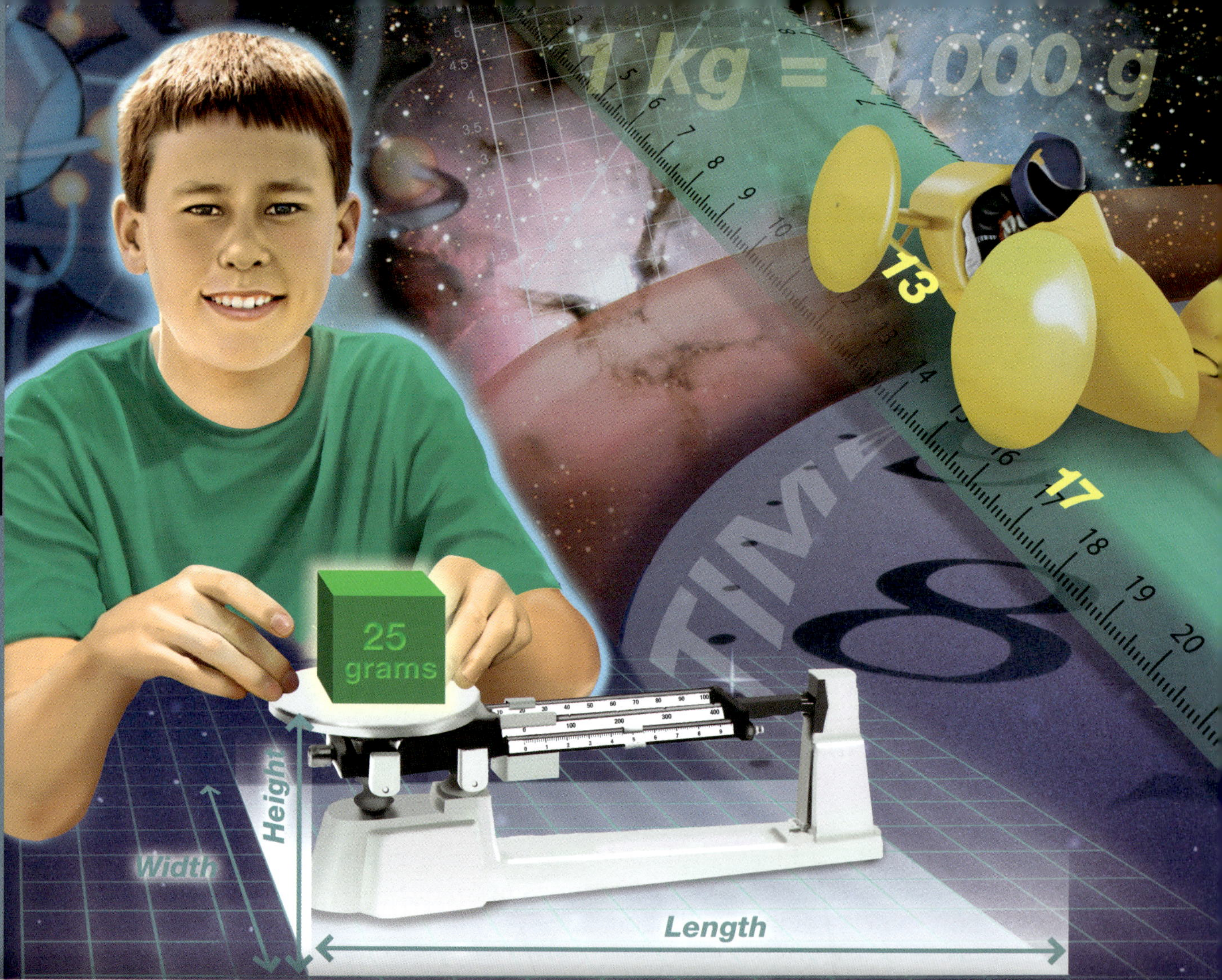

TRY THIS AT HOME

Do your own metric scavenger hunt! Find a ruler that is marked off in centimeters. If you don't have a ruler you can print a paper ruler from the Internet. Find ten different objects. The first should have a length, width, or thickness of one centimeter. The second should have a length, width, or thickness of two centimeters, and so on until you find an object that is ten centimeters long, wide, or thick. List and describe your ten objects.

Chapter 1

Studying Physics and Chemistry

Imagine humans when they first had fire, when they first made wheels, when they first established where they were going by looking at the stars. Today, when you get off the school bus, go home and turn on the lights, put a CD on the player, you are in a way as far from your 19th-century ancestors — from your own relatives who lived like everyone, without electricity and automobiles, without radio and telephones — as you are from those ancient groups from early human history. How did the things we consider basic — heat, light, navigation, transportation, entertainment — and the things your great great-grandparents before you considered basic, ever come to be?

The answers all touch on physics and chemistry. Physical science, the subject of this book, includes physics and chemistry. Physics tells us how and why things move. Chemistry tells us how to make things and what things are made of. Even the most brilliant scientists, the most imaginative inventors, are born not knowing any science at all. Science has been *learned* over many thousands of years. This chapter is also about how scientists use inquiry to learn how the world works.

Key Questions

1. ***What is physical science and why is it important to learn?***

2. ***How did people figure out the scientific knowledge that we know?***

1.1 The Physical Science in Your Life

In Unit One we will look at the 'big picture' of physical science. We will also review how people learned science in the first place. Each Unit after the first will then focus on one important area of physical science in more detail.

The physics of a car

Cars changed the way people live

Few inventions have changed the world as much as the car. Before cars, it was a three-week, dangerous adventure to travel 500 miles between San Diego and San Francisco. Today, you might drive the same trip in a day to visit your relatives. Let's take a look at the physics and chemistry in an ordinary car.

Forces are described by physics

When you push on the gas pedal at a stoplight the car starts moving. The car starts moving because the engine causes a force that pushes the car forward. If there were no force, the car would not start moving. Forces are described by physics. Physics tells us how much force it takes to get the car moving. Physics also tells us how much force it takes to make the car turn or stop.

Mass is also described by physics

If you apply the same amount of force to a small sports car and to a big garbage truck, which one will start moving faster? Of course you know the answer (Figure 1.1). The sports car starts moving faster because a sports car has less **mass** than a garbage truck. The more mass you have, the more force it takes to get you moving. The reverse is also true. If you apply the same force to two objects, the one with less mass will start moving faster. The relationship between mass and motion is described by physics.

VOCABULARY

mass - the amount of "stuff" (matter) an object contains. More about mass in chapter two.

Objects with greater mass respond more slowly to the same force.

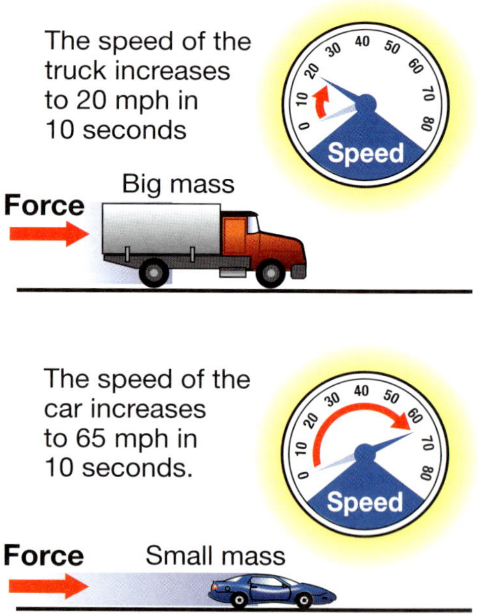

Figure 1.1: *When the same amount of force is applied to a sports car and a garbage truck, the sports car picks up speed much more rapidly because it has less mass.*

The chemistry of a car

Chemistry describes the different forms of matter

Look at a car and describe what the car is made of. At the most basic level a scientist might say a car is made of **matter**. Matter is everything that has mass and takes up space. This book is matter. You are matter. Even the air is matter. At the next level of detail, a car contains many different kinds of matter, such as steel, aluminum, plastic, and rubber. Chemistry is the science that concerns the different kinds of matter, how different kinds of matter are created and how matter can be changed from one kind to another.

Some of the many kinds of matter in a car

Matter appears as solid, liquid, and gas

The matter in a car comes in three fundamental **phases**: called solid liquid, and gas (Figure 1.2). Solid matter (like ice) is stiff, holds its shape, and may be strong. The frame of the car is made of solid steel. Liquid matter (like water) flows and does not hold its shape. Gasoline, oil, and windshield washer fluid are all liquids that are used in a car. Matter that is a gas (like air) can expand and contract. The air in the tires is a gas, as is the exhaust coming out the tailpipe.

Burning gasoline is a chemical reaction

Virtually all of today's cars and trucks run on gasoline. You put gasoline in the tank and it is *burned* in the engine. The burning gasoline releases energy that makes the engine turn and makes the car work. Burning is a **chemical change** that turns gasoline and oxygen into carbon dioxide, water and other kinds of matter. Chemistry describes what gasoline is, and how gasoline combines with oxygen in a chemical reaction that releases energy.

VOCABULARY

matter - everything that has mass and takes up space.

phases of matter - the different forms matter can take; commonly occur as solid, liquid, or gas.

chemical change - a chemical change transforms one kind of matter into another kind (or several) which may have different properties.

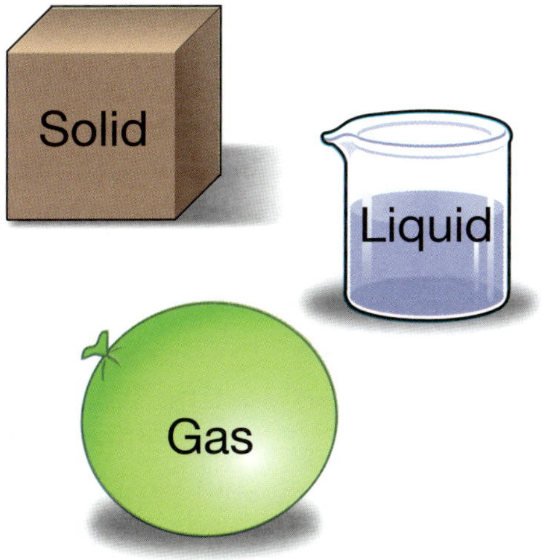

Figure 1.2: *Matter appears in three different phases that we call solid, liquid, and gas.*

Physical science and sea otters

Physical science and living things

A sea otter is a small furry mammal that spends all of its life in the water. Sea otters live in the giant kelp forests off the California coast. Like all living creatures, sea otters depend on physics and chemistry to exist.

The force of water friction

In order to swim, a sea otter has a streamlined body and webbed feet. The streamlined body shape reduces the friction of water against the sea otter's skin as it swims (Figure 1.3). The lower the force of water friction, the faster the sea otter can swim. The otter's webbed feet work in the opposite way. Having webs between its toes increases the force of water friction against the sea otter's outstretched feet. This makes the sea otter more effective in pushing against the water to swim. Friction and forces are described by physics.

Why a sea otter floats instead of sinking

To find the shellfish they eat, sea otters dive down to the sea floor at the bottom of the kelp forest. Once they get a tasty clam or scallop, the otter swims back up to the surface and floats there to eat it. When it is floating, the sea otter is not swimming, yet it does not sink! A sea otter floats because its body is less *dense* than water. Density is a property of all matter, not just sea otters. Objects that are less dense (like the sea otter) float in liquids that are more dense (like water).

Why living organisms need food to survive

Like most living organisms, a sea otter eats to survive. What is eating? Why do living organisms need food? The answer is provided by chemistry (Figure 1.4). Living organisms are made of special chemicals known as proteins, fats and carbohydrates. The sea otter needs these chemicals to provide energy to keep its body warm. The otter also needs these chemicals to grow. The otter gets the chemicals it needs by eating shellfish. *Biochemistry* is a branch of chemistry that explores exactly how plants and animals use chemicals and energy.

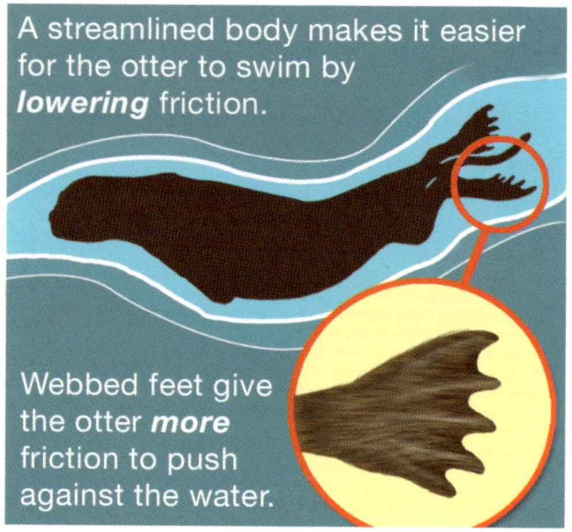

Figure 1.3: *The otter's streamlined body reduces the force of water friction when it swims. The otters webbed feet increase the force of water friction against its feet, making the otter more effective at pushing against the water.*

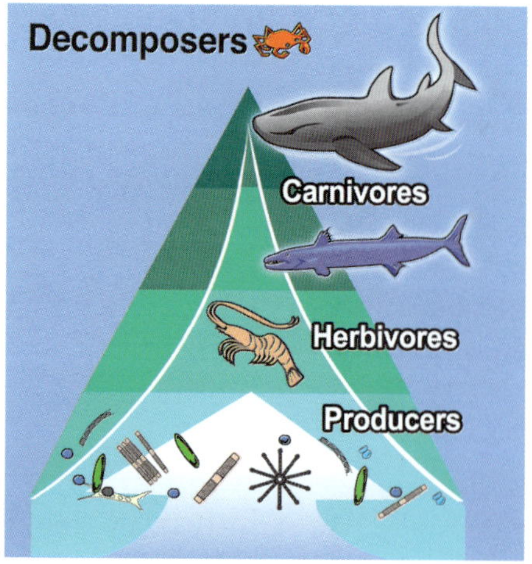

Figure 1.4: *Chemistry describes how living organisms use the food they eat.*

The sun is the source of energy for Earth

The sun and energy

Both living creatures and human technology derive virtually all of their energy from the sun. Without the Sun's energy, Earth would be a cold icy place with a temperature of -273 degrees Celsius. As well as warming the planet, the Sun's energy drives the entire food chain (Figure 1.5). Plants store the energy in carbohydrates, like sugar. Animals eat the plants to get energy. Other animals eat *those* animals for their energy. It all starts with the sun.

Understanding what energy really is

If energy is so important, what is it? The answer is slippery but essential to physics and chemistry. Energy measures the ability for things to change. Energy is exchanged when *anything* changes. Nothing changes when no energy is exchanged. For example, when you press its accelerator, the speed of a car changes. The change comes from the exchange of energy between the gasoline in the engine and the motion of the whole car. The sea otter's body is warmer inside than the surrounding ocean water. Because the otter is warmer, heat energy flows from the otter into the ocean water. The otter needs a constant supply of energy from its food to replenish the energy exchanged with the water. Energy is such an important idea that you will find examples of energy in every single Unit of this book!

Life on Mars and other planets

A very important question in science today is whether there is life on other planets, such as Mars. Mars is farther from the sun than Earth. For this reason, Mars receives less energy from the sun than does Earth. In fact, the average temperature on Mars is well below the freezing point of water. Can life exist on Mars? Recent research suggests that can. Scientists have found living bacteria in the Antarctic ice living at a temperature colder than the average temperature of Mars.

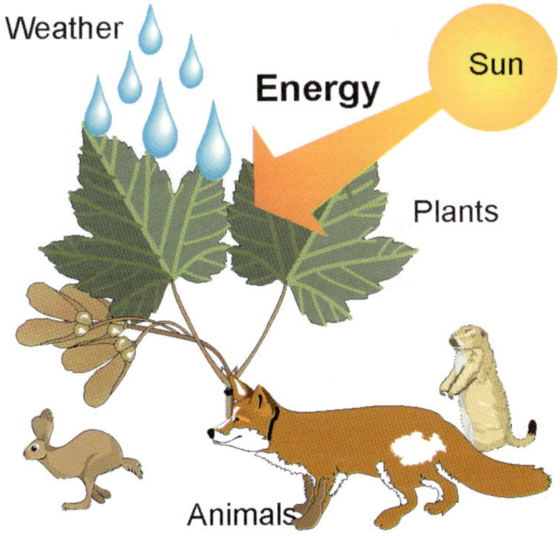

Figure 1.5: *The flow of energy from the sun supports all living things on Earth.*

CHALLENGE

The planet Venus is closer to the sun than Earth. Should this make Venus warmer or colder than Earth? Research your answer to see what scientists think Venus is like on the planet surface.

1.1 Section Review

Based on what you learned in this section, decide which of the following statements are *true* and which are *false*.

1. Understanding force and mass are part of the subject of physical science.
2. Physical science applies to mechanical inventions such as a car but not to living creatures.

Is this true?

Physical science applies

Physical science does not apply

3. Steel is matter because it is solid but water is not matter because water can flow and change its shape.
4. Matter can take the form of solid, liquid, or gas.
5. A sea otter floats because it eats a special kind of shellfish that it gets from the ocean floor.
6. The Earth receives a significant amount of its energy from the Moon and the other planets.
7. A car increases its speed primarily because energy is exchanged between the tires and the road. (careful, this is tricky to answer!)
8. Mars is warmer than Earth because it receives more energy from the sun than Earth does.
9. Proteins, fats, and carbohydrates are the three phases of matter in living creatures.
10. Gasoline is necessary for most cars because it is used in a chemical reaction inside the engine.

STUDY SKILLS

Using new words

Write two sentences about something you know using each new glossary word. Using a new word helps you to remember what it means and how it is used. In fact, if you do not use a new word right away you are very likely to forget it within a few hours!

Science assigns precise meanings to many common words

Many words in science are words you already know, like “force” or “energy”. However, science defines these words in very precise ways that may be different from how they are used in everyday conversation. When you see a common word used in science, do not assume it means the same thing as it would in casual conversation.

1.2 Time and Length

In science we often want to know what happens next based on what happened before. The concepts of *next* and *before* involve time. Time plays an essential role in science. We also need a way to describe the size of things from the tiniest bacteria to the entire solar system. In physical science, size is described by length. This section is about the way scientists measure and communicate information about time and length.

Two meanings for time

What time is it? Time has two important meetings (Figure 1.6). One meaning is to identify a particular moment in the past or in the future. For example, saying your 18th birthday party will be on January 1st, 2010 at 2:00 p.m. identifies a particular moment in the future for your party to start. This is the way "time" is usually used in everyday conversation. You may think of this meaning as *historical time*. If you ask, "What time is it?" you usually want to identify a moment in historical time. To answer the question, you would look at a calendar, clock or your watch.

How much time? The second meaning is to describe a *quantity* of time. For example, saying that a class lasts for 45 minutes is identifying a quantity of time; 45 minutes. If you ask, "How much time?" (did something take to occur, for instance), you are looking for a quantity of time. To answer, you need to measure an interval of time that has both a beginning and an end. For example, you might measure how much time has passed between the start of a race and when the first runner crosses the finish line. A quantity of time is often called a time interval. A microwave oven with a built-in clock (Figure 1.7) displays both kinds of time: historical time and time intervals. In physical science, the word "time" will usually mean a time interval instead of historical time.

Figure 1.6: *There are two different ways to understand time.*

Figure 1.7: *A microwave oven can understand time in either mixed units (minutes and seconds) or in a single unit (seconds). Both 1:30 and 0:90 will result in the same cooking time.*

Units for Time

Units for measuring time

You are probably familiar with the units for measuring time: seconds, minutes, hours, days, and years. But you may not know how they relate to each other. The table below gives some useful relationships between units of time.

Time unit	. . . in seconds . .	. . . and in days
1 second	1	0.0001157
1 minute	60	0.00694
1 hour	3,600	0.0417
1 day	86,400	1
1 year	31,557,600	365.25

Figure 1.8: *Digital timers have displays that show time in mixed units.*

SOLVE IT!

Suppose you won a contest that gave you two choices. You could have either:

a) 1 million dollars, or

b) $1 per second for a month.

Which choice is worth more money?

One second

The second (s) is the basic unit of time in both the English and metric systems. One second is about the time it takes to say "thousand." There are 60 seconds in a minute and 3,600 seconds in an hour. The second was originally defined in terms of one day: There are 86,400 seconds in an average day of 24 hours (86,400 = 3,600 x 24).

Mixed units

In everyday life, time is often expressed in mixed units. For example, the timer in Figure 1.8 shows the time for a race as 2 hours, 30 minutes and 45 seconds. This time is in mixed units (hours, minutes, and seconds). People are used to hearing time this way. However, to do scientific calculations, you must convert the time into a single unit.

Converting time to seconds

For most physical science problems, you will need to express time in seconds. When converting to seconds the first thing you do is convert each quantity of hours and minutes to seconds. Then you add up all the seconds to get the total.

For example, 2 hours = 7,200 seconds. 30 minutes = 1,800 seconds. Therefore, 2:30:45 = 7,200 + 1,800 + 45 = 9,045 seconds.

Length and Distance

What is distance? You can think of distance as the amount of space between two points, like the points of a pencil that has been sharpened on both ends. You probably have a good understanding of distance from everyday experiences, like the distance from one house to another, or the distance between Los Angeles and San Francisco. The concept of distance in science is the same, but the actual distances may be much larger and much smaller than anything you normally think about in your everyday life.

Distance is measured in units of length **Distance** is measured in units of length. Some of the commonly used units of length include inches, feet, miles, centimeters, kilometers, and meters. The metric (or SI) system uses millimeters (mm), centimeters (cm), meters (m), and kilometers (km). There are 10 millimeters in a centimeter, 100 centimeters in a meter, and 1,000 meters in a kilometer. The names of units in the metric system are based on multiplying by ten (Table 1.1). Almost all fields of science use metric units because they are easier to work with.

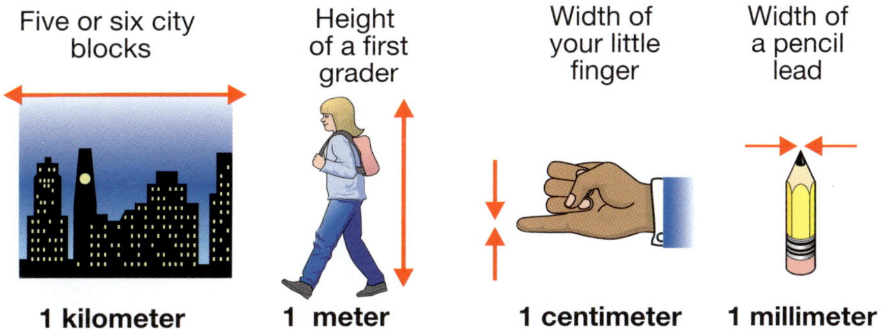

Always include units It is important to always specify which length unit you are using for a measurement. All measurements must have units. Without a unit, a measurement cannot be understood. For example, if you asked someone to walk 10, she would not know how far to go: 10 feet, 10 meters, 10 miles, 10 kilometers are all 10, but the units are different and therefore the distances are also different.

VOCABULARY

distance - the amount of space between two points.

Distance

Distance is the amount of space between two points.

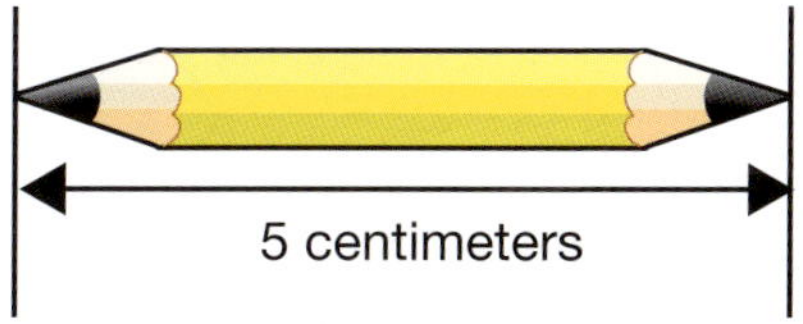

Figure 1.9: *The definition of distance.*

Table 1.1: Metric prefixes

Prefix	Meaning	
giga (G)	1 billion	1,000,000,000
mega (M)	1 million	1,000,000
kilo (k)	1 thousand	1,000
centi (c)	1 one-hundredth	0.01
milli (m)	1 one-thousandth	0.001
micro (μ)	1 one-millionth	0.000001

Measuring length

The meter stick For ordinary lengths you must measure in the laboratory a meter stick is the most convenient tool to use. A meter stick is one meter long and has divisions of millimeters and centimeters. The diagram in Figure 1.10 shows a meter stick along with several different length objects. Can you see how the meter stick is used to measure the length shown for each object?

Microscopic lengths

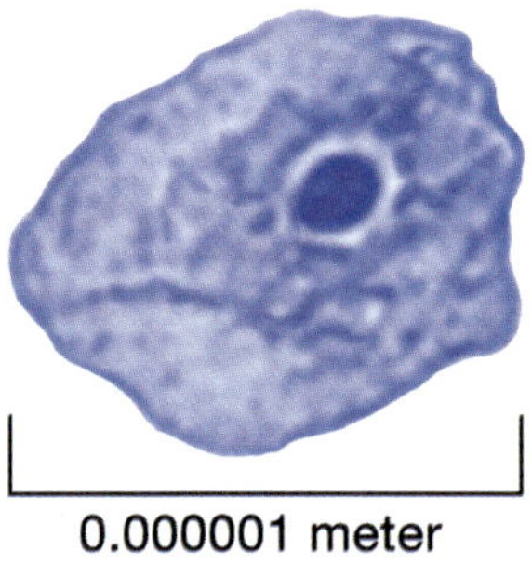

For very small lengths, such as the size of a bacteria, scientists use measuring instruments that fit into a microscope. The micron (μ) is a very tiny unit of length, appropriate to measuring tiny living creatures like bacteria. There are one million microns in a meter. The picture shows a single cell with a micron scale. Can you determine the length of the cell?

Geographic lengths For geographical lengths the most common metric unit is the kilometer. One kilometer is equal to one thousand meters (1 km = 1,000 m). A mile is longer than a kilometer. In fact, one mile is a bit more than one and a half kilometers (1 mile = 1.6 km).

Astronomical lengths

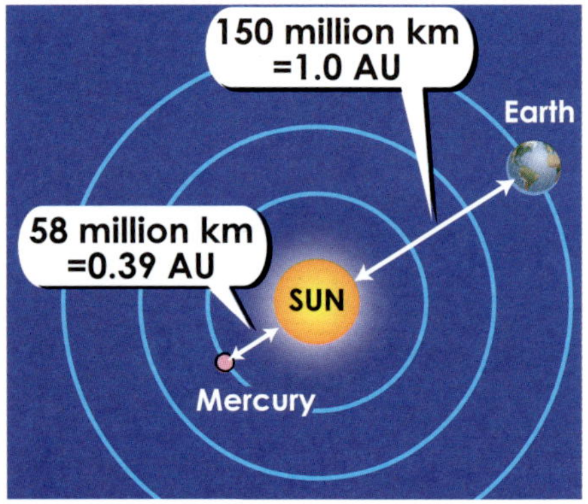

Science also considers very large lengths, such as the distance between planets and stars. One **astronomical unit (AU)** is the distance between Earth and the sun, about 150 million kilometers. One **light year** is the distance light travels in one year, which is 9.5 trillion kilometers. The nearest star is 4.3 light years away from us.

VOCABULARY

astronomical unit (AU) - the distance between Earth and the Sun (1 AU).

light year - the distance light travels in one year (9.5 trillion km).

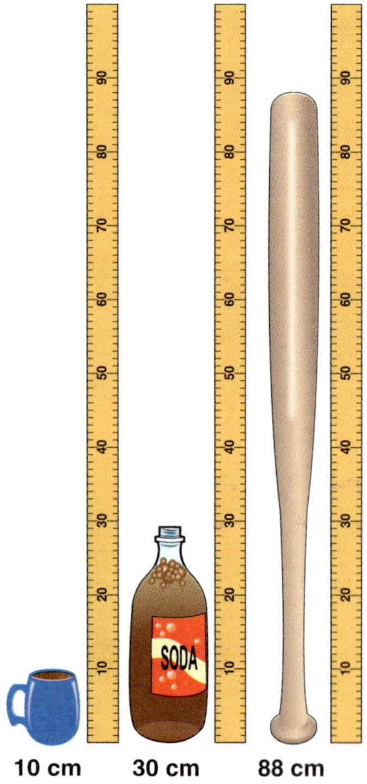

Figure 1.10: *How to read a meter stick.*

Converting between units of length

The problem of multiple units

It would be best (for science) if everyone always used the same unit for length, like the meter. Unfortunately, many different units of length are used for different things, and in different places. In California, you will find inches, feet and miles used more commonly than centimeters, meters and kilometers. In many problems you will need to translate a measurement in one unit into another unit.

Comparing feet and meters

For example, a backhoe on one tractor gives the maximum depth the tractor can dig as 1.83 meters (Figure 1.11). The contractor using the backhoe needs to dig a foundation for a house that is 8 feet deep. Can the backhoe do the job?

Doing units conversions

To answer the question you need to convert from feet to meters and then compare the distances in the same units. To do the conversion you multiply by *conversion factors*. A conversion factor is a ratio that has the value of one. That means the same length is on the top and bottom of the fraction. The trick is that the top and bottom are in different units. For example to convert 1.83 meters into feet you need a conversion factor that relates meters and feet. The table in Figure 1.12 gives two related conversion factors between feet and meters: 1 m / 3.28 ft and 1 foot / 0.305 m. The diagram below shows you what happens when you multiply 1.83 meters by both conversion factors. the correct answer is 6 feet. Can you tell why this answer is correct and the other one is not?

The wrong way

$$1.83\text{ m} \times \left(\frac{1\text{ m}}{3.28\text{ ft}}\right) = 0.56\ \frac{\text{m}^2}{\text{ft}}$$

conversion factor

The right way

$$1.83\ \cancel{\text{m}} \times \left(\frac{1\text{ ft}}{0.305\ \cancel{\text{m}}}\right) = 6\text{ feet}$$

conversion factor

Answer: This backhoe can only dig 6 feet deep so a larger machine is needed.

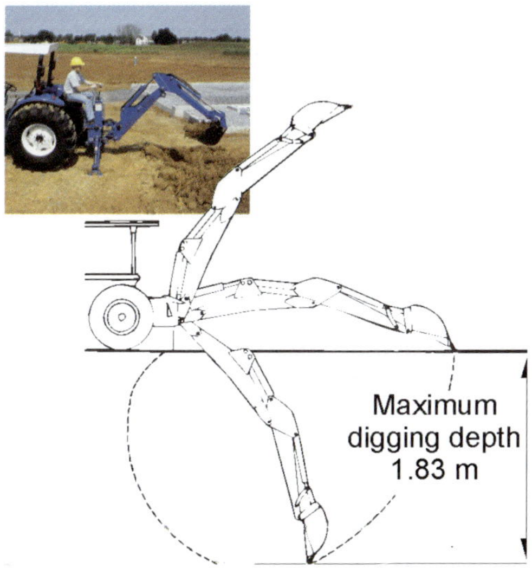

Figure 1.11: *This backhoe specification gives the maximum depth the machine can dig in meters.*

$\frac{1\text{ cm}}{0.01\text{ m}}$	$\frac{1\text{ cm}}{0.394\text{ in}}$	$\frac{1\text{ cm}}{10\text{ mm}}$
$\frac{1\text{ inch}}{0.0254\text{ m}}$	$\frac{1\text{ inch}}{2.54\text{ cm}}$	$\frac{1\text{ inch}}{25.4\text{ mm}}$
$\frac{1\text{ foot}}{0.305\text{ m}}$	$\frac{1\text{ foot}}{12\text{ in}}$	$\frac{1\text{ foot}}{30.5\text{ cm}}$
$\frac{1\text{ m}}{3.28\text{ ft}}$	$\frac{1\text{ m}}{39.4\text{ in}}$	$\frac{1\text{ m}}{100\text{ cm}}$
$\frac{1\text{ km}}{0.622\text{ mi}}$	$\frac{1\text{ km}}{1{,}000\text{ m}}$	$\frac{1\text{ km}}{3{,}280\text{ ft}}$
$\frac{1\text{ mile}}{1.61\text{ km}}$	$\frac{1\text{ mile}}{1{,}609\text{ m}}$	$\frac{1\text{ mile}}{5{,}280\text{ ft}}$

Figure 1.12: *Length conversion factors. Each fraction is equal to 1.*

1.2 Section Review

1. There are two ways to view time in science: as historical time and as time interval.
   a. Give an example in which the two meanings are similar.
   b. Give an example in which the two meanings are different.
2. Arrange the following intervals of time from shortest to longest.
   a. 160 seconds
   b. 2 minutes
   c. 2 minutes 50 seconds
3. A bicyclist completes a race in one hour, five minutes, and 27 seconds (Figure 1.13). How many seconds did it take for the bicyclist to finish the race?
4. The length of a sheet of standard (letter size) paper is closest to:
   a. 0.11 meters
   b. 11 centimeters
   c. 29 centimeters
   d. 279 millimeters
5. The height of an average person is closest to:
   a. 1.0 meters
   b. 1.8 meters
   c. 5.6 meters
   d. 10 meters
6. Someone who sells rope offers you 200 of rope for $10. What is wrong with this offer?
7. Which is longer, one kilometer or one mile?
8. Many home improvement stores sell plywood with a thickness of seven millimeters. How does this compare to standard plywood which has thicknesses of 1/4", 3/8", 1/2" and 3/4"? Which one of these thicknesses could be replaced with seven millimeter plywood?

Figure 1.13: *The bicyclist for question 3.*

1.3 Inquiry and the Scientific Method

We believe that the universe obeys a set of rules that we call **natural laws**. We believe that everything that happens everywhere obeys the same natural laws. Unfortunately, the natural laws are not written down nor are we born knowing them. *The primary goal of science is to discover what the natural laws are*. Over time, we have found the most reliable way to discover the natural laws is called *scientific inquiry*.

What inquiry means

Inquiry is learning through questions

Learning by asking questions is called **inquiry**. An inquiry is like a crime investigation in that there is a mystery to solve. With a crime, something illegal happened and the detective must figure out what it was. Solving the mystery means accurately describing what actually happened.

Deduction

One problem always is that the detective did not see what happened. The detective must **deduce** what happened in the past from information collected in the present.

Theories

In the process of inquiry, the detective asks lots of questions related to the mystery. The detective searches for evidence and clues that help answer the questions. Eventually, the detective comes up with a theory about what happened that is a description of the crime and what occured down to the smallest details.

How you know you have learned the truth

At first, the detective's theory is only one explanation among several of what *might have happened*. The detective must have proof that a theory describes what *did happen*. To be proven, a theory must pass three demanding tests. First, it must be supported by significant evidence. Second, there cannot be even a *single* piece of evidence that proves the theory is false. Third, the theory must be unique. If two theories both fit the facts equally well, you cannot tell which is correct. When the detective arrives at a theory that passes all three tests, he or she believes they have learned from their inquiry what happened. happened.

ă VOCABULARY

natural laws - the set of rules that are obeyed by every detail of everything that occurs in the universe, including living creatures and human technology.

inquiry - a process of learning that starts with questions and proceeds by seeking the answers to the questions.

deduce - to figure something out from known facts using logical thinking. For example, "this is what happened ... because ...".

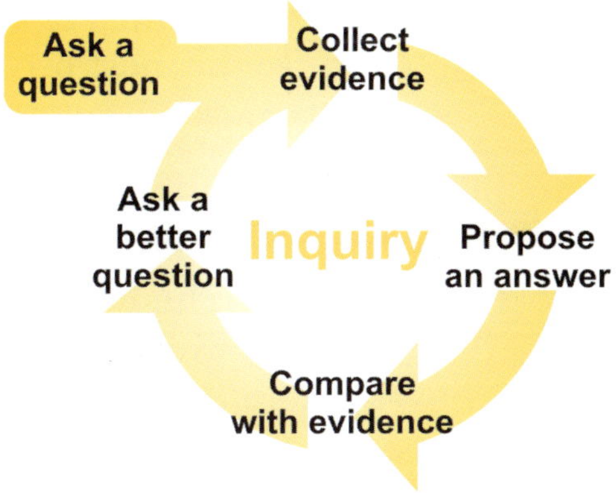

Figure 1.14: *The steps in learning through inquiry.*

Scientific Theories and Natural Laws

How theories are related to natural laws

A scientific theory is a human attempt to describe a natural law. For example, if you leave a hot cup of coffee on the table eventually it will cool down. Why? There must be some natural law that explains what causes the coffee to cool. A good place to start looking for the answer is by asking what it is about the coffee that makes it hot. Whatever quality creates "hot" must go away or weaken as the coffee gets cool (Figure 1.15). The question of what heat is — not how to create it or what it feels like but what it is — puzzled people for a long time.

The theory of caloric

Before 1843, scientists believed (a theory) that heat was a kind of fluid (like water) that flowed from hotter objects to colder objects. They called this fluid *caloric*. Hot objects had more caloric than cold objects. When you put a hot object in contact with a cold object, the caloric flowed between the two until the temperature was equal.

Testing the theory

This theory was at first supported by the evidence. However, problems with the theory came up as soon as people learned to measure weight accurately. If caloric flowed from a hot object to a cold object, a hot object should weigh slightly more than the same object when it was cold. Experiments showed that this was *not true*. To the most precise measurements that could be made, an object had the same weight, hot or cold. The theory had to be changed because new evidence proved it could not be correct.

How theories are tested against evidence

Scientists continually test theories against new evidence.

1. The current theory correctly explains the new evidence. This gives us more confidence that the current theory is the right one.
2. The current theory does *not* explain the new evidence. This means scientists must revise the theory or come up with a completely new one that explains the new evidence as well as all the previous evidence, too.

Figure 1.15: *A question that might begin an inquiry into what "heat" really is.*

CHALLENGE

Many things in science were unknown 1,000 years ago. That doesn't mean people didn't know about qualities like temperature! Ancient people certainly knew the difference between hot and cold. What they did not know is the explanation for qualities such as hot or cold. People believed different things about these qualities than they do today.

Research a theory about something in science that was believed in the past and is no longer believed today. What evidence convinced people to change their minds?

Theories and hypotheses

The hypothesis

In a criminal investigation, a good detective often proposes many possible but different theories for what might have happened. Each different theory is then compared with the evidence. The same is true in science, except that the word **theory** is reserved for a single explanation supported by lots of evidence collected over a long period of time. Instead of "theory" scientists use the word **hypothesis** to describe a possible explanation for a scientific mystery.

Theories in science are hypotheses that correctly explain every bit of evidence

Theories in science start out as hypotheses. The explanation of heat in terms of caloric is an incorrect hypothesis, one of many leading up to the modern theory of heat. The first hypothesis that heat was a form of energy was made by the German doctor, Julius Mayer in 1842, and confirmed by experiments done by James Joule in 1843. Energy has no weight so Mayer's hypothesis explained why an object's weight remained unchanged whether hot or cold. After many experiments, Mayer's hypothesis (that heat was a form of energy) became the accepted *theory* of heat we believe today.

Hypotheses must be testable to be scientific

A scientific hypothesis must be testable. That means it must be possible to collect evidence that proves whether the hypothesis is true or false. This requirement means *not all hypotheses* can be considered by science. For instance, it has been believed at times that creatures are alive because of an undetectable "life force." This is not a scientific hypothesis because there is no way to test it. If the "life force" is undetectable that means no evidence can be collected that would prove whether it existed or not. Science restricts itself only to those ideas which may be proved or disproved by actual evidence.

VOCABULARY

hypothesis - an unproved or preliminary explanation that can be tested by comparison with scientific evidence. Early hypotheses are rarely correct and are often modified as new evidence becomes available.

theory - a scientific explanation supported by much evidence collected over a long period of time.

Figure 1.16: *Eventually, scientists were able to deduce that heat is a form of energy. A hot cup of coffee has more heat energy than a cold cup of coffee. As coffee cools, it's heat energy is transferred to the room. As a result, air in the room is warmed.*

The Scientific Method

Learning by chance At first, humans learned about the world by trial and error, trying one thing at a time, like a small child attempting to open a jar. She will try what she knows: biting the lid, pulling on it, shaking the jar, dropping it ... until, by chance, she twists the lid. It comes off. She puts it back and tries twisting it again — and the lid comes off again. The child learns by trying many things and *remembering what works*.

Learning by the scientific method It takes a long time to learn by randomly trying everything. What is worse, you can never be sure you tried *everything*. The **scientific method** is a much more dependable way to learn.

The Scientific Method

1. Scientists observe nature, then invent or revise hypotheses about how things work.
2. The hypotheses are tested against evidence collected from observations and experiments.
3. Any hypothesis which correctly accounts for all of the evidence from the experiments is a potentially correct theory.
4. A theory is continually tested by collecting new and different evidence. Even a single piece of evidence that does not agree with a theory causes scientists to return to step one.

Why the scientific method works The scientific method is the underlying logic of science. It is basically a careful and cautious way to build a supportable, evidence-based understanding of our natural world. Each theory is continually tested against the results of observations and experiments. Such testing leads to continued development and refinement of theories to explain more and more different things. The way people learned about many things great and small, to the solar system and beyond, can be traced through many hypotheses (Figure 1.17).

VOCABULARY

scientific method - a process of learning that begins with a hypothesis and proceeds to prove or change the hypothesis by comparing it with scientific evidence.

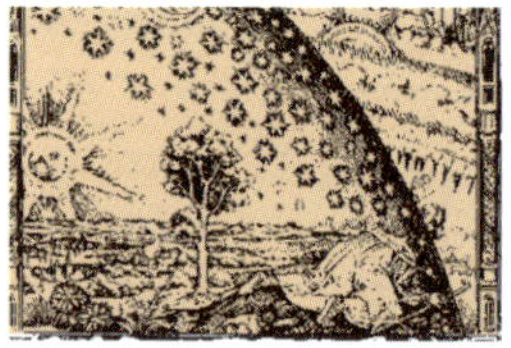

Early civilizations believed the Earth was covered by a dome on which the Sun, stars, and planets moved.

In the middle ages people thought the Sun, stars, and planets circled the Earth which sat in the center.

Today we know the Earth and planets orbit around the Sun and the stars are very far away.

Figure 1.17: *Three different models for Earth and the solar system that were believed at different times in history.*

Scientific evidence

What counts as scientific evidence? The scientific method tells us the only sure way to know you are right is to compare what you think against evidence. However, what types of evidence qualify as *scientific* evidence? Do feelings or opinions count as scientific evidence? Does what other people think qualify as scientific evidence? The answer is no. Because evidence is so important in science, there are careful rules defining what counts as scientific evidence.

An example of scientific evidence Scientific evidence may be measurements, data tables, graphs, observations, pictures, sound recordings, or any other information that describes what happens in the real world (Figure 1.18). Scientific evidence may be collected even without doing experiments in a laboratory. For example, Galileo used his telescope to observe the moon and recorded his observations by sketching what he saw. Galileo's sketches are considered scientific evidence.

When is evidence considered scientific? The two most important characteristics of scientific evidence are that it be **objective** and **repeatable**. "Objective" means the evidence should describe *only what actually happened* as exactly as possible. "Repeatable" means that others who repeat the same experiment the same way observe the same results. Scientific evidence must pass the tests of both objectivity and repeatability. Galileo's sketches describe in detail what he actually saw through the telescope, therefore they pass the test of objectivity. Others who looked through his telescope saw the same thing, therefore the sketches pass the test of repeatability. Gallileo's sketches convinced people that the Moon was actually a world like the Earth with mountains and valleys. This was not what people believed prior to Galileo's time.

Communicating scientific evidence It is important that scientific evidence be communicated clearly, with no room for misunderstanding. For this reason, in science we must define concepts like "force" and "weight" very clearly. Usually, the scientific definition is similar to the way you already use the word, but more exact. For example, your "weight" in science means the force of gravity pulling on the mass of your body.

VOCABULARY

objective - describes evidence that documents only what actually happened as exactly as possible.

repeatable - describes evidence that can be seen independently by others if they repeat the same experiment or observation in the same way.

Examples of scientific evidence

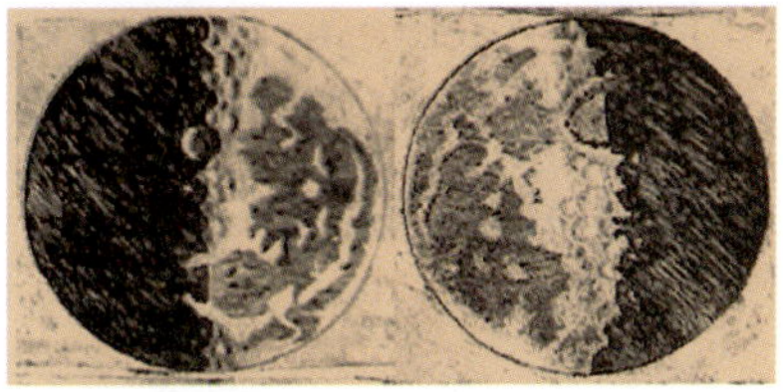

Pictures or sketches that show actual observations

time (sec)	speed (m/sec)	position (m)
0.0	0.00	0.00
0.2	0.83	0.08
0.4	1.66	0.33
0.6	2.50	0.75
0.8	3.33	1.33
1.0	4.16	2.08

Measurements and data

Figure 1.18: *Some examples of scientific evidence.*

Learning science through inquiry

What learning through inquiry means

The goal of this book is to help you learn science through inquiry. This means you will be asked questions instead of being given answers. You will be asked to propose explanations (hypotheses) for things that you see. And like your most ancient ancestors and the most modern detectives, and almost everyone in between, *not all your explanations will be correct*. In inquiry, getting the right answer immediately *is not important*. What is important is having a possible answer that can be tested to see whether it is right or not. You will be given ways (investigations) to collect evidence so you can decide which hypotheses are correct and which are not.

Why not just tell you the answers?

Of course, you could just read the answers in the book and skip all the thinking and testing. There are two major reasons why learning through inquiry is a better way. For one, just reading the answers in a book is boring and you are unlikely to remember them anyway.

Inquiry teaches you how to use science

The most important reason to learn through inquiry is that someday, maybe tomorrow, you will be stumped by something that does not work in your life outside the classroom. It is very unlikely this book or any book will have taught you to fix the exact thing that has you stumped. Inquiry teaches you how to learn how almost anything works. Once you know how something works, you can often figure out how to make it work for *you*. This is called problem-solving and it is an extraordinarily useful skill. In fact, *problem-solving is the most important thing you will learn in this course*. Simple answers, like the temperature of Mars can always be looked up in a book. Inquiry teaches you how to *use* the answers to solve *your problems* and become successful.

MY JOURNAL

You use inquiry and the scientific method every day, only you probably do not realize it. Write down three things you have learned in your life by being curious and testing your ideas for yourself.

SOLVE IT!

Problem-solving in the real world often involves things that are not written down. For example, suppose you have a flashlight. You turn on the switch, and nothing happens. There is no light.

List two potential reasons why the flashlight might not work. These are your hypotheses.

Give at least one way to test each hypothesis to see if it is the correct one. Each test is an experiment that produces scientific evidence that allows you to evaluate the hypothesis.

It is not hard to fix a flashlight by doing the things you just wrote down. This is the scientific method as applied to every day problem-solving.

1.3 Section Review

1. Which of the following is an example of deduction?
   a. Hector calls the weather service to find out if the temperature outside is below freezing.
   b. Caroline looks out the window and concludes the temperature is below freezing because she sees that the puddles in her neighbor's driveway are frozen.
2. Describe the relationship between a hypothesis, a theory, and a natural law.
3. To be correct, a scientific theory must be everything **except**
   a. supported by every part of a large collection of evidence,
   b. believed by a large number of reputable people,
   c. testable by comparison with scientific evidence,
   d. an explanation of something that actually occurs in the natural world or human technology
4. Julie, a third grade student, believes that the moon disappears on certain days every month. Explain why the following information is or is not scientific evidence which can be used to evaluate Julie's hypothesis.
   a. Julie sometimes cannot see the moon all night even though the sky is clear.
   b. Anne, Julie's older sister, thinks the phases of the moon are caused by the moon's position in its orbit around the Earth.
5. When describing scientific evidence, what is the meaning of the word "repeatable"?
6. Which of the following is an example of learning through inquiry?
   a. Miguel is told that hot objects, like a cup of coffee, cool off when left on the table in a cooler room.
   b. Enrique wonders what happens to hot objects if you remove them from the stove. He puts a thermometer in a cup of hot coffee and observes that the coffee cools off.

STUDY SKILLS

Keep your eyes and ears open to the world

A great many discoveries were made almost by accident! For example, paper used to be made of cotton or linen. Both of these are plant fibers and are very expensive. Many inventors searched for a better and less expensive way to make paper. In 1719 French scientist and inventor Rene de Reaumer was walking in the woods when he noticed that wasp nests were made from something a lot like paper! How did the wasps do it? Reaumer discovered that the wasps used wood fibers to make paper. Reaumer began experimenting with a way to make paper from wood, like the wasps. It was trickier than he realized and it was not until 1840 that Friedrich Keller made the first all wood paper. However, Reaumer's curiosity and alert eyes lead directly to the modern, wood-based paper we use today.

TECHNOLOGY CONNECTION

The Art and Sole of Scientific Thinking

When you are playing a sport, how often do you think about your shoes? If they are "working" properly, you shouldn't think about them at all. This is no accident, either. Your athletic shoes have been scientifically designed and engineered.

Not that long ago, athletic shoes were relatively crude. They were made of leather or canvas. They had cleats if they were used on grass and rubber soles if they were used on hard surfaces. What they did not have was much cushioning to protect athletes' feet and legs and knees and hips from the impact of running and jumping, starting and stopping.

Using the scientific method

Today's athletic shoes are high-tech. Scientists and engineers are constantly studying the forces that act on the feet and legs of athletes in various sports and using those studies to redesign and improve footwear.

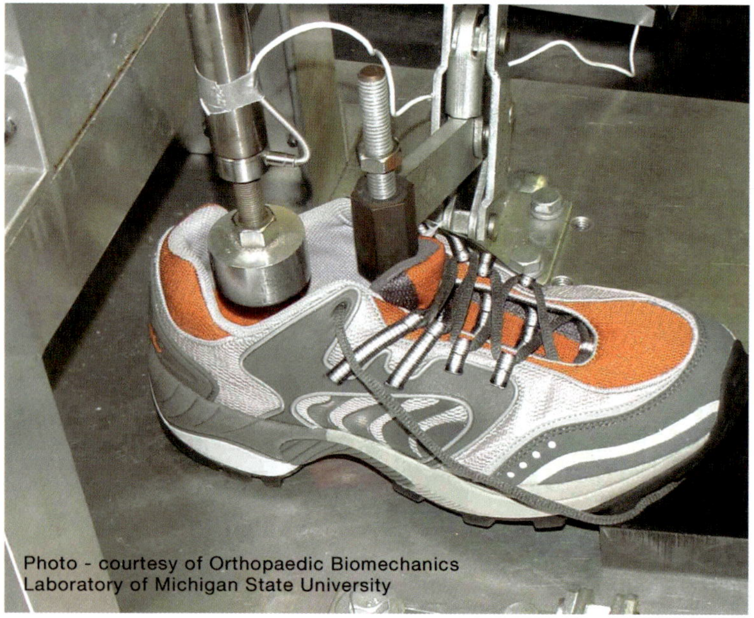

Photo - courtesy of Orthopaedic Biomechanics Laboratory of Michigan State University

So how do scientists and engineers work together to study and improve running shoes - or any other product? They use the scientific method. The scientific method helps ensure that new shoe designs actually improve performance.

Remember, the scientific method begins with observations that lead to questions. For example: "how can a shoe reduce the impact force on a runner's foot?" A design engineer proposes answers to the questions, such as "this new shape of wedge should lessen impact force". The designer's proposed answer is a testable hypothesis because it includes a prediction that can be tested by actual measurements. In the instance of a running shoe, the engineer does experiments that measure the impact forces with the new wedge shape. The measurements can then be compared to other shapes.

This process of scientific thinking is applied to all aspects of the research and testing of athletic shoes at the Orthopaedic Biomechanics Laboratories (OBL) at Michigan State University, in East Lansing, Mich. The OBL is a unique facility that has won recognition for its excellence in science research.

Building a better running shoe

Modern athletic shoes are specialized for various sports, and the running shoe may be the most specialized and highly engineered of all. Running puts great stress on the feet and legs because of the repeated motion and impact. Many runners develop stress injuries. As running has become more popular both as a way to exercise and in competition, millions of people are relying on - and investing in - shoes that will keep them going.

Is there a connection between the shoes runners wear and the injuries they develop? The OBL conducts experiments on athletic shoes in order to study their durability, comfort, and the protection they provide from injury. In terms of the scientific method, this is an observable fact. Researchers can hypothesize whether shoes can reduce the natural stresses of running on the feet and legs. By observing runners, they see that a runner's heel hits the ground first, then the foot rolls forward, and lastly the runner pushes off on the forefoot.

Early experiments measured the forces of these actions. It was calculated that the heel strikes the ground with a force equal to two or three times the body weight of the runner. Scientists hypothesized that the materials in a running shoe's sole would affect how a runner pushes off on the forefoot. Testing this prediction with different variables, they discovered that the forefoot pushed off with more spring when the sole of the shoe was flexible.

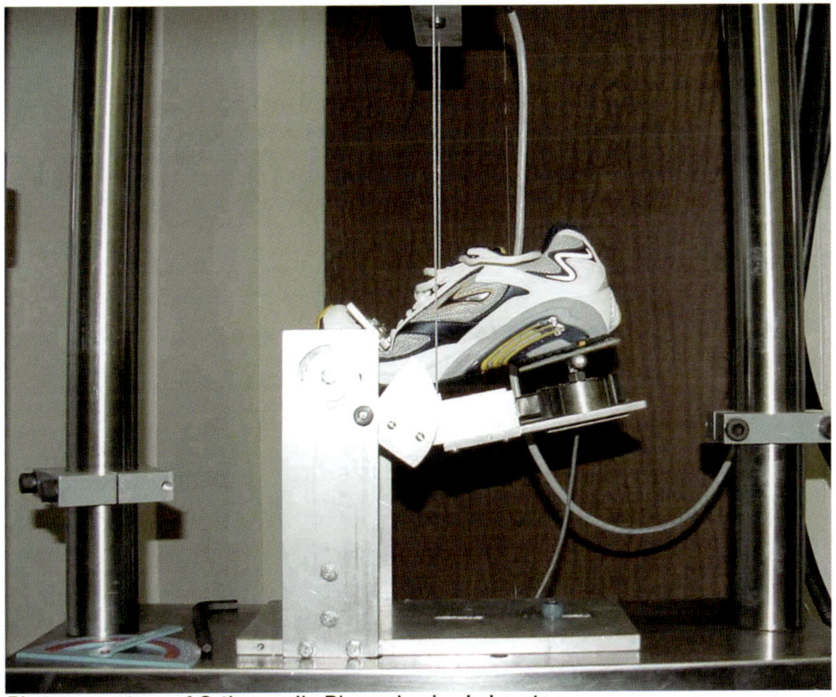

Photo - courtesy of Orthopaedic Biomechanics Laboratory of Michigan State University

Scientists predicted that if a shoe allowed a person to run more efficiently, stress-related injuries would be less probable. This led to the general hypothesis that shoes with well-cushioned heels and flexible soles would reduce injuries in runners.

From running shoes to turkey bones?

Because of their special expertise, the biologists, engineers, and clinicians at the OBL work on many projects involving the musculo-skeletal system of both humans and animals.

For example, Dr. Roger Haut, the labs' director, worked on turkeys! Turkeys raised on big farms develop so fast that they actually outgrow their bone structure. In an average week, 1% of farm turkeys break their femurs. Farmers can lose almost 20% of their flock as a result of this problem. Dr. Haut and his staff at OBL conducted a series of experiments to determine the strength of turkey femurs. They are are researching ways to improve the turkeys' bone strength.

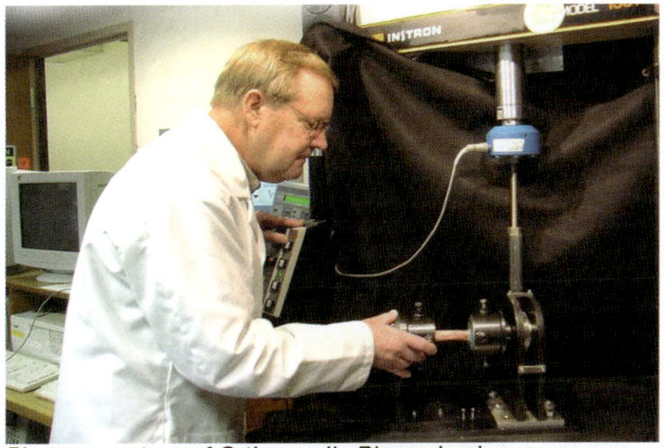

Photo - courtesy of Orthopaedic Biomechanics Laboratory of Michigan State University

Whether it be research on a runner's bones or a turkey's, the scientific method is the foundation. Now you know there is more to your athletic shoe than just good looks.

Questions:

1. What is the first step in the scientific method?
2. What are the main qualities of a good hypothesis?
3. What general hypothesis did scientists make about running shoes?
4. How do scientists and engineers test running shoes in the lab?

Chapter 1 Connection

CHAPTER ACTIVITY Make a Water Clock

A clock is a tool used to measure time. Inside a clock are parts that move with a constant repetition. We can record a quantity of time by counting how many "movements" of our clock occur during an interval of time that has a beginning and an end. The number of movements gives us our time measurement. In this activity, you and a partner will make your own clock and measure time with it.

Materials

Two empty 2-liter soda bottles
400 mL beaker
Water
Duct tape
Stopwatch

Making your clock

1. Put three beakers of water (1200 mL) into one of the empty soda bottles.
2. Attach the mouths of the two soda bottles together and seal them with duct tape.
3. Turn the two attached soda bottles upside down so that the water runs from one soda bottle to the other.
4. Seal your bottles with more duct tape if you have any leaks.

Measuring time with your clock

5. One partner should hold the stopwatch, while the other partner holds the soda bottle clock.

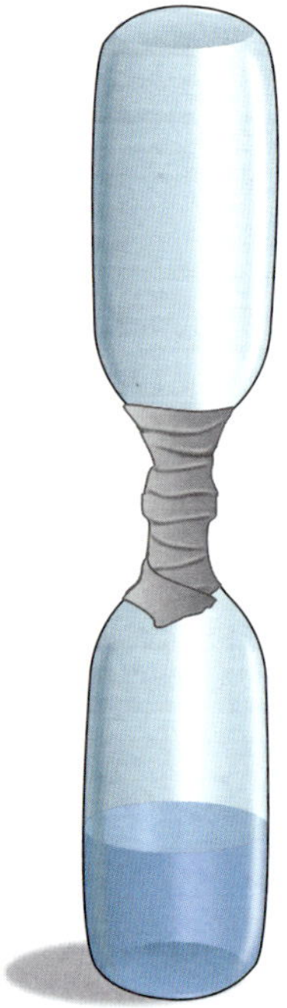

6. One partner turns the soda bottle clock upside down at the same time the other partner starts the stopwatch.
7. Stop the stopwatch as soon as all the water has emptied from the top bottle to the bottom bottle. Record your data in the table.
8. Repeat the procedure two more times, and record your data in the table.
9. Switch roles and repeat the activity three more times. Record your data in the table.

Trial	Time (seconds)
1	
2	
3	
4	
5	
6	
Average	

Discussion Questions

a. Clocks are described as tools that measure time using moving parts. What is the moving part of your soda bottle clock?
b. The new unit of time you created was the time it took the water to run from one bottle completely into the other. Give this unit of time a name.
c. Using the data from your six trials, find the average number of seconds in your new unit of time and record it in the table.
d. Maurice Green can run 100 meters in about 9.8 seconds. How many of your new units did it take?
e. What are some problems with the clock you made today? Why isn't this type of design used in many of our current clocks?

Chapter 1 Assessment

Vocabulary

Select the correct term to complete the sentences.

phases of matter	mass	energy
natural laws	chemical change	scientific method
theory	hypothesis	inquiry
objective		

Section 1.1

1. Solids, liquids and gases are three fundamental ____.
2. Changes in matter are caused by an exchange of ____.
3. The amount of matter an object contains is called ____.
4. When one kind of matter is changed into another kind of matter a(n) ____ has taken place.

Section 1.2

No vocabulary words in this section

Section 1.3

5. When evidence used in an investigation contains only factual information, the evidence is called ____.
6. Problem solving by moving from hypothesis to conclusion using information collected through inquiry, experimentation, and comparison is called the ____.
7. An unproven or preliminary idea that can be tested by scientific inquiry is a(n) ____.
8. A process of learning that starts with questions and arrives at answers to the questions is known as ____.
9. Rules that explain how all things in the entire universe always behave are called ____.
10. When an explanation of something in nature is verified by all known facts it may be called a(n) ____.

Concepts

Section 1.1

1. Describe one effect of chemistry and one effect of physics in each of the following actions.
   a. A birthday cake is made by heating a mixture of salt, flour, baking powder, sugar, eggs and oil.
   b. Iron, not silver, is used to build the frame of a car.
   c. A small wagon accelerates faster than a large pickup truck when the same force is applied to both.
   d. Sea otters can float in water on their backs while feeding.
2. Write the letters **L** (liquid), **S** (solid), or **G** (gas) to indicate the phase of matter of each of the following examples.
   a. ____ Helium in a balloon
   b. ____ A piece of bread
   c. ____ Steam
   d. ____ Cooking oil
3. The source of virtually all of the energy on Earth is:
   a. green plants.
   b. gasoline and oil.
   c. the sun.

Section 1.2

4. Explain the difference between historical time and a quantity of time. Give one example of each.
5. Most people in the United States use inches, yards, and miles while scientists and people in most other countries, including Canada, Mexico, Japan, Germany, China and England use the SI units of centimeters, meters, and kilometers. What would be the advantage of having people in the United States use SI units of length?

6. What unit would be most convenient for measuring each of the following lengths?
   a. The height of a door
   b. The distance between planets
   c. The length of tiny bacteria
   d. The distance between stars
   e. The distance between cities

Section 1.3

7. Write the letters **H** (hypothesis), **T** (theory), or **L** (law) to describe the following statements. Letters will be used more than once.
   a. ____ Complex animal life evolved from simpler forms of life.
   b. ____ An attractive force exists between two bodies. Its strength is determined by the mass of the bodies and the distance between them.
   c. ____ Atoms are the smallest particles of matter.
   d. ____ It is possible for a race car to jump across the Grand Canyon.
8. What three tests must a scientific theory pass to be accepted as the correct theory for a natural event?

Problems

Section 1.1

1. Describe a common substance that you have experienced in all three phases of matter (solid, liquid, and gas.)
2. The same force is applied to a ping-pong ball and a bowling ball. Both balls are free to roll along a level floor. Describe the differences between the motion of the two balls.
3. Select the correct description and examples for each phase of matter to fill in the table:

expands freely	wood, rock, ice
maintains shape	air, oxygen, helium
flows; has no definite shape	milk, water, gasoline

Phase	Description	Examples
Solid		
Liquid		
Gas		

Section 1.2

4. Convert 36,000 seconds to the units shown:
   a. _____ years
   b. _____ days
   c. _____ minutes
5. Convert the following distances to the units shown:
   a. 3.0 miles is equal to __________ kilometers
   b. 1.23 miles is equal to __________ meters
   c. 8.2 feet is equal to __________ meters

Section 1.3

6. A student notices that some plants in her class have grown faster than others and wants to know why. Unscramble the steps of the scientific method she might use to investigate. Place them in a logical order from the first step to the last.
   a. She thinks it might be light (a hypothesis).
   b. She wonders why (a question).
   c. She concludes that it is not light (a conclusion).
   d. She grows similar plants under different amounts of light (an experiment).
   e. She compares the plants growh (analyzes data).

Chapter 2

Experiments and Variables

On Aug. 21, 2003, on a specially built hill in Irvine, California, six adults climbed into cars with no motors and rolled downhill. Called the Extreme Gravity Race, the cars reached speeds of up to 60 miles per hour as they raced down the hill using nothing but gravity for energy. Fiercely competitive, each of the six cars represented a different design and engineering team. The race featured teams from five different automakers. Each team had created the slipperiest, low-friction car they could, using carbon fiber, titanium and many high-tech materials.

How did the cars reach such high speeds using nothing but gravity? How did each design team evaluate its car so that it would be as fast as possible? Answers to these questions involve experiments and variables. Read on, and you will find out how people learn to make things better, faster, or more efficient!

Key Questions

1. *How do scientists know when they are right?*
2. *How is an experiment designed?*
3. *What is the purpose for taking the average of many measurements?*

2.1 Variables and Relationships

Imagine you are in a submarine exploring the deepest part of the ocean. There is no light. You cannot see anything around you. How do you know where you are? How do you tell the rest of your exploration team where you are? What you need is a way to precisely describe where you are at every moment and communicate this information to your fellow explorers. When scientists need to communicate precise information they use *variables*. A variable is a quantity that can be precisely described, such as the depth of a submarine.

What is a variable?

Variables and values

A **variable** is a quantity that has a **value** which describes something. For example, *color* is a variable. If you are choosing a car to buy, color is often an important variable. For a particular car, the values of color could be red, blue, brown, black, or yellow. The variable *color* represents all possible choices and a value, red, is a specific choice.

When talking about the car, it is clearest to use the variable (color) to describe all possible values. For example, you would say,"What colors do you have?" It is much more difficult to list all possible colors one at a time: "Do you have silver?" "Do you have blue?" "Do you have brown?" "Do you have black?" "Do you ...?"

Variables often have values of numbers with units

In science, many variables represent numbers with units. In the example in Figure 2.1, the depth below the surface is a variable. The name of the variable is depth. The value can be any number between zero (at the surface) and the distance between the surface and the bottom. For example, you could specify your depth as 100 meters. Mathematically, you would write this as *depth = 100 m*. If your team calls on the radio and asks, "What is your depth?", they are asking for the value of the variable *depth* at that exact moment. You would respond, "My depth is 100 meters."

VOCABULARY

variable - a quantity that can be precisely specified, often with a numerical value. For example, position and speed are variables.

value - the particular number (with units) or choice that a variable may have.

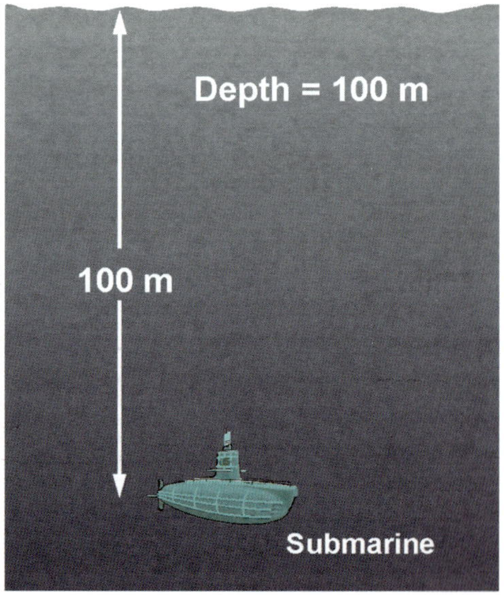

Figure 2.1: *The variable depth represents all possible values for the distance between the surface and the submarine. At any moment, the submarine has a specific value of depth. At the moment in the picture, the depth is 100 meters.*

Using variables in physical science

Variables in the car and track experiments Consider doing an experiment with a small car that rolls along a straight track (Figure 2.2). What variables affect the motion of the car in this system? What kinds of values can each variable take? What kinds of instruments are used to measure the variables?

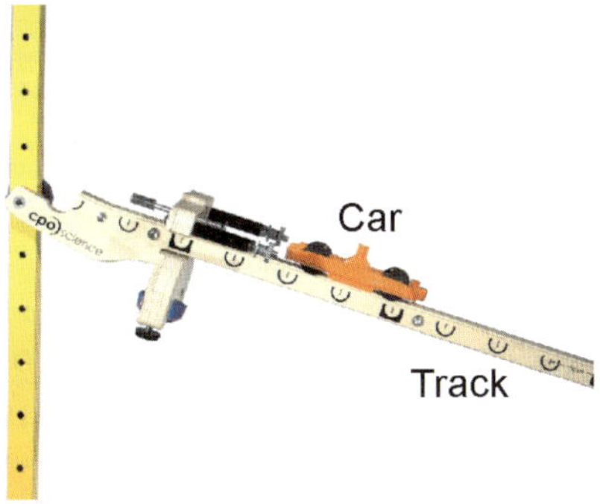

Figure 2.2: *The car and track used in many of the investigations.*

Mass Mass is an important variable in many experiments. Mass is measured with a balance and has units of grams or kilograms.

Time Time is an important variable in many experiments. In your investigations, time is measured in seconds using a stopwatch or photogate timer.

Position Position is another important variable. Position is measured in meters or centimeters (Figure 2.3).

Angle Angle is another important variable. The track may be set at different angles. The car will travel at different speeds depending on the angle at which you set the track.

Temperature In your investigations in chemistry, you also will measure different variables. For example, temperature is a variable that is important in many experiments. Temperature is measured with a thermometer and has units of degrees Celsius (°C) or degrees Fahrenheit (°F).

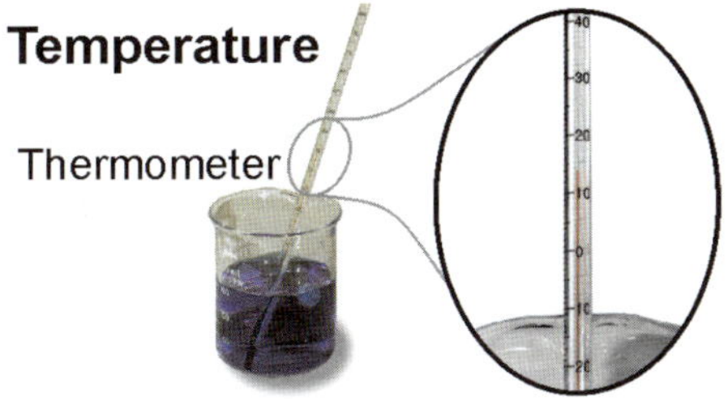

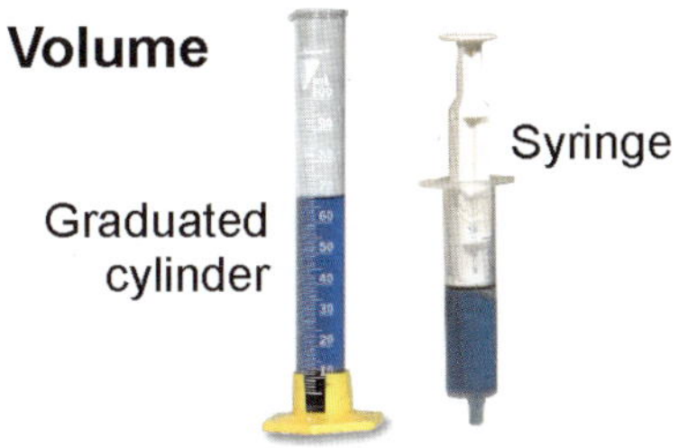

Volume Volume is a variable that you will use to measure liquids like water. Volume is measured in cubic centimeters (cc) or milliliters (mL). One cubic centimeter is the same as one milliliter. Graduated cylinders and syringes are used to measure volume in different experiments.

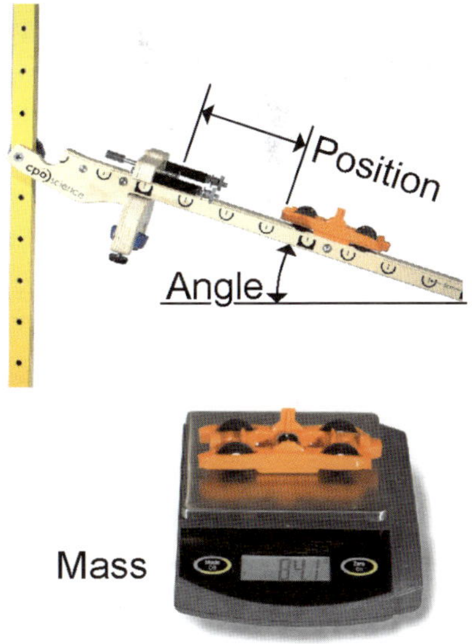

Figure 2.3: *Mass, position, and angle are important variables when doing experiments with the car and track.*

Graphs show relationships between variables

Relationships between variables

Physical science is all about relationships between variables. For example, you are rolling a car down the track. You suspect that increasing the angle would make the car go between points A and B in less time. How do you find out if your suspicion is correct? You need to know the relationship between the variables *angle* and *time*.

Patterns on a graph show relationships

A good way to show a relationship between two variables is to use a **graph**. A graph is a mathematical diagram that shows one variable on the vertical (or y) axis and a second variable on the horizontal (or x) axis. Each axis is labeled with the range of values the variable has. In Figure 2.4, the x-axis (angle) has data values between zero and 50 degrees. The y-axis (time) has data values between zero and 0.3286 seconds. You can tell there is a relationship because the graph is a curve that slopes down and to the right.

Recognizing a relationship from a graph

Two variables may have a strong relationship, a weak relationship, or no relationship at all. In a strong relationship, large changes in one variable make proportionately large changes in the other variable, like in Figure 2.4. When there is no relationship, the graph looks like scattered dots (below left). A weak relationship is in between strong and none, meaning large changes in one variable cause only small changes in the other.

No relationship between variables

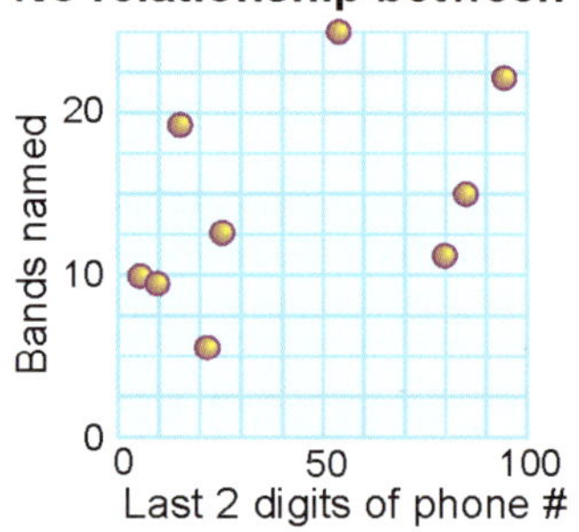

Bands named	Tel. # ends in
10	06
15	85
19	15
6	22
22	96
9	10
25	63
12	34
11	79

Weak relationship between variables

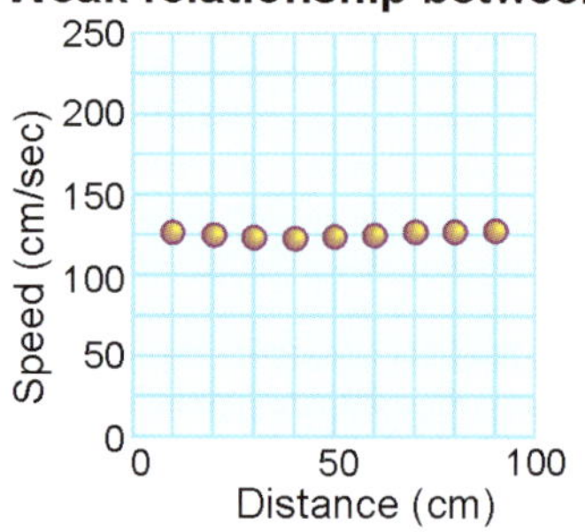

Distance (cm)	Speed (cm/sec)
10	126
20	125
30	123
40	122
50	123
60	124
70	126
80	127
90	128

VOCABULARY

graph - a mathematical diagram showing one variable on the vertical (y) axis and the second variable on the horizontal (x) axis.

Angle (degrees)	Time (seconds)
10	0.3286
20	0.2431
30	0.1937
40	0.1708
50	0.1565

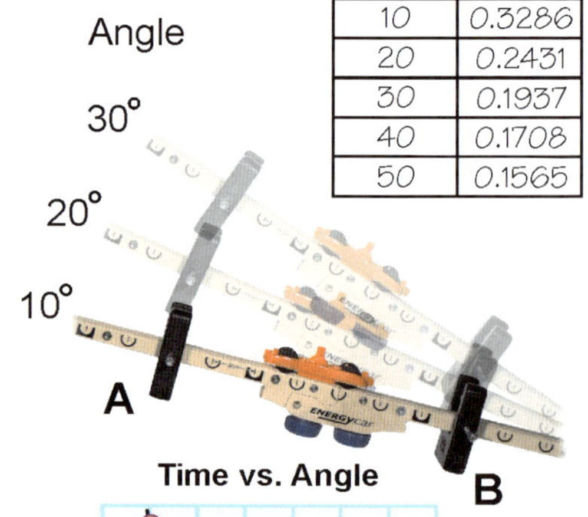

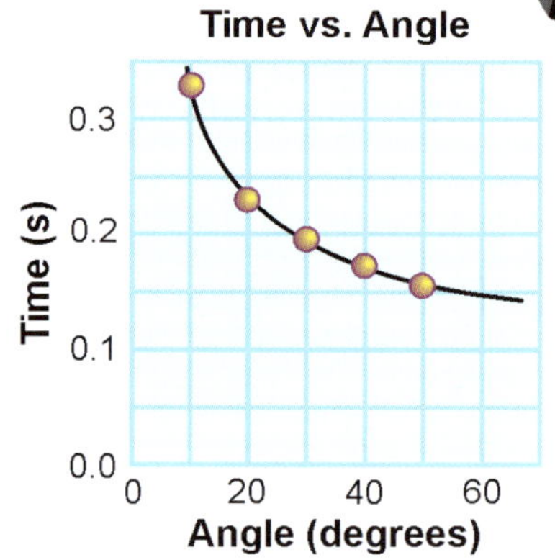

Figure 2.4: *Investigating how quickly the car gets from A to B as the angle of the track is changed. What kind of relationship does the graph show?*

Direct and inverse relationships

Direct relationships

In a direct relationship, when one variable increases, the other also increases. For example, if you add a volume of water to water already in a beaker, the mass of water increases. The relationship between mass and volume is a direct relationship. Graphs showing direct relationships usually slope up and to the right, like the graph in Figure 2.5. This is also called a *linear relationship*.

Inverse relationships

In an inverse relationship, when one variable increases, the other decreases. For example, suppose you put one kilogram of water on the stove for 10 minutes. The temperature of the water increases. If you put two kilograms of water on the same stove for the same time, the temperature would increase only half as much because the heat is spread out over twice the amount of water. The relationship between temperature and mass for this experiment is an inverse relationship. The graph of an inverse relationship often slopes down to the left, like the graph in Figure 2.6.

Relationships involving more than two variables

Many natural laws relate three or more variables. In fact, the example of the stove involves three variables: mass, time, and temperature. When more than two variables are involved you must do careful planning to figure out the relationships. For example, to figure out the relationship between mass and temperature, you should keep the time the same. It is easiest to figure out the relationships between individual variables by changing them only one at a time.

Complex relationships

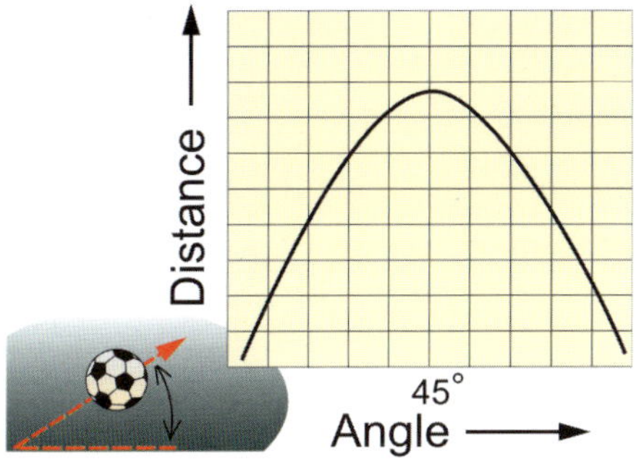

Some relationships are neither direct nor inverse. The graph on the left shows the distance a soccer ball flies when kicked at different angles. The maximum distance is near 45 degrees. Increasing the angle past 45 degrees decreases the distance. Decreasing the angle below 45 degrees also decreases the distance. This is a *nonlinear relationship*.

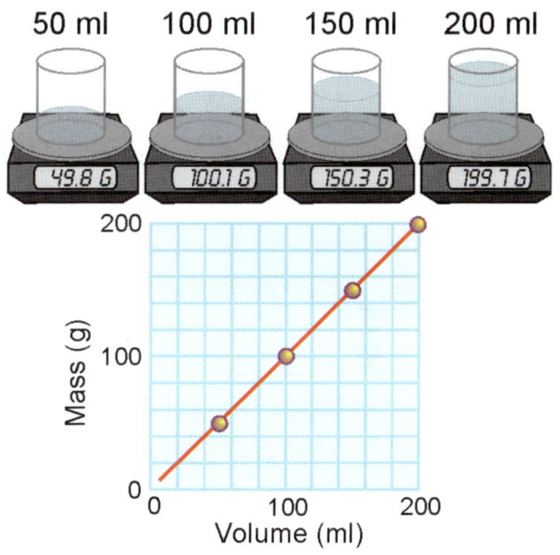

Figure 2.5: *The relationship between mass and volume of water is an example of a direct relationship.*

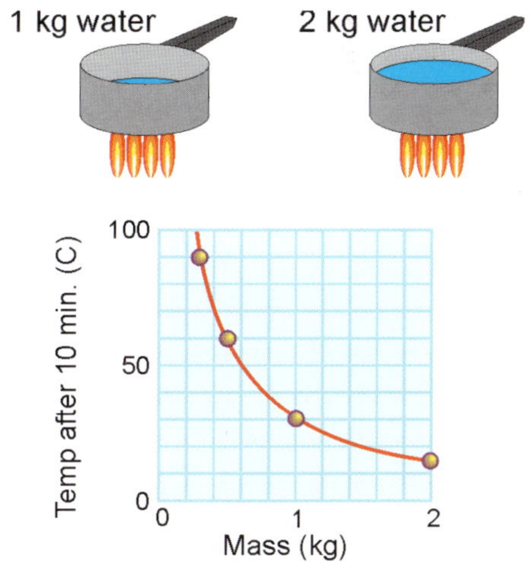

Figure 2.6: *If heating time is kept constant, the relationship between temperature change and mass is an inverse relationship.*

Designing a graph: dependent and independent variables

What to put on the *x*- and *y*-axes

To a scientist, a graph is a language that shows the relationship between two variables. Graphs are drawn a certain way just like words are spelled a certain way. The first rule in making a correct graph is to choose which variables to put on which axis.

The independent variable

Graphs are usually created to show a cause and effect relationship between two variables. A graph makes it easy to see if changes in one variable *cause* changes in the other variable (the *effect*). The variable that causes the change is called the **independent variable**. In an experiment, this is the variable that the experimenter is free to change. *By agreement among scientists, the independent variable goes on the x-axis.* In the example, mass is the independent variable so mass goes on the x-axis (horizontal).

The dependent variable

The **dependent variable** shows the effect of changes in the independent variable. *The dependent variable goes on the y-axis.* In the example, temperature is the dependent variable and therefore goes on the y-axis (vertical).

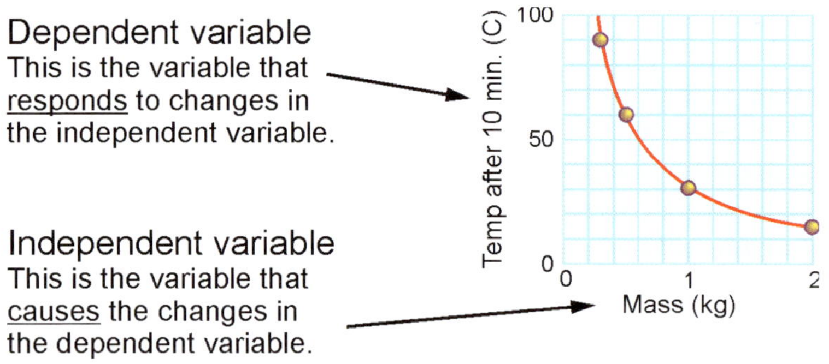

If time is a variable

Time is often an exception to the rule about which variable goes on which axis. Time usually goes on the x-axis even though you may not think of time as an independent variable.

VOCABULARY

independent variable - in an experiment, a variable that is changed by the experimenter and/or causes changes in the dependent variable.

dependent variable - in an experiment, a variable that responds to changes in the independent variable.

SOLVE IT!

4 steps to making a graph

Step 1: Choose which will be the dependent and independent variables. The dependent variable goes on the y-axis and the independent variable goes on the x-axis.

Step 2: Make a scale for each axis by counting boxes to fit your largest value. Count by multiples of 1, 2, 5, or 10.

Step 3: Plot each point by finding the x-value and drawing a line upward until you get to the right y-value.

Step 4: Draw a smooth curve that shows the pattern of the points. Do not just connect the dots.

Reading a graph

Using a graph to make a prediction

Suppose you want to find out what the temperature of 1.5 kilograms of water would be after 10 minutes on the stove. But you only measured the temperature with 1 and 2 kilograms, not with 1.5 kilograms. The graph can give you an accurate answer even without your doing the experiment.

1) Start by finding 1.5 kg on the x-axis.
2) Draw a line vertically upward from 1.5 kg until it hits the curve representing what you actually measured.
3) Draw a line across horizontally to the y-axis.
4) Use the scale on the y-axis to read the predicted temperature.

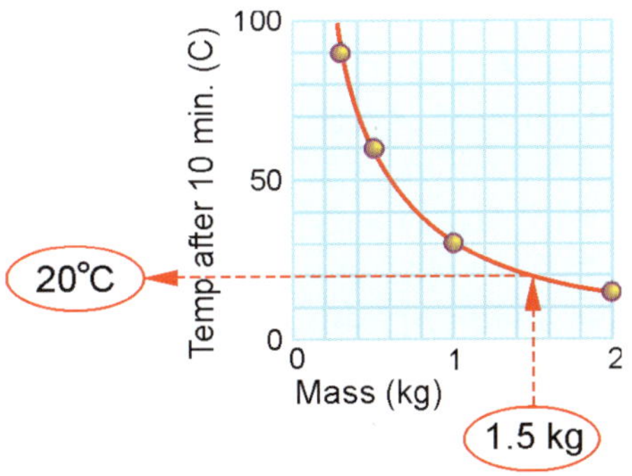

Large graphs are more precise to read

For this example, the graph predicts the temperature to be 20°C. Predictions are more accurate when the graph is large enough to read precisely. That is why it is a good idea to make your graphs fill as much of the page as you can.

A graph is a form of model

A graph is a simple form of model. Remember, a model is a relationship that connects two or more variables. Scientists use models to make and test predictions.

SOLVE IT!

A student measures the mass of water collected every five minutes on a rainy day. Design a graph to show the student's data. Once you have the graph, estimate when 20 grams of water was collected.

Time (min)	Mass (g)
0	0
5	17
10	26
15	38
20	49

Time is the independent variable, therefore mass is the dependent variable. The mass axis should go from 0 to at least 50 grams. The time axis should go from 0 to at least 20 minutes. The graph shows that 20 grams of rainwater fell in the first 7.5 minutes.

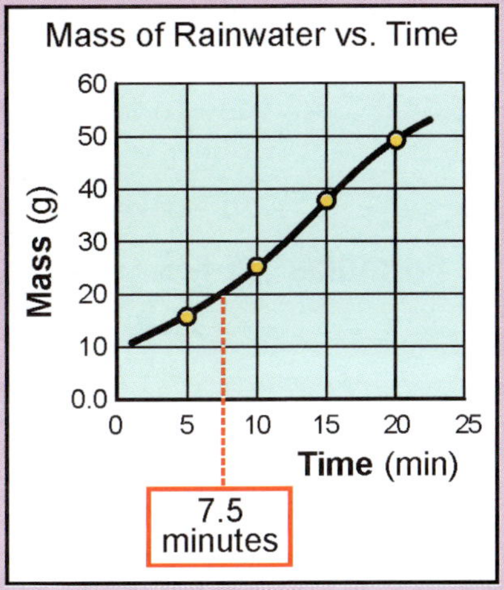

Using math to describe relationships between variables

Math is the language of variables

Math is the best language to describe relationships between variables. A graph is good for getting a quick picture of a relationship. However, math is more accurate and more useful. Like any foreign language, you have to learn to read and write math before it can be useful to you. Math allows us to solve problems as routine as balancing a checkbook or as complicated as landing a robot on Mars.

Using letters to represent variables

When you write out a relationship in math, you use a single letter or symbol to represent each variable. For example, what is the relationship between volume and size for a box? The diagram below shows the relationship written out two ways.

Volume is length x width x height

$$V = lwh$$

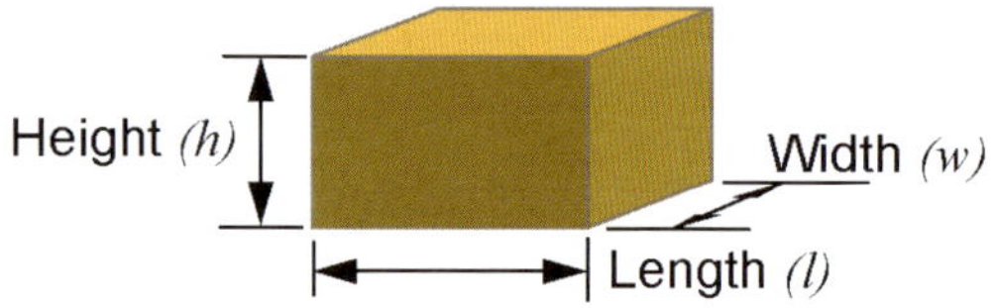

Operations

The relationships between variables are represented in math by *operations*. Four operations you know are: add, subtract, multiply, and divide. To show addition or subtraction you put the symbol between the variables. However, there are two ways to represent multiplication and division. In fact, you will rarely see the multiply or divide symbol. Instead, two variables written right next to each other mean they should be multiplied together. Two variables on top of each other mean the top one is divided by the bottom one. All four operations are shown in Figure 2.7.

Formulas

A formula is a relationship that gives one variable in terms of other variables. Think of each variable as a box to hold the value of that variable. To use the formula, you substitute actual values for the corresponding variables. Then add, subtract, multiply, or divide the numbers as directed by the formula. In this book, important formulas will be shown as "cards" with the variables and their units identified (Figure 2.8).

Add a and b	$a + b$
Subtract a from b	$b - a$
Multiply a and b	$a \times b$ or ab
Divide a by b	$a \div b$ or $\frac{a}{b}$

Figure 2.7: *The basic math operations and how they are written and read.*

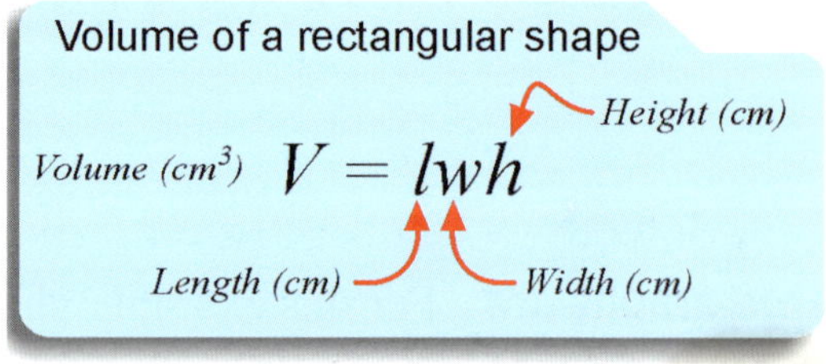

Figure 2.8: *An example of a formula for the volume of a rectangular box.*

Solving for one variable in terms of the others

Using a formula A formula is an exact description of the relationship between variables. Formulas allow you to calculate any one variable if you know the values of the others. This is easiest when the variable you want is by itself on one side of the equal sign (=). For example, the volume (V) of a box is length (l) × width (w) × height (h). If you know the length, width, and height, you plug values in for the variables and do the multiplication as directed by the formula (Figure 2.9).

Solving an equation In many problems the variable you want is not by itself. Getting the variable you want by itself is known as *solving* the equation (equation being another word for formula). Suppose the box must have a volume of 1,500 cubic centimeters — about the right size to hold a coffee cup. You want the box to be 10 centimeters wide and 15 centimeters long. How high should the box be?

Solving an equation with values Let's do the problem with numbers first. To calculate the required height, write down the formula substituting the values you know.

$$1500 = 15 \times 10 \times h$$

Next, do the operations you can. For example, multiply 15 times 10. Last, divide both sides by 150. Dividing by 150 cancels out the 150 on the right-hand side of the formula, leaving only the variable h (height), which gives you the solution: $h = 10$ cm.

$$1500 = 150h \quad \rightarrow \quad 10 = h$$

Solving an equation with algebra We can also solve the problem using algebra to rearrange the formula. Once you have the formula in the form "h = ..." then you can plug in the numbers and find the required height. Many problems must be solved this way because the relationships are too complex to calculate using just numbers.

$$V = lwh \quad \rightarrow \quad h = \frac{V}{lw} \quad \rightarrow \quad h = \frac{1500}{15 \times 10} = 10 \text{ cm}$$

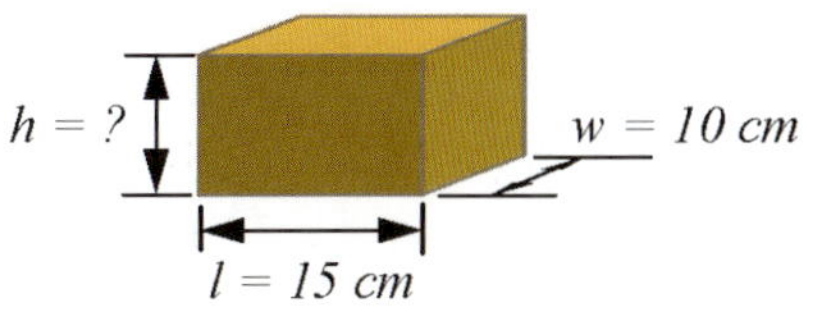

Formula:	$V = lwh$
Known values:	$1{,}500 = (15)(10)h$

Figure 2.9: *Using a formula to calculate the volume of a rectangular box.*

STUDY SKILLS

Solving math problems is a lot like playing a game. As in all games, there are rules you have to follow. You win the game when you get the variable you want all by itself on one side of the equals sign.

Here are a few of the rules:

Rule 1: Anything you do on the left of the equals sign (=) you must also do on the right. If you subtract 10 from the left, you must also subtract 10 from the right. If you divide the left by w, you must also divide the right by w.

Rule 2: Anything divided by itself equals one

Rule 3: Given three variables a, b, and c, the following are true:

$$ab + ac = a(b + c)$$

$$\frac{a}{c} + \frac{b}{c} = \frac{a+b}{c}$$

2.1 Section Review

1. An engineer designing a house needs to know how much force it takes to break different size beams of wood (Figure 2.10). List at least five variables important for figuring out this information.
2. Let the variable m be the mass of a student. Which of the following is NOT a possible value that m can have?
   a) 51 kilograms. b) 51,000 grams. c) 51 meters.
3. Figure 2.11 shows four different graphs. Which one shows a direct relationship? Which one shows an inverse relationship? Which one shows a complex relationship? Which one shows no relationship?
4. A student does an experiment that measures the temperature at which water freezes when different amounts of salt are mixed in. What is the independent variable in the experiment? What is the dependent variable?
5. A student collects data on the distance between people's outstretched fingertips and their height. The data are shown in the graph in Figure 2.12. If a person's height is 1.7 meters, what is the most probable distance between his outstretched fingertips?
6. On a car trip, the distance you travel is equal to the speed you go multiplied by the time that you are driving. Write this relationship as a formula, using d to represent distance, v to represent speed, and t to represent time.
7. Which TWO of the following formulas are mathematically correct? Being correct means what is on the left of the equals sign *always* has the same value as what is on the right.

   a) $ab = a + b$ c) $a + a + b = 2a + b$

   b) $\frac{a-b}{b} = \frac{a}{b} - 1$ d) $a + b - b = a - 2b$

8. Solve the equation $f = ab - c$ for the variable c.

Figure 2.10: *The wood beam in question 1.*

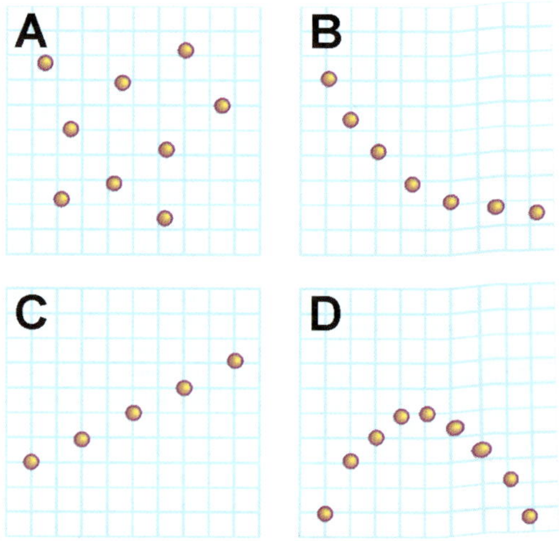

Figure 2.11: *Four different graphs.*

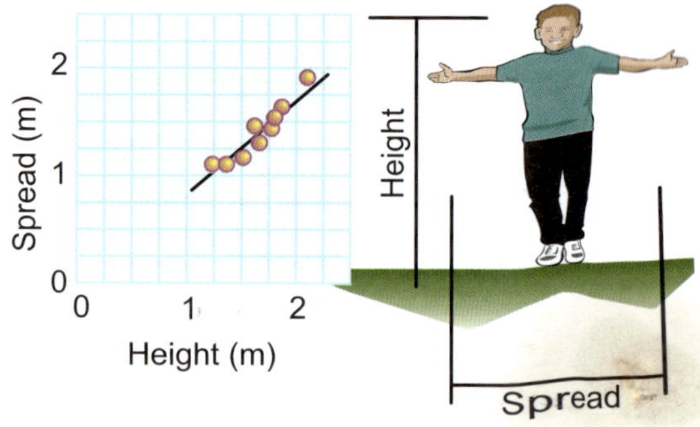

Figure 2.12: *A graph showing the spread of people's arms compared with their height.*

2.2 Experiments and Data

If you wish to know someone's name, it's easy enough to ask, "What is your name?", and get a simple answer: "Sandra." If you want to know how something works in science, the questions are not so easy because the answers are not so easy. Earlier we talked about the importance of inquiry. Inquiry is how science is learned. Inquiry is built around questions, and doing *experiments* is how scientists learn the answers to questions about nature and technology.

Experiments provide the test of truth

Experiments

An **experiment** is a situation specially set up to investigate the relationships between specific variables. Scientists do experiments to gather evidence so they can evaluate their ideas. An example of an idea (or hypothesis) could be: "The temperature of water increases proportional to the amount of time you apply heat." One way a scientist could test this idea is with an experiment that measured the temperature of water at different times while heat was applied.

The importance of experiments

Experiments are important in science because *experiments are the test of whether a hypothesis is correct or not.*

In science, a hypothesis is correct ONLY if it agrees with the results of actual experiments.

Something is true in science ONLY if it agrees with what actually happens. People's opinions of what *did or should or could* happen are NOT scientific tests of what is true and what is not true. The best way to test a theory or hypothesis is to do experiments that provide scientific evidence.

Observations may substitute for experiments

There are times when experiments cannot be done, such as in astronomy. Then a theory may be tested by comparing predictions of the theory with observations of what occurs in nature (Figure 2.14).

VOCABULARY

experiment - a situation specially set up to investigate relationships between variables.

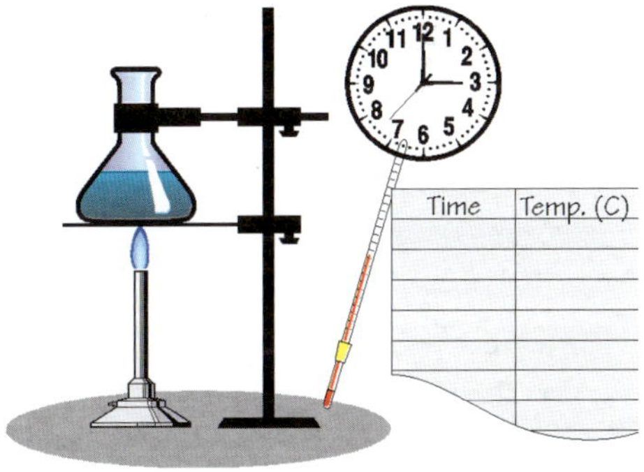

Figure 2.13: *An experiment to test the hypothesis that the temperature of water increases proportional to the amount of time you apply the heat.*

Figure 2.14: *There are several competing theories for how the solar system formed. Solar systems cannot be re-created in experiments so the theories are tested by comparing them with observations of the actual solar system.*

Designing a good experiment

The importance of experiment design A good experiment is carefully planned and carried out. If the experiment is well designed, the results will help you learn what you want to know. If the experiment is poorly designed, the results may fool you into believing the wrong things. Many people have been mistaken and confused by results of poorly designed experiments.

The hypothesis The first step in designing a good experiment is to clearly state what you want to test (the *hypothesis*). Remember, a hypothesis is a tentative statement about how or why something happens. For example, "Cars rolling down steep ramps go faster than those rolling down shallow ramps" is a good hypothesis. It is good because it may be tested by doing an experiment (Figure 2.15).

The procedure The second step is to plan how you will test your hypothesis. This part of experiment design is called the **procedure**. A good procedure describes the equipment you use, the techniques you use, and the data you collect from the experiment. For example, you could measure the time it takes a car to roll between two photogates as you vary the angle of the ramp. The procedure should contain enough information for someone else to repeat your experiment the exact same way you did.

Data The purpose of experiments is to collect scientific evidence. In many experiments the scientific evidence will be **data**. Data are usually the values of variables that you measure during an experiment. For example, the ramp angle and time between photogates are data.

Analysis and conclusion Once you collect your data, you need to **analyze** it. Analyzing includes thinking about, graphing, or doing calculations with the data. If your experiment was successful, your analysis will lead you to a **conclusion**. A conclusion is a statement of what you learned from the experiment. A conclusion must always be supported by data and analysis. For example, that cars roll faster down steeper ramps is a conclusion supported by data that shows what the cars did on ramps of various heights in the experiment.

VOCABULARY

procedure - a description of an experiment that details the equipment used, the techniques used, and the data collected.

data - information collected during an experiment or other scientific inquiry. Data are often values of variables measured in an experiment.

analysis - the process of evaluating data. Analysis may include thinking, creating graphs, doing calculations, and discussing ideas with others.

conclusion - a statement of what was learned in an experiment or observation.

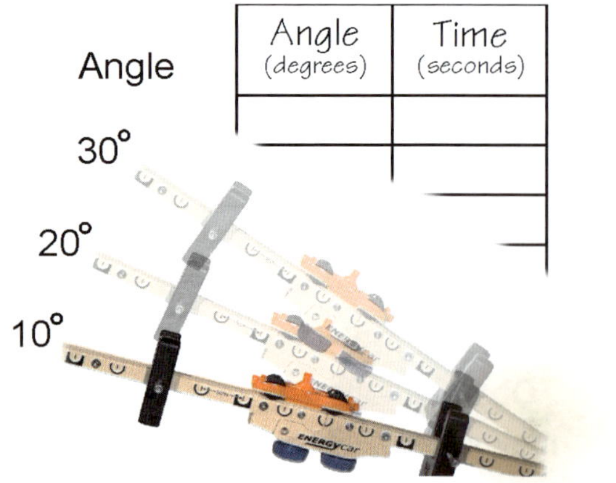

Figure 2.15: *An experiment to test whether cars rolling down steeper ramps go faster.*

Experimental and control variables

Only one variable is changed in an ideal experiment

In an ideal experiment only a single variable is changed at a time. If anything happens in the experiment, you can then assume what happened was caused by the variable you changed. On the other hand, if you make many changes at once, you will never be able to sort out which one caused what happened. The experiment will still *work*, you just won't be able to understand its results.

The experimental variable

The variable you allow to change in an experiment is called the **experimental variable**. This is usually the variable that you have direct control over. Going back to the experiment with a car on a ramp, the angle of the ramp is the experimental variable.

Control variables

In real experiments there are often many variables that matter. If only one variable is allowed to change, then the others must be kept constant. Constant means NOT allowed to change. Variables that are kept constant are called **control variables**. Many variables affect the car on the ramp. For example, changing the mass of the car, pushing it, or releasing it from a different position all change how quickly the car moves between the photogates. If you want to test the effect of steeper ramps you need to ensure that the ramp angle is *the only variable that changes*. The mass and release position should stay the same for every angle you test. Mass and release position are control variables in this experiment.

Identifying the experimental variable

In some tests you may be asked to identify the experimental variable by looking at data collected in an experiment. The most reliable way to do this has two steps:

Step 1) Eliminate all the variables that remain constant since these are controlled variables.

Step 2) Of the remaining variables, the experimental variable is the one the experimenter has changed — and often in equal steps, such as every 2 minutes or every 5 centimeters.

VOCABULARY

experimental variable - a variable that changes in an experiment.

control variable - a variable that is kept constant in an experiment.

Control variables
Mass, Position, Starting point, Distance between photogates, Force applied at start

Experimental variable
Angle

Figure 2.16: *Control and experimental variables for the effect of angle on the time between photogates.*

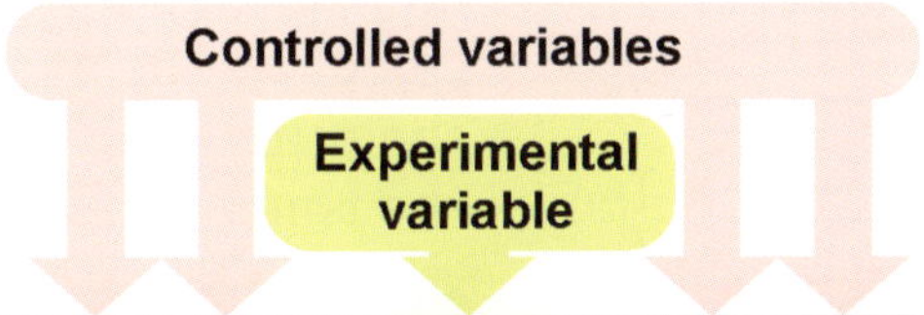

Mass	Position	Angle	Start	Push
58 g	40 cm	10 deg.	10 cm	no
58 g	40 cm	20 deg	10 cm	no
58 g	40 cm	30 deg	10 cm	no
58 g	40 cm	40 deg	10 cm	no

Figure 2.17: *The angle is the experimental variable because it changes and is set by the experimenter.*

Accuracy, errors, and averages

Accuracy In science, the word **accuracy** means how close a measurement is to the true value of what is being measured. For example, suppose it takes a car exactly 0.125 seconds to pass between two photogates. *You have no way to know the exact true time.* The best you can do is make a measurement and *assume* that your measurement is the true time. The table in Figure 2.18 shows several measurements of the time made by different students using different techniques. Which is the most accurate? Which is the least accurate?

Error In your experiments with motion you will use a photogate timer to make measurements to the nearest 0.0001 seconds. You will notice that you rarely get the exact same time three times in a row *even if you do the exact same thing*. This is because any measurement always contains some **error**. Error is the difference between a measurement and the true value of what you are trying to measure. The errors are small if the experiment is accurate.

The average When you make many measurements of the same thing you will notice that they cluster around an **average** value. Some measurements are more than the average and some are less. To calculate the average you add up all the measurements and divide by the number of measurements you have. For example, the average of the times in Figure 2.18 is 0.1253 seconds.

Why taking the average is useful The average of several measurements is usually more accurate than a single measurement. The average is more accurate because errors in the negative direction partially cancel errors in the positive direction.

$$\text{Avg} = \frac{.1255 + .1248 + .1252 + .1256 + .1252}{5}$$

$$= 0.1253 \text{ s}$$

Measurement	Error	− +
0.1255 s	+.0002 s	
0.1248 s	-.0005 s	
0.1252 s	-.0001 s	
0.1256 s	+.0003 s	
0.1252 s	-.0001 s	

VOCABULARY

accuracy - describes how close a measurement is to the true value.

error - the difference between a measurement and the true value.

average - a mathematical process in which you add up all the values, then divide the result by the number of values.

Average

$$\text{Avg} = \frac{\text{sum of values}}{\text{number of values}}$$

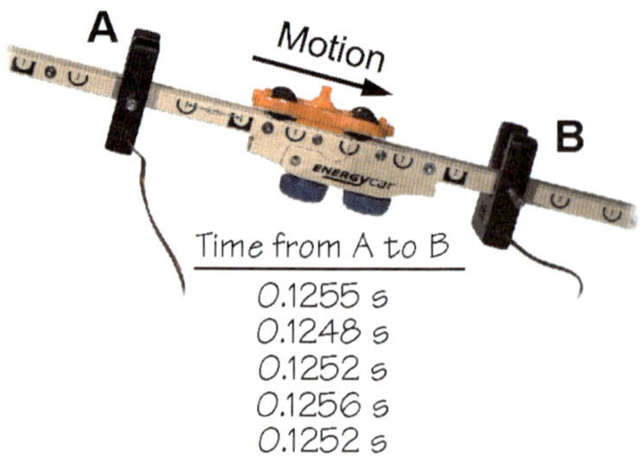

Figure 2.18: *Some data collected by different students using different techniques. Each measurement is the time for a car to roll between two photogates.*

Reproducibility: When are two results the same?

A new condition for reproducibility

In Chapter 1 we said one of the criteria for scientific evidence was that it be *reproducible*. Reproducibility means *two* things that we can now discuss more carefully.

1. Others who repeat the same experiment get the same result (which is what we said in Chapter 1).
2. If you repeat the experiment the same way, you always get the same result.

When different numbers are "the same"

According to rule number 2, an experiment is most reproducible when errors are small. However, even small errors *are still errors*. This brings up an important question: How can you tell if two results are the same when both results have errors in them? The word "same" when applied to experimental measurements does not mean what it does in ordinary conversation. Two measurements are considered the same if their difference is less than or equal to the amount of error. *This is important to remember.*

Two measurements are considered the same if their difference is less than or equal to the amount of error.

How to estimate the error

How can we know the error if we do not know the true value of a measurement? The way scientists estimate the error is to *assume the average is the true value*. The error is roughly the largest difference between the average and a measured value. The number line below shows Group 1's (sidebar) data with the estimated error.

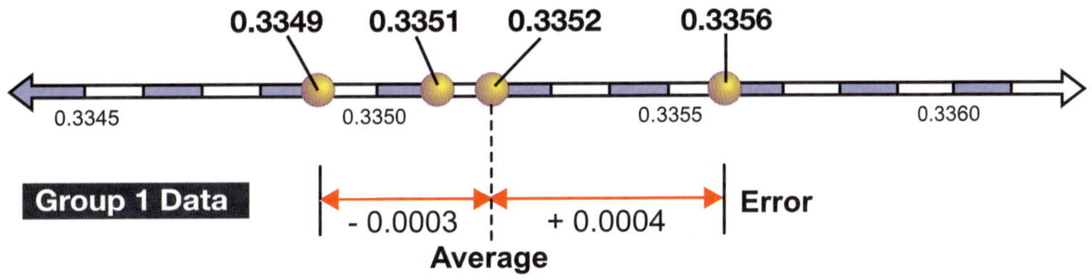

SOLVE IT!

Two groups of students do an experiment in which they measure the time it takes a car to roll between two photogates. Each group makes four measurements and takes the average. One group claims its results are the same. The other group claims its results are different. Who is right?

Group 1 Time (s)	Group 2 Time (s)
0.3356	0.3346
0.3351	0.3353
0.3349	0.3350
0.3352	0.3347
Average	**Average**
0.3352	0.3349
Error	**Error**
± 0.0004	± 0.0004

You need to know the size of the error. For both groups, the largest difference beween the average and a measured value is 0.0004. Therefore, the error is approximately +/- 0.0004 seconds.

The difference between the two groups' results is only 0.0003 seconds. This difference is NOT greater than the error, therefore, scientifically, **the results are the same**.

Drawing conclusions from data

We do experiments to reach conclusions

The point of experiments is to produce data that allows a scientist (like you) to come to a conclusion. The conclusion tells whether your idea, the hypothesis, is right or not. *You need to know about errors before you can make a conclusion.* For example, suppose you think that pointy cars roll faster down a hill. An experiment tests cars with different shapes by measuring the time between two photogates. Your hypothesis (pointy cars roll faster) predicts that the time between photogates should be less for the pointy car.

A sloppy experiment

One group rolls each car once and records the time in a data table (Figure 2.19). Their results show a shorter time for the pointy car. This seems to confirm the hypothesis. Are they right? Have they proved that pointy cars go faster?

A better experiment

Scientifically there is no way to tell if the time difference for Group 1 is **significant** or not. A difference is only significant if the difference is greater than the amount of error. The same experiment is also done by a second group. The second group did five identical trials for each shape. Doing identical trials allowed the second group to estimate its errors. The largest difference between the average and a measured value is 0.0007 for the blunt car and 0.0006 for the pointy car.

When differences are significant

The second group can say that, scientifically, the data do NOT show that pointy cars are faster. At the speeds in their experiment, the shape of the car made no *significant* difference. In fact, shape only starts to become an important factor at speeds at least five times higher than those you will observe in your investigations.

How numbers can be different but still the same

Two results are significantly different only when the difference between them is greater than the error. This is an important consideration because experiments rarely produce exactly the same numbers twice in a row. Numbers that are "different" in a mathematical sense may not be *significantly* different in a scientific sense.

ă VOCABULARY

significant - a difference between two measured results is significant if the difference is greater than the error in measurement.

Which is faster?

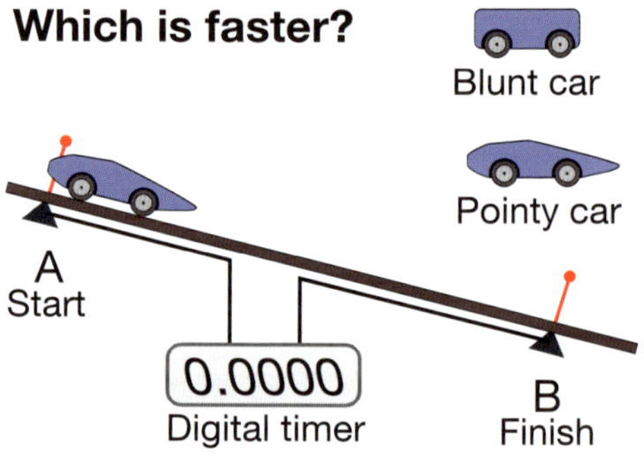

Group 1

Blunt	Pointy
0.3545 seconds	0.3542 seconds

Group 2

Blunt	Pointy
0.3540 seconds	0.3542 seconds
0.3550 seconds	0.3548 seconds
0.3541 seconds	0.3538 seconds
0.3539 seconds	0.3540 seconds
0.3546 seconds	0.3541 seconds
Averages	
0.3543 ±.0007	0.3542 ±.0006

Figure 2.19: *An experiment to test whether the shape of a car has an effect on its speed going downhill.*

2.2 Section Review

1. Georgiana has the idea that salt water heats up faster than fresh water. Which of the following would be acceptable scientific proof of her hypothesis?
   a. Three of her friends believe it.
   b. She read on an Internet website that it was so.
   c. She did an experiment that showed it.
2. A careful description of how an experiment is conducted is called
   a. a data table.
   b. a procedure.
   c. an analysis.
   d. a conclusion.
3. Explain why all variables but one should be controlled in a well designed experiment.
4. Is it possible to measure the true value of a physical property, such as mass? Explain why or why not.
5. Three groups do the same experiment rolling a car through two photogates. The track is set at the same angle and all other variables are kept the same for each group. The data is shown in Figure 2.20. The first group does the experiment in the daylight. The second group does it under a magnet. The third group does the experiment in the dark. Group 2 claims that its experiment shows that magnets make the car go faster. Is their claim supported by the evidence? Explain why or why not.
6. Which of these sets of data has an average of 10.5?
   a. 8.5, 9.5, 10.5, 11.5.
   b. 9.0, 10.0, 11.0, 12.0.
   c. 10.0, 10.5, 11.0, 11.5.
   d. 10.5, 10.6, 10.7, 10.8.
7. What does it mean when two values are different but not *significantly* different?

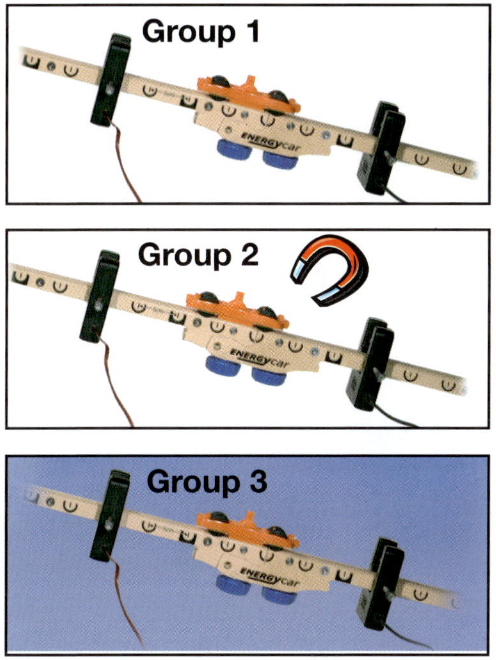

Group 1	Group 2	Group 3
Time A to B	Time A to B	Time A to B
0.343	0.339	0.341
0.346	0.338	0.337
0.341	0.345	0.344
0.340	0.341	0.340
0.349	0.339	0.339

Figure 2.20: *The three experiments of question 5.*

HEALTH CONNECTION

Testing the Power of Suggestion

Have you ever bought one brand of a common item instead of another, just because you heard a name in an ad? We all do that. Ads sometimes influence us through the power of suggestion. We recognize the name of the product. We may not be sure why. And then we buy the product.

When the power of suggestion occurs in medicine, it is called the "placebo effect." A placebo is a substance that has no healing properties. Yet this same substance causes a patient's condition to improve. Why? Because he or she believes it has the power to do so. This is the placebo effect.

Is it real? One way to try to find out is through a scientific experiment. This will test the hypothesis. A hypothesis is an idea that is based on evidence and that, most importantly, can be tested. In this case, the hypothesis is that the placebo effect is real. People who take a placebo for an illness will get well.

Let's imagine an experiment that will test the placebo effect. The hypothesis will be tested on a group of people that, or "subjects." All of the people in the experiment need to believe that they are receiving the same treatment for the same problem. They need to believe that they are all being treated in exactly the same way. This is called a "blind" test. The subjects do not know if they are receiving a medicine or other treatment or a placebo.

For our experiment, let's assume our subjects are concerned about catching a colds. They all believe that taking high doses of vitamin C helps prevent colds. Some of the subjects will be given a placebo, but they will believe it is vitamin C. If the placebo effect is real, everyone in the experiment should catch fewer colds. It should not matter if a subject is taking vitamin C or not.

Test the hypothesis

Half of the subjects will receive vitamin C pills. The other half will take pills that look the same but are placebos. For example, they might be made of powdered sugar.

To test the effect, all of the subjects take the "vitamin" pills for a month. We record how all of them respond to the pills. This becomes our experimental data. At the end of the experiment, which group caught fewer colds? If there is no difference, the placebo effect may be real.

To test the hypothesis further, we would also follow up. When the experiment ends and the powdered sugar is no longer being given to half of the test subjects, do the number of colds caught by the subjects go up?

If so, then we have more evidence to support our hypothesis. What were the results when the subjects believed they were taking vitamin C? How do those results compare with what happened when they stopped taking the pills? This is the point of comparison.

Keep conditions the same

For a good scientific experiment, other factors must also be considered. As experimenters, we need to be sure that conditions are the same at every point in the experiment. For example, how did the weather change? Did any subject's overall well-being change (with, for instance, a death in the family, a broken arm, a job loss, for instance)? Were the subjects getting the same amount of rest before and after you gave them the pills? Were the subjects eating the same types of food before and after? These factors are called "variables." For the experiment to be a success, the variables need to be controlled.

Let's assume everything was controlled in our experiment. The placebo effect is real. A person believes that a treatment will work. Then it is suggested that this type of treatment will work. The hypothesis is tested using controlled, observable, scientific methods and the results are measured.

Before a new drug can be sold, it must prove to be more effective against a disease than a placebo.

Real or imagined?

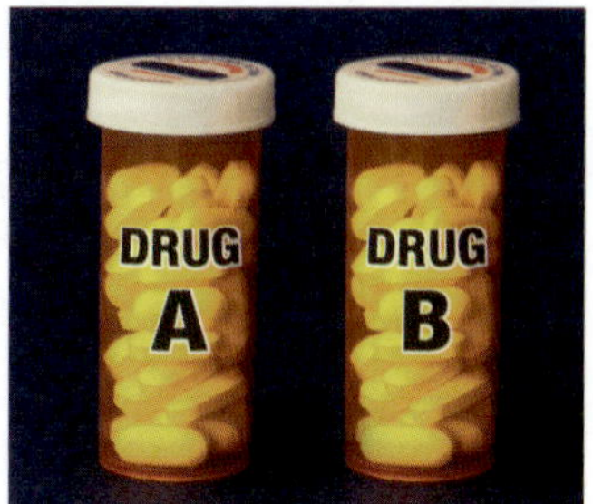

A doctor's attitude about a medication can affect how it works in a patient. Double-blind studies overcome this problem.

Some scientists believe the placebo effect is a myth. Simple experiments like one discussed cannot prove that the effect is real. This is because all of the variables cannot be controlled. Does taking the placebo reduce a subject's anxiety? Does it bring down stress levels? Does the extra attention and care the subject receives affect the results? Do encouraging messages trigger the subject's response to the placebo?

Variables play an important role in the outcome of any experiment. Most often, even the doctor conducting the test does not know which subjects are receiving a placebo. This type of test is known as a double-blind study. It means both the subject and the experimenter are "blind" about who is receiving a placebo. This is how new drugs are tested before they are sold.

Whether or not you believe in the placebo effect, it is important for all experiments to be objective. Using a placebo helps keep experiments with human subjects objective because it separates the actual effects of a medicine from effects caused by how a patient *feels* about a medicine. A doctor's attitude about a medication can also effect how it works in a patient. Using the double-blind technique is a second way of maintaining objectivity because it separates the doctor's own opinions form the results of the experiment.

Questions:

1. Why are drugs tested in double-blind studies?
2. What is the placebo effect?
3. What is a hypothesis?
4. How is a "double-blind" study set up?

CHAPTER ACTIVITY Guess the Gram

Mass is the measure of the amount of matter an object contains. An object's mass is related to its size (volume) and the density of the material from which it is made. Common units for measuring mass include the gram and kilogram. One kilogram is equal to 1000 grams. In this activity you will be using a triple beam balance to measure mass and estimate the masses of everyday objects.

Materials:

Balance capable of measuring 300 grams +/- 0.1 grams

Assorted items such as coins, paperclips, rubber bands, washers, marbles, popcorn kernels, etc.

Two kinds of mechanical balances

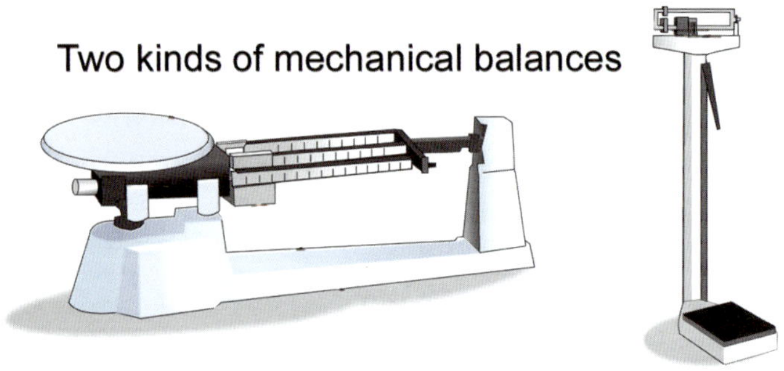

What you will do

1. Make sure your balance is calibrated properly and shows a reading of zero when it is empty. Make adjustments to the thumb screw if necessary.
2. Copy the table shown to the right.
3. Choose an object from your assortment that you believe has a mass close to one gram. Record the name of the object in the first row of the table.
4. Use the balance to measure the mass of the object to the nearest tenth of a gram. Record the mass in the table in the first try column.
5. Unless your mass was exactly one gram, make a second attempt to find one or more objects with a mass of one gram. Record the object(s) you selected, measure the mass, and record the mass.
6. Repeat with all the other masses in the table.

Mass (g)	Object(s)	Mass (g) first try	Object(s)	Mass (g) second try
1				
5				
10				
25				
50				
100				
250				

Applying your knowledge

a. Did the accuracy of your estimations increase as the activity went on?

b. Calculate your error in estimating the mass of the 10 gram object on your first try. Calculate your error in estimating on your second try.

c. Estimate the mass of the following objects in grams:
- your textbook
- an unsharpened pencil
- a sheet of notebook paper
- a compact disc

d. Estimate your mass in grams and in kilograms.

e. If you wanted to estimate the mass of a car, would you use grams or kilograms?

Chapter 2 Assessment

Vocabulary

Select the correct term to complete the sentences.

variable	dependent variable	graph
independent variable	experiment	control variable
procedure	conclusion	experimental variable
average		

Section 2.1

1. A mathematical picture that may show a pattern between two variables is a ____.
2. On a graph of two variables, the variable that causes changes in the other is the ____.
3. A quantity which can have many values is a ____.
4. The variable on a graph which is most often represented on the y-axis is the ____.

Section 2.2

5. A situation set up to investigate the relationship between certain variables is called a(n) ____.
6. A statement of what was learned in an experiment is called a(n) ____.
7. The description of how an experiment is done, including equipment, techniques used and the type of data collected is the ____.
8. The variable that you change in an experiment is called the ____ .
9. In an experiment, variables that are NOT allowed to change are ____ variables.
10. The sum of all measured values divided by the number of measurements is called the ______________ value.

Concepts

Section 2.1

1. Name four types of variables you might commonly use in doing an experiment. For each variable, name a unit and an instrument or tool that could be used to make that measurement. An example is *force-newton-spring scale*; where *force* is the variable, *newton* is the unit, and *spring scale* is the measuring instrument.
2. Identify the relationship for each graph pictured below as either direct, inverse, or no relationship.

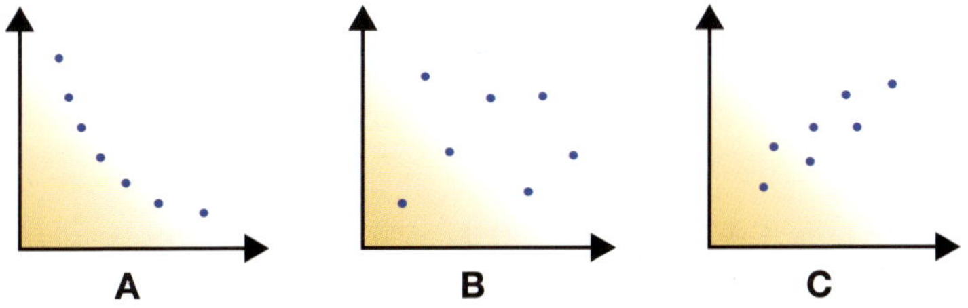

3. You have designed an experiment to test whether stretching a spring farther will cause it to shoot a marble a greater distance. You record the distances the spring is stretched, and the distances the marble travels with each release. Answer the following questions about your experiment.
   a. Name the independent variable.
   b. Name the dependent variable.
   c. Name the variable you would graph on the x-axis.
   d. Name the variable you would graph on the y-axis.

Section 2.2

4. Write an **X** next to each hypothesis that could be tested by experiment.
   a. ____ Steeper ramps result in higher speeds.
   b. ____ Red apples taste better than green apples.
   c. ____ Alien life forms are hiding on Earth.
   d. ____ A parallel universe exists that cannot be detected.

5. Celia designs an experiment to test if the speed of a wagon changes as masses are added when a constant force is applied for 2 meters. As masses are added to the wagon, she measures decreasing values of speed. She draws a graph that shows that the mass and speed are inversely related. Based on the description of Celia's experiment:

   a. Identify a control variable mentioned in the procedure.
   b. Identify the experimental variable in the procedure.
   c. Write a hypothesis for this experiment.
   d. Identify the data in this experiment.
   e. State a conclusion for this experiment.

Problems

Section 2.1

1. A graph is 20 boxes by 20 boxes. Time is plotted on the x-axis and its range is 0 to 40 minutes. Position is plotted on the y-axis and the range is 0 to 20 meters. What should the scale be for each axis of this graph?

2. The density of a material is calculated by dividing the mass of the material by its volume. If d represents density, m represents mass, and V represents volume, write a formula to represent this relationship. Is the relationship between density and volume direct or inverse?

3. Which of the following graphs shows an inverse relationship?

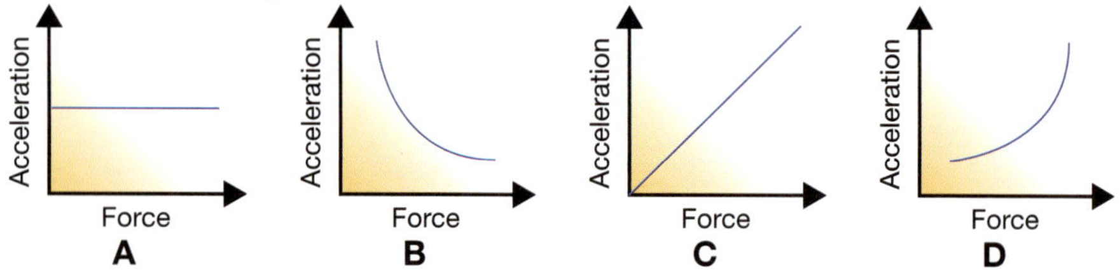

4. The graph to the right represents the position of a rocket above the ground as time passes.

   a. What is the position of the rocket after 3.0 seconds?
   b. Does this graph show a direct or inverse relationship?

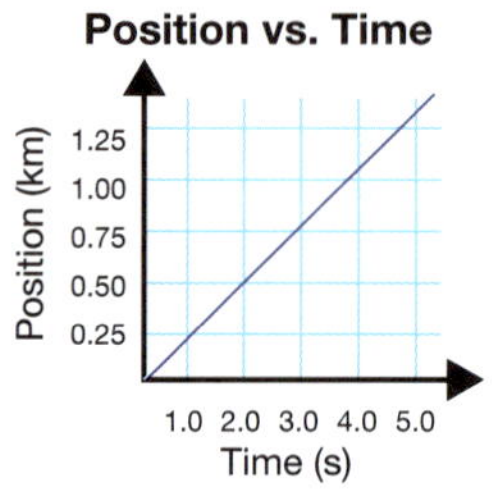

Section 2.2

5. If the width of a machine part should be 4.00 cm with an acceptable error of +/- 0.03 cm, which of the following lengths is significantly different?
   a) 4.04 b) 3.98 c) 4.29

6. Jaden measured the mass of a 2002 penny on a mass balance 4 different times. Kylie repeated the same measurement 4 times using the same penny and the same mass balance. Their results are shown below:

Trial	Jaden	Kylie
1	2.45 g	2.51 g
2	2.49 g	2.50 g
3	2.55 g	2.49 g
4	2.51 g	1.70 g

   a. What is the average of Jaden's measurements? Of Kylie's measurements?
   b. Who had the most accurate measurements? Explain your reasoning.
   c. What is the "true" mass of the penny?

Chapter 3

Key Concepts in Physical Science

What would you answer if someone asked you "what is everything?" In science "everything" means all the matter and energy in the universe. Matter and energy are key concepts in physical science because they make up the natural world. Einstein's famous equation $E = mc^2$ is often used to represent science in cartoons, movies, and popular culture. The "E" in the equation stands for energy, and the "m" in the equation represents matter. Einstein used this equation to relate energy and matter.

Almost all the matter you can touch, taste, or feel is made of incredibly tiny particles called atoms. An atom is so small that if one atom were the size of a marble, you would be about the size of the entire planet Earth! In this chapter, you will learn about atoms, matter and energy – some of the most important ideas in science!

Key Questions

1. *What is matter made of?*
2. *What is energy and where does it come from?*
3. *Are temperature and heat the same thing or are they different?*

3.1 Mass and the Atomic Theory of Matter

Mass describes the amount of matter in an object. A car has more mass than a bicycle because the car contains more matter (Figure 3.1). Steel, plastic, and rubber are different kinds of matter and a car has a lot more of each kind than a bicycle. Ordinary matter is made of small particles called atoms. Atoms are so small they cannot be seen even with a powerful microscope. To imagine how atoms could be real and yet unseen, look at a sugar cube. Held in your hand, a sugar cube looks like a single piece of matter. But up close, you can tell it is made up of tiny, individual crystals of sugar fused together. Matter is made up of atoms in a similar way.

Grams and kilograms

Kilograms Mass is measured in kilograms (kg). Most of the world uses kilograms for daily measurement, such as buying food (the U.S. is an exception). A bunch of bananas or a 1-liter bottle of soda each have a mass of about 1 kilogram. People have a mass of around 55 kilograms. Common machines range in mass from a bicycle (about 12 kg) to a motorcycle (about 200 kg) or car (1,000 - 2,000 kg). You should try to develop an intuitive sense for how much mass there is in one kilogram since this is a basic quantity in science.

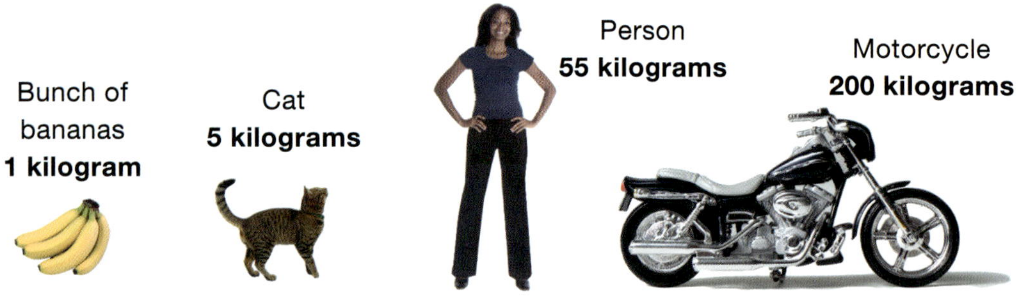

Grams For small amounts of mass, the kilogram is too large a unit to be convenient. One gram (g) is one-thousandth of a kilogram. One grain of rice has a mass of about a gram, so a bag of 1,000 grains of rice has a mass of approximately 1 kilogram.

VOCABULARY

kilogram (kg) - the basic metric (SI) unit of mass.

gram (g) - a unit of mass smaller than a kilogram. One kg equals 1,000 g.

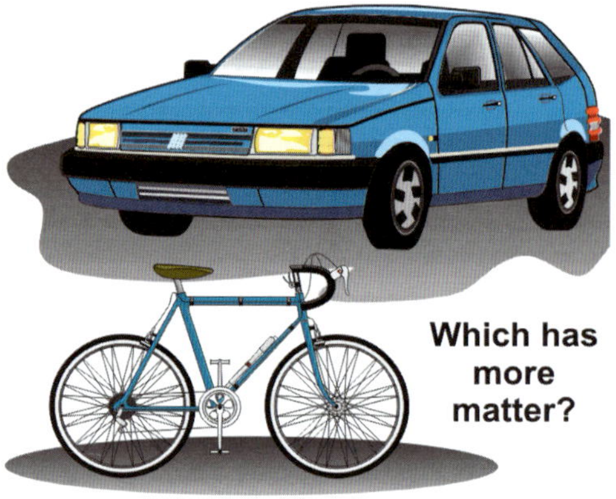

Figure 3.1: *A car contains more matter than a bicycle, therefore it has more mass.*

CHALLENGE

Where did the unit of kilograms come from? Research the origin of the metric system to discover how the kilogram was first defined and what other units of mass it replaced.

Measuring mass in the laboratory

Using a mass balance

In the laboratory you will usually measure mass with a balance. The balance displays mass in grams. For example, the balance in Figure 3.2 shows the mass of six steel nuts to be 96.2 grams. A single gram is a tiny mass and balances are therefore sensitive (and quite delicate). Never drop things onto a balance! Instead, set things gently on the balance.

Converting grams to kilograms

For many calculations you will need to convert masses from grams to kilograms. To convert a mass in grams to kilograms, you need to divide by 1,000 since there are 1,000 grams in a kilogram (g/kg).

Masses you will consider in science

Ordinary objects tend to have masses between a few grams and a few hundred kilograms. You will encounter a much wider range of masses in science. A bacteria has a mass of 0.000000001 kg. That seems small — but then an atom has a mass a thousand billion times smaller. Science also concerns large masses, such as planets and stars. A star like our sun has a mass of 2 million trillion trillion kilograms.

Electronic balance

Converting from grams to kilograms

$$96.2\ \cancel{g} \times \frac{1\ \text{kg}}{1{,}000\ \cancel{g}} = 0.0962\ \text{kg}$$

$$96.2\ \text{g} = 0.0962\ \text{kg}$$

Figure 3.2: *A balance displays mass in grams. You may need to convert grams to kilograms when doing calculations.*

Converting mass units

A laboratory balance shows the mass of a banana is 175.5 grams. How much is this in kilograms?

1. Looking for: You are asked for the mass in kilograms.
2. Given: You are given the mass in grams.
3. Relationships: There are 1,000 grams in one kilogram.
4. Solution: $175.5 \div 1{,}000 = 0.1755$ kg

Your turn...

a. A sack of onions has a mass of 5 kilograms. Can this sack be measured using a balance that reads up to 500 grams? **Answer: No; 5 kg = 5,000 g**

b. Convert 1.77 kilograms to grams. **Answer:** 1,770 g

Atoms

The smallest piece of matter Suppose you wanted to make the smallest possible piece of gold. You cut a pure gold coin into smaller and smaller pieces until you can't cut it any smaller. That smallest possible piece is one atom. A single **atom** is the smallest amount of gold (or any element) you can have.

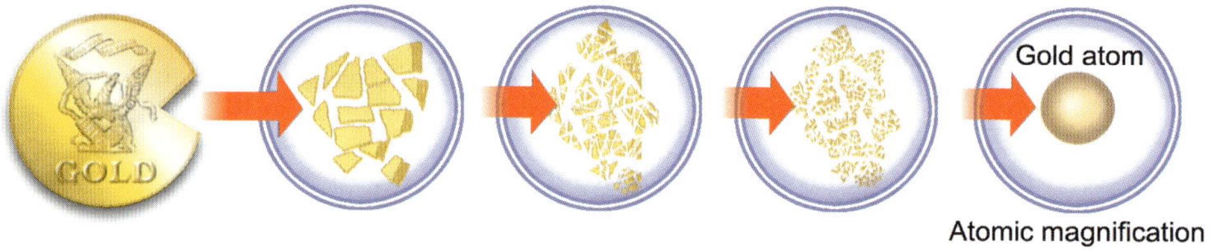

Atoms We know ordinary matter is made up of atoms. Atoms make up everything that we see, hear, feel, smell, and touch. We cannot experience atoms directly because they are so small. Aluminum foil is thin but is still more than 200,000 atoms thick (Figure 3.3).

Elements An **element** is a pure substance (like gold) that cannot be broken down into other elements. All of the matter you are ever likely to experience is made from one or more of 92 naturally-occuring elements. Each of those 92 elements has a unique type of atom. All atoms of a given element are similar to each other. If you could examine a million atoms of carbon, you would find them all to be similar. But carbon atoms are different from sodium, aluminum, or oxygen atoms. The atoms of an element are similar to atoms of the same element but different from atoms of other elements.

VOCABULARY

atom - the smallest particle of matter that retains the identity of its element, such as an atom of gold.

element - a pure substance that cannot be broken down into other elements.

Figure 3.3: *Even a thin sheet of aluminum foil is 200,000 atoms thick. NOTE: In the illustration, the atoms are drawn much larger than they are in reality. You could never see an atom with a magnifying glass.*

How atoms and elements explain chemical changes

Compounds The properties of matter depend mostly on how atoms of different elements are combined in *compounds*. A compound is a substance that is made up of more than one element. Pure elements are rare. Most matter exists in the form of compounds.

Properties depend more on compounds than on elements The properties of a compound are usually different from the properties of the pure elements that make up the compound. For example, salt is a compound of the elements sodium and chlorine (Figure 3.4). The properties of salt are quite different from the properties of sodium or chlorine. The pure element sodium is a soft, silvery metal. Pure chlorine is a yellow-green, toxic gas. Yet the compound of the two (salt) is a hard white crystal used to flavor food!

Salt crystal

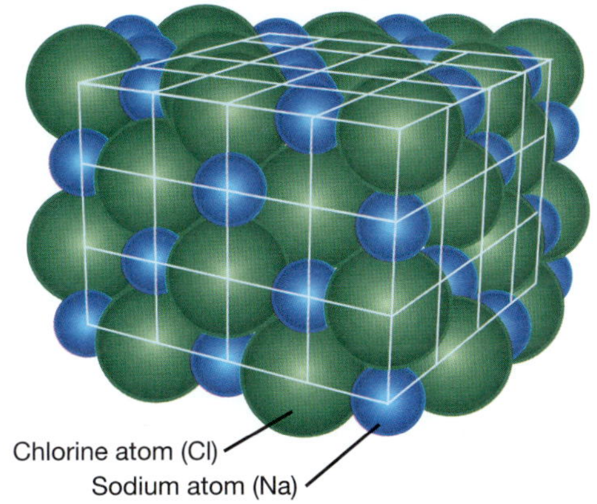

Figure 3.4: *Table salt is a compound made of sodium and chlorine atoms.*

Matter can change its properties

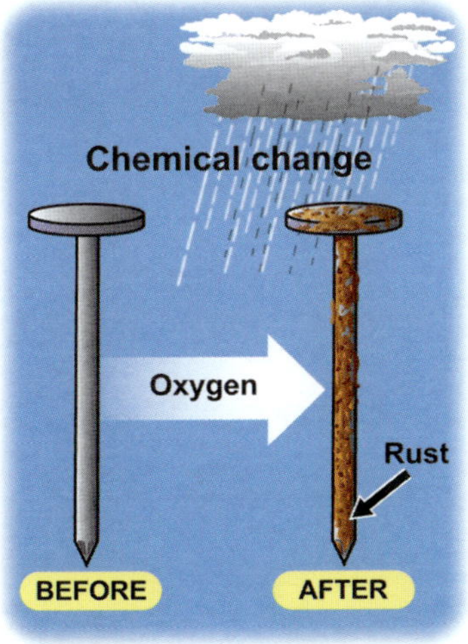

One kind of matter can change into another kind of matter that has different properties than the original matter. For example, if you leave an iron nail out in the rain, the silver-colored surface soon turns brown with rust. Scaly, brown rust is so different from hard, silvery steel. How does one turn into the other?

Chemical changes can rearrange atoms Rust forms through a chemical change between iron in the nail and oxygen in the air. Chemical changes rearrange atoms into different molecules and compounds. Remember, the properties of matter depend on the arrangement of atoms in a compound. The iron atoms in the nail combine with oxygen atoms from the air and water to make rust. Rearranging atoms into new compounds is how one kind of matter changes into another kind with different properties.

How atoms explain solids, liquids, and gases

Atoms in a solid The concept of atoms explains how the same substance can be a solid, a liquid, or even a gas. In a gold ring, every atom of gold is attached to its neighboring atoms and cannot easily move (Figure 3.5). At room temperature, gold is solid because the individual atoms are bound firmly to each other. They still wiggle around, but do not have enough energy to break away from their neighbors.

Atoms in a liquid A liquid holds its volume, but not its shape — it flows. Liquids flow because the atoms can move around. If you heat gold with a torch, it finally melts and becomes liquid when the temperature reaches 1,064°C (1,947°F). The atoms in a liquid have enough energy to temporarily break their attachments to their neighbors. The atoms are about as close together as they are in a solid, but they continually change neighbors, like changing partners in a fast dance.

Atoms in a gas A gas flows like a liquid, but can also expand or contract to fill a container. A gas can expand or contract because the atoms are completely "unbonded" from each other and are relatively far away from each other. If you kept heating the molten gold up to 2,856 °C (5,173°F), it would finally boil and turn into gold gas. Like other gases, gold gas expands to fill any container because the atoms are free to move around. Atoms in a gas are much farther apart than atoms in a liquid or solid.

Temperature Atoms are never still. Atoms are always moving, vibrating around like drops of water on a hot griddle. You can tell from a substance's temperature the amount of energy of motion each atom has. When the temperature is low, each atom has very little energy of motion. That is why solids form at low temperatures. When the temperature is high, each atom has a lot of energy of motion. Gases form at high temperatures because the atoms have enough energy to fly away from each other instead of sticking together.

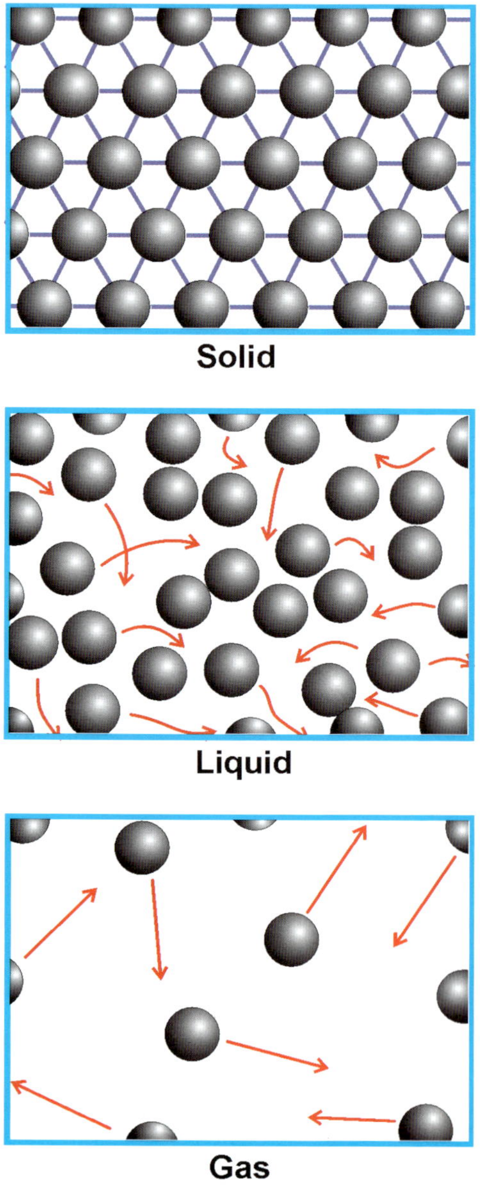

Figure 3.5: *Atoms in a solid, a liquid, and a gas.*

3.1 Section Review

1. Which of the following objects has a mass of about one kilogram?
    a. A golf ball.
    b. A one-liter bottle of soda.
    c. A medium-size dog.
    d. A motorcycle.
2. How many grams are there in 2.2 kilograms?
3. The objects on the right have the same mass. That means they:
    a. have the same size.
    b. have the same elements in them.
    c. are all the same phase of matter.
    d. contain the same amount of matter.
4. Name one substance that is solid, one substance that is liquid, and one substance that is a gas at ordinary room temperature.
5. Which of the three phases of matter (solid, liquid, gas) would be the best to use to build a bridge? Explain.
6. Is water a compound or an element? Explain.
7. Atoms of iron are much more tightly attached to each other than are atoms of lead. That means it takes more energy to separate iron atoms from one another than to separate lead atoms. Based on this information, which of the following statements is probably true?
    a. An iron bar is harder to bend than a lead bar of the same size.
    b. An iron bar is easier to bend than a lead bar of the same size.
    c. An iron bar has more mass than a lead bar of the same size.
    d. An iron bar has less mass than a lead bar of the same size.

Figure 3.6: *The four objects for question 3.*

STUDY SKILLS

In the metric system, you can convert from one unit to another by shifting the decimal point.
For example:

0.056 kg = 56 grams

2.55 kg = 2,550 grams

125 grams = 0.125 kg

8 grams = 0.008 kg

Moving the decimal point to the left three digits is the same as dividing by 1,000. Moving the decimal point to the right three digits is the same as multiplying by 1,000.

3.2 Temperature and Energy

If you look around, you are reminded that nothing stays the same for long. Things are always changing. Motion, chemical reactions, and a cup of coffee cooling down are all examples of change. Why do these changes occur? How do the changes occur? This section is about how things change through the exchange of energy. The flow of energy connects matter, time, and space and causes things to change.

Two important concepts: systems and energy

What is a system?

To learn science, we have to break up the complex universe into pieces small enough to understand. We call these pieces **systems**. In science, the word "system" means a small group of related things that work together. For example, your eye is a system that contains many parts, including a lens, pupil, and retina (Figure 3.7).

We learn science one system at a time

There are big, complex systems like the solar system. Gravity ties the sun and planets together so that they all interact with one another. There are also little systems within the big systems, like the weather on Earth (which is part of the solar system). Your body is a big, complex system. All the organs in your body (like your eye) are smaller systems.

The parts of a system change by exchanging energy

The parts of a system interact with each other by exchanging **energy**. Energy is a quantity that measures the ability to cause change. Anything with energy can change itself or cause change in other objects or systems. Energy can cause changes in temperature, speed, position, mass, or other physical variables. Energy can cause change in materials, such as burning wood changing into smoke and ashes.

- A gust of wind has energy because it can move objects in its path.
- A piece of wood in a fireplace has energy because it can change into smoke and ashes, releasing heat and light.
- You have energy because you can change the motion of your own body.

VOCABULARY

system - a small group of related things that work together.

energy - a quantity that measures the ability to cause change in the physical system. Energy is measured in joules.

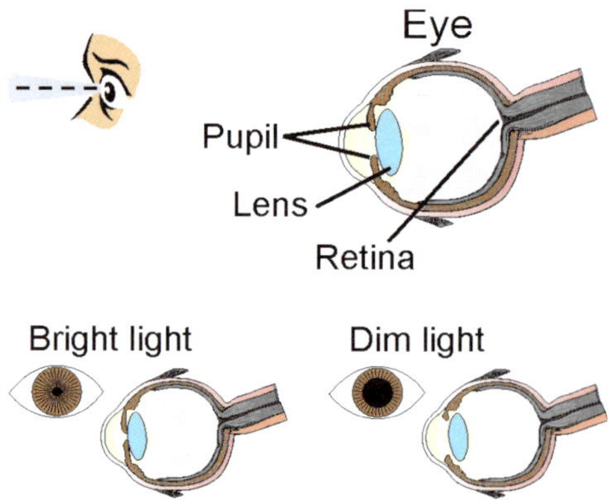

Figure 3.7: *Your eye is a system that includes the lens, pupil, and retina. All the parts of the system work together to create your sense of vision. The pupil opens in dim light to allow more light into the eye. The pupil closes partly in bright light.*

Forces and energy

Force is measured in newtons

If you want to move a box, you must apply a **force** to it. Force is a fundamental quantity in physical science. In the metric system, force is measured in **newtons** (N). A newton is a small unit of force. To lift the average person takes a force of 500 to 1,000 newtons.

Energy is measured in joules

The unit of energy is related to the units for force and distance. One **joule (J)** of energy is enough to push with a force of one newton for a distance of one meter. So a joule is about as much energy as it takes to pull a small book across a table. If you have more energy, you can pull with more force or for a greater distance — or both.

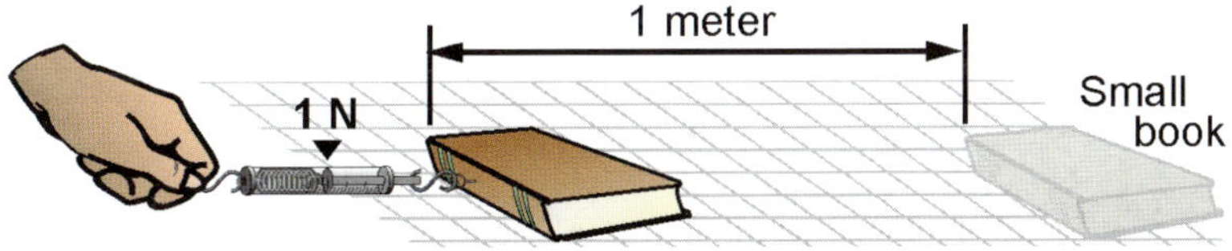

One joule of energy is enough to apply a force of 1 N over a distance of 1 meter

Forces can transfer energy from one object to another

Energy often moves through the action of forces. In fact, one important property of energy is the *stored ability to create force*. An object which has energy may exert forces to transfer its energy to another object. If an object has absolutely *zero energy,* then no forces may be produced. Another way to think about force is as an action that may transfer energy from one object to another.

Potential energy

Objects may have energy due to their height. For example, a book high on a shelf has more energy than a book on the floor (Figure 3.8). This kind of energy is called potential energy and comes from Earth's gravity.

Kinetic energy

Objects in motion have energy due to their speed. Energy of motion is called kinetic energy. The faster an object is moving, the more kinetic energy it has. If an object has kinetic energy, it may transfer that energy to other objects by applying forces. For example, if you catch a falling book, the book exerts a force against your hand. The energy of motion of the book is transferred to your hand through that force.

VOCABULARY

force - a push, pull, or any action which has the ability to change motion. Force is measured in newtons.

newton (N) - the unit of force in the metric (SI) system.

joule (J) - the unit of energy in the metric (SI) system. One joule is enough energy to push with a force of one newton for a distance of one meter.

Figure 3.8: *Energy of motion (kinetic energy) and energy of height (potential energy).*

Temperature

Hot and cold You have seen a thermometer hung outside someplace to register the **temperature**. Temperature is the measurement we use to make the sensations of hot and cold more precise. Experience has taught you to wear a coat if the weather forecast predicts a high of 40°F, and to wear shorts if a high of 95°F is predicted. But what is different about the air that makes it feel 95 degrees rather than 40?

Thermometers A **thermometer** is an instrument used to accurately measure temperature. The common alcohol thermometer uses the expansion of colored liquid alcohol to show temperature. As the temperature increases, the alcohol expands and rises up a long, thin tube. You read the temperature by looking at the mark the alcohol reaches.

Fahrenheit scale You are probably most familiar with the English system of measuring temperature, known as the **Fahrenheit scale**. It was developed in 1714 by Gabriel Fahrenheit, a German physicist who was the first person to use a mercury thermometer. He chose the lowest temperature he could create in his lab (using water, salt, and ice) to be the zero point of his scale. For the other end of the scale he used the temperature of the human body as 100 degrees. Eventually the Fahrenheit scale was standardized so that the freezing point of water is 32 degrees and the boiling point is 212 degrees.

Celsius scale In 1742, Anders Celsius, a Swedish astronomer, invented a temperature scale in which there were 100 degrees between freezing and boiling. He called it the centigrade scale. In 1948 this official scale of the metric system was named the **Celsius scale** in his honor. Most countries use the Celsius scale. Figure 3.9 shows how the two temperature scales compare.

VOCABULARY

temperature - a measurement of hot or cold that depends on the thermal energy in a material.

thermometer - an instrument used to measure temperature.

Fahrenheit scale - temperature scale in which water freezes at 32 degrees and boils at 212 degrees.

Celsius scale - temperature scale in which water freezes at 0 degrees and boils at 100 degrees.

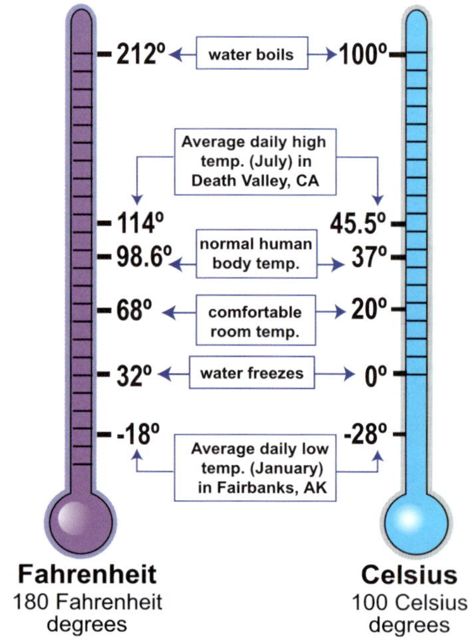

Figure 3.9: *The Celsius and Fahrenheit temperature scales.*

Converting between Fahrenheit and Celsius

Converting between the two scales

A weather report forecasting 21°C in Barcelona, Spain, predicts a pleasant day, suitable for shorts and a T-shirt. A weather report predicting 21°F in Minneapolis, Minnesota, suggests wearing a heavy winter coat, gloves, and a hat. Because the United States is one of the few countries that still use the Fahrenheit scale, it is useful to know how to convert between the two temperature scales.

CONVERTING BETWEEN FAHRENHEIT AND CELSIUS

$$T_{Fahrenheit} = \frac{9}{5}T_{Celsius} + 32 \qquad T_{Celsius} = \frac{5}{9}(T_{Fahrenheit} - 32)$$

STUDY SKILLS

Using the proper units

Temperatures in Fahrenheit and Celsius are easy to confuse. Science usually works in Celsius, as do most countries. However, the United States uses Fahrenheit for everyday communication. Get in the habit of writing the units whenever you write down a number in science. For example, write 10°C instead of just 10 or 25.5 grams instead of just 25.5. This will keep you from getting confused later on.

+ − × ÷

Converting between temperature scales

A friend in Paris sends you a recipe for a French cake. The recipe says to bake the cake at a temperature of 200°C for 45 minutes. At what temperature should you set your oven, which uses the Fahrenheit scale?

1. Looking for: You are asked for the temperature in degrees Fahrenheit.
2. Given: You are given the temperature in degrees Celsius.
3. Relationships: Use the conversion formula: $T_F = {}^9/_5 T_C + 32$.
4. Solution: $T_F = ({}^9/_5)(200) + 32 = 392$ °F.

Your turn...

a. You are planning a trip to Iceland this summer. You find out that the average July temperature in Iceland is 11.2°C. What is the average July temperature in degrees Fahrenheit? **Answer:** 52.2°F

b. You are doing a science experiment with a Fahrenheit thermometer. Your data must be in degrees Celsius. If you measure a temperature of 125°F, what is this temperature in degrees Celsius? **Answer:** 51.7°C

Temperature extremes

Absolute zero

There is a limit to how cold matter can get. **Absolute zero** is -273°C (-459°F). You cannot have a temperature lower than absolute zero. As temperature is reduced, all atoms move more and more slowly. At -273°C atoms have the lowest energy they can have and the temperature cannot get any lower. Think of absolute zero as the temperature at which atoms are "frozen." Technically, molecules never completely stop moving, but at absolute zero their energy is so small they might as well be stopped.

High temperatures

Unlike with absolute zero, there is no maximum temperature for matter. As temperature increases, exotic forms of matter appear. For example, at 10,000°C atoms start to come apart and become a *plasma*. In a plasma, the atoms are broken apart into separate positive ions and negative electrons. Plasma conducts electricity and is formed in lightning and inside stars.

Temperatures in the Solar System

When compared with temperatures in other parts of the universe, temperatures on Earth fall in a small range (Figure 3.10). In the solar system, temperatures can range from a low of approximately -270°C (-454°F) between the outer planets to a high of 15 million °C at the center of the sun.

Temperatures on Earth

The coldest place on Earth is Antarctica. The lowest temperature ever recorded there was -89°C (-129°F) in 1983. The hottest temperature recorded on Earth was in Libya, in North Africa, in 1922 when it was 58°C (136°F). The hottest temperature recorded in California is in Death Valley where it reached 57°C (135°F). Much colder and hotter temperatures have been produced in laboratories. Scientists have experimented with temperatures as low as billionths of a degree above absolute zero and as high as 100 million million degrees Celsius.

VOCABULARY

absolute zero - the lowest temperature there can be, equal to -273.3°C or -460°F.

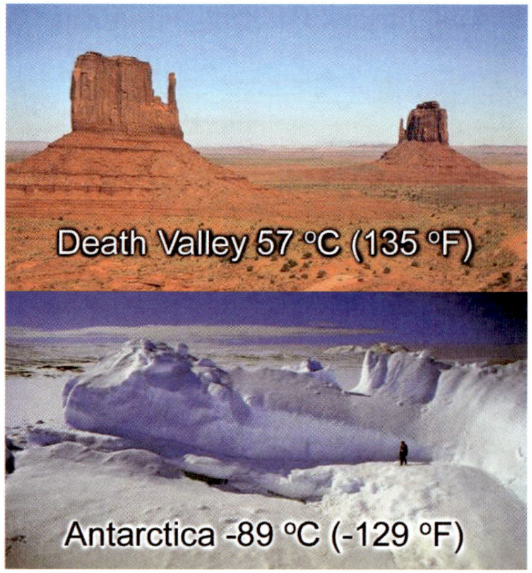

Figure 3.10: *Some extremes of temperature on Earth.*

CHALLENGE

Can you think of a reason the temperature cannot go lower than absolute zero? Hint: Think about the energy of molecules as they get colder and colder.

Heat and thermal energy

Temperature measures thermal energy

Temperature measures a kind of energy called **thermal energy**. The higher the temperature, the more thermal energy is present. A hot object has more thermal energy than the same object when it is cold. On the molecular level, thermal energy comes from the motion of atoms in matter. When matter is hot, atoms are vigorously moving with lots of kinetic energy. When objects are cold, the atoms move slower and have less kinetic energy. Fundamentally, thermal energy is the total energy stored in the kinetic energy of individual atoms.

Heat is moving thermal energy

When you put something hot in a cold room, the temperature of the hot object decreases. The temperature of the cold air in the room increases. This is because thermal energy flows from hot to cold, or from higher temperature to lower temperature. We call thermal energy that is flowing **heat**. Heat flows from hot to cold whenever two quantities of matter at different temperatures are allowed to interact with each other (Figure 3.11).

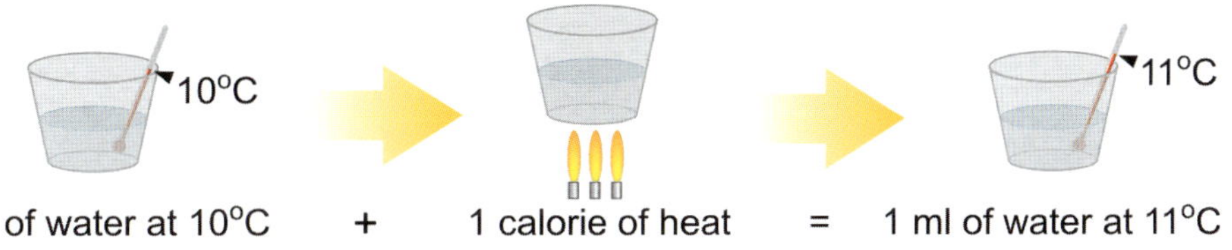

Calories

Thermal energy is often measured in **calories**. One calorie is the amount of energy it takes to raise the temperature of one milliliter of water by one degree Celsius. The calorie is a slightly larger unit of energy than the joule. One calorie is 4.184 joules. There are two different units because scientists started measuring heat in calories before they knew heat was a form of energy.

Food energy is measured in kilocalories

Joules and calories are tiny amounts of energy compared with the energy used around you every day. Because the joule is so small, the energy in foods is measured in kilocalories, also called food calories, or just Calories with a capital "C." One Calorie (kilocalorie) is equal to 4,184 joules. The energy in a single jelly doughnut (200 Calories) is an astounding 837,000 joules.

VOCABULARY

thermal energy - energy that is due to difference in temperature. Thermal energy comes from kinetic energy of individual atoms.

heat - thermal energy that is moving.

calorie - a unit of energy equal to 4.184 joules or the energy needed to heat 1 gram of water by 1°C.

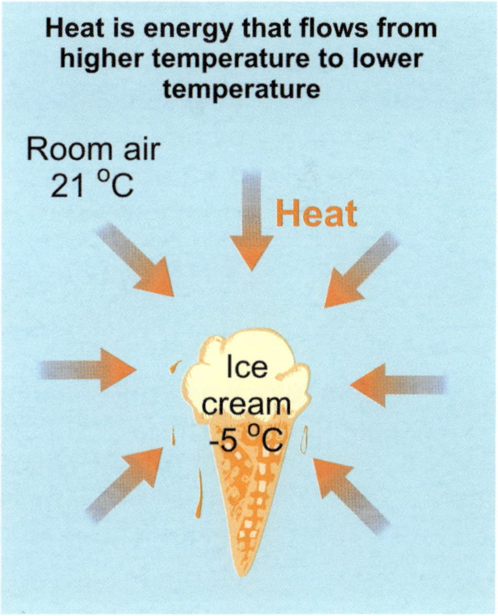

Figure 3.11: *Heat is the flow of thermal energy from higher temperature to lower temperature.*

Specific heat

Temperature and mass

If you add heat to an object, how much will its temperature increase? It depends in part on the mass of the object. If you double the mass of the object you are going to heat, you need twice as much energy to increase the temperature.

Temperature and type of material

The amount of temperature increase also depends on the kind of material you are heating. It takes different amounts of energy to raise the temperature of different materials. You need to add 4,184 joules of heat to 1 kilogram of water to raise the temperature by 1°C. (Figure 3.12). You only need to add 470 joules of heat to raise the temperature of a kilogram of steel by 1°C. It takes nine times more energy to raise the temperature of water by 1 degree than it does to raise the temperature of the same mass of steel by 1 degree.

Specific heat

The **specific heat** is a property of a substance that tells us how much heat is needed to raise the temperature of one kilogram of a material by one degree Celsius. A large specific heat means you have to put in a lot of energy for each degree increase in temperature. Specific heat is measured in joules per kilogram per degree Celsius (joule/kg°C).

The specific heat is the energy that will raise the temperature of one kilogram by 1°C.

Uses for specific heat

Knowing the specific heat tells you how quickly the temperature of a material will change as it gains or loses energy. If the specific heat is *low* (like it is for steel), then temperature will change relatively quickly because each degree of change takes less energy. If the specific heat is *high* (like it is for water), then temperature will change relatively slowly because each degree of change takes more energy. Next time you have some for dessert, remember that your apple pie filling will stay hot for a long time because it is mostly water and water has a high specific heat.

VOCABULARY

specific heat - a material property that tells how much energy is needed to change the temperature by one degree.

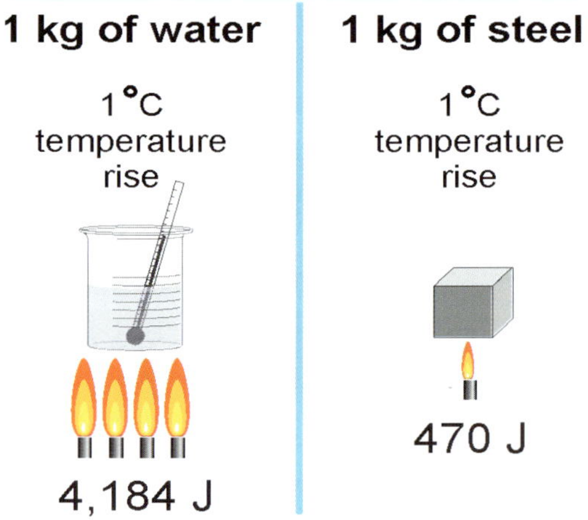

Figure 3.12: *It takes 4,184 joules to raise the temperature of 1 kilogram of water by 1°C but only 470 joules for a kilo of steel.*

Material	Specific heat (J/kg°C)
water	4,184
aluminum	900
steel	470
oil	1,900
concrete	880
glass	800
wood	2,500

Figure 3.13: *Specific heat values of some common materials.*

Conservation of energy

Law of conservation of energy One of the most important of all the natural laws concerns energy. The **law of conservation of energy** says that energy can never be created or destroyed, just converted from one form into another. The law of conservation of energy applies to all forms of energy, including potential energy, kinetic energy, and chemical energy.

Energy can never be created or destroyed, just converted from one form into another.

Energy transformations Although it cannot be destroyed, energy can be transformed easily between its different forms. The diagram in Figure 3.14 shows the energy transformations that produce the electrical energy you use.

Energy from water and wind Many parts of California get energy from water and wind. In hydroelectric power, energy is released by falling water. The potential energy of the water is converted into electrical energy. In wind power, the kinetic energy in moving air is converted into electrical energy.

Chemical energy Chemical energy is energy that is stored in the bonding forces between atoms. Chemical reactions may use or release chemical energy. A car engine converts about 13 percent of the chemical energy in gasoline into kinetic energy of the moving car. Much of the remaining 87 percent of the energy in the gasoline becomes heat in the exhaust gases and radiator.

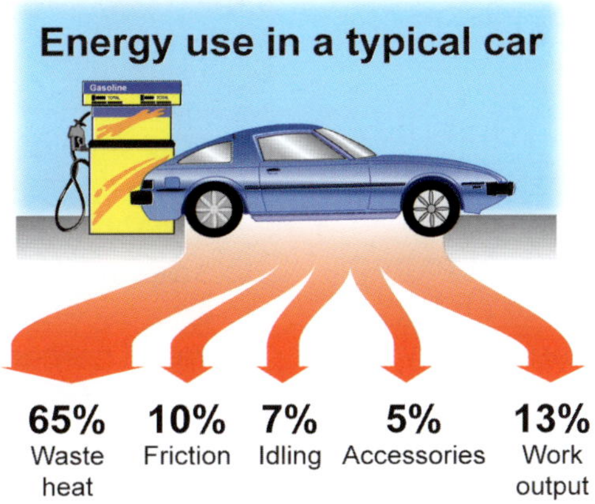

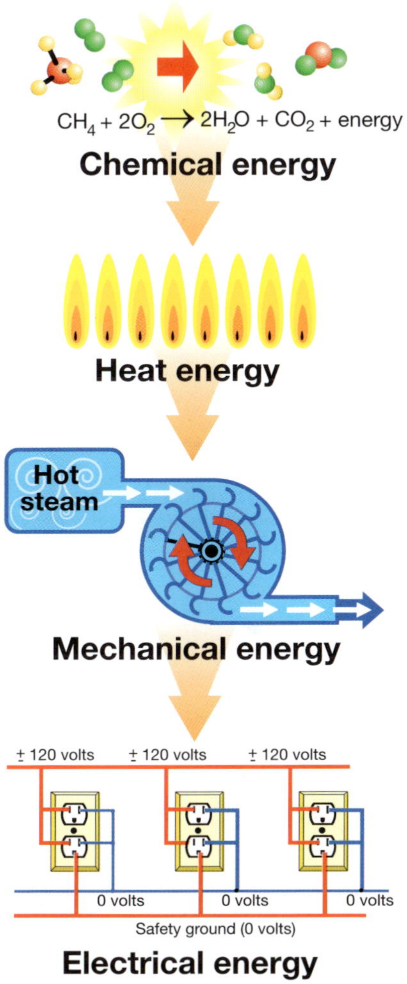

Figure 3.14: *.The electricity in your wall outlets at home carries energy that has been transformed several times before becoming electrical energy.*

'Using' and 'conserving' energy in the everyday sense

Conserving energy

We have all heard that it's good to "conserve energy" and not waste it. This is good advice because energy from gasoline and electricity costs money and uses resources. But what does it mean to "use energy" in the everyday sense? If energy can never be created or destroyed, how can it be "used up"? Why worry about "running out" of energy?

Using energy

When you "use" energy by turning on a light you are converting energy from one form (electricity) to other forms (light and heat). What gets "used up" is the amount of energy *in the form of electricity*. Electricity is a valuable form of energy because it is easy to move long distances (through wires). So the energy is not truly "used up" but is just converted into other forms. The total amount of energy stays constant.

Power plants

Electric power plants do not *make* electrical energy. Energy cannot be created. What power plants do is convert other forms of energy (chemical, solar, nuclear) into usable electrical energy. When someone advises you to turn out the lights to conserve energy, they are asking you to use less *electrical energy*. If people used less electrical energy, power plants would have to burn less oil, gas, or other fuels to "produce" the electrical energy.

Running out of energy

Many people are concerned about "running out" of energy. What they mean is running out of certain *forms* of energy that are easy to use, such as oil and gas. When you use gasoline in a car, the chemical energy in the gas mostly becomes heat energy. It is impractical to put the heat energy back into the form of gasoline, so we say the energy has been "used up" even though the energy itself is still there, only in a different form. Earth contains a limited amount of oil and gas. When what we have is gone — that is, converted from gas and oil into other forms — there will be no more. There will still be plenty of energy, just not in the form of oil or gas. Those two are valuable because much of our economy has been developed around using the energy contained in them.

CHALLENGE

Please turn out the lights when you leave!

There are about 285,000,000 people living in the United States. If an average house has four light bulbs per person, it adds up to 1,140,000,000 light bulbs. The average bulb uses 100 joules of electrical energy each second. Multiplying it out gives an estimate of 114,000,000,000 joules every second — just for light bulbs.

A big electric power plant puts out 2 billion joules each second. That means 67 big power plants are burning up resources just to run your light bulbs. If everyone were to switch their incandescent bulbs to fluorescent lights, we would save 75 percent of this electricity. That means we could save 50 big power plants' worth of pollution and wasted resources.

3.2 Section Review

1. Write a paragraph about a system inside your home or school building. Describe what the system does as a whole. Describe at least three parts of the system. For each part, describe how it contributes to the function of the whole system.
2. Scientists would like to understand many things that are large and complex, like the ecology of Earth. Scientists divide complex things into smaller groups called systems because:
   a. It is easier to understand a small group than a large complex thing.
   b. There is not enough money to study the entire complex thing.
3. Which is the higher temperature: 30°C or 60°F?
4. A cook warms up 1 bowl of soup from 20°C to 60°C (Figure 3.15). Another cook warms up 10 identical bowls of soup from 20°C to 60°C. Which of the following statements is FALSE?
   a. Both cooks use the same amount of heat (energy) because the temperatures are the same.
   b. Both cooks use different amounts of heat (energy) even though the temperatures are the same.
5. Describe the flow of energy as a cup of hot coffee cools down as it sits on a table.
6. True or false: If the same amount of heat is applied to water, steel, and wood, the temperature of each one will rise by the same amount.
7. Imagine you are the teacher of a science class. A student brings in a newspaper article that claims the world will run out of energy by the year 2050 because all the oil will be pumped out of the planet. The student is confused because she has learned in science that energy can never be created or destroyed. How would you explain to her what "running out of energy" means in the article.

CHALLENGE

Research what is going on in your community around energy conservation. Write about a project that is anticipated to save energy. How much energy might be saved?

CHALLENGE

Every month your family pays an electric bill for energy you have used. Research the cost of electricity in your area. How much does it cost for 1 million joules? This is the amount of energy used by a single electric light bulb in three hours.

Figure 3.15: *Question 4*

ECOLOGY CONNECTION

A Mighty Energizing Wind

There is a new kind of farm that is unlike any other - it reaps the wind. These farms can help solve the energy crisis by generating electricity from the powerful forces in wind.

Not that long ago, most farms in the United States had a windmill. It was used to pump water from a well to supply the farm's needs. These days an electric motor pumps the water, and the old windmill is gone or just an antique.

New windmills, however, are going strong. Tower-mounted wind turbines that are far larger and more efficient have replaced the old models. When these big turbines are grouped, they form a wind farm. They are being built on land that is still used for farming. With support from industry and the government, wind farms are sprouting across the country. Researchers are finding ways to improve windmill efficiency and solve the issue of low wind speed.

A wind turbine is almost the opposite of a fan. A fan uses electricity to make wind; the turbine uses wind to make electricity. The operation is quite simple: Wind turns the turbine's blades, which spins a shaft that connects to a generator, which produces electricity. The old farm windmills had several blades on a small metal or even wooden tower. Today's wind turbines have two or three blades mounted on towers that may be hundreds of feet tall.

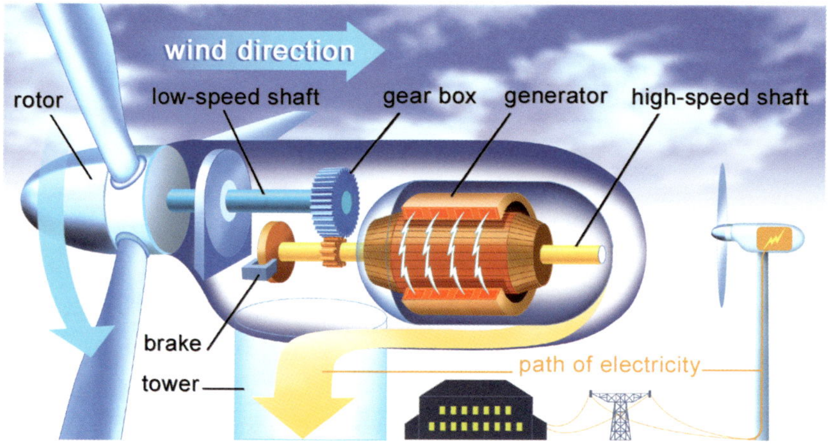

The promise of wind's power

According to the U.S. Department of Energy, wind power costs 4 to 6 cents per kilowatt-hour. Coal-fired power costs 4.8 to 5.5 cents, and natural gas can cost as little as 4 cents, so wind power is competitive. And it has notable advantages.

- A clean fuel source that does not pollute the air like coal- or gas-burning power plants.
- Does not need to be imported.
- Abundant resource - it will never run out.
- Requires no mining or drilling.
- One of the lowest-priced renewable energy technologies available.

Wind power can also benefit the economies of rural areas. The power companies pay rent to landowners. Since the turbines occupy only a fraction of the land on a wind farm, the landowner can farm around the towers.

Obstacles, naturally

The biggest problem with wind power is obvious: Wind is intermittent. It cannot be counted on to blow when electricity is needed. It does not blow at a steady rate. Another problem is that electricity from wind cannot be stored, except in batteries.

Also, the best sites for wind farms are often in remote locations, in mountains or deserts, far from cities where the most electricity is needed. The map of the United States shows wind energy potential. Find your state to see how windy it is compared with other states.

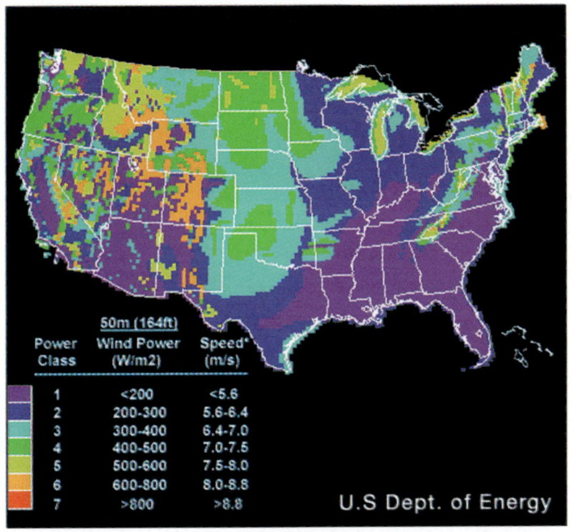

According to the Department of Energy, 6 percent of the nation's land mass has wind energy resources. This area has the potential to supply more than one and half times the electricity being consumed today. Yet obstacles stand in the way of harvesting this natural resource.

- Wind farms are not always welcome in communities, for a variety of reasons.
- As the turbines spin, rotor blades produce a certain amount of noise.
- Some people dislike the industrial look of the wind-farm towers.
- Concern for the fact that some birds and bats are killed when they fly into the rotors.

Searching for solutions

There needs to be more research and better methods of harvesting in areas with less wind speed. Wind industry scientists and engineers, in partnership with the Department of Energy, are designing, analyzing, and testing equipment and methods in order to improve performance.

Progress in research requires test after test. Before a new product such as an improved wind turbine is placed on the market, a single model is made and tested repeatedly.

Not all wind farms are on land. Offshore wind energy projects such as the Nantucket Sound wind farm are being looked at more closely. Research is underway on floating turbines to be tested in US coastal waters and the lower Great Lakes. Such sites would be one way to solve the drawback of distance from large cities that need electricity.

Questions:

1. How does a wind turbine operate?
2. Compare and contrast wind energy with fossil fuel.
3. What are the disadvantages to wind power?
4. Why is it important to research and study wind energy?

CHAPTER ACTIVITY Your Own Science Experiment

The scientific method sets science apart from other types of study. The method begins with observation of some phenomenon. After making observations, scientists propose a hypothesis. A hypothesis is an informed guess about the outcome of the experiment. The hypothesis is then tested to find out whether or not it is true. During this activity, you will be designing and conducting your own experiment.

Materials:

String, assorted sizes of balls, newspaper, meter stick, stopwatch, balance, paper cups, macaroni, rubber bands, paper clips

What you will do

1. Brainstorm with your group to decide on an experiment you could make up and carry out with the above materials. Working together, brainstorm what questions your group would like to find answers to.
2. Once you decide on an experiment you would like to design, identify the independent variable, dependent variable, and control variables. Think about how changing the different variables or keeping them constant might affect your experiment.
3. Write a procedure for your experiment. List which of the materials you are using. Think about what types of measurements and observations you need to make. Could any of the materials could be used for making measurements?
4. Make an educated guess, using your experience and knowledge, as to what you predict will happen in this experiment. Think about the relationship between the independent and dependent variables. Your educated guess is your hypothesis.
5. Carry out your experiment. Take your measurements and make your observations. Record all your data.
6. Analyze your data to determine whether your hypothesis was correct. This may involve making a graph or other type of model.
7. Write a conclusion to explain your results.

Applying your knowledge

a. Did your results support or contradict your hypothesis? If your hypothesis turned out to be wrong, what do you think might be a correct hypothesis?

b. Did you have any problems with your experiment? Do you need to change anything in the procedure and try the experiment again?

Chapter 3 Assessment

Vocabulary

Select the correct term to complete the sentences.

kilogram	gram	atom
element	compound	system
energy	Calorie	newton
heat	thermometer	Fahrenheit
Celsius	force	conservation of energy

Section 3.1

1. The basic SI unit of mass, approximately equal to 1000 grains of rice, is the ____.
2. A pure substance that cannot be broken down into other substances is known as a(n) ____.
3. A combination of elements that most often has properties different than the elements from which it is made is called a(n) ____.
4. The smallest particle of matter which retains the identity of the element it comes from is called a(n) ____.
5. An amount of mass equal to one-thousandth of a kilogram is a(n) ____.

Section 3.2

6. A small group of related things that work together may be called a ____.
7. A quantity that measures the ability to cause change in a physical system in units of joules is known as ____.
8. A unit of energy larger than a joule used to measure the energy available in food is the ____.
9. An SI unit used to measure force, equal to less than one-quarter of a pound, is the ____.
10. A natural law that says energy cannot be created or destroyed is the law of ____.
11. An action, measured in newtons, that has the ability to transer energy or change motion is called a(n) ____.
12. A(n) ____ is an instrument used to measure temperature.
13. The temperature scale on which the freezing point of water is 32 degrees and boiling point of water is 212 degrees is the ____ scale.
14. The temperature scale on which the freezing point of water is 0 degrees and boiling point of water is 100 degrees is the ____ scale.
15. When thermal energy flows from hot to cold it is called ____.

Concepts

Section 3.1

1. Scientists find it useful to use both grams and kilograms for measuring mass. Why is it necessary to have two different SI units of mass?
2. Mass describes:
   a. an object's size.
   b. the amount of matter in an object.
   c. the type of elements in an object.
3. What laboratory instrument is most often used to measure mass?
4. Identify the following substances as an *element* or a *compound*:
   a. ____ table salt (sodium chloride)
   b. ____ oxygen gas
   c. ____ rust
   d. ____ iron

5. If a scientist were to cut a piece of the element iron into the smallest possible piece of iron, what would be the name given to that smallest piece of iron?

6. Describe the difference between the arrangement of atoms in solids, liquids, and gases.

7. Which has the lowest energy of motion?

   a. solid

   b. liquid

   c. gas

8. Which has its atoms farthest apart?

   a. solid

   b. liquid

   c. gas

9. Which of the following is a true statement?

   a. In the gas state, atoms move around freely.

   b. Liquids do not change shape easily.

   c. Gas molecules move more slowly as they are heated.

Section 3.2

10. Name 3 natural or man-made systems.

11. Why do scientists organize nature into systems?

12. Explain how force and energy are related.

13. Energy takes many forms. Compare potential energy to kinetic energy and give two examples of each.

14. Write the letters **KE** (kinetic energy) or **PE** (potential energy) to indicate which type of energy is illustrated by each of the following examples.

   a. ____ a car moving down the street

   b. ____ a bicycle stopped at the top of a hill

   c. ____ a box sitting on a table

   d. ____ a ball rolling across the floor

15. Which has the higher kinetic energy, ice or steam? Why?

16. Describe what happens when matter is heated to extreme temperatures greater than 10,000°C.

17. Explain the difference between thermal energy, heat, and temperature.

18. Heat always flows from ____ temperature to ____ temperature.

19. What did Gabriel Fahrenheit use to set 100°F for his Fahrenheit temperature scale?

20. What happens to atoms at absolute zero?

21. Energy is measured in:

   a. joules.

   b. newtons.

   c. kilograms.

22. People who say that "using less electrical energy will conserve energy" are not really saying (scientifically) what they mean. Explain in a scientifically accurate way what these people are trying to say.

23. What happens to the electrical energy used to turn on the lights in your home?

   a. It is used up.

   b. It is destroyed.

   c. It is converted to light and heat energy.

24. Do power plants create electrical energy from empty space?

25. When energy transformations occur in a system, the total amount of energy in the system:

   a. increases.

   b. decreases.

   c. stays the same.

Problems

Section 3.1

1. Kyela has a mass of 45 kg. What is her mass in grams?
2. An average baseball has a mass of 150 grams. What is its mass in kilograms?
3. What is the mass in grams of a 0.454 kilogram soccer ball?
4. Which graph best represents the relative energy of the atoms of substances in the solid, liquid, and gas phases?

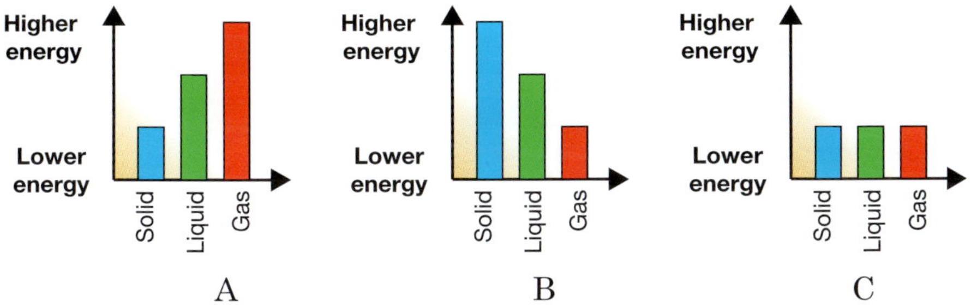

Section 3.2

5. For each of the following examples, which has more thermal energy?
   a. 1 kg of water at 50°C or 2 kg of water at 50°C
   b. 1 kg of ice at 0°C or 1 kg of water at 5°C
   c. 1 kg of water at 5°C or 1 kg of steam at 105°C
6. Which of the following will have the highest temperature when 1,000 joules of energy are added to it? Assume each starts at the same temperature.
   a. 1 kg of water (specific heat = 4184 J/kg°C)
   b. 1 kg of wood (specific heat = 2500 J/kg°C)
   c. 1 kg of glass (specific heat = 800 J/kg°C)
7. How much heat in joules would you need to raise the temperature of 1 kg of water by 2°C?
8. If a single light bulb uses 100 joules of energy every second, how many Calories of energy are used each second if there are 4,184 joules in one Calorie?
9. A very bright light bulb uses 150 joules of energy every second. An average fast-food burger contains 350 Calories of energy. If the energy of the burger could be converted to electricity without any energy loss, how long would the energy from the burger light the bulb?

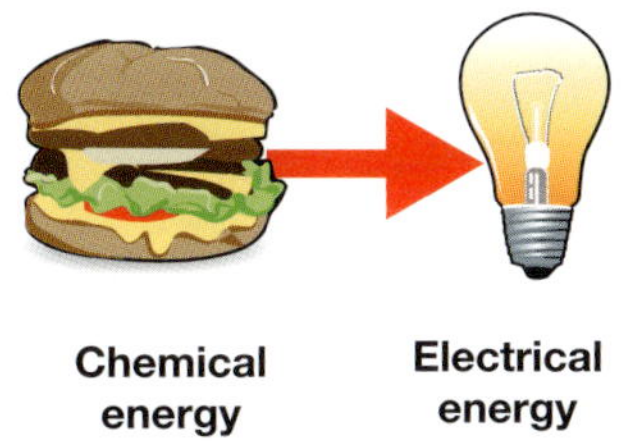

10. One pancake contains about 80 Calories of energy. One Calorie contains 4,184 joules. What is the amount of energy in joules in one pancake?
11. A pizza restaurant advertises that their brick oven cooks their pizza at 800°F. What is this temperature in degrees Celsius?
12. Earth's core is estimated to be 7,000°C. What is this in degrees Fahrenheit?
13. Which temperature scale, Fahrenheit or Celsius, has the greatest change in temperature for one degree?
14. Describe the energy transformations or conversions for each of the diagrams below:

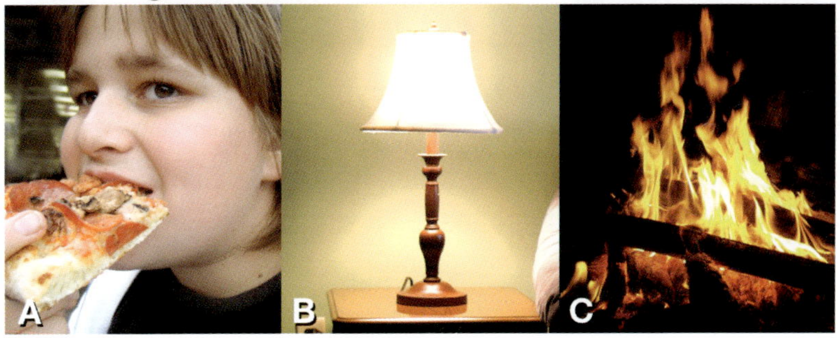

UNIT 2

Properties of Matter

TRY THIS AT HOME

Can you make an egg float? Place an egg in a glass that is half full of fresh water. The egg will sink because it is denser than the water. Now mix about 8 to 10 teaspoons of salt into the fresh water.

Place the egg in the saltwater mixture. Can you explain the result? Can you figure out how to make the egg sink halfway into a glass of water and then stop before hitting the bottom?

Chapter 4

Density and Buoyancy

Will it float or will it sink? If you are designing ships this is a very important question. The largest ship in the world is the *Jahre Viking*, an oil-carrying tanker. This super-sized ship is 1,504 feet long and 264 feet wide, longer than 5 football fields laid end-to-end. If the Empire State building was laid on its side, the Jahre Viking would be longer by 253 feet! Crew members use bicycles to get from place to place on the ship. The ship is too large to fit through the Panama or Suez Canal, and it cannot dock in many of the world's seaports. The Jahre Viking is largely constructed of steel – so how can a big, heavy ship like this actually float? By the time you finish studying this chapter on density and buoyancy, you will be able to explain how ships and boats of all shapes and sizes can float.

Key Questions

1. ***What is density and how can you measure it?***
2. ***What two things does density depend on?***
3. ***How does a steel ship float, when a steel marble sinks?***

4.1 Density

When you think about the many kinds of matter you come into contact with every day, what properties come to mind? Some matter is solid and hard, like steel and wood. Some matter is liquid like water, or a gas like air. And in the category of solid matter, there are big differences. A block of wood and a block of steel may be the same size but one has a lot more mass than the other. Because of that difference in mass, wood floats in water and steel sinks. Whether an object floats or sinks in water is related to its density. This chapter will explain density, a property of all matter.

VOCABULARY

density - the mass of matter per unit volume; density is typically expressed in units of grams per milliliter (g/mL), grams per cubic centimeter (g/cm^3), or kilograms per cubic meter (kg/m^3).

Density is a property of matter

Density is mass per unit volume

Density describes how much mass is in a given volume of a material. Steel has high density; it contains 7.8 grams of mass per cubic centimeter. Aluminum, as you well might predict, has a lower density; a one-centimeter cube has a mass of only 2.7 grams.

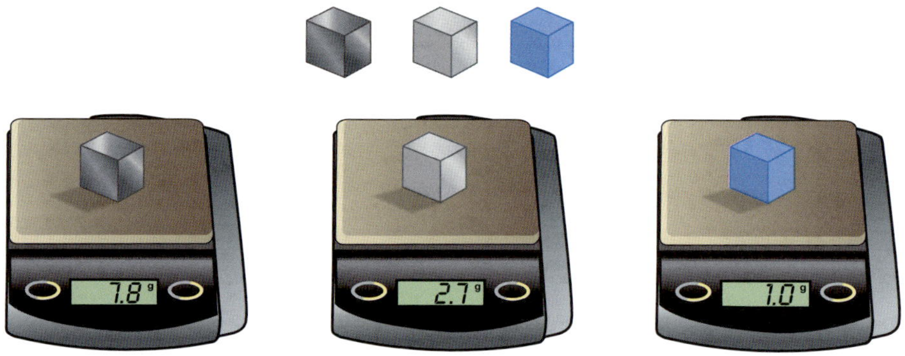

Steel 7.8 g   Aluminum 2.7 g   Water 1.0 g

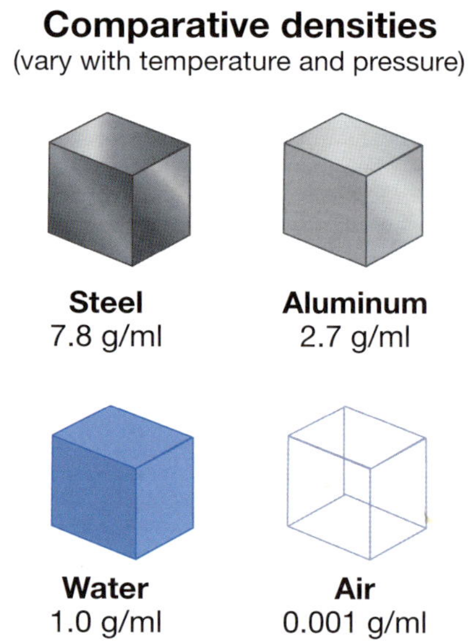

Figure 4.1: *The density of steel, aluminum, water, and air expressed in different units.*

The density of water and air

Liquids and gases are matter and have density. The density of water is about one gram per cubic centimeter. The density of air is lower, of course — much lower. The air in your classroom has a density of about 0.001 grams per cubic centimeter.

Units of density

Density in units of grams per milliliter Your laboratory investigations will typically use density in units of grams per milliliter (g/mL). The density of water is one gram per milliliter. That means one milliliter of water has a mass of one gram, and 100 milliliters of water a mass of 100 grams.

Density in g/cm^3 and kg/m^3 Some problems use density in units of grams per cubic centimeter (g/cm^3). Since one milliliter is exactly the same volume as one cubic centimeter, the units of g/cm^3 and g/mL are actually the same. For applications using large objects, it is more convenient to use density in units of kilograms per cubic meter (kg/m^3). Table 4.1 gives the densities of some common substances in both units.

Converting units of density To convert from one to the other, remember that 1 g/cm^3 is equal to 1000 kg/m^3. To go from g/cm^3 to kg/m^3 you multiply by 1,000. For example, the density of ice is 0.92 g/cm^3. This is the same as 920 kg/m^3.

To go from kg/m^3 to g/cm^3 you divide by 1,000. For example, the density of aluminum is 2,700 kg/m^3. Dividing by 1,000 gives a density of 2.7 g/cm^3.

Convert between units of density

A reference book lists the density of ceramic tile as 2,650 kg/m^3. Estimate the mass of one cubic centimeter of tile.

1. Looking for: Mass of 1 cm^3, which is the density in g/cm^3
2. Given: Density of 2,650 kg/m^3
3. Relationships: 1 g/cm^3 = 1,000 kg/m^3
4. Solution: Divide by 1,000 to get the density in g/cm^3.
   2,650 ÷ 1,000 = 2.65 g

Your turn...

a. A bronze statue has a density of 6,000 kg/m^3. What is the density in g/mL?
Answer: 6 g/mL

Table 4.1: Densities of common substances

Material	(kg/m^3)	(g/cm^3)
Platinum	21,500	21.5
Lead	11,300	11.3
Steel	7,800	7.8
Titanium	4,500	4.5
Aluminum	2,700	2.7
Glass	2,700	2.7
Granite	2,600	2.6
Concrete	2,300	2.3
Plastic	2,000	2.0
Rubber	1,200	1.2
Liquid water	1,000	1.0
Ice	920	0.92
Oak (wood)	600	0.60
Pine (wood)	440	0.44
Cork	120	0.12
Air (avg.)	0.9	0.0009

Figure 4.2: *The densities of some common materials.*

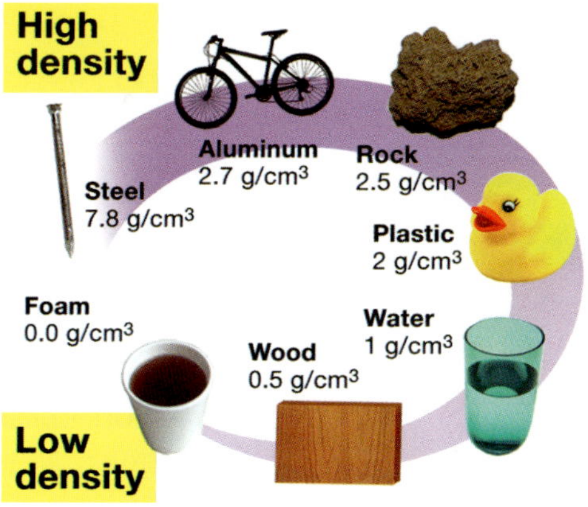

Figure 4.3: *The density range of common materials.*

Solving problems

A four-step technique

The method for solving problems has four steps. Follow these steps and you will be able to see a way to the answer most of the time and will at least make progress toward the answer almost every time. There is often more than one way to solve a problem. Sometimes you will have to use creativity to work with the given information to find the answer. Figure 4.4 shows the steps for solving problems.

Solved example problems are provided

Throughout this book you will find example problems that have been solved for you. Following each solved example, there are often practice problems. The answers to the practice problems are provided so that you can check your work while practicing your problem-solving skills. Always remember to write out the steps when you are solving problems on your own. If you make a mistake, you will be able to look at your work and figure out where you went wrong. Here is the format for example problems:

Calculating density

A wax candle has a volume of 1,000 ml. The candle has a mass of 1.4 kg (1,400 g). What is the density of the candle?

1. Looking for:	You are asked for the density.
2. Given:	You are given the mass and volume.
3. Relationships:	Density is mass divided by volume:
4. Solution:	density = (1,400 g) ÷ (1,000 ml) = 1.4 g/ml

Physics problems

Problem solving is important in all careers. For example, financial analysts are expected to look at information about businesses and figure out which companies are succeeding. Doctors collect information about patients and must figure out what is causing pain or an illness. Mechanics gather information about a car and have to figure out how to fix the engine. All these examples use problem-solving skills.

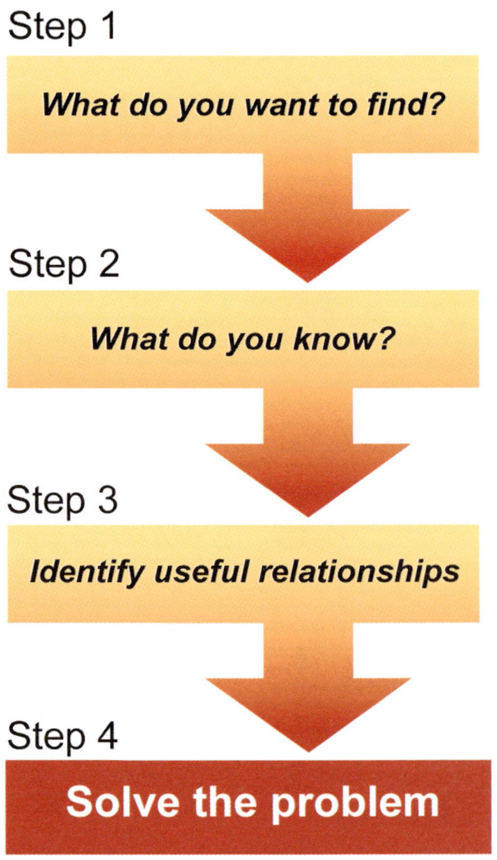

Figure 4.4: *Follow these steps and you will be able to see a way to the answer most of the time.*

Density of common materials

Material density is independent of shape

Density is a property of material independent of quantity or shape. For example, a steel nail and a steel cube have different amounts of matter and therefore different masses (Figure 4.5). They also have different volumes. However, if you calculate density by dividing mass by volume, the result is the same for both the nail and the cube.

Strong solids typically have higher density

Solids that are strong, such as steel, typically have high density. High density means there are many atoms per cubic centimeter. Many atoms in a material means many bonds between atoms, and those bonds are what ultimately create strength. There are exceptions, however. Lead is very dense but not very strong because the bonds between lead atoms are relatively weak.

Soft materials typically have lower density

Solids with low density, such as cork or foam, are often used as cushioning material. Low density means there are relatively large spaces between atoms. That means materials may be compressed relatively easily, which is why foam and other low-density substances make good packing materials.

Liquids tend to be less dense than solids of the same material

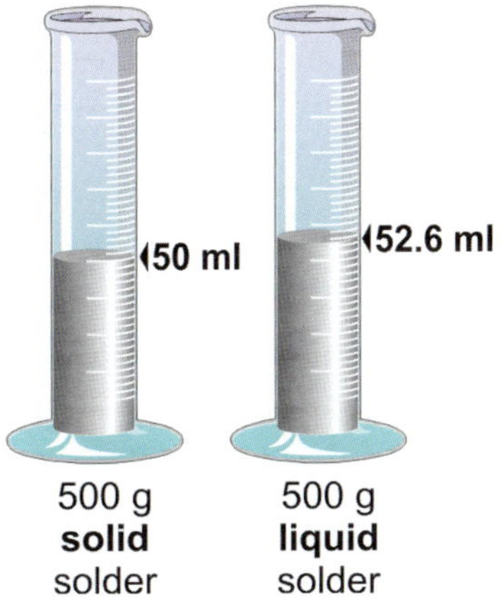

The density of a liquid is usually close to solid density, surprisingly enough, but less. For example, the density of solder is 10 g/ml. The density of liquid solder is 9.5 g/ml. The liquid density is lower because the atoms are not packed as uniformly as they are in the solid. Picture a brand-new box of toy blocks. When you open the box, you see the blocks tightly packed in a repeating pattern, like the atoms in a solid. Now imagine dumping the blocks out of the box, and then trying to pour them back into the original tight packing pattern. The jumbled blocks take up more space, like the atoms in a liquid. Water is an exception to this rule. The density of solid water, or ice, is *less* than the density of liquid water.

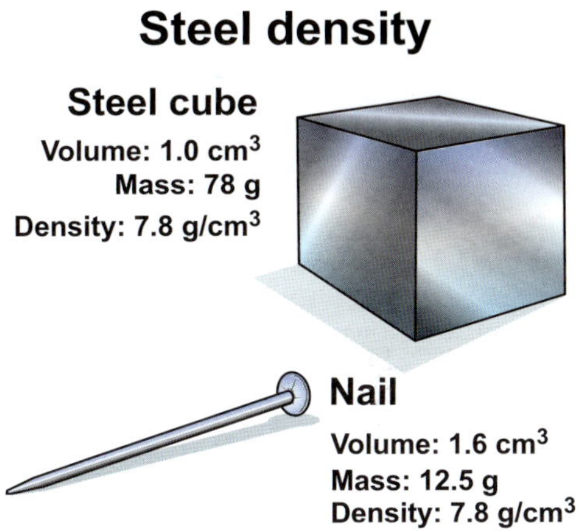

Figure 4.5: *The density of a steel nail is the same as the density of a solid cube of steel.*

Figure 4.6: *The same number (or mass) of blocks arranged in a tight, repeating pattern take up less space than when they are jumbled up.*

Volume

Volume Volume is the amount of space an object takes up. The units used to measure volume depend on whether the object is solid or liquid. The volume of solids is measured in cubic centimeters (cm^3) or cubic meters (m^3). The volume of liquids is measured in milliliters (mL) or liters (L). One cubic centimeter is the same volume as one milliliter.

Measuring the volume of liquids Measuring the volume of liquids is easy. Pour the liquid into a marked container called a *graduated cylinder* and read the volume. There are two things to keep in mind to measure accurately. First, read the mark at eye level. Second, you will notice that the surface of the liquid forms a curve rather than a straight line (Figure 4.7). This curve is called the meniscus. Read the volume at the center of the meniscus.

Volume of solids You have probably already learned to measure the volume of some solid shapes. The volume of a rectangular solid (a shoebox shape), for example, is found by multiplying length times width times height. The volume of a sphere is $4/3\pi r^3$, with r equal to the radius of the sphere.

The displacement method You can find the volume of an irregular shape using a technique called displacement. To displace means to "take the place of" or to "push aside." You can find the volume of an irregularly shaped object by putting it in water and measuring the amount of water displaced.

How you make the measurement You can use the displacement method to find the volume of an ordinary item like a house key. Fill a 100-milliliter graduated cylinder with 50 mL of water (Figure 4.7). Gently slide the key into the water. The water level in the container will rise, because the key displaced, or pushed aside, some water. If the level now reads 53.0 mL, you know that the key displaced 3.0 mL of water. The volume of the key, or of any object you measured in this way, is equal to the volume of the water it displaced. The key has a volume of 3.0 milliliters, or 3.0 cubic centimeters (cm^3).

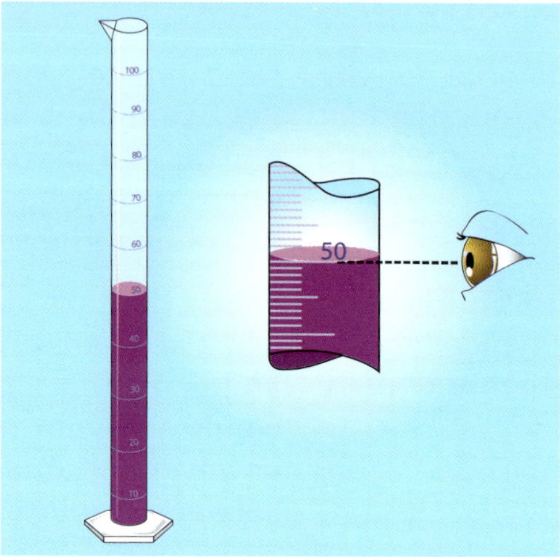

Figure 4.7: *The meniscus of water has a concave shape. Read the mark at the bottom of the curve.*

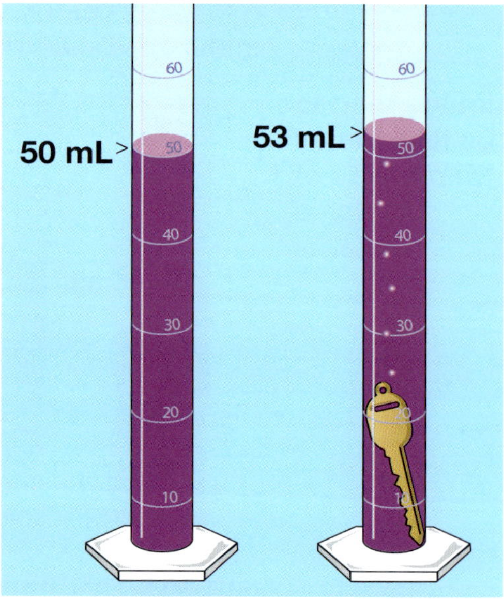

Figure 4.8: *The key displaced 3.0 milliliters of water.*

Determining density

Measuring density To find the density of a material, you need to know the mass and volume of a solid sample of the material. You can calculate the density from the formula below.

DENSITY

$$D = \frac{m}{V}$$

Density (kg/m³ **or** g/cm³) → D; m ← *Mass* (kg **or** g); V ← *Volume* (m³ **or** cm³)

Mass is measured with a balance or scale. For irregular objects (Figure 4.10) the displacement method can be used to find the volume.

Calculating volume For simple shapes you can calculate the volume. The volume of spheres, cylinders, and rectangular solids is given in the diagram in Figure 4.9. When calculating volume, all of the units of length involved in the calculation must be the same. For example, if you want volume in cubic centimeters, all of the measurements in the calculation must be in centimeters.

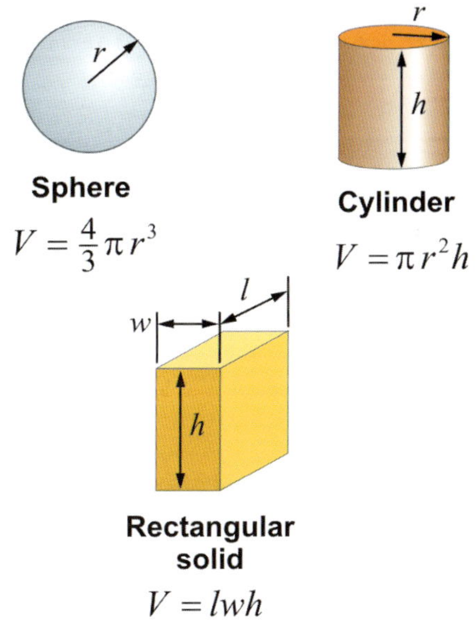

Figure 4.9: *The volume of some simple geometric shapes.*

Figure 4.10: *Use the displacement method to find the volume of irregular objects. Use a scale to find the mass.*

Calculate density from mass and volume

A student measures the mass of five steel nuts to be 96.2 grams. The nuts displace 13 mL of water. Calculate the density of the steel in the nuts.

1. Looking for:	Density
2. Given:	Mass (96.2 g) and volume (13 mL)
3. Relationships:	Density = mass ÷ volume
4. Solution:	D = 96.2g ÷ 13 mL = 7.4 g/mL

Your turn...

a. A solid brass block measures 2 cm × 2 cm × 3 cm and has a mass of 48 g. What is its density? **Answer**: 4 g/cm^3

Why density varies

Atoms have different masses

The density of a material depends on two things. First is the individual mass of each atom or molecule. Solid lead is denser than solid aluminum mostly because a single atom of lead has 7.7 times more mass than a single aluminum atom.

Atoms may be "packed" tightly or loosely

Second, density depends on how tightly the atoms are packed. A diamond is made of carbon atoms and has a density of 3,500 kg/m^3. The carbon atoms in diamonds are closely packed. Paraffin wax is mostly carbon, but the density of paraffin is only 870 kg/m^3. The density of paraffin is low because the carbon atoms are mixed with hydrogen atoms in long molecules that take up a lot of space.

Solving density problems

Density problems usually ask you to find one of the three variables (mass, volume, density), given the other two. Figure 4.12 shows three forms of the density equation you can use. Which one you choose depends on what you are asked to find.

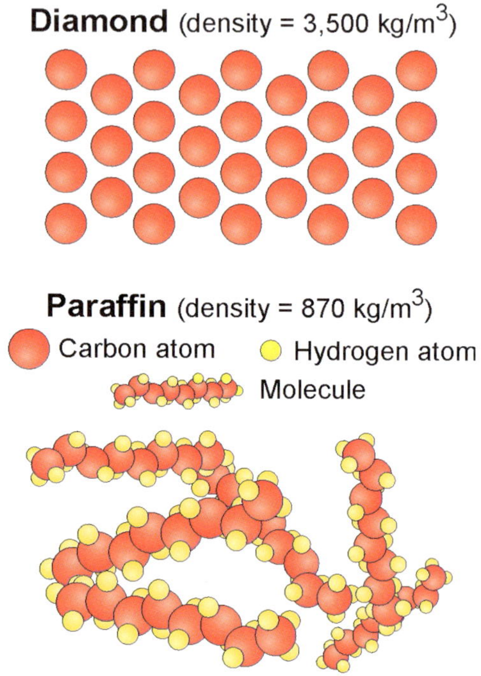

Figure 4.11: *The carbon atoms in diamonds are packed tightly while the carbon atoms in paraffin are not.*

Using the density equation

A 4,500-gram cube of titanium is 10 centimeters on each side. Calculate its volume in cm^3, and then calculate its density in g/cm^3.

1. Looking for: You are looking for the volume of a solid and its density.
2. Given: You are given the size and mass.
3. *Relationships:* V = *length x width x height* $\quad D = m/V$
4. Solution: V = 10 cm x 10 cm x 10 cm = 1,000 cm^3
   D = 4,500 g ÷ 1,000 cm^3 = 4.5 g/cm^3

Your turn...

a. Calculate the volume and density of a block that has the dimensions 10 cm x 5 cm x 4 cm and a mass of 400 grams. **Answer:** 200 cm^3, 2 g/cm^3

b. A 6-gram marble put in a graduated cylinder raises the water from 30 mL to 32 mL. Calculate the marble's volume and density. **Answer:** 2 cm^3, 3 g/cm^3

Use. . .	. . . if you know . . .	. . . and want to find . . .
$D = m \div V$	mass and volume	density
$m = D \times V$	volume and density	mass
$V = m \div D$	mass and density	volume

Figure 4.12: *Relationships to use when solving density problems.*

4.1 Section Review

1. List three physical properties of a piece of wood.
2. One cubic centimeter is the same volume as one _______.
3. Explain two ways to find the volume of a plastic cube.
4. A certain material has a density of 0.2 g/mL. Is this material good for building a bridge or for making cushions for a couch?
5. The density of a gas is lower than the density of a liquid because:
   a. gas atoms are smaller than liquid atoms.
   b. gas atoms are larger than liquid atoms
   c. atoms in a gas are farther apart than atoms in a liquid.
   d. atoms in a gas are closer together than atoms in a liquid
6. Density is a _______ property of matter.
7. Density is the _______ per unit volume of a substance.
8. What measurements must be known in order to find the density of a substance?

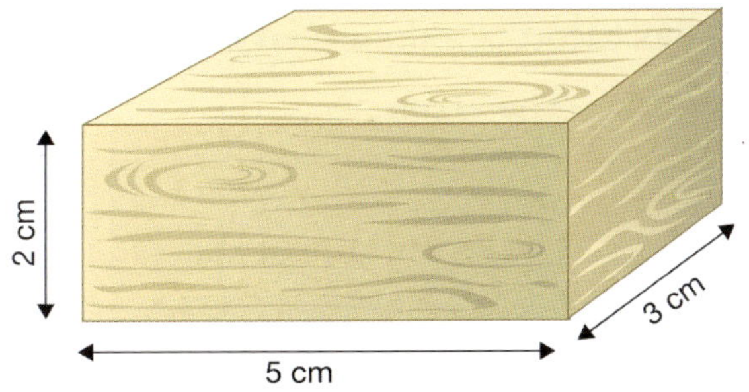

9. The piece of wood shown above has a mass of 18 grams. Calculate it volume and density. Then use the table in Figure 4.2 to determine which type of wood it is. What are the two factors that determine a material's density?

CHALLENGE

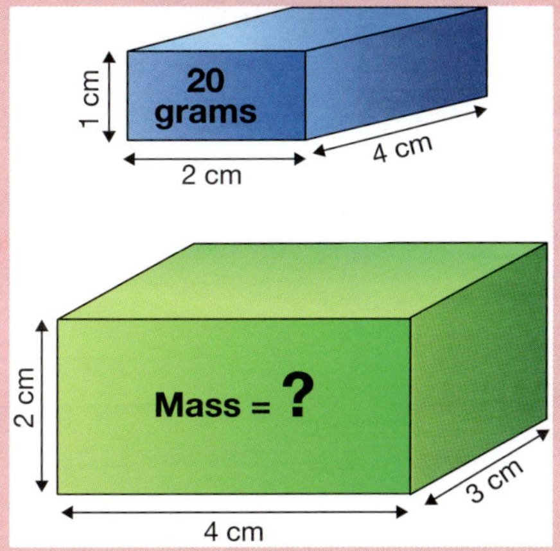

Two toy blocks are made of the same type of material. One has a mass of 20 grams and its dimensions are 2 centimeters x 4 centimeters x 1 centimeter. The other is 4 cm x 3 cm x 2 cm. Calculate the mass of the second block.

4.2 Buoyancy

If you drop a steel marble into a glass of water, it sinks to the bottom. The steel does not float because it has a greater density than the water. And yet many ships are made of steel. How does a steel ship float when a steel marble sinks? The answer has to do with gravity and weight.

Weight and buoyancy

Weight and mass are not the same We all tend to use the terms *weight* and *mass* interchangeably. In science however, *weight and mass are not the same thing*. Mass is a fundamental property of matter. **Weight** is a force, like any other pushing or pulling force, and is caused by Earth's gravity. It is easy to confuse mass and weight because heavy objects (more weight) have lots of mass and light objects (less weight) have little mass.

Buoyancy is a force It is much easier to lift yourself in a swimming pool than to lift yourself on land. That is because the water in the pool exerts an upward force on you that acts in a direction opposite to your weight (Figure 4.13). We call this force **buoyancy**. Buoyancy is a measure of the upward force a fluid exerts on an object that is submerged.

Pushing a ball underwater

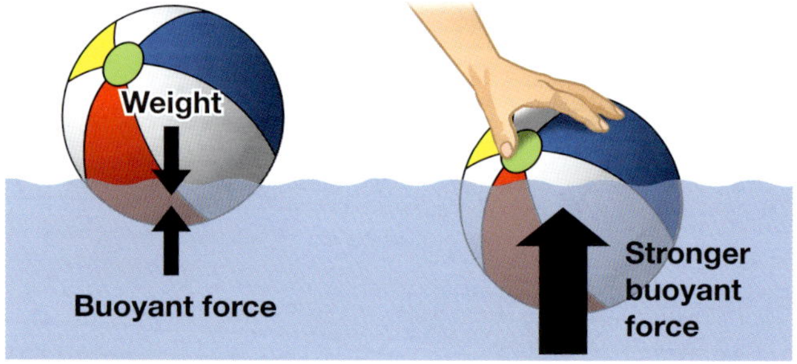

The strength of the buoyant force on an object in water depends on the volume of the object that is underwater. Suppose you have a large beach ball you want to submerge in a pool. As you keep pushing downward on the ball, you notice the buoyant force getting stronger and stronger. The greater the part of the ball you manage to push underwater, the stronger the force trying to push it back up. The strength of the buoyant force is proportional to the volume of the part of the ball that is submerged.

VOCABULARY

weight - the downward force of gravity acting on mass.

buoyancy - the measure of the upward force a fluid exerts on an object that is submerged.

Figure 4.13: *The water in the pool exerts an upward force on your body, so the net force on you is lessened.*

Archimedes' principle

Archimedes' principle

In the third century BC, a Greek mathematician named Archimedes realized that buoyant force is equal to the weight of fluid displaced by an object. We call this relationship **Archimedes' principle**. For example, suppose a rock with a volume of 1,000 cubic centimeters is dropped into water (Figure 4.14). The rock displaces 1,000 cm^3 of water, which has a mass of 1 kilogram. The buoyant force on the rock is the weight of 1 kilogram of water, which is 9.8 newtons.

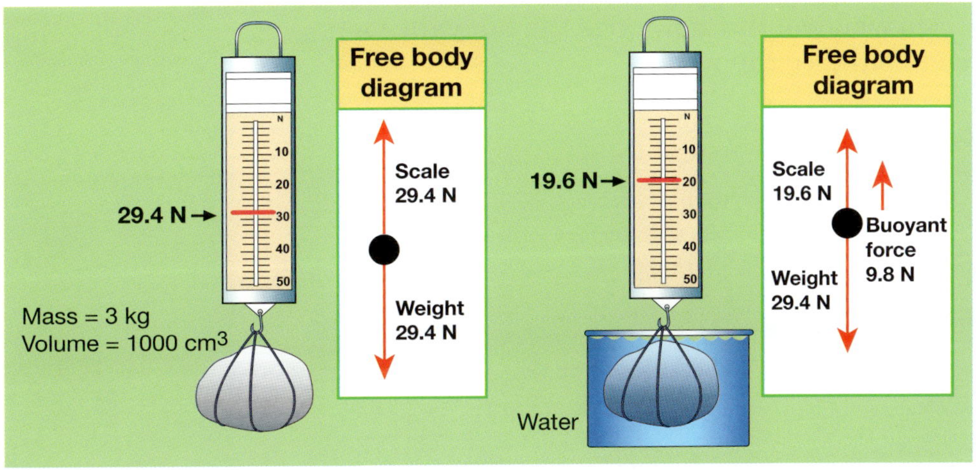

A simple buoyancy experiment

A simple experiment can be done to measure the buoyant force on a rock (or any object) with a spring scale. Suppose you have a rock with a volume of 1,000 cubic centimeters and a mass of three kilograms. In air, the scale shows the rock's weight as 29.4 newtons. The rock is then gradually immersed in water, but not allowed to touch the bottom or sides of the container. As the rock enters the water, the reading on the scale decreases. When the rock is completely submerged, the scale reads 19.6 newtons.

Calculating the buoyant force

Subtracting the two scale readings, 29.4 newtons and 19.6 newtons, results in a difference of 9.8 newtons. This is the buoyant force exerted on the rock, and it is the same as the weight of the 1,000 cubic centimeters of water the rock displaced.

VOCABULARY

Archimedes' principle - states that the buoyant force is equal to the weight of the fluid displaced by an object.

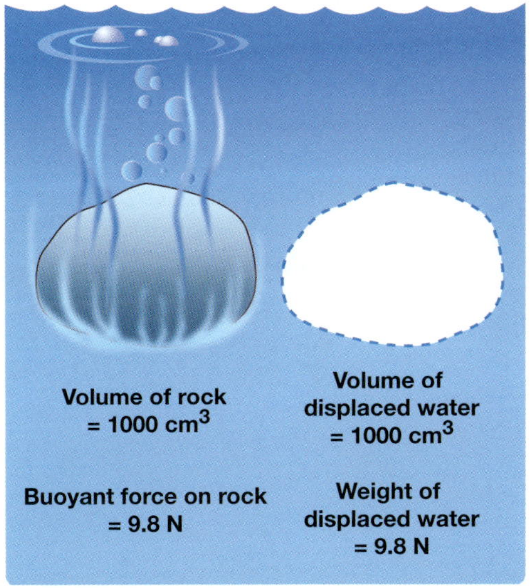

Figure 4.14: *A rock with a volume of 1,000 cm^3 experiences a buoyant force of 9.8 newtons.*

Sinking and floating

Comparing buoyant force and weight Buoyancy explains why some objects sink and others float. A submerged object floats to the surface if the buoyant force is greater than its weight (Figure 4.15). If the buoyant force is less than its weight, then the object sinks.

Equilibrium Suppose you place a block of foam in a tub of water. The block sinks partially below the surface. Then it floats without sinking any farther. The upward buoyant force perfectly balances the downward force of gravity (the block's weight). But how does the buoyant force "know" how strong to be to balance the weight?

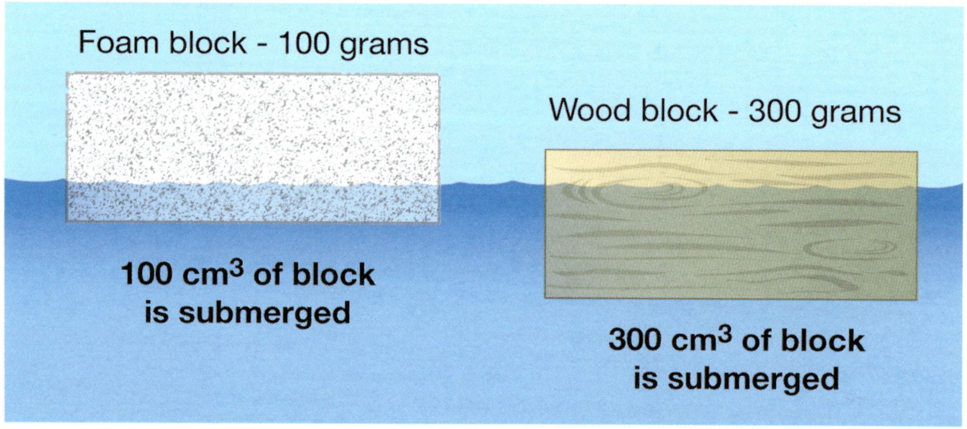

Denser objects float lower in the water You can see the answer to this question in the pictures above. If a foam block and a wood block of the same size are both floating, the wood block sinks farther into the water. Wood has a greater density, so the wood block weighs more. A greater buoyant force is needed to balance the wood block's weight, *so the wood block displaces more water*. The foam block has to sink only slightly to displace water with a weight equal to the block's weight. A floating object displaces just enough water to make the buoyant force equal to the object's weight.

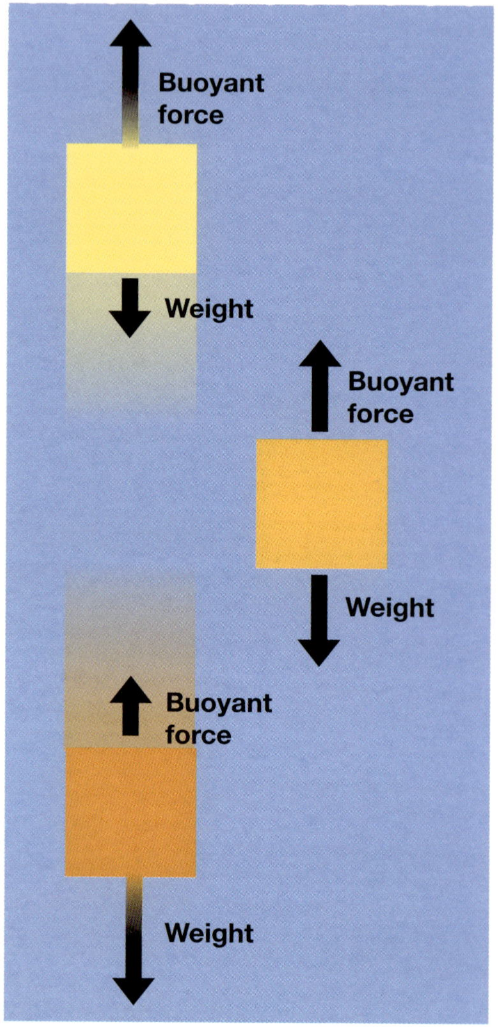

Figure 4.15: *Whether an object sinks or floats depends on how the buoyant force compares with the weight.*

Density and buoyancy

Comparing densities

If you know an object's density, you can immediately predict whether it will sink or float — without measuring its weight. An object sinks if its density is greater than that of the liquid. It floats if its density is less than that of the liquid.

Two balls with the same volume but different density

To see why, picture dropping two balls in a pool of water. The balls have the same size and volume but have different densities. The steel ball has a density of 7.8 g/ml which is greater than the density of water (1.0 g/ml). The wood ball has a density of 0.75 g/ml, which is less than the density of water.

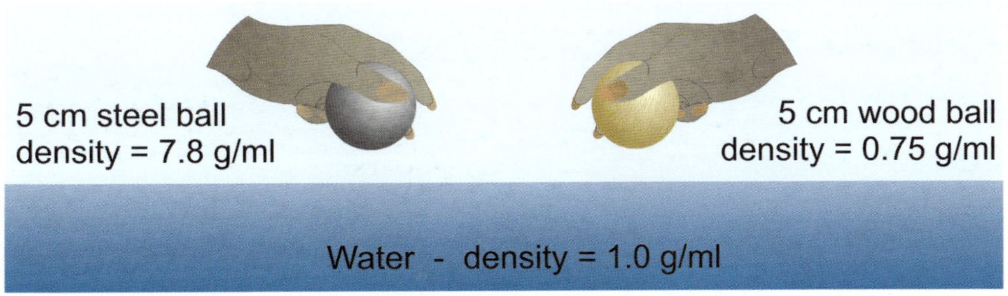

Why one sinks and the other floats

When they are completely underwater, both balls have the same buoyant force because they displace the same volume of water. However, the steel ball has more weight since it has a higher density. The steel ball sinks because steel's higher density makes the ball heavier than the same volume of water. The wood ball floats because wood's lower density makes the wood ball lighter than the same volume of displaced water.

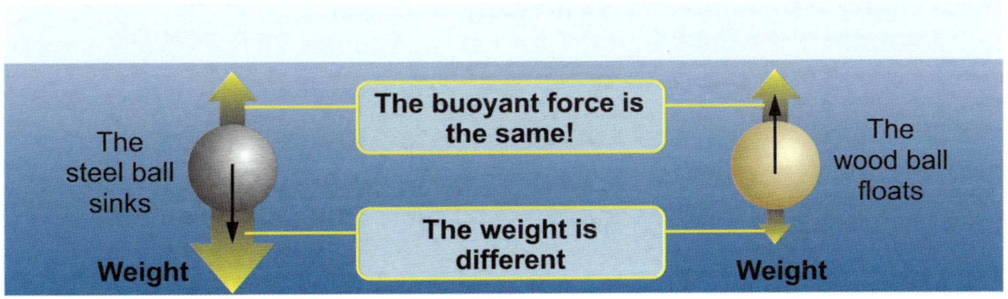

An object with an average density GREATER than the density of water will sink

An object with an average density LESS than the density of water will float.

Average density

Average density is the total mass divided by the total volume.

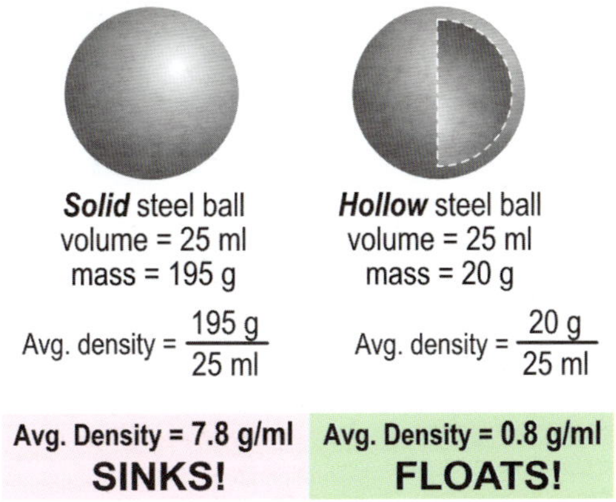

Figure 4.16: *The meaning of average density.*

Boats and average density

How do steel boats float?

If you place a solid chunk of steel in water, it immediately sinks because the density of steel (7.8 g/cm^3) is much greater than the density of water (1.0 g/cm^3). So how is it that thousands of huge ships made of steel are floating around the world? The answer is that it is the *average density* that determines whether the object sinks or floats.

Solid steel sinks because it is denser than water

To make steel float, you have to reduce the *average* density somehow. Making the steel hollow does exactly that. Making a boat hollow expands its volume a tremendous amount without changing its mass. Steel is so strong that it is quite easy to reduce the average density of a boat to 10 percent of the density of water by making the shell of the boat relatively thin.

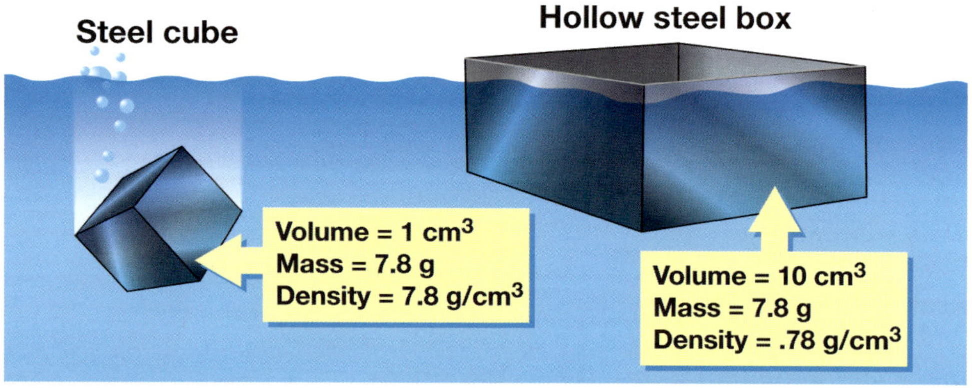

Increasing volume decreases density

Ah, you say, but that is an empty ship. True, so the density of a new ship must be designed to be under 1.0 g/cm^3 to allow for cargo. When objects are placed in a boat, of course its average density increases. The boat must sink deeper to displace more water and increase the buoyant force (Figure 4.17). If you have seen a loaded cargo ship, you might have noticed that it sat lower in the water than an unloaded ship nearby. In fact, the limit to how much a ship can carry is set by how low in the water the ship can get before rough seas cause waves to break over the side of the sip.

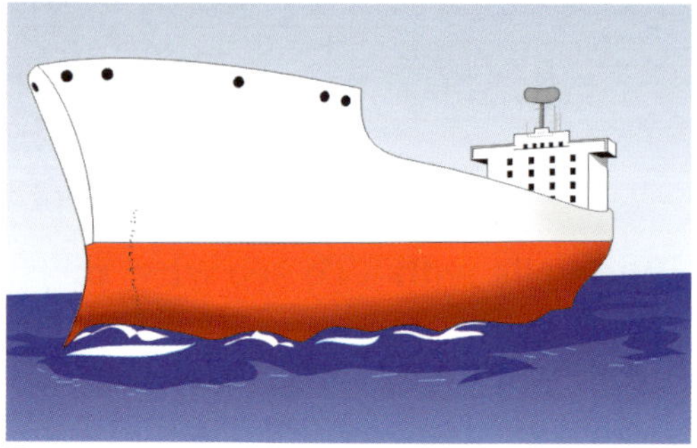

Empty cargo ship - less displaced water

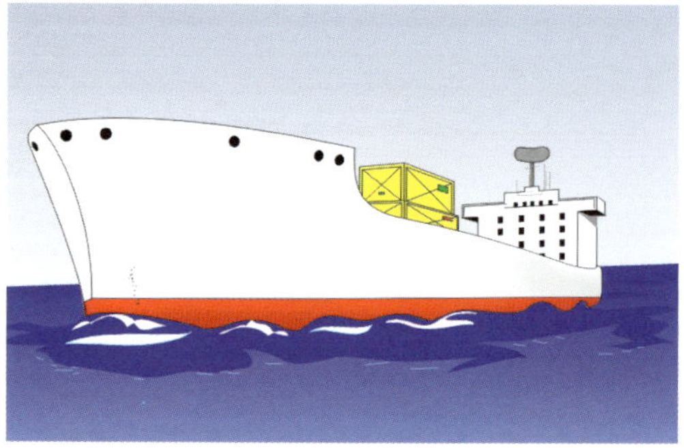

Full cargo ship - more displaced water

Figure 4.17: *A full ship has more mass than an empty ship. This means a full ship must displace more water (sink deeper) to make the buoyant force large enough to balance the ship's weight.*

4.2 Section Review

1. The buoyant force on an object depends on the _______ of the object that is underwater.
2. What happens to the buoyant force on an object as it is lowered into water? Why?
3. The buoyant force on an object is equal to the weight of the water it _________________.
4. When the buoyant force on an object is greater than its weight, the object ___________.

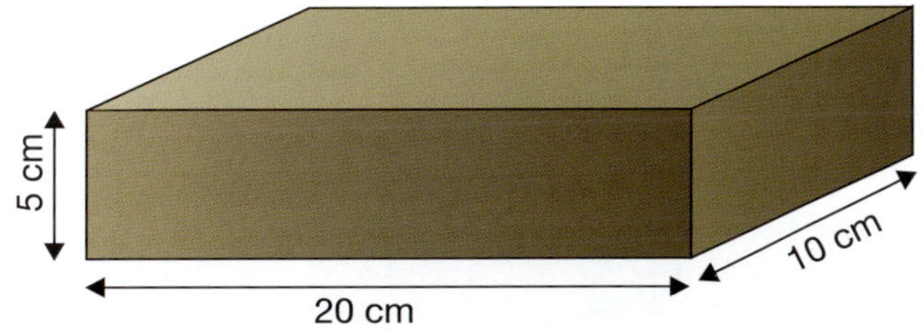

5. A rectangular object is 10 centimeters long, 5 centimeters high, and 20 centimeters wide. Its mass is 800 grams.
   a. Calculate the object's volume in cubic centimeters.
   b. Calculate the object's density in g/cm^3.
   c. Will the object float or sink in water? Explain.
6. Solid iron has a density of 7.9 g/cm3. Liquid mercury has a density of 13.6 g/cm^3. Will iron float or sink in mercury? Explain.
7. Why is it incorrect to say that heavy objects sink in water?
8. Steel is denser than water and yet steel ships float. Explain.

CHALLENGE

Legend has it that Archimedes added to his fame by using the concepts of volume and density to figure out whether a goldsmith had cheated Hiero II, the king of Syracuse. The goldsmith had been given a piece of gold of a known weight to make the crown. Hiero suspected the goldsmith had kept some of the gold for himself and replaced it with an equal weight of another metal. Explain the steps you could follow to determine whether or not the crown was pure gold.

PHYSICS CONNECTION

Airships in the Sky

What do televised football, baseball, and auto racing all have in common? You might think of sports, dedicated fans, and athletes. While these similarities are true, one common feature is that blimps provide video coverage for each of these events. Large, oval shaped airships hover over stadiums providing aerial shots of the action. Television cameras attached to the blimp's underside take wide-angle pictures. In addition, blimps provide advertising for major corporations. You have probably seen one in a sports telecast.

Hot air balloons

Anything that flies is an aircraft. In fact, the hot air balloon was the first aircraft invented, dating back to the early 1700s almost 200 years before the Wright brothers tested their first successful powered airplane. A hot air balloon depends on buoyancy forces for lift. Air in the balloon is kept hot by a burner. Hot air has a lower density than cold air. As long as the air inside the balloon is significantly hotter than the air outside, the balloon floats.

The density of air decreases with altitude. A hot air balloon floats upward until its average density matches the average density of the surrounding air. To go up father, the balloon pilot makes the air hotter inside the balloon, decreasing the average density. To go down, the pilot releases some hot air, making the balloon smaller and decreasing its average density.

The problem with hot air balloons is how to steer! With a hot air balloon you can only control motion up or down. The only way to steer is to change your altitude in hopes that the wind will be blowing the way you want to go at *some* altitude. This is all right if you are ballooning for fun. However, the direction of the wind is not reliable enough to use hot air balloons for travel or for shipping.

Blimps

A blimp is a type of *airship*, like a hot air balloon. Like a hot air balloon, a blimp also gets its lift from buoyancy of surrounding air. However, a blimp is filled with helium, a gas that is lighter than air. Unlike a balloon, a blimp keeps its helium gas at the same temperature as the surrounding air.

A blimp has four major parts: the envelope, gondola, engines, and controls. The envelope is the large cavity filled with helium. Made of polyester fabric similar to spacesuits, the envelope is a cigar-shaped, aerodynamic design. The gondola carries passengers and pilots and contains the engines and controls. Gasoline powered engines move blimps an average of 35 miles per hour.

Because the blimp has a motor, it can steer with a rudder, like a boat. Controls allow the pilot to steer the blimp up, down, right, or left. This is a big advantage over a hot air balloon.

Henri Giffard invented the first powered airship in 1852. In 1900, Count Ferdinand von Zeppelin invented the first rigid airship with a metal structure providing its shape. For buoyancy, early blimps used hydrogen. Hydrogen is very light bit also very explosive! In fact, the famous Hindenburg blimp, filled with hydrogen, exploded and burned during an accident in New Jersey in 1937.

Today, blimps are filled with helium instead of hydrogen, and do not have a rigid steel structure. Instead, modern blimps are nothing more than a large, strong, gas balloons. The aerodynamic shape is maintained by keeping the pressure of the helium in the blimp greater than the pressure of the air outside, just like the balloons you get at a party.

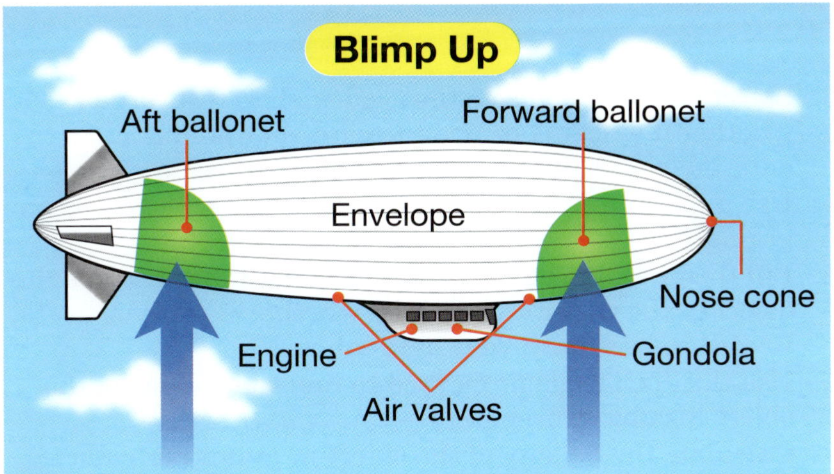

1. Air is released from the ballonets through air vents controlled by air valves
2. Ballonets deflate causing the envelope to fill with **HELIUM**
3. Envelope air becomes lighter and less dense
4. Outside air becomes heavier and more dense
5. Positive buoyancy causes blimp to rise

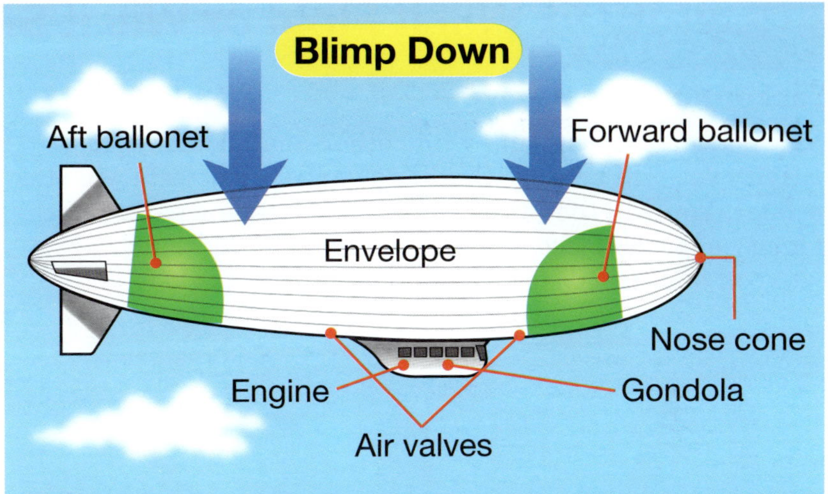

1. Air fills ballonets through air vents controlled by air valves
2. Ballonets inflate causing the envelope to fill with **AIR**
3. Envelope air becomes heavier and denser
4. Outside air becomes lighter and less dense
5. Negative buoyancy causes blimp to descend

Controlling altitude in a blimp

Inside a blimp's envelope there are two compartments called ballonets. These are bags filled with relatively heavy air (not helium). The pilot controls the amount of air in the ballonets through air valves. Since air is heavier than helium, the pilot deflates or inflates the ballonets to make the blimp rise or fall. Changing the proportions of air and helium changes the blimp's average density. Just like a hot air balloon, the blimp rises or falls until its average density matches the surrounding air.

Blimps in science

Today blimps are used for more than just sporting events and advertising. The U.S. Geological Survey has used blimps to fly over volcanoes. A blimp is less likely to be damaged by ash than a helicopter or plane.

Blimps are also used to study whales and their behavior. One research blimp is outfitted with cameras and equipment for tracking whales. Picture taking is much more stable from a blimp than from an airplane. The stability allows cameras with high-power zoom lenses to take highly detailed pictures from far above the ocean surface.

Questions:

1. What is a blimp?
2. What is buoyancy and how does it affect a blimp
3. What are negative, positive, and neutral buoyancy, and how do they affect a blimp?

CHAPTER ACTIVITY Will it Float or Will it Sink?

Background

Why do some objects sink while other float? During this activity you will predict whether a variety of objects will sink or float. You will then test your predictions and calculate the density of each object. You will look at the relationship between a material's density and its ability to float.

Materials:

Graduated cylinder, balance or scale, plastic cup
An assortment of small objects made of materials such as wood, various types of plastic, foam, steel, glass, cork, and aluminum

What you will do

1. Find the mass of an empty plastic cup. Pour exactly 50 mL (50 cm^3) of water into the cup. Measure the mass of the water and cup together, and subtract to find the mass of the water alone. Record your result in the data table below.

Description of object	Prediction: Sink or float?	Result: Sink or float?	Mass (g)	Volume (cm^3)	Density (g/cm^3)
water	X	X		50	

2. Calculate the density of the water.
3. Select one of the objects that you wish to test. Predict whether it will sink or float. Record your prediction in the data table.
4. Use a balance or scale to record the mass of the object in grams.
5. Fill the graduated cylinder halfway with water. Place the object in the water and observe whether it sinks or floats. Record the result in your data table.
6. Use the displacement method to find the volume of the object in cubic centimeters.
7. Calculate the density of the object in grams per cubic centimeter.
8. Repeat steps 3-7 using the other objects.

Applying your knowledge

a. Explain how you found the volume of the various objects you tested.
b. List the objects in order of density, starting with the least dense material. Circle each object that floats.
c. What is the pattern between the density of an object and the ability of the object to sink or float?
d. Suppose you were to do this experiment using other liquids instead of water. Which of the objects would float in each of the liquids listed below? Explain how you found your answer.

Liquid	Density (g/cm^3)
corn syrup	1.4
mercury	13.6
cooking oil	0.91

Chapter 4 Assessment

Vocabulary

Select the correct term to complete the sentences.

density	buoyancy	Archimedes' principle
weight		

Section 4.1

1. The mass of matter per unit of volume is known as ____.

Section 4.2

2. The upward force exerted by a fluid on an object submerged in the fluid is called ____.
3. The idea that the buoyant force exerted on an object is equal to the weight of the fluid displaced by the object is known as ____.
4. The force exerted on an object by the gravity of Earth is called ____.

Concepts

Section 4.1

1. Which has a greater weight, 1.0 gram of steel or 1.0 gram of aluminum?
2. Which has a greater density, 1.0 gram of steel or 1.0 kilogram of aluminum?
3. Which has a greater volume, 1.0 gram of steel or 1.0 gram of aluminum?
4. Name three units that can be used to represent density.
5. When comparing solids, liquids, and gases:
   a. Which phase generally has the greatest density?
   b. Which phase has the lowest density?
6. The density of ice is less than the density of water. How does this affect life in a pond over a long, cold winter?
7. Using Table 4.1 on page 75, which material might be used to make an object with a density of 0.60 g/cm^3?
8. A glass block has a density of 2.7 g/cm^3. If the same type of glass was used to make a 2-liter water pitcher, what would be the density of the pitcher?
9. By adding more lead to a bar of lead you:
   a. increase the bar's density.
   b. decrease the bar's density.
   c. do not change the bar's density.
10. If you know the mass and density of a material, how would you find its volume?
11. Based upon the diagram to the right, arrange the three materials, cork, water, and lead in order from most to least dense.

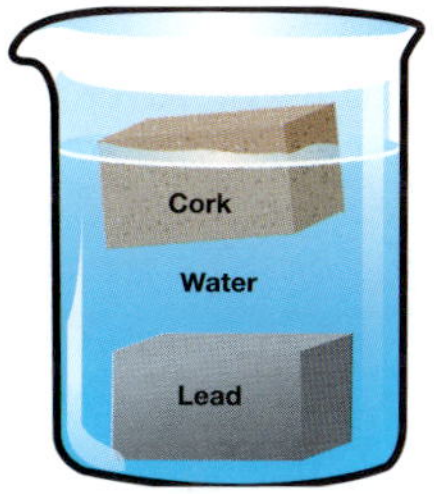

12. A graduated cylinder contains 25 mL of water. An object placed in the cylinder causes the water level to rise to 43 mL. What is the volume of the object?
13. What is the name of the dark, curved "band" at the top of the column of water pictured in the graduated cylinder to the right?

Section 4.2

14. How does the buoyant force of a rock submerged in water compare to the weight of the water displaced by the rock?
15. Why does ice float in a glass of water? Explain in terms of density and buoyancy.
16. What property of an object determines the strength of buoyant force that will be exerted on it when submerged in water?
17. A cargo barge weighs 200,000 N when empty. Cargo weighing 50,000 N is loaded onto the barge. What is the total weight of water displaced by the loaded barge?
18. What is the maximum average density that a fully loaded cargo ship may have?
19. Why do helium balloons float in air? Use the terms *buoyancy* and *density* in your answer.

Problems

Section 4.1

1. A piece of granite has a mass of 26 grams. The granite is placed in a graduated cylinder containing 10 mL of water. What is the reading of the water level in the graduated cylinder after the granite is fully submerged? Granite has a density of 2.6 g/cm^3.
2. Convert a density of 123 kg/m^3 to units of g/cm^3.
3. What is the volume of the liquid in the graduated cylinder pictured in the diagram?

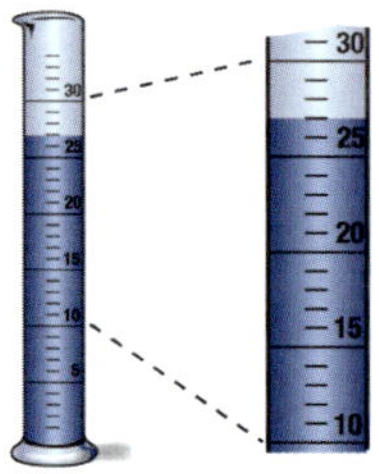

4. A rubber ball has a radius of 2.5 cm. The density of rubber is 1.2 g/cm^3.
   a. What is the volume of the ball?
   b. What is the mass of the ball?
5. The density of ice is 0.92 g/cm^3. What is the volume of 1 kg of ice? If that 1 kg of ice completely melted, what would the volume of water be? The density of water is 1 g/cm^3.
6. Your teacher gives you two stainless steel balls. The larger has a mass of 25 grams and a volume of 3.2 cm^3. The smaller has a mass of 10 grams. Calculate the volume of the smaller ball.
7. The cube in the diagram has a mass of 7.8 grams and measures 1.2 centimeters on an edge.
   a. Find the density of the cube. Show your work, including an equation.
   b. Will the cube float in water? Explain.

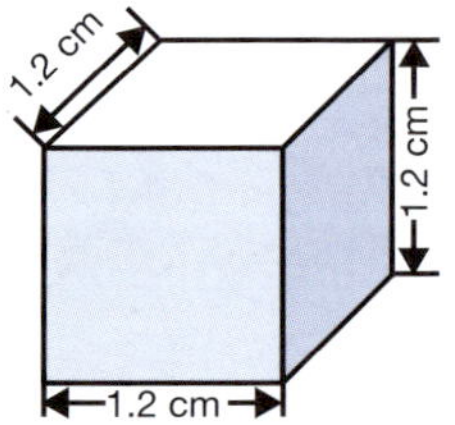

Section 4.2

8. An object weighing 45 newtons in air is suspended from a spring scale. The spring scale reads 22 newtons when the object is fullly submerged. Calculate the buoyant force on the object.
9. A stone that weighs 6.5 newtons in air weighs only 5.0 newtons when submerged in water. What is the buoyant force exerted on the rock by the water?
10. A 100 mL oak object is placed in water. What volume of water is displaced by the oak object? The density of oak is 0.60 g/cm^3.
11. A 100 mL steel object is placed in water. What volume of water is displaced by the steel object? The density of steel is 7.8 g/cm^3.

Chapter 5

States of Matter

Imagine a liquid substance that can harden rapidly and form an exact replica of any container that holds it. The material does not burn, boil, melt, or dissolve in any commonly available acid or solvent. Once it hardens, it won't change under normal circumstances. What sort of material is this? This substance was created in 1907 by a chemist named Leo Baekeland. He called the substance *Bakelite*, and it was the first useful plastic. Since then, many more plastics have been developed for many different uses. It is hard to imagine life today without plastic! Read this chapter to learn about interesting properties of plastics and other matter.

Key Questions

1. ***How are liquids and gases alike, and how are they different?***

2. ***Why does a frozen pond have ice on the top and not the bottom?***

3. ***Why are solids solid?***

5.1 Liquids and Gases

A **fluid** is a form of matter that flows when any force is applied, no matter how small. Liquids are one kind of fluid, gases are another. You have seen water flow from a faucet (or overflow a sink) and felt cool air flow through an open window (or carry the aroma of cooking food into your room). What are some other properties of fluids?

Atoms and molecules in liquids and gases

Molecules in a liquid The molecules in liquid water have more energy and move around much more than do the molecules in ice. In a **liquid**, molecules can slide over and around each other. This is how liquids flow and change shape. But the atoms do not have enough energy to completely break their bonds with one another. That is why liquids have constant volume even though the shape may change.

Molecules in a gas As in liquids, molecules in a **gas** are free to move around and so gas flows. However, molecules in a gas have much more energy than molecules in a liquid. On average, each molecule has enough energy to completely break away from its neighbors. That is why gas expands to fill any container (Figure 5.1).

Density of liquids and gases In general, a liquid material is a little less dense than the same material in a solid form. This is because the molecules in a liquid move around more and take up a little more space. A gas is usually much less dense than either a liquid or solid. This is because the molecules in a gas are spread out with comparatively large spaces between them.

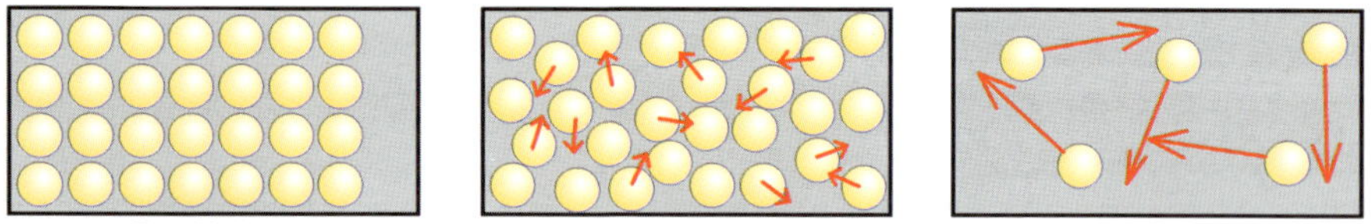

VOCABULARY

fluid - a form of matter that flows when any force is applied, no matter how small. Liquids and gases are fluids.

liquid - phase of matter that can flow and change shape but has constant volume.

gas - phase of matter with high energy molecules that can expand to fill a container.

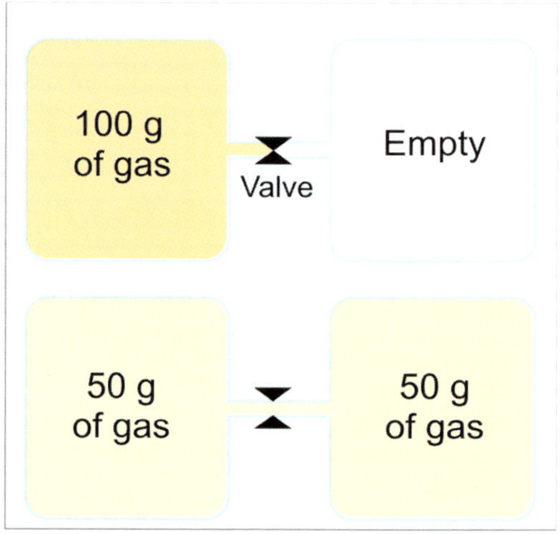

Figure 5.1: *Gases flow like liquids, but they also may expand or contract to completely fill any container.*

Pressure

Forces in fluids

Think about what happens when you push down on an inflated balloon. The downward force you apply creates forces that act sideways as well as down. This is very different from what happens when you push down on a bowling ball. The ball transmits the force directly down. Because fluids change shape, forces in fluids are more complicated than forces in solids.

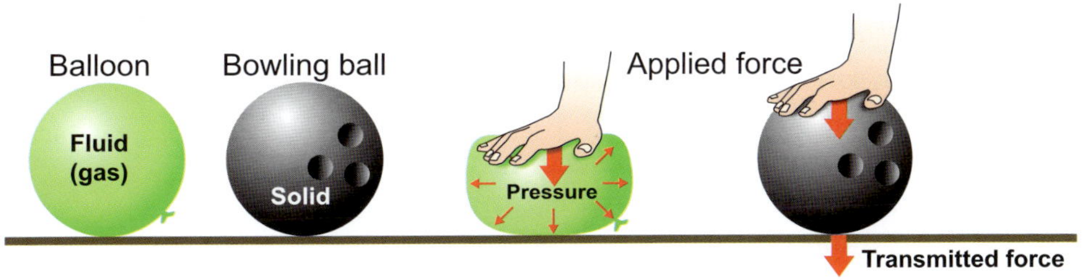

Pressure

A force applied to a fluid creates **pressure**. Pressure acts in all directions, not just the direction of the applied force. When you inflate a car tire, you are increasing the pressure in the tire. A pressure of 40 pounds per square inch means every square inch of the inside of the tire feels a force of 40 pounds. This force acts up, down, and sideways in all directions inside the tire. The downward portion of the pressure force is what holds the body of the car up (Figure 5.2).

The molecular explanation

What causes pressure? On the microscopic level, pressure comes from collisions between atoms. If you had a jar of water, and if there were such as thing as an atomic-magnification video camera, you would see trillions of atoms bounce off each other and the walls of the jar every second (Figure 5.3). Every square centimeter of the inside surface of the jar feels a force from the constant impact of atoms. That force is what we feel as pressure. Pressure comes from the constant collisions of many, many atoms.

VOCABULARY

pressure - a distributed force per unit area that acts within a fluid.

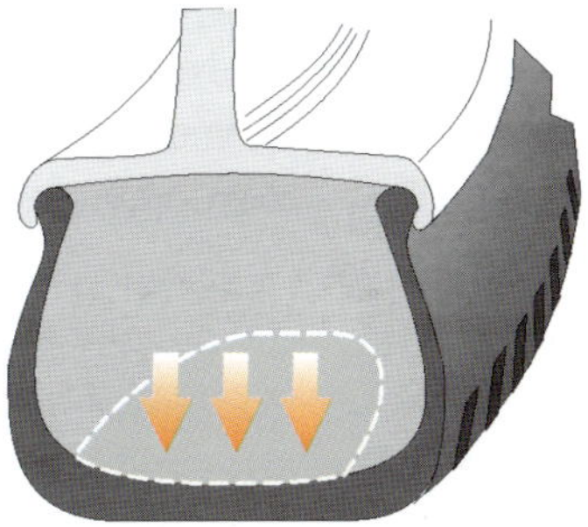

Figure 5.2: *The pressure inside your tire is what holds your car up.*

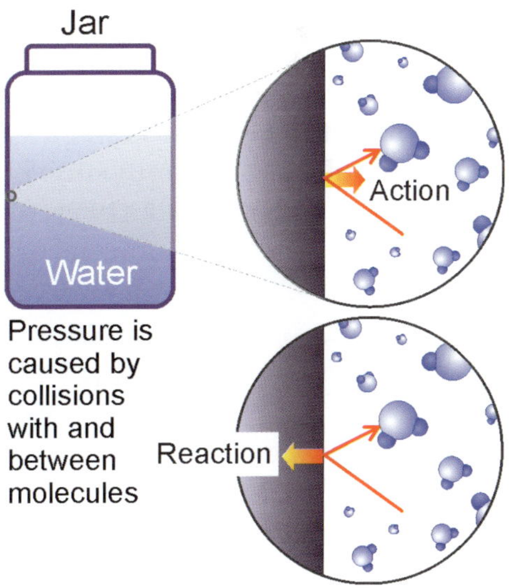

Figure 5.3: *The molecular explanation of pressure.*

Intermolecular forces

What intermolecular forces do

There are two types of forces that act between atoms. The strongest forces are between atoms that are bonded together into molecules and compounds. These forces act *within* molecules, such as the forces that hold the hydrogen and oxygen atoms together in a water molecule. A weaker type of force acts *between* molecules, or between atoms that are not bound together in molecules. These in-between forces are called **intermolecular forces**. For example, intermolecular forces hold water molecules together in liquid water and in ice.

Properties of intermolecular forces

At distances greater than the size of the molecule, intermolecular forces are attractive and pull molecules together. Once molecules become close enough to touch, intermolecular forces become repulsive. This is what prevents one molecule from overlapping another. Intermolecular forces pull molecules together at long range and hold them apart at short range.

The role of thermal energy

The phases of matter — solid, liquid, gas — exist because of competition between thermal energy and intermolecular forces. Intermolecular forces always try to bring molecules close. Thermal energy causes molecules to vibrate and spread apart.

Explaining the phases of matter

When molecules have a lot of thermal energy (high temperature), intermolecular forces are completely overcome and the molecules spread apart, as in a gas. When molecules have a medium amount of thermal energy, they come together to form a liquid. In a liquid, the molecules have enough thermal energy to partially overcome intermolecular forces and move around, but not enough energy to completely escape. When molecules have a low amount of thermal energy, the intermolecular forces dominate and molecules become fixed in place as a solid.

VOCABULARY

intermolecular forces - forces between separate atoms and molecules that are attractive at a distance but repulsive at close range.

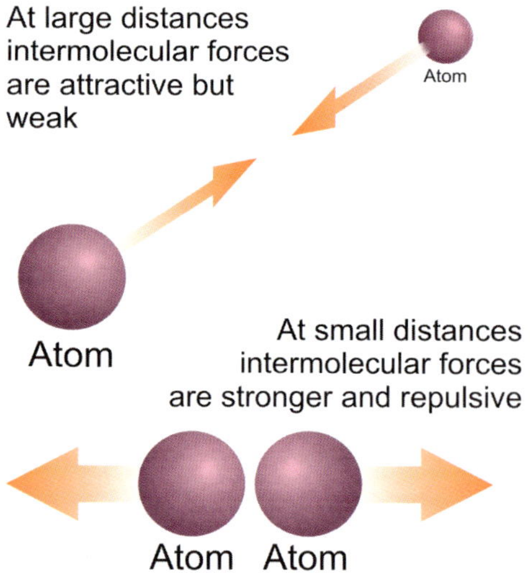

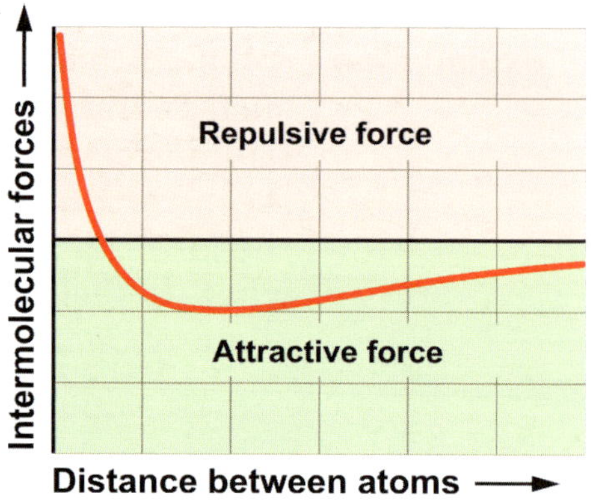

Melting and boiling

Melting point The **melting point** is the temperature at which a substance changes from a solid to a liquid. For example, the melting point of water is zero degrees Celsius. Different substances have different melting points because intermolecular forces have different strengths in different substances. Stronger forces require more energy to break. For example, iron melts at a much higher temperature than water, about 1,500°C. The difference in melting points tells us that the intermolecular forces in iron are stronger than they are in water.

Boiling When enough thermal energy is added, intermolecular forces are completely overcome and a liquid becomes a gas. The temperature at which a liquid becomes a gas is called the **boiling point**. For water, the boiling point is 100 degrees Celsius. That is the temperature at which liquid water becomes a gas (steam). Boiling takes place within a liquid as bubbles of gas particles form and rise to the surface (Figure 5.4).

Changes in state require energy It takes energy to overcome intermolecular forces. This explains a peculiar thing that happens when a substance melts or boils. As heat energy is added to ice, the temperature increases until it reaches 0°C. Then *the temperature stops increasing*. As you add more heat, more ice becomes liquid water but the temperature stays the same. This is because the added energy is being used to break the intermolecular forces and change solid into liquid. Once all the ice has become liquid, the temperature starts to rise again if more energy is added.

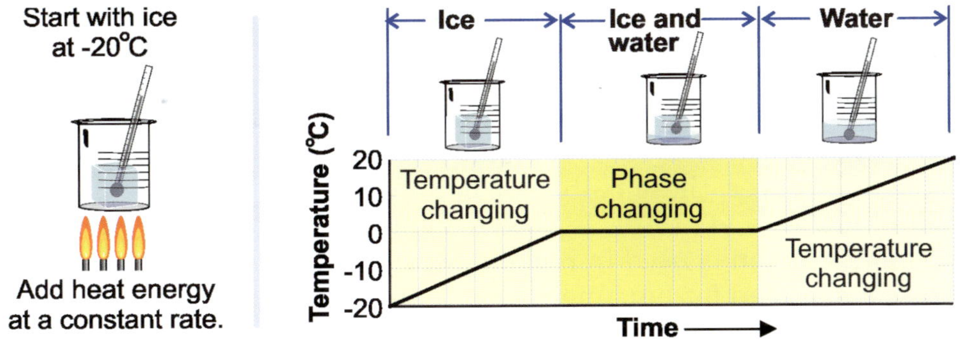

VOCABULARY

melting point - the temperature at which a substance changes from a solid to a liquid.

boiling point - the temperature at which a substance changes from a liquid to a gas.

Water melts at 0° C (32°F)

Water boils at 100°C (212°F)

Figure 5.4: *Melting and boiling.*

Melting and boiling points of common substances

Range of melting and boiling points

Materials have a wide range of melting and boiling points. This is essential for life because we need some materials to be solid at room temperature, others to be liquid, and still others to be gas. Table 5.1 gives the melting and boiling points for some ordinary materials.

Table 5.1:

Material	Melting point (°C)	Boiling point (°C)
Tungsten	3,422	5,555
Iron	1,538	2,861
Copper	1,085	2,562
Aluminum	660	2,519
Lead	327	1,749
Hard plastic	240	300
Candle wax	50	400
Water	0	100
Alcohol	-108	78
Nitrogen	-210	-196
Oxygen	-219	-183
Helium	none	-269

Water is *less dense* in solid form

Most materials have a higher density as a solid than as a liquid. Water is a notable exception. Solid water has an open crystal structure that resembles a honeycomb, where each water molecule forms intermolecular bonds with four other water molecules. This creates a six-sided arrangement of molecules. The six-sided crystal form explains the six-way symmetry you see when you examine snowflakes with a magnifying lens.

Decreasing density

As water freezes, molecules of water separate slightly from each other because of the honeycomb structure. This causes the volume to increase slightly, while the mass stays the same. As a result the density decreases. This explains why water expands when it is frozen and also floats. The density of ice is about 0.92 g/cm^3 whereas the density of water is about 1.0 g/cm^3.

Water's density and living organisms

Because ice is less dense than liquid water, it floats on the surface of lakes and ponds when they freeze over in winter. When this occurs, the temperature of the water below the ice layer remains above freezing. This is one factor that helps fish and other aquatic organisms to survive over long, cold winters (Figure 5.5).

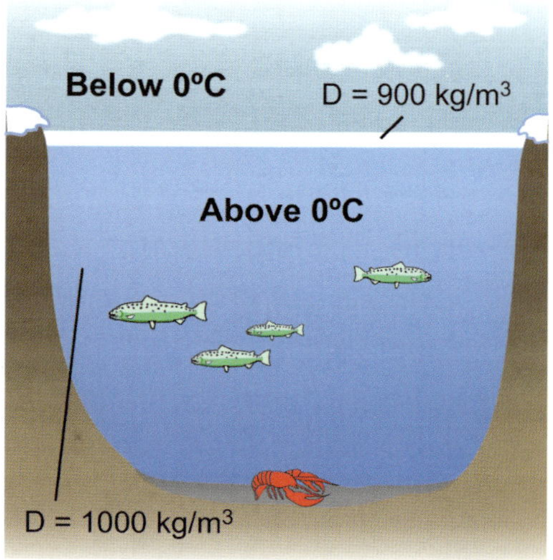

Figure 5.5: *Ice floats on the surface of a pond, keeping the water beneath it from reaching freezing temperatures.*

Oxygen and nitrogen are ordinarily gases at room temperature. If the temperature gets low enough, however, these materials become liquid and even solid. Liquid nitrogen at -196°C is used to rapidly freeze or cool materials. Liquid oxygen is used in rockets because in outer space there is no gaseous oxygen for burning rocket fuel.

Evaporation and condensation

Evaporation **Evaporation** occurs when molecules go from liquid to gas at temperatures *below* the boiling point. Evaporation happens because temperature measures the *average* kinetic energy of molecules. Some have energy above the average and some below the average. Some of the highest-energy molecules have enough energy to break bonds with their neighbors and become a gas if they are near the surface. These high-energy molecules are the source of evaporation.

Evaporation cools liquids Evaporation takes energy away from a liquid because the molecules that escape are the ones with the most energy. The average energy of the molecules left behind is lowered. That is why your body sweats on a hot day. The evaporation of sweat from your skin cools your body by carrying away energy (Figure 5.6).

Condensation **Condensation** occurs when molecules go from gas to liquid at temperatures below the boiling point (Figure 5.7). Condensation occurs because water vapor molecules with less than the average energy stick to a cool surface forming drops of liquid water. Condensation raises the temperature of a gas because atoms in a gas have more energy than atoms in a liquid. Low-energy atoms condense into liquid, leaving the higher-energy (warmer) atoms in the gas.

Relative humidity Ordinary air contains some water vapor. Evaporation adds water vapor to the air. Condensation removes water vapor. The percentage of water vapor in the air is a balance between evaporation and condensation. When air is *saturated*, it means evaporation and condensation are exactly balanced. If you try to add more water vapor to saturated air, it condenses immediately back into liquid again. The *relative humidity* tells how close the air is to saturation. When the relative humidity is 100 percent, the air is completely saturated. That means any water vapor that evaporates from your skin is immediately condensed again, which is why you feel hot and sticky when the humidity is high.

VOCABULARY

evaporation - change from liquid to gas at a temperature below the boiling point.

condensation - change from gas to liquid at a temperature below the boiling point.

Figure 5.6: *Sweat evaporating from skin removes energy and cools the body.*

Figure 5.7: *Dew forms when water vapor in air condenses into droplets.*

Convection

What is convection? Convection is the transfer of heat through the motion of fluids such as air and water. If you have warmed your hands over a fire, you have felt convection. Heat from the flame was transferred to your hand by the upward movement of air.

Natural convection Convection occurs because fluids expand when they heat up. Since expansion increases volume, but not mass, the density of a warm fluid becomes lower than the density of surrounding cooler fluid, causing the warmer fluid to float upward. In a pot on the stove, hot water circulates to the top and cooler water sinks to the bottom. This circulating flow is called *natural convection.*

Forced convection In many buildings and houses, a boiler heats water that is then pumped throughout the structure to distribute the heat. Since the heat is being carried by a moving fluid, this is another example of convection. However, the flow is created by pumps, which makes this an example of *forced convection*. Natural convection also occurs in the same system. The heat from a hot radiator warms the air in a room by natural convection. The warmer air rises and cooler air is drawn from the far side of the room. The cooler air is then warmed and rises, creating circulation that spreads heat through the room.

VOCABULARY

convection - the transfer of heat through the motion of fluids such as air and water.

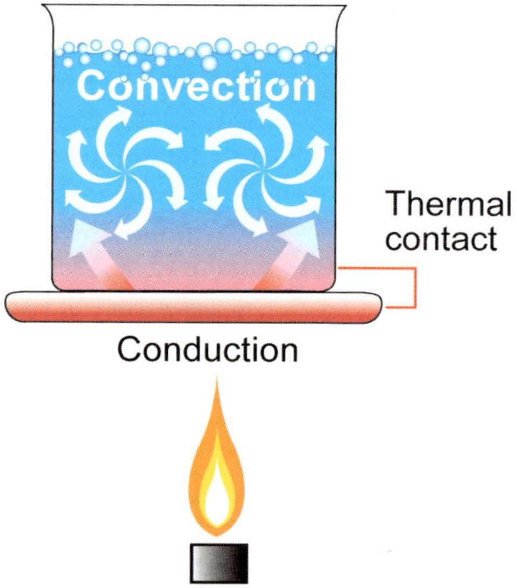

Figure 5.8: *Convection currents in water. The hot water at the bottom of the pot rises to the top and replaces the cold water.*

The atmosphere of Earth

Air is not "nothing" Air feels "light" because it is 1,000 times less dense than water. Air may seem like "nothing" but all the oxygen our bodies need and all the carbon needed by plants comes from air. As a tree grows, you will not see soil disappear to provide mass for the tree. After oxygen and hydrogen (from water), the most abundant element in a tree is carbon. All of those carbon atoms come from carbon dioxide (CO_2) in the air.

Air is a mixture of gases Air is the most important gas to living things on the Earth. The atmosphere of Earth is a mixture of gases (Figure 5.9). Molecular nitrogen (N_2) and oxygen (O_2) account for 97.2 percent of the mass of air. Argon and water vapor make up most of the rest.

Atmospheric pressure Gravity creates pressure because fluids have mass and therefore weight. The Earth's atmosphere has a pressure due to the weight of air. The density of air is low, but then the atmosphere is more than 80,000 meters deep (Figure 5.10).

Weather

Earth's weather is created by gigantic convection currents in the atmosphere. Energy from the sun mostly passes through the atmosphere to warm the ground. Air near the ground becomes warm and expands. Warmer air is less dense than cold air and therefore the warm air near the ground rises.

How rain forms Over the oceans, the warm air may be nearly saturated with water vapor. At high altitude, the temperature of the atmosphere drops rapidly. As the temperature drops, the ability of the air to hold water vapor also decreases. That excess water vapor condenses to create rain and other forms of precipitation.

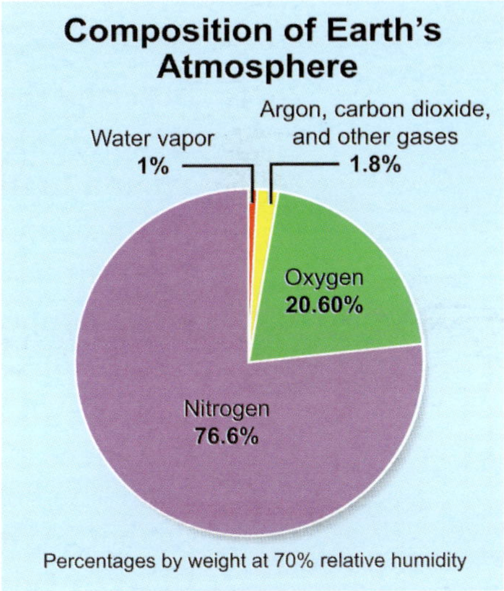

Figure 5.9: *Air is a mixture of gases.*

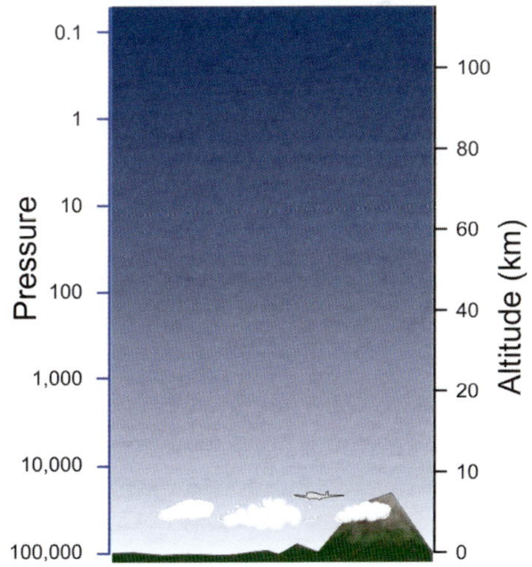

Figure 5.10: *The change in pressure with altitude in the atmosphere.*

5.1 Section Review

1. Describe the movement of the atoms or molecules in a gas.
2. A liquid takes the shape of its container, but why doesn't a liquid expand to fill the container completely?
3. When you push *down* on a confined fluid, you create pressure. In what direction does the pressure act?
4. What happens to the temperature of ice at its melting point while you add heat? While it is melting, does it gain or lose energy?
5. What is evaporation? How is it different from boiling?
6. You place 1 liter of a substance into a 2-liter bottle and tightly cover the bottle. The substance expands until it completely fills the bottle. What state is the substance in?
7. Describe how the density of ice affects our daily lives. Explain why ice forms on the top of ponds and lakes, and not the bottom. Use the following terms in your explanation: density, organized structure, and water molecules. How does this property of water help support life in lakes and ponds?
8. Why doesn't convection occur in a solid material?
9. Why is it more comfortable to exercise on a day when the relative humidity is low?
10. Convection creates circulating currents in a pot of boiling water because ____ water rises and ____ water sinks.
11. Describe how water can be present in all three states at the same time in the atmosphere?
12. Would you expect a higher atmospheric pressure at the top of a mountain in Alaska's Denali national park or near sea level in Florida's Everglades national park?

CHALLENGE

Visitors to high-altitude regions may suffer from Acute Mountain Sickness (AMS) if they do not allow their bodies to acclimate to the new surroundings. Do some research to find a set of guidelines for preventing this condition. Design a brochure for travelers that describes symptoms of AMS and provides recommendations for preventing and/or treating them.

5.2 Solid Matter

You have learned that matter is made up of atoms and molecules. In a solid the atoms or molecules are closely packed and stay in place, which is why solids hold their shape. In this section you will learn how the properties of solids result from the behavior of atoms and molecules.

The molecular structure of solids

Why solids are solid

In a **solid**, thermal energy is not enough to overcome intermolecular forces of attraction. Individual molecules are bound together tightly enough that they do not change their positions as they do in liquids and gases. Imagine that the molecules in a solid are connected by springs (Figure 5.11) that represent the intermolecular forces. Thermal energy keeps the molecules moving, but because of those intermolecular forces, they only "spring" back and forth around the same average position. That is why solid materials hold their shape.

Solids hold their shape

Because the molecules are bound to each other, all solids have some ability to hold their shape when forces are applied. Some solids, like steel, can hold their shape under much greater forces than others, like rubber (Figure 5.12). Many solids, like plastic, have properties between the softness of rubber and the hardness of steel. Engineers design the molecular structures of solids to have the properties that are needed for given applications.

Physical properties of solids

Some important physical properties of solids are:

- Density: mass per-unit volume.
- Strength: the ability to maintain shape under great force.
- Elasticity: the ability to stretch and return to the same shape.
- Ductility: the ability to bend without breaking.
- Thermal conductivity: the ability to transmit heat energy.
- Electrical conductivity: the ability to allow electricity to flow.

VOCABULARY

solid - a phase of matter with a definite shape and constant volume.

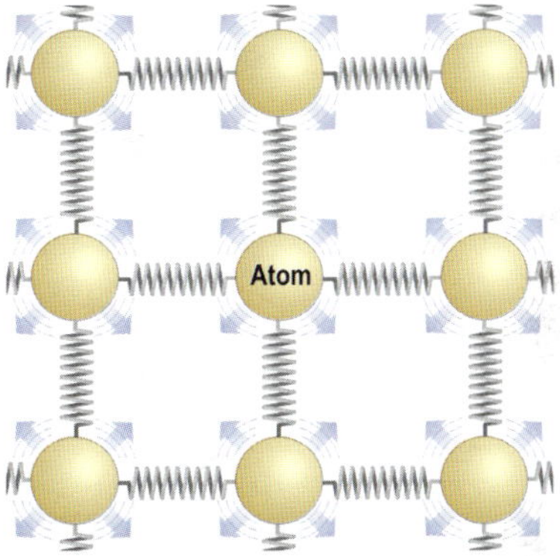

Figure 5.11: *Atoms in a solid are connected by bonds that act like springs. The atoms still vibrate but stay in the same average position relative to each other.*

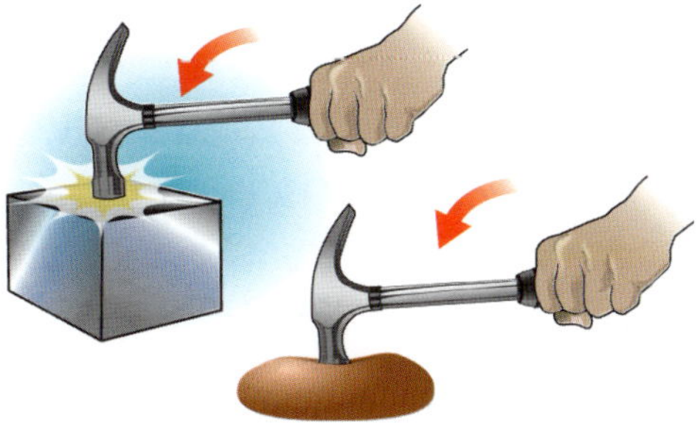

Figure 5.12: *Steel and rubber are both solids but they have different strengths, or abilities to hold their shape under force.*

Mechanical properties

The meaning of "strength"

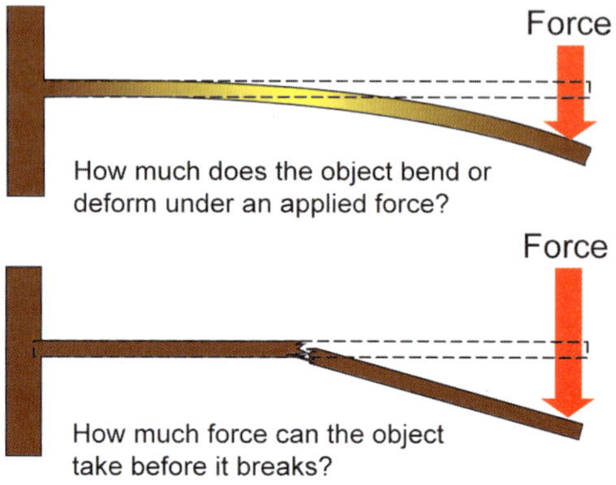

When you apply a force to an object, the object may change its size, shape, or both. The concept of "strength" describes the ability of a solid object to maintain its shape even when force is applied. The **strength** of an object depends on the answers to the two questions in the diagram.

Elasticity If you pull on a rubber band, its shape changes. If you let it go, the rubber band returns to its original shape. Rubber bands can stretch many times their original length before breaking, a property called elasticity. **Elasticity** describes a solid's ability to be stretched and then return to its original size. This property also gives objects the ability to bounce and to withstand impact without breaking.

Brittleness **Brittleness** is defined as the tendency of a solid to crack or break before stretching very much. Glass is a good example of a brittle material. You cannot stretch glass even one-tenth of a percent (0.001) before it breaks. To stretch or shape glass you need to heat the glass until it is almost melted. Heating causes molecules to move faster, temporarily breaking the forces that hold them together.

Ductility One of the most useful properties of metals is that they are ductile. A ductile material can be bent a relatively large amount without breaking. For example, a steel fork can be bent in half and the steel does not break. A plastic fork cracks when it is bent only a small amount. Steel's high **ductility** means steel can be formed into useful shapes by pounding, rolling, and bending. These processes would destroy a brittle material like glass.

VOCABULARY

strength - the ability to maintain shape under the application of forces.

elasticity - the ability to be stretched or compressed and then return to original size.

brittleness - the tendency to crack or break; the opposite of elasticity.

ductility - the ability to bend without breaking.

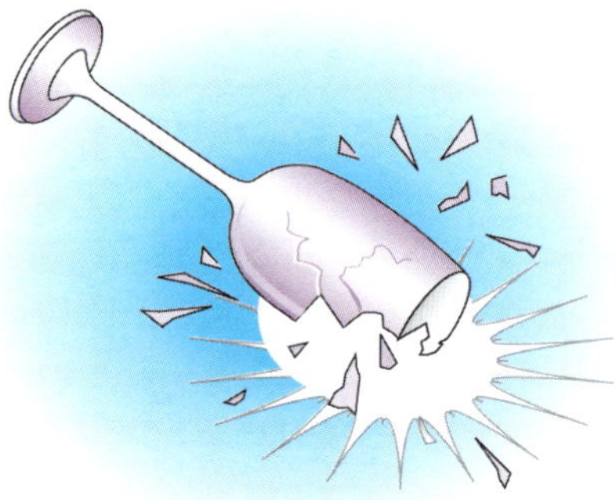

Figure 5.13: *Brittleness is the tendency of a solid to crack when force is applied.*

Heat conduction in solids

What is conduction? **Heat conduction** is the transfer of heat by the direct contact of particles of matter. When you hold a warm mug of tea or cocoa, you experience conduction. Heat is transferred from the mug to your hand. Conduction occurs between two materials at different temperatures when they are touching each other. Heat can also be transferred by conduction *through* materials. If you stir hot cocoa with a metal spoon, heat is transferred *from* the cocoa *through* the spoon and *to* your hand.

How does conduction work? Picture yourself placing a spoon into a mug of hot cocoa. The molecules in the cocoa have a higher average kinetic energy than those of the spoon. The molecules in the spoon exchange energy with the molecules in the cocoa through collisions. The molecules in the spoon spread the energy up the handle of the spoon through the intermolecular forces between them. Conduction works through collisions and through the intermolecular forces between molecules.

Thermal equilibrium As the collisions continue, the molecules of the hotter material (the cocoa) lose energy and the molecules of the cooler material (the spoon) gain energy. The kinetic energy of the hotter material is transferred, one collision at a time, to the cooler material. Eventually, both materials are at the same temperature. When this happens, they are in thermal equilibrium. **Thermal equilibrium** occurs when two bodies have the same temperature. No heat flows in thermal equilibrium because the temperature is the same in the two materials.

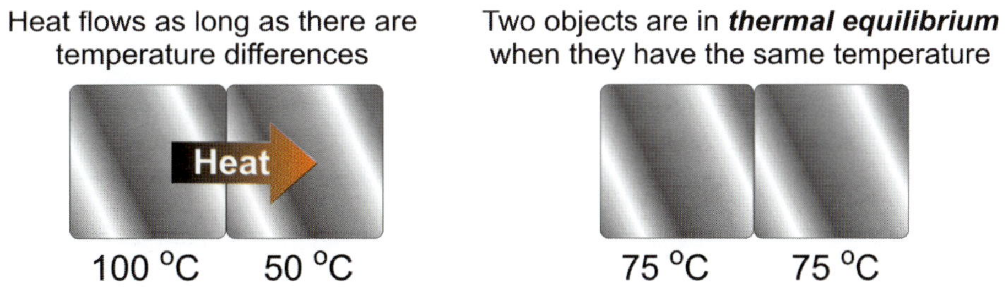

VOCABULARY

heat conduction - the transfer of heat by the direct contact of particles of matter.

thermal equilibrium - a condition where temperatures are the same and no heat flows.

Figure 5.17: *Heat flows by conduction from the hot cocoa into the spoon and up its handle.*

Thermal conductors and insulators

Which state of matter conducts best?

Although conduction also occurs in liquids and gases, solids make the best conductors because the molecules in a solid are packed close together. Because molecules in a gas are spread so far apart, relatively few collisions occur, making air, for instance, a poor conductor of heat. This explains why materials used to keep things warm, such as fiberglass insulation and down jackets, have thousands of air tiny spaces inside (Figure 5.18).

Figure 5.18: *Because air is a poor conductor of heat, a down jacket keeps you warm in cold winter.*

Thermal conductors and insulators

Materials that conduct heat easily are called thermal conductors and those that conduct heat poorly are called thermal insulators. For example, metal is a thermal conductor, and a foam cup is a thermal insulator. The words *conductor* and *insulator* are also used to describe a material's ability to conduct electrical current. In general, good electrical conductors like silver, copper, gold, and aluminum are also good thermal conductors.

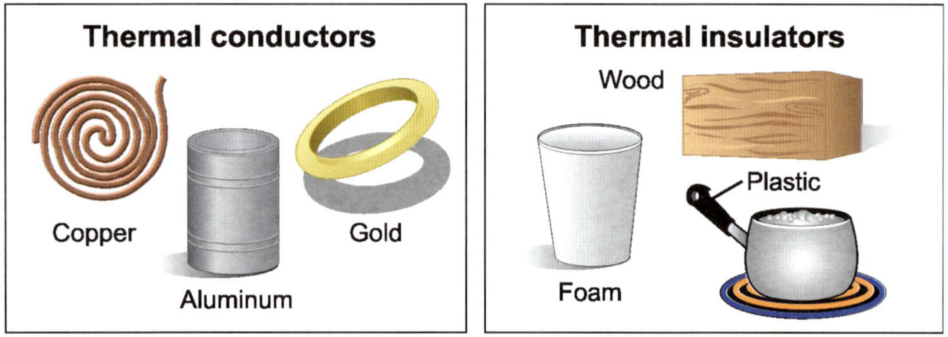

Heat conduction cannot occur through a vacuum

Conduction cannot occur in the vacuum of space where there is no matter. A thermos bottle keeps liquids hot for hours using a vacuum. A thermos is a container consisting of a bottle surrounded by a slightly larger bottle. Air molecules have been removed from the space between the bottles to create a vacuum. This prevents heat transfer by conduction. A small amount of heat is conducted through the cap and the glass where the two walls meet, so eventually the contents will cool off (Figure 5.19).

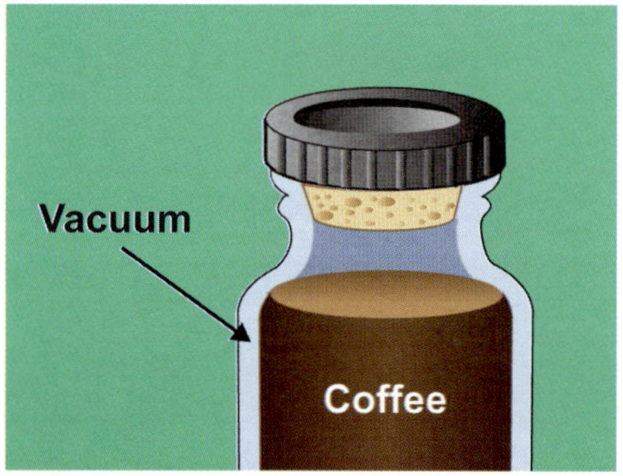

Figure 5.19: *A thermos bottle uses a vacuum to prevent heat transfer by conduction or convection.*

5.2 Section Review

1. Observe the world around you and find a useful object made of an elastic material. Would this object work if it was made of a brittle material? Why does the elasticity of the material allow this object to work so well?
2. Golf balls are made with a rubber core. Why does it make no sense to make the core of glass?
3. What property of copper allows it to be pulled into thin wire?
4. Rubber and steel are both elastic, yet engineers do not design bridges out of rubber. Explain why.
5. Name one example of a material for each set of properties:
   a. high elasticity and ductile.
   b. amorphous and brittle.
   c. crystalline and brittle.
   d. crystalline and elastic.
6. Describe how the arrangement of the atoms and molecules in a sugar crystal differ from those in a piece of plastic.
7. You are an engineer who must choose a type of plastic to use for the infant car seat you are designing. Name two properties of solids that would help you decide, and explain why each is important.
8. In nature, heat will always flow from a:
   a. cold object to a warm object.
   b. small object to a large object.
   c. warm object to a cold object.
9. Why do you think pots and pans for cooking are made out of metal?
10. What properties make a material a good thermal insulator? Give three examples of good thermal insulator.
11. Air spaces between the feathers of a down-filled coat cause the coat to be a good thermal ____.
12. Name one example of heat transfer through conduction.

CHALLENGE

Find out how much insulation is recommended for homes in your community. Where is the most insulation recommended: in the ceiling, walls, or floors? Using what you know about heat transfer, explain why.

CHEMISTRY CONNECTION

Silly Putty®: Solid or Liquid?

Silly Putty—it's been a popular party favor for more than fifty years. Your parents probably played with it when they were kids. Some people call it America's longest lasting fad.

It's easy to understand why people like Silly Putty. Roll it into a ball, and you can bounce it around the room. Pull on it slowly and it will stretch out like a long lazy snake. Give it a quick yank and it will break with a satisfying *snap*.

Have you ever tried to smash a ball of Silly Putty with a hammer? It keeps its shape every time. However, if you gently press on it with your thumb, you can flatten it easily. If you leave a ball of Silly Putty on your dresser overnight, in the morning you'll see that it flattened out by itself while you were sleeping.

What's going on here?

Silly Putty isn't easy to categorize. It holds it shape when hammered, yet flows into a puddle when left alone overnight. No wonder the people who make Silly Putty call it "a real solid liquid."

Rheologists (scientists who study how matter flows and/or deforms) have another term for Silly Putty: it's a *viscoelastic* liquid.

Viscoelastic is a compound word (like *snowman*). The *visco-* part comes from the word *viscous*, which means "resistant to flow." Thick, gooey, slow-flowing liquids like hot fudge sauce are *viscous*. Silly Putty is like that.

You're probably already familiar with the second half of the word. *Elastic*, in physics terms, describes a material that returns to its original shape when deformed.

So, rheologists describe Silly Putty as a slow-flowing, elastic liquid.

How did it get that way?

It's not too surprising that Silly Putty bounces, because it was accidentally invented by a chemist looking for a substitute for rubber. In 1943, James Wright, a researcher for General Electric, dropped some boric acid into silicone oil, creating a gooey compound.

This compound, first called "nutty putty," was sent to engineers around the world—but no practical uses were found. In 1949, a man named Peter Hodgson decided to sell it as a toy. He borrowed $147 to buy a batch from General Electric, divided the batch into one-ounce lumps, and placed each lump into a plastic egg. He renamed the compound "Silly Putty" after the main ingredient, silicone.

A *New Yorker* magazine reporter wrote an article about Silly Putty in 1950, and afterward Hodgson received 250,000 orders in three days. Silly Putty was a hit!

Inside Silly Putty

The silicone oil used to make Silly Putty is known to chemists as polydimethylsilo xane, or PDMS. PDMS is a

polymer, which means each molecule is made up of long chain of identical smaller molecules.

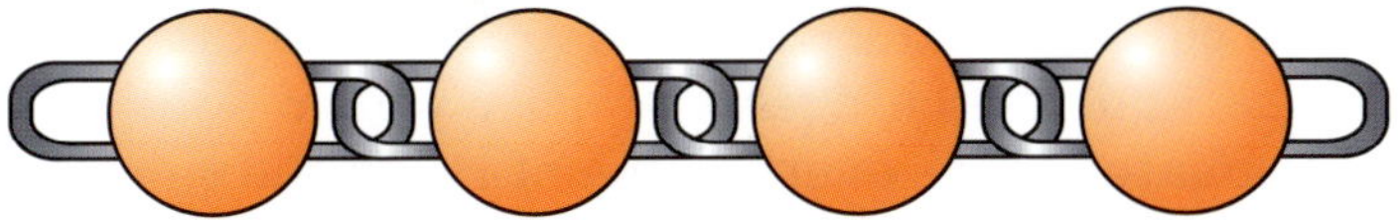

When boric acid is added to the long chains of PDMS, boron crosslinks begin to form. This means that the boron hooks chains of PDMS molecules together like this:

These boron crosslinks are not very strong. Remember that molecules in solids and liquids are always in motion. This motion breaks boron crosslinks, but over time new crosslinks form. This action is called *dynamic* (changing) *crosslinking*.

Because of this dynamic crosslinking, Silly Putty reacts one way to quick forces and another way to long-acting forces.

When you strike Silly Putty with a hammer, the Silly Putty reacts like an elastic solid: it bounces back. That's because most of the boron crosslinks remain in place during the split second of the hammer's strike.

When you leave a ball of Silly Putty untouched overnight, the boron crosslinks that help Silly Putty hold its shape have about eight hours to break down. Over that time, molecular motion breaks many of the original crosslinks. Gravitational force constantly pulls the PDMS molecules downward, and in the morning you're left with a Silly Putty puddle.

Questions:

1. Silly Putty does have some practical uses, despite the fact that engineers in the 1940's couldn't think of any. Find out about these using the Internet, or come up with one on your own.
2. Use the Internet to find out about a man named Earl Warrick. What was his role in the invention of Silly Putty?
3. The crew of Apollo 8 took some Silly Putty to the moon. Use the Internet to find out how the astronauts used it.

*Permission granted by Binney and Smith to publish trademark named Silly Putty.

CHAPTER ACTIVITY Make Your Own Viscoelastic Liquid

The exact recipe for Silly Putty is kept secret, but you can make your own viscoelastic liquid with ingredients you may have around the house. The homemade compound uses different molecules to form the polymer chains, but the boron crosslinks work the same way.

What you will need

White glue and water solution made in a 1:1 ratio

Borax and water solution: mix 5 mL of Borax in 60 mL of water (Borax powder is found in supermarket laundry detergent aisles)

8-ounce paper cup

Stirring stick (A tongue depressor works well)

What you will do

1. Pour 60 mL of the white glue solution into the cup.
2. Add 30 mL of the borax solution.
3. Stir the mixture for 2-3 minutes.
4. Remove the mixture from the cup and knead it with your hands. It will be sticky at first. Keep kneading until it is easy to pull the Putty away from your hands in a single lump.

Applying your knowledge

a. Develop a class procedure for measuring the Putty's bounciness and stretchiness. Compare your results with your classmates'. Was every batch of Putty the same? If not, can you suggest reasons for the differences?

b. There are lots of experiments you could do with your home-made Putty. Here are a few examples:
   a. How does temperature affect bounciness?
   b. Does stretchiness change over time?
   Choose one of these questions or make up your own question to answer about your Putty.

c. State your hypothesis.

d. Develop a procedure for testing your hypothesis. Remember, only one variable can be changed!

e. Create a data table to record your results. Here's a sample:

Temperature	Bounce height when dropped 50 cm
-10°C	
5°C	
20°C	
35°C	
50°C	

f. Carry out your experiment and record your results. What conclusion(s) can you draw?

g. Share your results with your classmates.

Chapter 5 Assessment

Vocabulary

Select the correct term to complete the sentences.

melting point	boiling point	evaporation
intermolecular forces	convection	pressure
condensation	fluid	strength
heat conduction	thermal equilibrium	ductility
crystalline	amorphous	brittleness
elasticity		

Section 5.1

1. When a substance changes from gas to liquid at a temperature below its boiling point, ____ has taken place.
2. The temperature at which a substance changes from liquid to gas is its ____.
3. The transfer of heat through the motion of fluids such as water or air is known as ____.
4. A scientist would call the change of a substance from liquid to gas at a temperature below its boiling point ____.
5. The temperature at which a substance changes from solid to liquid is its ____.
6. A form of matter that flows when any force is applied to it is called a(n) ____.
7. A force that acts in all directions and comes from the constant collisions of many atoms is ____.
8. ____ are what hold water molecules together in liquid water and in ice.

Section 5.2

9. Transfer of heat by direct contact between particles of matter is called ____.
10. A solid whose atoms are arranged in an orderly, repeating pattern would be called a ____ solid.
11. When heat does not transfer from one object to another because both objects are at the same temperature, the condition is called ____.
12. The ability to bend without breaking is known as ____.
13. Solids whose atoms or molecules have **no** orderly, repeating pattern are called ____ solids.
14. ____ is a solid's ability to be stretched and then return to its original size.
15. The tendency of a solid to crack or break before stretching very much is known as ____.
16. The ability of an object to maintain its shape even when force is applied is ____.

Concepts

Section 5.1

1. For each phase or form, identify the matter as liquid (**L**), gas (**G**), or both (**B**):
    a. ____ definite volume but changes shape to fit the shape of the container.
    b. ____ generally has the lower density of the two forms.
    c. ____ expands to completely fill any container.
    d. ____ bonds between atoms are not completely broken.
    e. ____ may be called a fluid.
    f. ____ molecules of this form have more energy.
    g. ____ force exerted on this form is transmitted as pressure in all directions.
2. Explain what causes pressure in a fluid on a microscopic level. In what direction does the pressure act on the fluid?

3. Describe the two types of forces that act between atoms.
4. Use the words *attractive* and *repulsive* to make the following statements true:
   a. At distances greater than the size of the molecules, intermolecular forces are ____.
   b. Once molecules are close enough to touch, intermolecular forces become ____.
5. What is the result of intermolecular forces being repulsive and attractive at varying distances?
6. How do thermal energy and intermolecular forces behave with each other?
7. What phase of matter has a low amount of thermal energy, which allows the intermolecular forces to dominate?
8. The solid, liquid, and gaseous phases of a material each have different strengths of intermolecular force compared to the amount of thermal energy. For each diagram below, rank as low, medium, or high:
   a. the amount of intermolecular force, and
   b. the amount of thermal energy

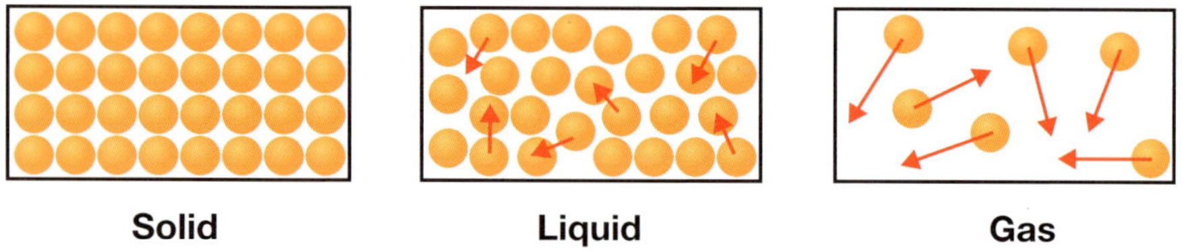

9. Name one factor that causes iron to have a higher boiling point than water.
10. As heat energy is added to ice, the temperature increases until it reaches 0°C. What happens at this point and why?
11. Why is ice less dense than water?
12. Why is oxygen transported as a liquid in rocket ships instead of as a gas?
13. How does the evaporation of sweat on a hot day help to cool your body?
14. If a meterologist describes the air as saturated, what does he or she mean?
15. Give one example of natural convection.
16. What type of heat transfer is represented in the diagram below?

17. From where does the carbon in a tree come from?
18. What two gases make up the majority of the atmosphere?
19. How does rain form?

Section 5.2

20. Why can solid materials hold their shape?
21. How does glass behave differently when it is solid versus when it is heated? Why?
22. What crystal forms the basis for much of the microelectronics industry? What makes it so valuable?
23. What physical properties make plastics such a valuable material for manufacturing goods?

24. Describe an example using the terms *conduction* and *thermal equilibrium*.
Identify each of the following as a thermal conductor or insulator:
   a. copper pipe
   b. styrofoam cup
   c. wooden spoon
   d. a vacuum space
   e. aluminum pot

Problems

Section 5.1

1. According to the diagram, most of our weather occurs below what altitude?

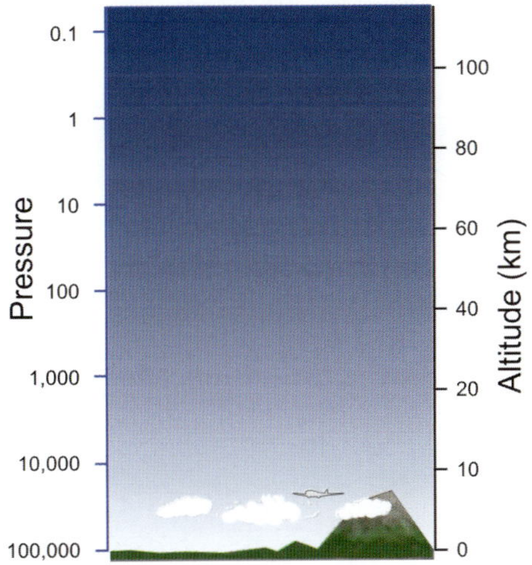

2. The air in your school can hold 20 g/m^3 of water when it is saturated at 70°F. What is the *relative humidity* of the air in your school at 70°F if the moisture content is 5 g/m^3?

3. The diagram to the right shows a graph of temperature vs. time for a material which starts as a solid. Heat is added at a constant rate. Using the diagram, answer the following questions:

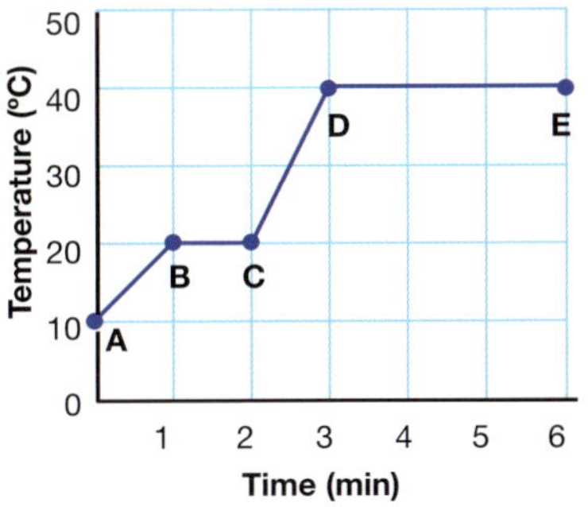

   a. During which time interval does the solid melt?
   b. During which time interval is the material all liquid?
   c. What is the boiling point of the substance?
   d. Does it take more heat energy to melt the solid or boil the liquid?

Section 5.2

4. Based on the definition of a vacuum, if you were to put an alarm clock inside a vacuum and set it to ring, what would happen when the alarm went off?

5. The diagram shows a cup of cocoa at 65°C. The arrows show the direction of heat conduction as a cold spoon is placed into the cup. What could the temperature of the spoon be?

   a. 75°C
   b. 65°C
   c. 55°C

6. Which make the best thermal conductor: solids, liquids, or gases? Why?

UNIT 3

Atoms and the Periodic Table

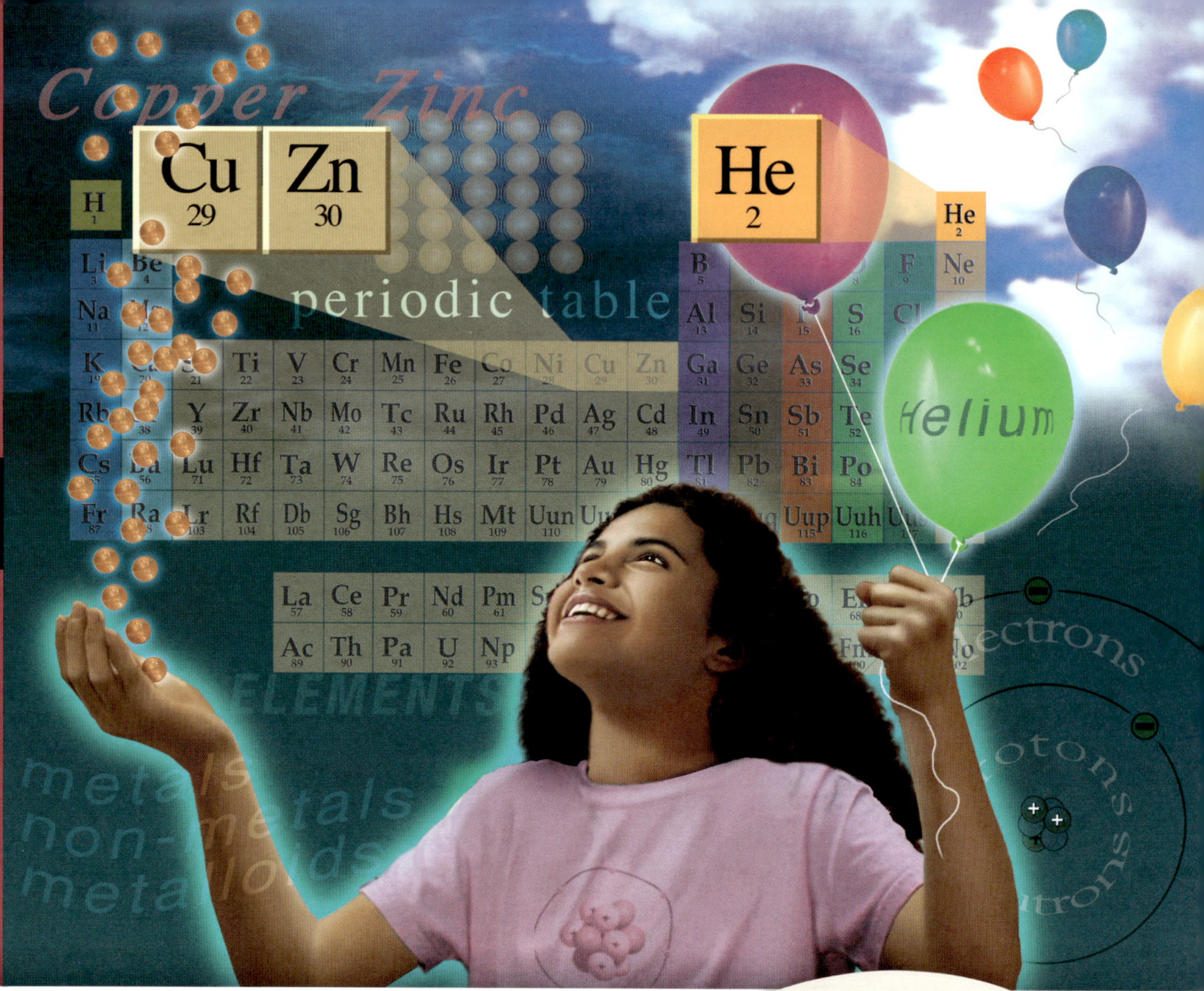

TRY THIS AT HOME

Matter contains positive and negative charges. Put a small handful of puffed cereal or puffed rice into a saucer. Inflate and tie a balloon. Rub the balloon back and forth on your hair, a sweater, or a sock a dozen times. Hold the balloon near the cereal. What happens? Try this with another kind of puffed cereal and compare to the first. Write a paragraph about your observations.

Chapter 6
The Atom

There is something more to wintergreen-flavored candy (the kind with the hole in the middle) than the refreshing taste. When you bite and crush one of these candies, blue sparks jump out of your mouth! You can only see the sparks if you hold a mirror up to your mouth in a very dark place, like a closet. You will be able to see the light even better if you crush one of the candies with a pair of pliers (no mirror required). To understand why the blue sparks appear, you must know what an atom is and what it is made of. After reading this chapter on atoms, you can do an Internet search on the term triboluminescence to find out why this candy sparks when you crush it.

Key Questions

1. *How did scientists figure out what atoms are like if they couldn't see them?*
2. *What makes atoms of different elements different?*
3. *What are atoms themselves made of?*

6.1 Fundamental Particles and Forces

Scientists once believed that atoms were the smallest particles of matter. With the advancement of technology, it became clear that atoms themselves are made of simpler particles. Today, we believe all atoms are made of three basic particles: the proton, electron, and neutron. Astonishingly, the incredible variety of matter in universe can be constructed using just these three subatomic particles!

Electric charge

Electric charge is a property of matter

Along with mass and volume, matter has another fundamental property that we call **electric charge**. In order to understand atoms, we need to understand electric charge because one of the forces that hold atoms together comes from electric charge.

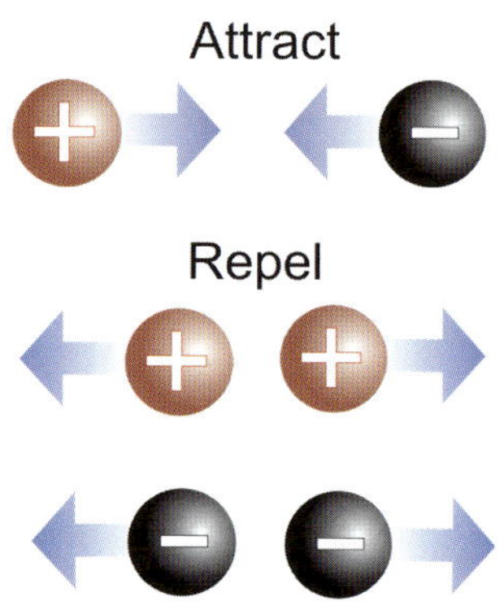

We know of two different kinds of electric charge and we call them *positive* and *negative*. Because there are two kinds of charge, the force between electric charges can be either attractive or repulsive. A positive and a negative charge will attract each other. Two positive charges will repel each other. Two negative charges will also repel each other.

The elementary charge

We use the letter *e* to represent the **elementary charge**. On the atomic scale, electric charge always comes in units of +*e* or -*e*. It is *only* possible to have charges that are multiples of e, such as +e, +2e, -e, -2e, -3e, and so on. Scientists believe it is impossible for ordinary matter to have charges that are fractions of e. For example, a charge of +0.5e is impossible in ordinary matter. Electric charge only appears in units of the elementary charge.

VOCABULARY

electric charge - a fundamental property of matter that comes in two types called positive and negative.

elementary charge - the smallest unit of electric charge that is possible in ordinary matter; represented by the lowercase letter *e*.

Electric charge only appears in multiples of the elementary charge, e.

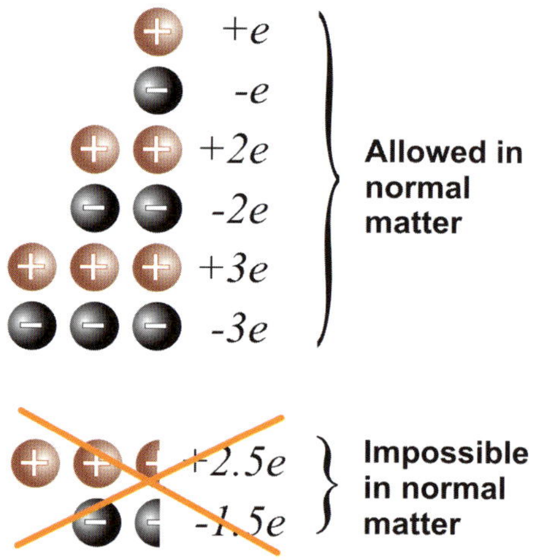

Figure 6.1: *Just as normal matter is divided into atoms, electric charge appears only in units of the elementary charge, e.*

Static electricity

Neutral means zero charge

We say an object is electrically **neutral** when its total electric charge is zero (Figure 6.2). Your pencil, your textbook, even your body are electrically neutral, at least most of the time. These forms of matter are made up of charged particles, which you don't usually notice because there is perfect cancellation between positive and negative, leaving a net charge of precisely zero.

Charged objects

An object is **charged** when its total electric charge is *not* zero. Objects become charged when they have an excess of either positive or negative electric charge.

Static electricity and charge

A tiny imbalance of positive or negative charge is the cause of **static electricity.** If two neutral objects are rubbed together, the friction often pulls some charge off one object and puts it temporarily on the other. This is what happens to clothes in the dryer and to your socks when you walk on a carpet. The static electricity you feel when taking clothes from a dryer or scuffing your socks on carpet typically results from an excess charge of less than one part in a hundred trillion!

What causes shocks

Static electricity

The forces between electric charges are incredibly strong. That is why charged objects do not stay charged very long. An object with excess positive charge strongly attracts negative charge until the object becomes neutral again. When you walk across a carpet on a dry day, your body picks up excess negative charge. If you touch a neutral door knob your negative charge repels negative charge in the door knob causing the doorknob to become slightly positive. The negative charge on your skin is now attracted to the positive charge on the doorknob. The shock you feel is the energy released when your excess negative charge jumps the gap between your skin and the door knob.

ă VOCABULARY

neutral - a condition where the total positive charge is canceled by the total negative charge. Matter is neutral most of the time.

charged - a condition where there is an excess of positive or negative charge.

static electricity - the buildup of either positive or negative charge; made up of isolated, motionless charges.

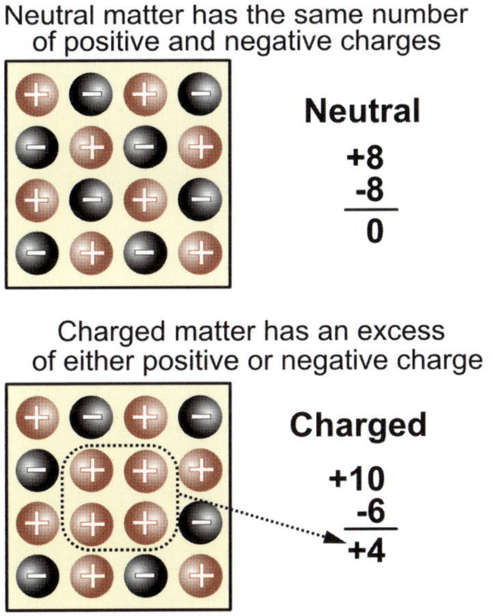

Figure 6.2: *An object is neutral if it has an equal number of positive and negative charges.*

Inside an atom: solving the puzzle

The electron identified

The first strong evidence that something existed smaller than an atom came in 1897. English physicist J. J. Thomson discovered that electricity passing through a gas caused the gas to give off particles that were too small to be atoms. Thomson's new particles also had negative electric charge while atoms have zero electric charge. Thomson called his particles *corpuscles*, which were eventually named **electrons,** and proposed that they came from the inside of atoms.

The proton and the nucleus discovered

In 1911, Ernest Rutherford, Hans Geiger, and Ernest Marsden did a clever experiment to test Thomson's model of the atom. They launched positively-charged helium ions (a charged atom is called an *ion*) at extremely thin gold foil (Figure 6.3). They expected the helium ions to be deflected a small amount as they passed through the foil. However, a few bounced back in the direction they came! The unexpected result prompted Rutherford to remark "*it was as if you fired a five inch (artillery) shell at a piece of tissue paper and it came back and hit you!*"

The nuclear model of the atom

The best way to explain the pass-through result was if the gold atoms were mostly empty space, allowing most of the helium ions to go through virtually undeflected. The best way to explain the bounce-back result was if nearly all the mass of a gold atom were concentrated in a tiny, hard core at the center. Further experiments confirmed Rutherford's ideas and we know that every atom has a tiny **nucleus**, which contains more than 99% of the atom's mass.

The neutron

The positively charged **proton** was soon discovered and shown to be the particle in the nucleus. But there still was a serious problem with the atomic model. Protons could only account for about half the observed mass. This problem was solved in 1932 by James Chadwick. Chadwick's experiments revealed another particle in the nucleus which has no electric charge and similar mass as the proton. Chadwicks neutral particle was named the **neutron**.

VOCABULARY

electron - a particle with an electric charge (-e) found inside of atoms but outside the nucleus.

proton - a particle with an electric charge (+e) found in the nucleus of atoms.

neutron - a particle with zero charge found in the nucleus of atoms.

nucleus - the tiny core at the center of an atom containing most of the atom's mass and all of its positive charge.

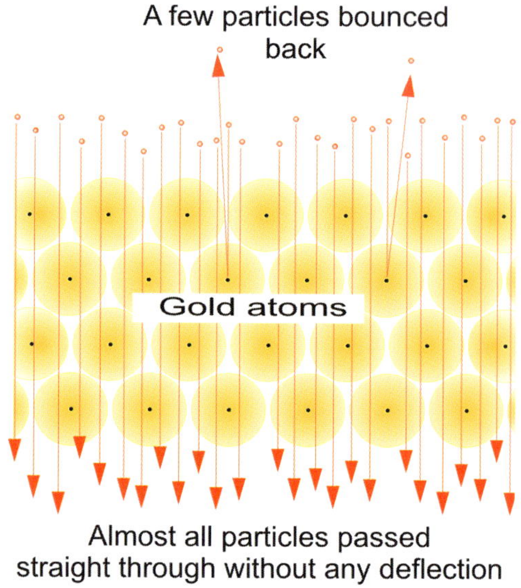

Figure 6.3: *Rutherford's famous experiment led to the discovery of the nucleus.*

Three subatomic particles make up an atom

Protons, neutrons, and electrons Today we know that atoms are made of three tiny *subatomic* particles: protons, neutrons, and electrons. Protons have positive charge. Electrons have negative charge. Neutrons add mass but have zero charge. The charge on a proton (+e) and an electron (-e) are exactly equal and opposite. Atoms that have the same number of protons and electrons have a total charge of precisely zero.

The nucleus The protons and neutrons are grouped together in the nucleus, which is at the center of the atom. The mass of the nucleus determines the mass of an atom because protons and neutrons are much larger and more massive than electrons (Figure 6.4). In fact, a proton is 1,836 times heavier than an electron. All atoms have both protons and neutrons in their nuclei except the simplest type of hydrogen, which only has one proton and no neutrons. The chart below compares electrons, protons, and neutrons in terms of charge and mass.

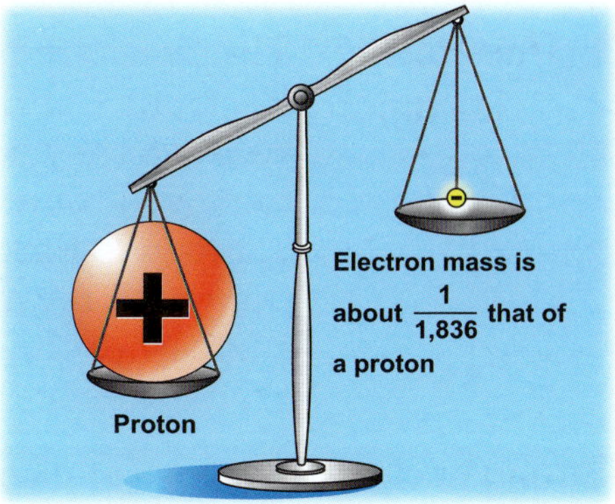

Figure 6.4: *The mass of at atom is mostly in the nucleus because protons and neutrons are much heavier than electrons.*

	Occurrence	Charge	Mass (g)	Relative Mass
⊖ Electron	found outside of nucleus	-1	9.109×10^{-28}	1
⊕ Proton	found in all nuclei	+1	1.673×10^{-24}	1,836
Neutron	found in almost all nuclei (exception: most H nuclei)	0	1.675×10^{-24}	1,839

Electrons define the volume of an atom Electrons take up the region *outside* the nucleus in a region called the *electron cloud*. The diameter of an atom is really the diameter of the electron cloud (Figure 6.5). Compared to the tiny nucleus, the electron cloud is enormous, more than 10,000 times larger than the nucleus. As a comparison, if an atom were the size of a football stadium, the nucleus would be the size of a pea, and the electrons would be equivalent to a small swarm of gnats buzzing around the stadium at extremely high speed. Can you visualize how much empty space there would be in the stadium? The atom is mostly empty space!

Size and Structure of the Atom

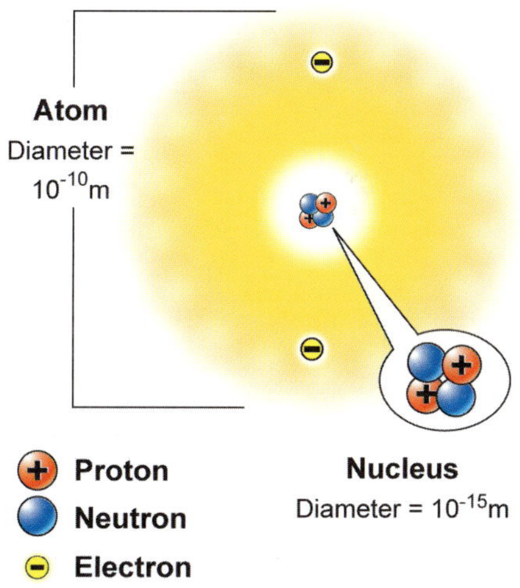

Figure 6.5: *The overall size of an atom is the size of its electron cloud. The nucleus is much, much smaller.*

Forces inside atoms

Electromagnetic forces Electrons are bound to the nucleus by the attractive force between electrons (-) and protons (+). The electrons don't fall into the nucleus because they have kinetic energy. The energy of an electron causes it to move around the nucleus instead of falling in (Figure 6.6). A good analogy is Earth orbiting the sun. Gravity creates a force that pulls the Earth toward the sun. Earth's kinetic energy causes it to orbit the sun rather than fall straight in. While electrons don't really move in orbits, the energy analogy is approximately right.

Strong nuclear force Because of electric force, all the positively charged protons in the nucleus *repel* each other. What holds the nucleus together? There is another force that is even stronger than the electric force. We call it the *strong nuclear force*. The strong nuclear force is the strongest force known to science (Figure 6.7). This force attracts neutrons and protons to each other and works only at the extremely small distances inside the nucleus. If there are enough neutrons, the attraction from the strong nuclear force wins out over repulsion from the electromagnetic force and the nucleus stays together. In every atom heavier than helium, there is at least one neutron for every proton in the nucleus.

Weak force There is another nuclear force called the *weak force*. The weak force is weaker than both the electric force and the strong nuclear force. If you leave a single neutron outside the nucleus, the weak force eventually causes it to break down into a proton and an electron. The weak force does not play an important role in a stable atom, but comes into action in certain special cases when atoms break apart.

Gravity The force of gravity inside the atom is much weaker even than the weak force. It takes a relatively large mass to create enough gravity to make a significant force. We know that particles inside an atom do not have enough mass for gravity to be an important force on the scale of atoms. But there are many unanswered questions. Understanding how gravity works inside atoms is an unsolved mystery in science.

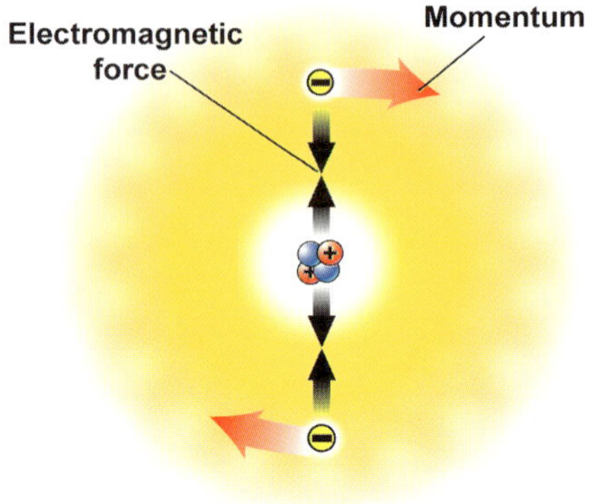

Figure 6.6: *The negative electrons are attracted to the positive protons in the nucleus, but their momentum keeps them from falling in.*

Figure 6.7: *When enough neutrons are present the strong nuclear force wins out over the repulsion between positively charged protons and pulls the nucleus together tightly. The strong nuclear force is the strongest force in the universe that we know.*

How atoms of different elements are different

The atomic number is the number of protons

How is an atom of one element different from an atom of another element? The atoms of different elements contain different numbers of protons in the nucleus. For example, all atoms of carbon have six protons in the nucleus and all atoms of hydrogen have one proton in the nucleus (Figure 6.8). Because the number of protons is so important, it is called the **atomic number.** The atomic number of an element is the number of protons in the nucleus of every atom of that element.

Atoms of the same element always have the same number of protons in the nucleus.

Elements have unique atomic numbers

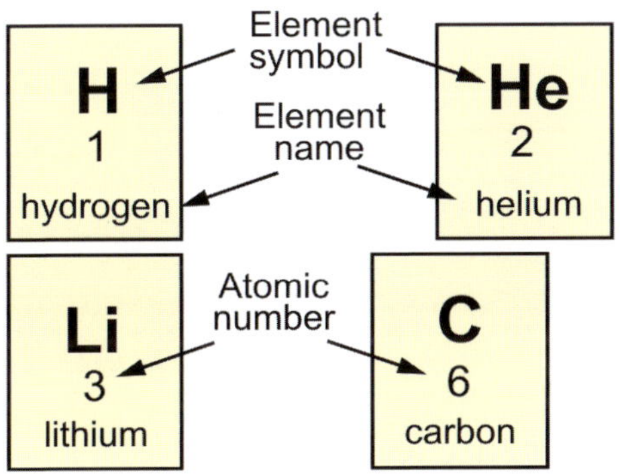

Each element has a unique atomic number. On a periodic table of elements, the atomic number is usually written above or below the atomic symbol. An atom with only one proton in its nucleus is the element hydrogen, atomic number 1. An atom with six protons is the element carbon, atomic number 6. Atoms with seven protons are nitrogen, atoms with eight protons are oxygen, and so on.

Complete atoms are electrically neutral

Because protons and electrons attract each other with very large forces, the number of protons and electrons in a complete atom is always equal. For example, hydrogen has one proton in its nucleus and one electron outside the nucleus. The total electric charge of a hydrogen atom is zero because the negative charge of the electron cancels the positive charge of the proton. Each carbon atom has six electrons, one for each of carbon's six protons. Like hydrogen, a complete carbon atom is electrically neutral.

VOCABULARY

atomic number - the number of protons in the nucleus. The atomic number determines what element the atom represents.

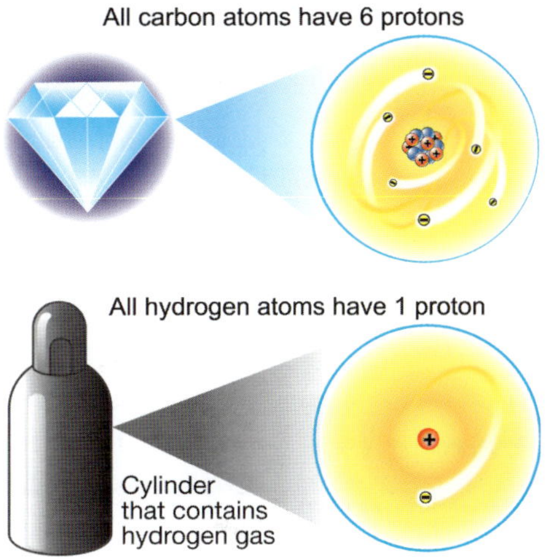

Figure 6.8: *Atoms of the same element always have the same number of protons.*

Isotopes

Isotopes All atoms of the same element have the same number of protons in the nucleus. However, atoms of the same element may have different numbers of neutrons in the nucleus. **Isotopes** are atoms of the *same* element that have different numbers of neutrons.

The isotopes of carbon Figure 6.9 shows three isotopes of carbon that exist in nature. Most carbon atoms have six protons and six neutrons in the nucleus. However, some carbon atoms have seven or eight neutrons. They are all carbon atoms because they all contain six protons, but they are different *isotopes* of carbon. The isotopes of carbon are called carbon-12, carbon-13, and carbon-14. The number after the name is called the mass number. The **mass number** of an isotope tells you the number of protons plus the number of neutrons.

VOCABULARY

isotopes - atoms of the same element that have different numbers of neutrons in the nucleus.

mass number - the number of protons plus the number of neutrons in the nucleus.

Calculating the number of neutrons in a nucleus

How many neutrons are present in an aluminum atom that has an atomic number of 13 and a mass number of 27?

1. Looking for: You are asked to find the number of neutrons.
2. Given: You are given the atomic number and the mass number.
3. Relationships: Use the relationship: protons + neutrons = mass number.
4. Solution: Plug in and solve: neutrons = 27 - 13 = 14
   The aluminum atom has 14 neutrons.

Your turn...

a. How many neutrons are present in a magnesium atom with a mass number of 24? *Answer: 12*

b. Find the number of neutrons in a calcium atom that has a mass number of 40. *Answer: 20*

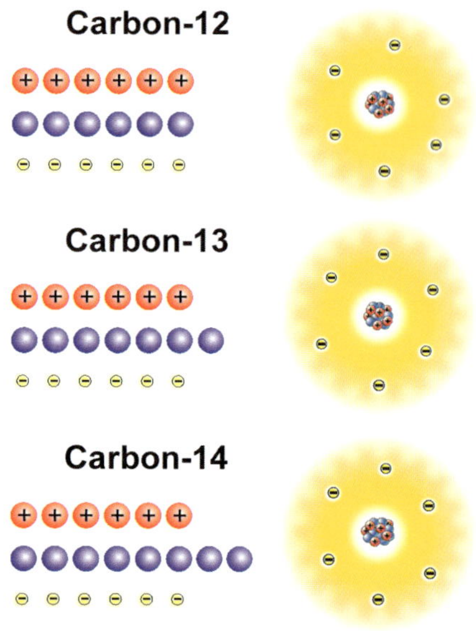

Figure 6.9: *The isotopes of carbon.*

Radioactivity

What if there are too many neutrons?

Almost all elements have one or more isotopes that are **stable**. "Stable" means the nucleus stays together. For complex reasons, the nucleus of an atom becomes unstable if it contains too many or too few neutrons relative to the number of protons. If the nucleus is unstable, it breaks apart. Carbon has two stable isotopes, carbon-12 and carbon-13. Carbon-14 is **radioactive** because it has an unstable nucleus. An atom of carbon-14 eventually changes into an atom of nitrogen-14.

Radioactivity

If an atomic nucleus is unstable for any reason, the atom eventually changes into a more stable form. Radioactivity is a process in which the nucleus spontaneously emits particles or energy as it changes into a more stable isotope. Radioactivity can change one element into a completely different element. For example carbon 14 is radioactive and eventually becomes nitrogen 14.

Alpha decay

In *alpha decay*, the nucleus ejects two protons and two neutrons (Figure 6.10). Check the periodic table and you can quickly show that two protons and two neutrons are the nucleus of a helium-4 (^{4}He) atom. Alpha radiation is actually fast-moving ^{4}He nuclei. When alpha decay occurs, the atomic number is reduced by two because two protons are removed. The atomic mass is reduced by four because two neutrons go along with the two protons. For example, uranium-238 undergoes alpha decay to become thorium-234.

Beta decay

Beta decay occurs when a neutron in the nucleus splits into a proton and an electron. The proton stays in the nucleus, but the high energy electron is ejected and is called beta radiation. During beta decay, the atomic number increases by one because one new proton is created. The mass number stays the same because the atom lost a neutron but gained a proton.

Gamma decay

Gamma decay is how the nucleus gets rid of excess energy. In gamma decay the nucleus emits pure energy in the form of gamma rays. The number of protons and neutrons stays the same.

VOCABULARY

stable - a nucleus is stable if it stays together.

radioactive - a nucleus is radioactive if it spontaneously breaks up, emitting particles or energy in the process.

Alpha decay

Nucleus ejects a helium-4 nucleus

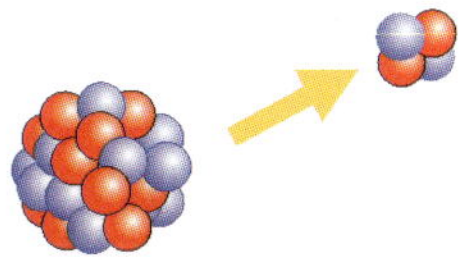

Protons	Decrease by 2
Neutrons	Decrease by 2
Atomic number	Decrease by 2
Mass number	Decrease by 4

Beta decay

Nucleus converts a neutron to a proton and electron, ejecting the electron.

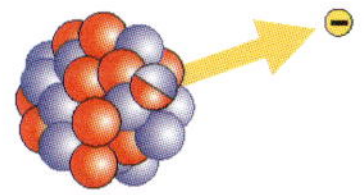

Protons	Increase by 1
Neutrons	Decrease by 1
Atomic number	Increase by 1
Mass number	Stays the same

Figure 6.10: *Two common radioactive decay reactions.*

6.1 Section Review

1. Which of the following statements regarding electric charge is TRUE?
   a. A positive charge repels a negative charge and attracts other positive charges.
   b. A positive charge attracts a negative charge and repels other positive charges.
2. Is electric charge a property of just electricity or is charge a property of all atoms?
3. Which of the drawings in Figure 6.11 is the most accurate model of the interior of an atom?
4. There are four forces in nature. Name the four forces and rank them strongest first, second strongest second, and so on.
5. There are three particles inside an atom. One of them has zero electric charge. Which one is it?
6. All atoms of the same element have (choose one)
   a. the same number of neutrons,
   b. the same number of protons,
   c. the same mass.
7. The atomic number is
   a. the number of protons in the nucleus,
   b. the number of neutrons in the nucleus,
   c. the number of neutrons plus protons.
8. The diagram in Figure 6.12 shows three isotopes of the element carbon. Which one is radioactive?
9. Radioactivity means
   a. an atom gives off radio waves,
   b. the nucleus of an atom is unstable and will eventually change,
   c. the electrons in an atom have too much energy.

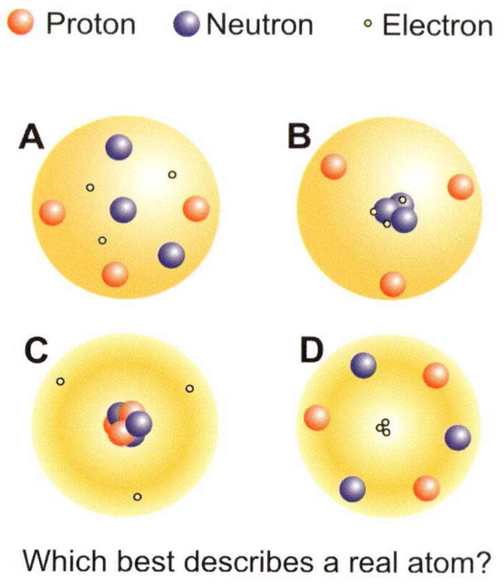

Figure 6.11: *Question 3*

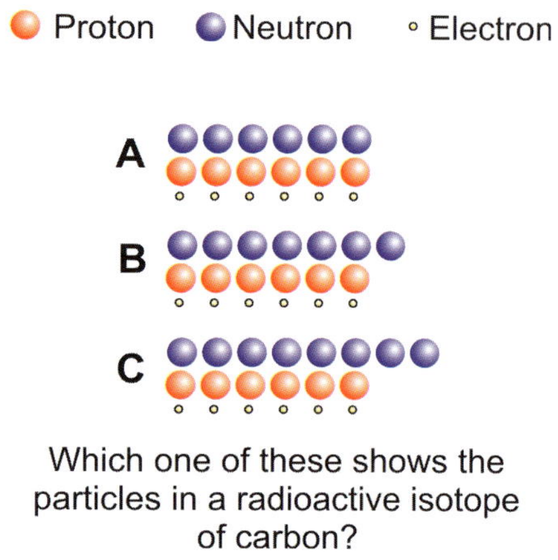

Figure 6.12: *Question 8*

6.2 Electrons in the Atom

Virtually all the properties of the elements (except mass) are due to the electrons outside the nucleus. Atoms interact with each other through their electrons. Chemical bonds involve only electrons so electrons determine how atoms combine into compounds. The rich variety of matter we experience comes directly from the complex behavior of electrons inside atoms. Exactly how electrons create the properies of matter was a puzzle that took bright scientists a long time to figure out!

The spectrum

The spectrum Almost all the light you see comes from atoms. For example, light is given off when electricity passes through the gas in a fluorescent bulb or neon sign. When scientists look carefully at the light given off by a pure element, they find that the light does not include all colors. Instead, they see a few very specific colors, and the colors are different for different elements (Figure 6.13). Hydrogen has a red line, a green line, a blue and a violet line in a characteristic pattern. Helium and lithium have different colors and patterns. Each different element has its own characteristic pattern of colors called a **spectrum**. The colors of clothes, paint, and everything else around you come from this property of elements to emit or absorb light of only certain colors.

Spectroscopes and spectral lines Each individual color in a spectrum is called a **spectral line** because each color appears as a line in a **spectroscope**. A spectroscope is a device that spreads light into its different colors. The diagram below shows a spectroscope made with a prism. The spectral lines appear on the screen on the right.

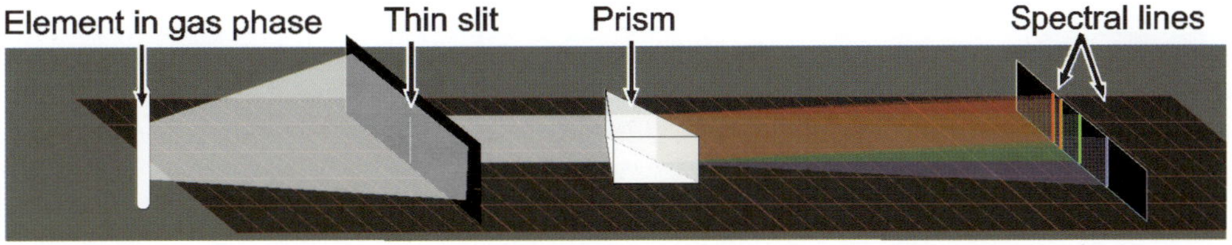

VOCABULARY

spectrum - the characteristic colors of light given off or absorbed by an element.

spectroscope - an instrument that separates light into a spectrum.

spectral line - a bright colored line in a spectroscope.

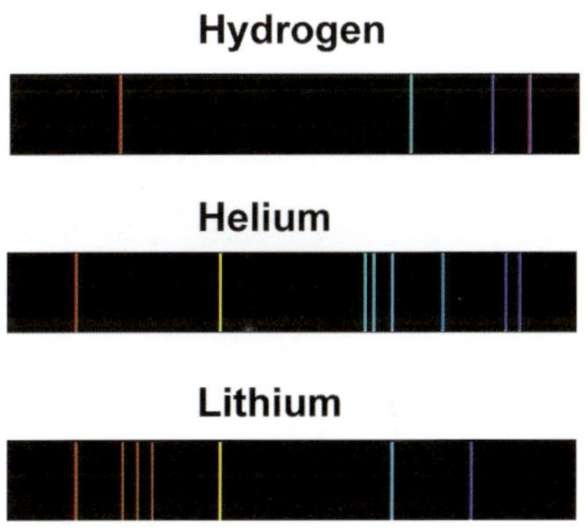

Figure 6.13: *When light from energized atoms is directed through a prism, spectral lines are observed. Each element has its own distinct pattern of spectral lines.*

The Bohr model of the atom

Energy and color

Light is a form of pure energy. The amount of energy determines the color of the light. Red light has low energy and blue light has higher energy. Green and yellow light have energy between red and blue. The fact that atoms only emit certain colors of light tells us that something inside an atom can only have certain values of energy.

Neils Bohr

Danish physicist Neils Bohr proposed the concept of energy levels to explain the spectrum of hydrogen. In Bohr's model, the electron in a hydrogen atom must be in a specific energy level. You can think of energy levels like steps on a staircase. You can be on one step or another, but you cannot be between steps except in passing. Electrons must be in one energy level or another and cannot remain in between energy levels. Electrons change energy levels by absorbing or emitting light (Figure 6.14).

Explaining the spectrum

When an electron moves from a higher energy level to a lower one, the atom gives up the energy difference between the two levels. The energy comes out as different colors of light. The specific colors of the spectral lines correspond to the differences in energy between the energy levels. The diagram below shows how the spectral lines of hydrogen come from electrons falling from the 3rd, 4th, 5th, and 6th energy levels down to the 2nd energy level.

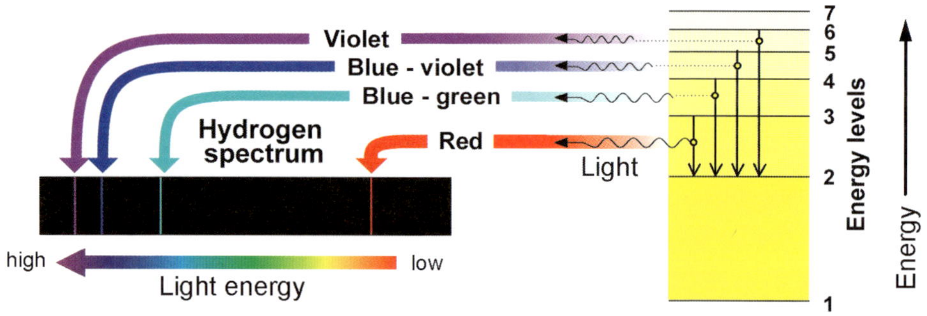

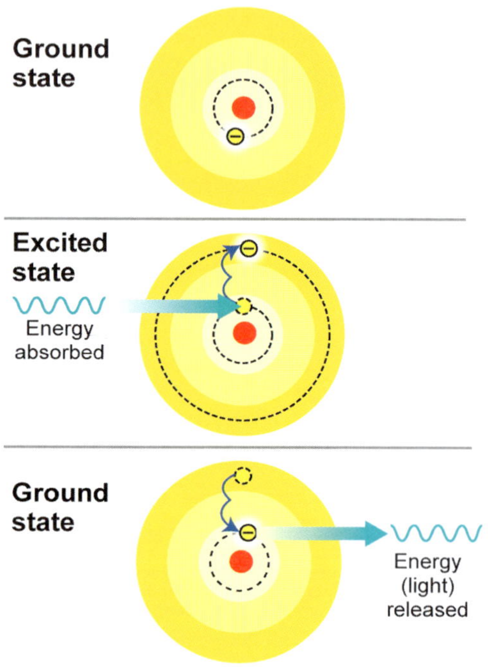

Figure 6.14: *When the right amount of energy is absorbed, an electron in a hydrogen atom jumps to a higher energy level. When the electron falls back to the lower energy, the atom releases the same amount of energy it absorbed. The energy comes out as light of a specific color.*

Electrons and energy levels

The electron cloud

Bohr's model of electron energy levels was incomplete. Electrons are so fast and light that their exact position within an atom cannot be defined. In the current model of the atom, we think of the electrons in an atom as moving around the nucleus in an area called an *electron cloud*. The energy levels occur because electrons in the cloud are at different average distances from the nucleus.

The energy levels are at different distances from the nucleus

The positive nucleus attracts negative electrons like gravity attracts a ball down a hill. The farther down the "hill" an electron slides, the less energy it has. Conversely, electrons have more energy farther up the hill, and away from the nucleus. The higher energy levels are farther from the nucleus and the lower energy levels are closer to the nucleus.

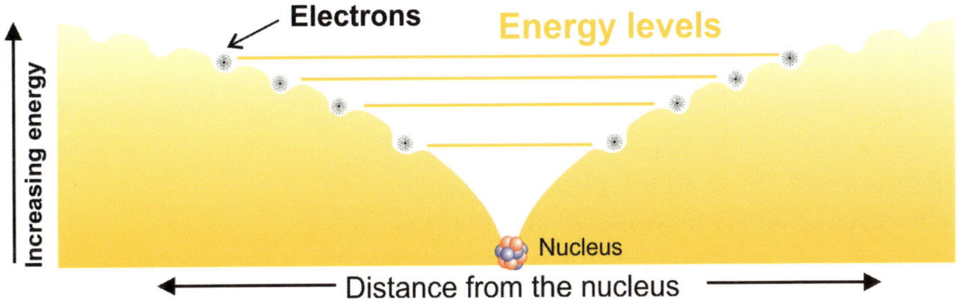

Rules for energy levels

Inside an atom, electrons behave in certain ways:

- The energy of an electron must match one of the energy levels in the atom.
- Each energy level can hold only a certain number of electrons, and no more.
- As electrons are added to an atom, they settle into the lowest unfilled energy level.

Quantum mechanics

Energy levels are predicted by *quantum mechanics*, the branch of physics that deals with the microscopic world of atoms. While quantum mechanics is outside the scope of this book, you should know that it is a very accurate theory and it explains energy levels.

Orbitals

The energy levels in an atom are grouped into different shapes called *orbitals*.

The s-orbital

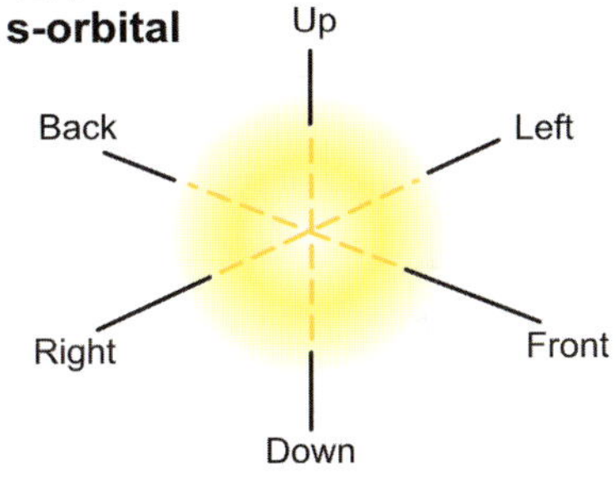

The s-orbital is spherical and holds two electrons. The first two electrons in each energy level are in the s-orbital.

The p-orbitals

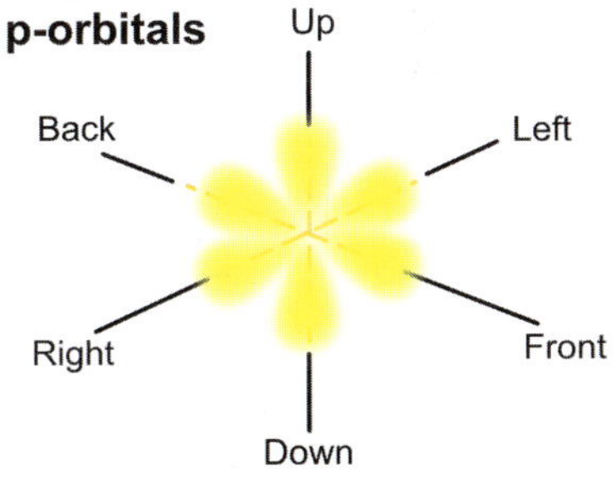

The p-orbitals hold 6 electrons and are aligned along the three directions on a 3-D graph.

The energy levels in an atom

How electrons fill in the energy levels

The first energy level can accept up to two electrons. The second and third energy levels hold up to eight electrons each. The fourth and fifth energy levels hold 18 electrons (Figure 6.15). A good analogy is to think of the electron cloud like a parking garage in a crowded city. The first level of the garage only has spaces for two cars, just as the first energy level only has spaces for two electrons. The second level of the garage can hold eight cars just as the second energy level can hold eight electrons. Each new car that enters the garage parks in the lowest unfilled space, just as each additional electron occupies the lowest unfilled energy level.

How the energy levels fill

The number of electrons in an atom depends on the atomic number because the number of electrons equals the number of protons. That means each element has a different number of electrons and therefore fills the energy levels to different point. For example, a helium atom has two electrons (Figure 6.16). The two electrons completely fill up the first energy level (diagram below). The next element is lithium with three electrons. Since the first energy level only holds two electrons, the third electron must go into the second energy level. The diagram shows the first 10 elements which fill the first and second energy levels.

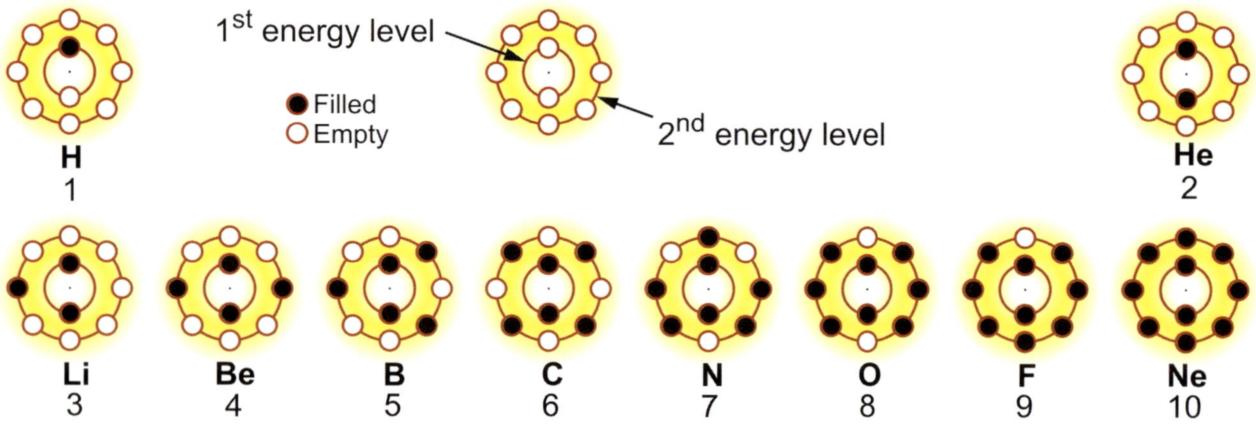

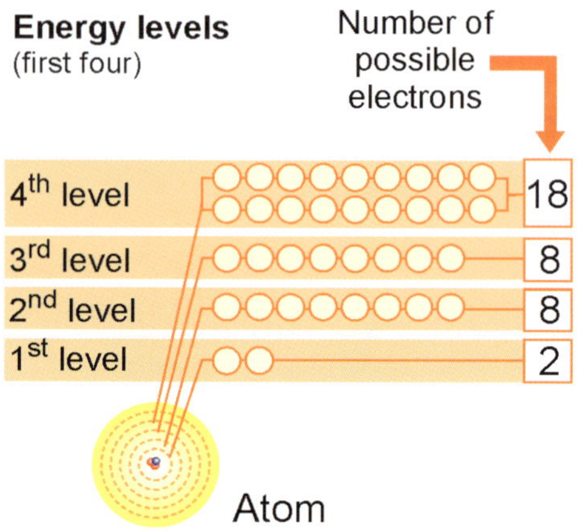

Figure 6.15: *Electrons occupy energy levels around the nucleus. The farther away an electron is from the nucleus, the higher the energy it possesses.*

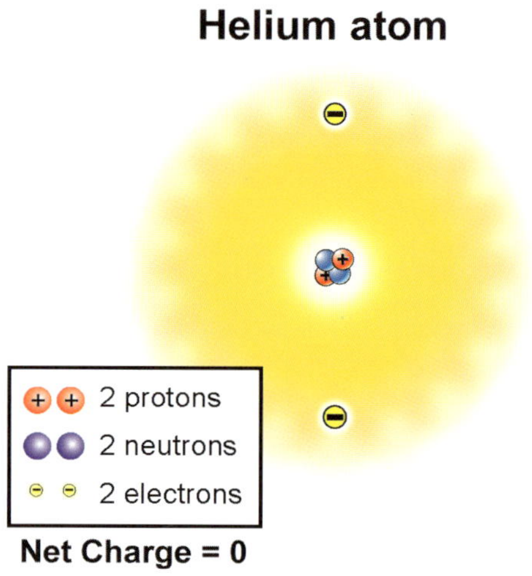

Figure 6.16: *A helium atom has two protons in its nucleus and two electrons.*

6.2 Section Review

1. The pattern of colors given off by a particular atom is called
   a. an orbital,
   b. an energy level,
   c. a spectrum.
2. Which of the diagrams in Figure 6.17 corresponds to the element lithium?
3. When an electron moves from a lower energy level to a higher energy level the atom
   a. absorbs light,
   b. gives off light,
   c. becomes a new isotope.
4. Two of the energy levels can hold eight electrons each. Which energy levels are these?
5. How many electrons can fit in the fourth energy level?
6. The element beryllium has four electrons. Which diagram in Figure 6.18 shows how beryllium's electrons are arranged in the first four energy levels?
7. Which two elements have electrons only in the first energy level?
   a. hydrogen and lithium
   b. helium and neon
   c. hydrogen and helium
   d. carbon and oxygen
8. On average, electrons in the fourth energy level are
   a. farther away from the nucleus than electrons in the second energy level,
   b. closer to the nucleus than electrons in the second energy level,
   c. about the same distance from the nucleus as electrons in the second energy level.

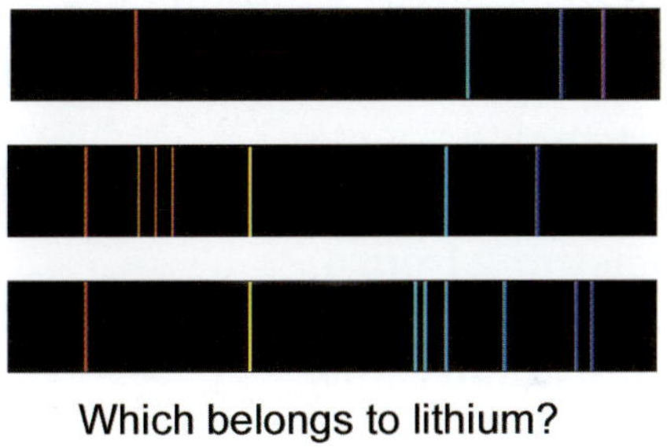

Figure 6.17: *Question 2*

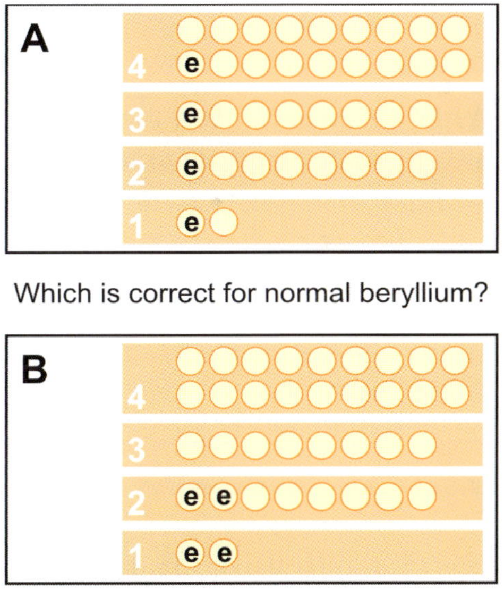

Figure 6.18: *Question 6*

BIOLOGY CONNECTION

Bioluminescence

Imagine you could make your hands glow like living flashlights. No more fumbling around for candles when the power goes out! You could read in bed all night, or get a job directing airplanes to their runways.

Although a glowing hand might sound like something from a science fiction movie, , many living things can make their own light. On warm summer evenings, fireflies flash signals to attract a mate. A fungus known as "foxfire" glows in decaying wood.

Photos by Garth Fletcher

While there are only a few kinds of glowing creatures that live on land, about *90 percent* of the animals that live in the deep parts of the ocean make their own light!

How do they do that?

Light is a form of energy. To make light, an atom must first absorb enough energy to promote an electron up to a higher-energy state. When the electron falls back to its original state, the energy is given off as light.

Atoms can absorb energy from a number of sources. Electrical energy is used in ordinary light bulbs. Mechanical energy can be used, too. Hit two quartz rocks together in a dark room, and you'll see flashes of light as the electrons you pumped up fall back down. You can also use the energy from a chemical reaction. When you bend a glow stick, you break a vial inside so that two chemicals can combine. When they react, energy is released and used to make light.

Bioluminescence

Like a glow stick, living things produce their own light using a chemical reaction. We call this process *bioluminescence* (*bio-* means "living" and *luminesce* means "to glow"). Bioluminescence is "cold light" because it doesn't produce a lot of heat. While it takes a lot of energy for a living thing to produce light, almost 100% of the energy becomes visible light. In contrast, only 10 percent of the used by an "incandescent" electric light bulb is converted to visible light. 90 percent of the energy is wasted as heat.

The chemical reaction

Three ingredients are usually needed for a bioluminescent reaction to occur: An organic chemical known as *luciferin*, a source of oxygen, and an enzyme called *luciferase*.

Luciferin and luciferase are categories of chemicals with certain characteristics. Luciferin in a firefly is not exactly the same as the luciferin in "foxfire" fungus. However, both luciferin chemicals are carbon-based and have the ability to give off light under certain conditions.

Firefly light

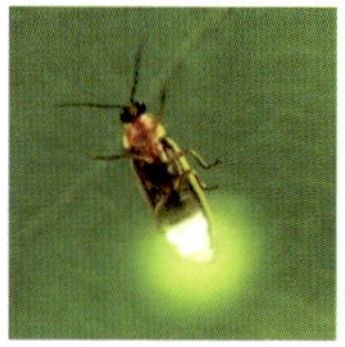

In a firefly, luciferin and luciferase are stored in special cells in the abdomen called "photocytes." To create light, fireflies push oxygen into the photocytes. When the luciferin and luciferase are exposed to oxygen, they combine with ATP (a chemical source of energy) and magnesium. This chemical reaction drives some of the luciferin electrons into a higher energy state. As they fall back down to their "ground state," energy is given off in the form of visible light.

Why make light?

Living creatures don't have an endless supply of energy. Since it takes a lot of energy to make light, there must be good reasons for doing it.

Fireflies flash their lights in patterns to attract a mate. The lights also warn predators to stay away, because the light-producing chemicals taste bitter. They can also be used as a distress signal, warning others of their species that there is danger nearby. The female of one firefly species has learned to mimic the signal of other types of fireflies. She uses her light to attract males of other species and then she eats them!

It's a little harder to figure out why foxfire fungus glows. Some scientists think that the glow attracts insects that help spread around the fungus spores.

Photo by E. Widder

Bioluminescent ocean creatures use their lights in amazing ways. The deep-sea angler fish has a glowing lure attached to its head. When a smaller fish comes to munch on the lure, it instead is gobbled up by the angler fish.

Comb jellies (shown below) are some of the ocean's most beautiful glowing creatures. When threatened, they release a cloud of bioluminescent particles into the water, temporarily blinding the attacker.

Photo by E. Widder

So far we know that living creatures use bioluminescence to attract mates, to communicate, to find food, and to ward off attackers. Perhaps someday you will be part of a research team that discovers even more uses for bioluminescence.

Questions:

1. Find out more about what is inside a glow stick. Make a poster to explain how glow sticks work, or prepare a demonstration for your classmates
2. Bioluminescence is found in a wide range of living organisms, including bacteria, fungi, insects, crustaceans, and fish. However, no examples have been found among flowering plants, birds, reptiles, amphibians, or mammals. Why do you think this is so?
3. Use the Internet or a library to find out more about bioluminescent sea creatures. Here are some questions to pursue: What is the most common color of light produced? What other colors of bioluminescence have been found?

CHAPTER ACTIVITY Atoms and Radioactivity

Radioactivity is how we describe any process where the nucleus of an atom emits particles or energy. All radioactive elements have a half-life. This means that there is a certain length of time after which half of the radioactive element has decayed. Radioactive elements have an unstable nucleus, which decays into an different type of atom with a more stable nucleus. As it decays, it releases radiation.

Materials:

can of pennies, graph paper

What you will do

Your teacher has given you a can of pennies to represent the atoms of a sample of a newly discovered, radioactive element. You will use the pennies to simulate the process of radioactive decay. Upon completion of the simulation, you will construct a graph of your data.

Shake your can of pennies and spill them out onto a tray or table.

1. Remove all pennies that are "heads" up and count them.
2. Record these as decayed atoms in a table like the one below.

Trial	Sample Number	# of decayed atoms
1		
2		
3		
4		
5		
6		
7		
8		
9		
10		

3. Put the rest of the pennies back into the can and shake them again.
4. Spill them out onto the tray or table, and again, remove and count the "heads".
5. Repeat this process until there are no pennies left.

Applying your knowledge

a. Graph your data. The sample number will be on the x-axis and the number of decayed atoms per sample will be on the y-axis. Label the axes clearly and provide a title for the graph.

b. Describe what your graph looks like.

c. How many trials did it take for half of your original number of pennies to decay to "heads up"?

d. How many trials did it take for all your pennies to decay?

e. Would it make a difference if you graphed the number of "tails" up instead?

f. If you were to put a sticker on one of the pennies and repeat the activity, could you predict in which trial the marked penny would decay?

g. Another student did this activity, and on the third shake 12 pennies decayed. Can you tell how many pennies the other student started with?

Chapter 6 Assessment

Vocabulary

Select the correct term to complete the sentences.

nucleus	spectroscope	neutral
spectral line	atomic number	isotopes
spectrum	neutron	radioactive
elementary charge		

Section 6.1

1. A particle with zero charge, found in the nucleus of an atom, is the ____.
2. The smallest unit of electric charge that occurs in ordinary matter is a(n) ____.
3. The tiny core of an atom containing most of the atom's mass and all of its positive charge is called the ____.
4. Atoms having the same atomic number but different mass numbers are called ____.
5. When an object has a net charge of zero, it is described as ____.
6. The number of protons in the nucleus of an atom is known as the ____.
7. A nucleus that spontaneously breaks up or emits particles may be called ____.

Section 6.2

8. A bright colored line viewed in a spectroscope is called a ____.
9. The characteristic pattern of colors iven off by an element is called a(n) ____.
10. The instrument used to separate the light from electrons into different colors is called a(n) ____.

Concepts

Section 6.1

1. When matter is observed closely, different electric charges may be seen. Of the following charges, which are NOT observed in matter on the atomic scale? There may be more than one answer.
   a. +2.3e
   b. -6.0e
   c. -3.5e
   d. +9.0e
2. How do forces between electrical charges differ from forces between masses?
3. Most objects are electrically neutral. Describe how an object may become charged.
4. Charges are attracted or repelled by one another. Use the words **attract** or **repel** to describe how the following combinations would react.

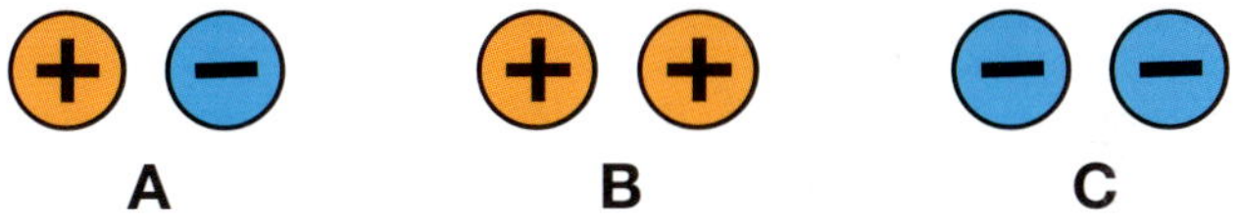

5. Which two particles are electrically attracted to each other?
   a. proton and neutron
   b. electron and neutron
   c. proton and electron
6. A neutral atom has:
   a. zero electrons.
   b. the same number of protons and neutrons.
   c. the same number of protons and electrons.
7. Why would an atom become radioactive?

8. Describe the three subatomic particles that make up an atom by completing the chart below.

Particle	Place in Atom	Charge	Relative Mass
electron			
proton			
neutron			

9. List the four fundamental forces of nature in order from strongest to weakest.
10. Describe the terms:
    a. alpha decay
    b. beta decay
    c. gamma decay

Section 6.2

11. When electricity passes through a gas it may become hot enough to give off light. How could a scientist determine the element(s) from which the gas is made?
12. How does the energy of an electron relate to its distance from the nucleus?
13. Correctly complete the following statements concerning electrons and energy levels:
    a. The energy of an electron in an atom must be ______________ (the same as, more than, less than) one of the energy levels in the atom.
    b. As electrons are added to an atom they must occupy the ______________ (highest, lowest) unfilled energy level.
    c. Each energy level can hold ___________________ (one, a specific number of, all available) electrons.

Problems

Section 6.1

1. One of the isotopes of the element of carbon is an atom with 6 electrons, 6 protons and 6 neutrons.
    a. What is the atomic number of this isotope?
    b. What is the mass number of this isotope?
    c. What is the charge on this isotope?
2. One atom has 12 protons and 12 neutrons. Another has 13 protons and 12 neutrons. Are they the same or different elements?
3. Of the following atoms, which are isotopes of the same element? There may be more than one correct answer.
    a. An atom with 11 protons, 11 electrons and 13 neutrons
    b. An atom with 12 protons, 11 electrons and 12 neutrons
    c. An atom with 13 protons, 13 electrons and 13 neutrons
    d. An atom with 13 protons, 13 electrons and 14 neutrons
4. Identify the following nuclear changes as alpha decay or beta decay.
    a. A radium atom with 88 protons and 138 neutrons becomes a radon atom with 86 protons and 136 neutrons.
    b. A sodium atom with 11 protons and 13 neutrons becomes a magnesium atom with 12 protons and 12 neutrons.

Section 6.2

5. Electrons "X" and "Y" are temporarily on different energy levels. As they fall back toward ground state, X emits red light and Y emits blue light. Which electron had more energy before falling?

Chapter 7

Elements and the Periodic Table

What are metals like? Think of things that are made with metals like aluminum, copper, iron, and gold. What do they have in common? They are usually shiny, and they can often be bent into different shapes without breaking. Did you know there is a metal that is shiny, but is so soft it can be cut with a knife? This metal is very reactive. If you place a piece of this metal in water, it will race around the surface, and the heat given off is often enough to melt the metal and ignite the hydrogen gas that is produced! This strange metal is called sodium. You can look at the periodic table of elements to find other metals that behave like sodium. In this chapter, you will become familiar with how you can predict the properties of different elements by their location on the periodic table.

Key Questions

1. ***Why are the elements arranged in the periodic table?***

2. ***What sort of information can the periodic table of elements give you?***

3. ***Why does the periodic table have the shape that it does?***

7.1 The Periodic Table of the Elements

Before scientists understood how atoms were put together they were able to identify elements by their chemical properties. In this section, you learn how the elements are organized in the *periodic table*, and how an element's chemical properties are related to the arrangement of electrons.

Physical and chemical properties

Physical properties Characteristics that you can see through direct observation are called **physical properties**. For example, water is a colorless, odorless substance that exists as a liquid at room temperature. Gold is shiny, exists as a solid at room temperature, and can be hammered into very thin sheets. Physical properties include color, texture, density, brittleness, and state (solid, liquid, or gas). Melting point, boiling point, and specific heat are also physical properties.

Physical changes are reversible Physical changes, such as melting, boiling, or bending are sometimes *reversible*, and no new substances are formed. When water freezes, it undergoes a physical change from a liquid to a solid. This does not change the water into a new substance. It is still water, only in solid form. The change can easily be reversed by melting the water. Bending a steel bar is another physical change.

Chemical properties Properties that can only be observed when one substance changes into a different substance are called **chemical properties**. For example, if you leave an iron nail outside, it will eventually rust (Figure 7.1). A chemical property of iron is that it reacts with oxygen in the air to form iron oxide (rust).

Chemical changes are hard to reverse Any change that transforms one substance into a different substance is called a *chemical change*. The transformation of iron into rust is a chemical change. Chemical changes are not easily reversible. Rusted iron will not turn shiny again even if you take it away from oxygen in the air.

VOCABULARY

physical properties - characteristics of matter that can be seen through direct observation such as density, melting point, and boiling point.

chemical properties - characteristics of matter that can only be observed when one substance changes into a different substance, such as iron into rust.

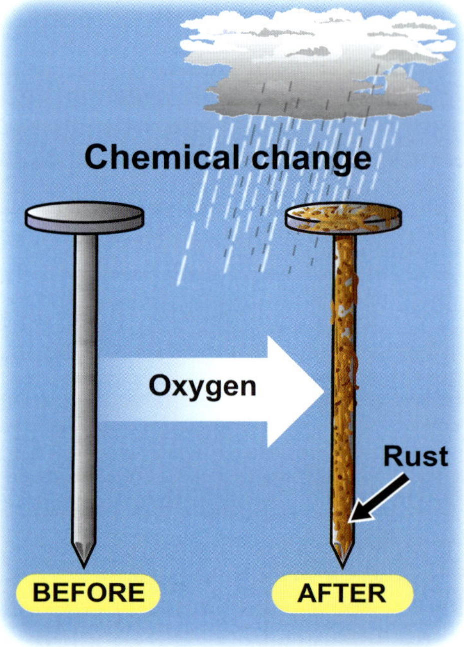

Figure 7.1: *Rusting is an example of a chemical change.*

The periodic table

How many elements are there?

In the 18th through 20th centuries, scientists tried to find and catalog all the elements that make up our universe. To do so, they had to carefully observe substances in order to identify them, and then try to break them apart by any possible means. If a substance could be chemically broken apart it could not be an element. As of this writing, scientists have identified 113 different elements, and five more are expected to be confirmed in the near future. Only 88 elements occur naturally. The others are made in laboratories.

The modern periodic table

As chemists worked on determining which substances were elements, they noticed that some elements acted like other elements. The soft metals lithium, sodium, and potassium always combine with oxygen in a ratio of two atoms of metal per atom of oxygen (Figure 7.2). By keeping track of how each element combined with other elements, scientists began to recognize repeating patterns. From this data, they developed the first *periodic table of the elements*. The **periodic table** organizes the elements according to how they combine with other elements (chemical properties).

Organization of the periodic table

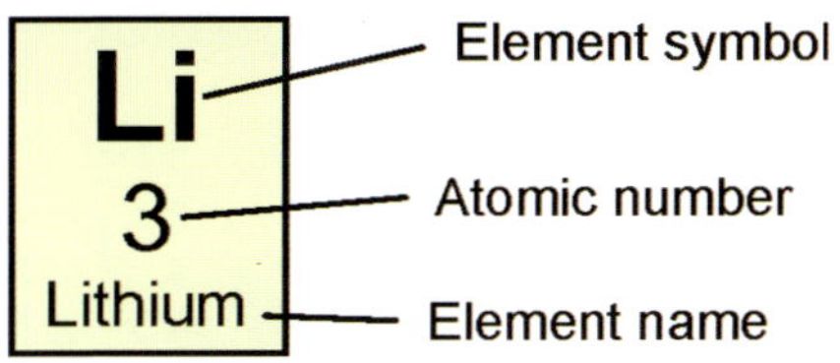

The periodic table is organized in order of increasing atomic number. The lightest element (hydrogen) is at the upper left. The heaviest (#118) is on the lower right. Each element corresponds to one box in the periodic table identified with the element symbol.

The periodic table is further divided into *periods* and *groups*. Each horizontal row is called a **period**. Across any period, the properties of the elements gradually change. Each vertical column is called a **group**. Groups of elements have similar properties. The *main group elements* are Groups 1-2 and 13-18 (the tall columns of the periodic table) Elements in Groups 3 through 12 are called the *transition elements*. The inner transition elements, called lanthanides and actinides, are usually put below to fit on a page.

ă VOCABULARY

periodic table - a chart that organizes the elements by their chemical properties and increasing atomic number.

period - a row of the periodic table is called a period.

group - a column of the periodic table is called a group.

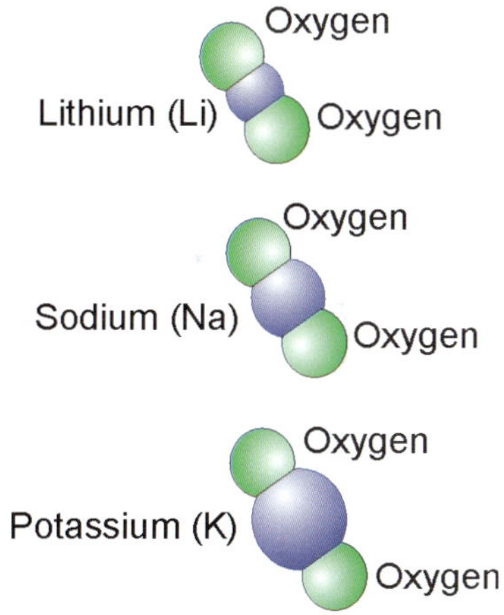

Figure 7.2: *The metals lithium, sodium, and potassium all form compounds with two atoms of oxygen. All the elements in group one of the periodic table form compounds with two oxygen atoms.*

Reading the periodic table

Metals, nonmetals, and metalloids

Most of the elements are metals. A **metal** is typically shiny, opaque, and a good conductor of heat and electricity as a pure element. Metals are also ductile, which means they can be bent into different shapes without breaking. With the exception of hydrogen, the nonmetals are on the right side of the periodic table. **Nonmetals** are poor conductors of heat and electricity. Solid nonmetals are brittle and appear dull. The elements on the border between metals and nonmetals are called *Metalloids*. Silicon is an example of a metalloid element with properties in between those of metals and nonmetals.

VOCABULARY

metal - elements that are typically shiny and good conductors of heat and electricity.

nonmetal - elements that are poor conductors of heat and electricity.

Periodic Table of the Elements

Main Group Elements | Non metals | Metalloids

Transition Elements | Metals

ROWS = PERIODS COLUMNS = GROUPS

1	2	3	4	5	6	7	8	9	10	11	12	13	14	15	16	17	18
H 1 hydrogen																	He 2 helium
Li 3 lithium	Be 4 beryllium											B 5 boron	C 6 carbon	N 7 nitrogen	O 8 oxygen	F 9 fluorine	Ne 10 neon
Na 11 sodium	Mg 12 magnesium											Al 13 aluminum	Si 14 silicon	P 15 phosphorus	S 16 silicon	Cl 17 chlorine	Ar 18 argon
K 19 potassium	Ca 20 calcium	Sc 21 scandium	Ti 22 titanium	V 23 vanadium	Cr 24 chromium	Mn 25 manganese	Fe 26 iron	Co 27 cobalt	Ni 28 nickel	Cu 29 copper	Zn 30 zinc	Ga 31 gallium	Ge 32 germanium	As 33 arsenic	Se 34 selenium	Br 35 bromine	Kr 36 krypton
Rb 37 rubidium	Sr 38 strontium	Y 39 yttrium	Zr 40 zirconium	Nb 41 niobium	Mo 42 molybdenum	Tc 43 technetium	Ru 44 ruthenium	Rh 45 rhodium	Pd 46 palladium	Ag 47 silver	Cd 48 cadmium	In 49 indium	Sn 50 tin	Sb 51 antimony	Te 52 tellurium	I 53 iodine	Xe 54 xenon
Cs 55 cesium	Ba 56 barium		Hf 72 hafnium	Ta 73 tantalum	W 74 tungsten	Re 75 rhenium	Os 76 osmium	Ir 77 iridium	Pt 78 platinum	Au 79 gold	Hg 80 mercury	Tl 81 thallium	Pb 82 lead	Bi 83 bismuth	Po 84 polonium	At 85 astatine	Rn 86 radon
Fr 87 francium	Ra 88 radium		Rf 104 rutherfordium	Db 105 dubnium	Sg 106 seaborgium	Bh 107 bohrium	Hs 108 hassium	Mt 109 meitnerium	Uun 110 ununnilium	Uuu 111 unununium	Uub 112 ununbium	113	Uuq 114 ununquadium	115	116	117	118

La 57 lanthanum	Ce 58 cerium	Pr 59 praseodymium	Nd 60 neodymium	Pm 61 promethium	Sm 62 samarium	Eu 63 europium	Gd 64 gadolinium	Tb 65 terbium	Dy 66 dysprosium	Ho 67 holmium	Er 68 erbium	Tm 69 thulium	Yb 70 ytterbium	Lu 71 lutetium
Ac 89 actinium	Th 90 thorium	Pa 91 protactinium	U 92 uranium	Np 93 neptunium	Pu 94 plutonium	Am 95 americium	Cm 96 curium	Bk 97 berkelium	Cf 98 californium	Es 99 einsteinium	Fm 100 fermium	Md 101 mendelvium	No 102 nobelium	Lr 103 lawrencium

Atomic mass

Atomic mass units

The mass of individual atoms is so small that the numbers are difficult to work with. To make calculations easier scientists define the **atomic mass unit (amu)**. One atomic mass unit is about the mass of a single proton (or a neutron). In laboratory units, one amu is 1.66×10^{-24} grams. That is 0.00000000000000000000000166 grams!

Atomic mass and isotopes

The **atomic mass** is the *average* mass (in amu) of an atom of each element (chart below). Atomic masses differ from mass numbers because most elements in nature contain more than one isotope (see chart below). For example, the atomic mass of lithium is 6.94 amu. That does NOT mean there are 3 protons and 3.94 neutrons in a lithium atom! On average, out of every 100 g of lithium, 94 grams are Li-7 and 6 grams are Li-6 (Figure 7.3). The *average* atomic mass of lithium is 6.94 because of the mixture of isotopes.

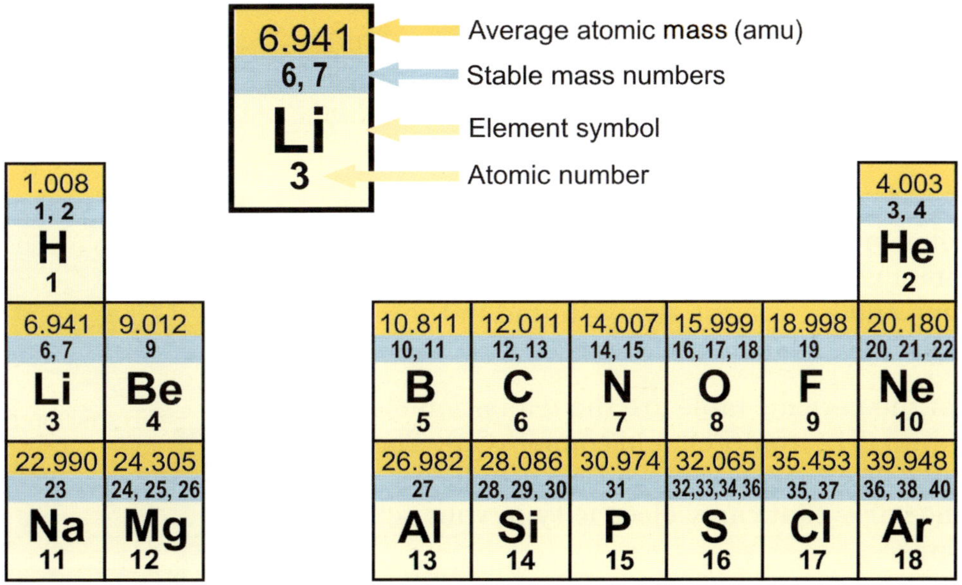

VOCABULARY

atomic mass unit (amu) - a unit of mass equal to 1.66×10^{-24} grams, which is one twelfth the mass of the isotope carbon-12.

atomic mass - the average mass of all the known isotopes of an element, expressed in amu.

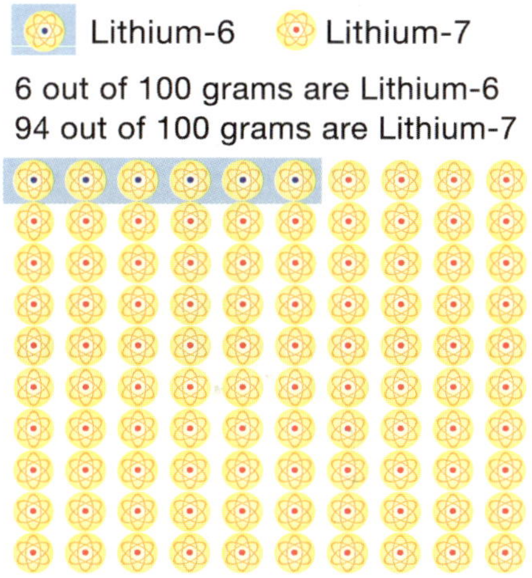

Figure 7.3: *Naturally occuring elements have a mixture of isotopes.*

Groups of the periodic table

Alkali metals

The different groups of the periodic table have similar chemical properties. For example the first group is known as the **alkali metals**. This group includes the elements lithium (Li), sodium (Na), and potassium (K). The alkali metals are soft and silvery in their pure form and are highly reactive. Each of them combines in ratio of two to one with oxygen. For example lithium oxide has two atoms of lithium per atom of oxygen.

Group 2 metals

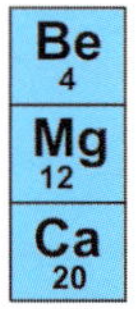

The group two metals include beryllium (Be), magnesium (Mg), and calcium (Ca). These metals also form oxides however they combine one-to-one with oxygen. For example, beryllium oxide has one beryllium atom per oxygen atom.

Halogens

The **halogens** are on the opposite side of the periodic table. These elements tend to be toxic gases or liquids in their pure form. Some examples are fluorine (F), chlorine (Cl), and bromine (Br). The halogens are also very reactive and are rarely found in pure form. When combined with alkali metals, they form salts such as sodium chloride (NaCl) and potassium chloride (KCl).

Noble gases

On the far right of the periodic table are the **noble gases**, including the elements helium (He), neon (Ne), and argon (Ar). These elements do not naturally form chemical bonds with other atoms and are almost always found in their pure state. They are sometimes called *inert gases* for this reason.

Transition metals

In the middle of the periodic table are the transition metals, including titanium (Ti), iron (Fe), and copper (Cu). These elements are usually good conductors of heat and electricity. For example the wires that carry electricity in your school are made of copper.

VOCABULARY

alkali metals - elements in the first group of the periodic table.

halogens - elements in the group containing fluorine, chlorine, and bromine, among others.

noble gases - elements in the group containing helium, neon, and argon, among others.

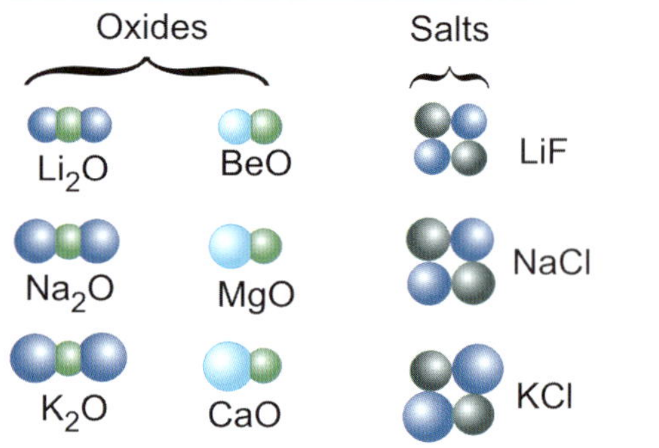

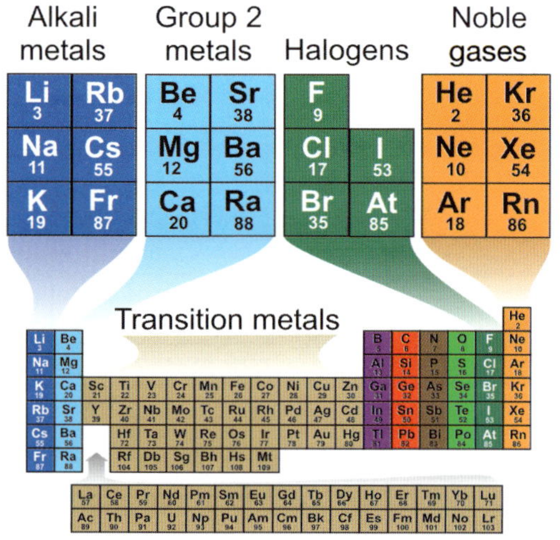

Energy levels and the periodic table

The periodic table The periods (rows) of the periodic table correspond to the energy levels in the atom (Figure 7.4). The first energy level can accept up to two electrons. Hydrogen has one electron and helium has two. These two elements complete the first period.

Row 2 is the second energy level The next element, lithium (Li), has three electrons. Lithium begins the second period because the third electron goes into the second energy level. The second energy level can hold eight electrons so there are eight elements in the second row of the periodic table, ending with neon. Neon (Ne) has 10 electrons, which completely fills the second energy level.

Row 3 is the third energy level Potassium (K) has 11 electrons, and starts the third period because the eleventh electron goes into the third energy level. We know of elements with up to 118 electrons. These elements have their outermost electrons in the seventh energy level.

Outer electrons As we will see in the next chapter, the outermost electrons in an atom are the ones that interact with other atoms. The outer electrons are the ones in the highest energy level. Electrons in the completely filled inner energy levels do not participate in forming chemical bonds.

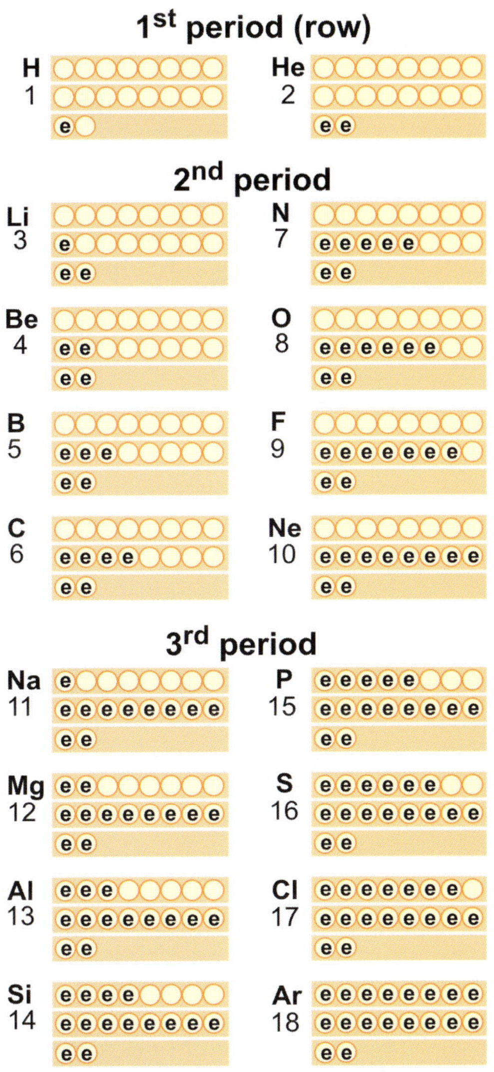

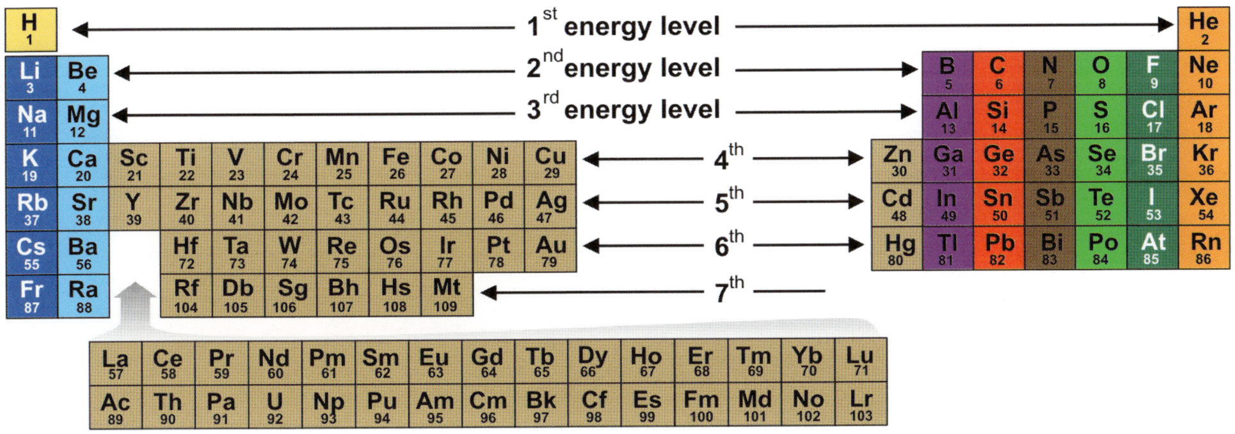

Figure 7.4: *The rows (periods) of the periodic table correspond to the energy levels for the electrons in an atom.*

7.1 Section Review

1. Which of the following (pick 2) are physical properties of matter and NOT chemical properties?
   a. melts at 650°C
   b. density of 1.0 g/ml
   c. forms molecules with two oxygen atoms
2. Groups of the periodic table correspond to elements with
   a. the same color
   b. the same atomic number
   c. similar chemical properties
   d. similar numbers of neutrons
3. Which element is the atom in Figure 7.5?
4. Name three elements which have similar chemical properties to oxygen.
5. The atomic mass unit (amu) is
   a. the mass of a single atom of carbon
   b. one millionth of a gram
   c. approximately the mass of a proton
   d. approximately the mass of electron
6. What element belongs in the empty space in Figure 7.6?
7. The outermost electrons of the element vanadium (atomic #23) are in which energy level of the atom? How do you know?
8. The elements fluorine, chlorine and, bromine are in which group of the periodic table?
   a. the alkali metals
   b. the oxygen-like elements
   c. the halogens
   d. the noble gases
9. What three metals are in the third period (row) of the periodic table?

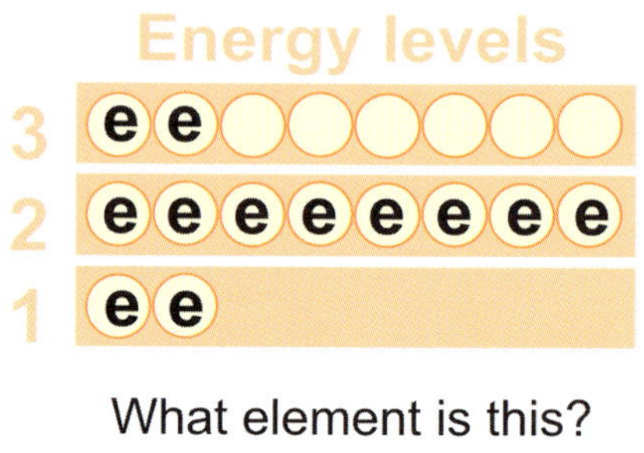

Figure 7.5: *Question 3*

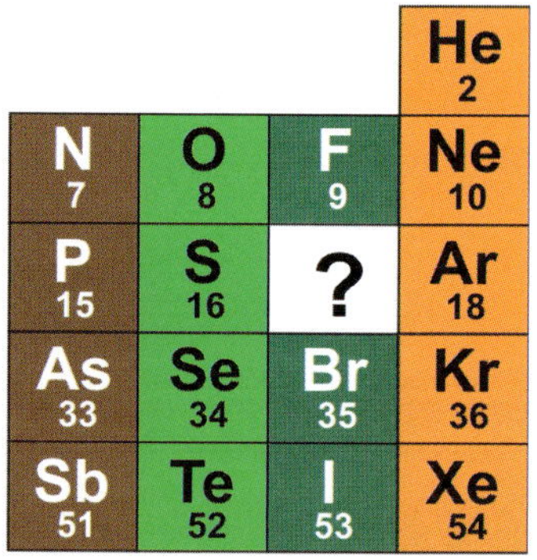

Figure 7.6: *Which element belongs in the empty space (question 6)?*

7.2 Properties of the elements

The elements have a wide variety of chemical and physical properties. Some are solid at room temperature, like copper. Others are liquid (like bromine) or gas (like oxygen). Some solid elements (like zinc) melt at very low temperatures and some melt at very high temperatures (like titanium). Chemically, there is an equally wide variety of properties. Some elements, like sodium, form salts that dissolve easily in water. Other elements, like neon do not form compounds with any other elements.

Room temperature appearance

Most elements are solid at room temperature Most of the pure elements are solid at room temperature. Only 11 of the 92 naturally occurring elements are a gas, and 10 of the 11 are found on the far right of the periodic table. Only 2 elements (Br and Hg) are liquid at room temperature.

What this tells us about intermolecular forces An element is solid when intermolecular forces are strong enough to overcome the thermal motion of atoms. At room temperature, this is true for most of the elements. The noble gases and elements to the far right of the periodic table are the exception. *These elements have completely filled or nearly filled energy levels* (Figure 7.7). When an energy level is completely filled, the electrons do not interact strongly with electrons in other atoms, reducing intermolecular forces.

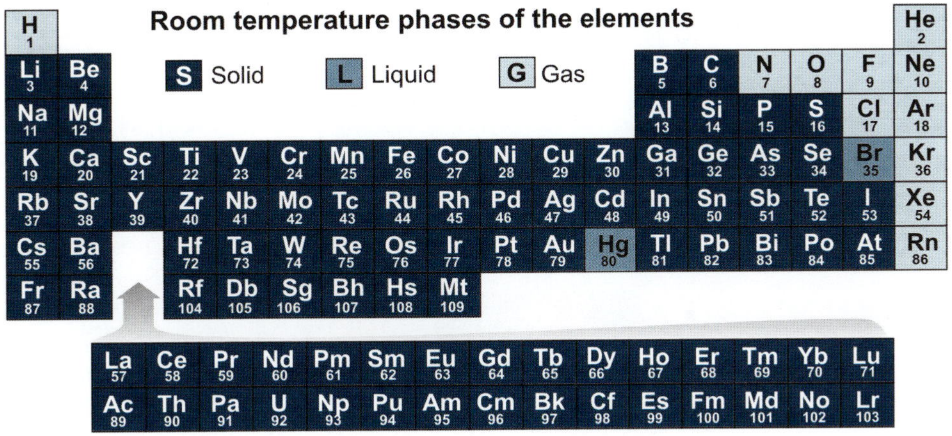

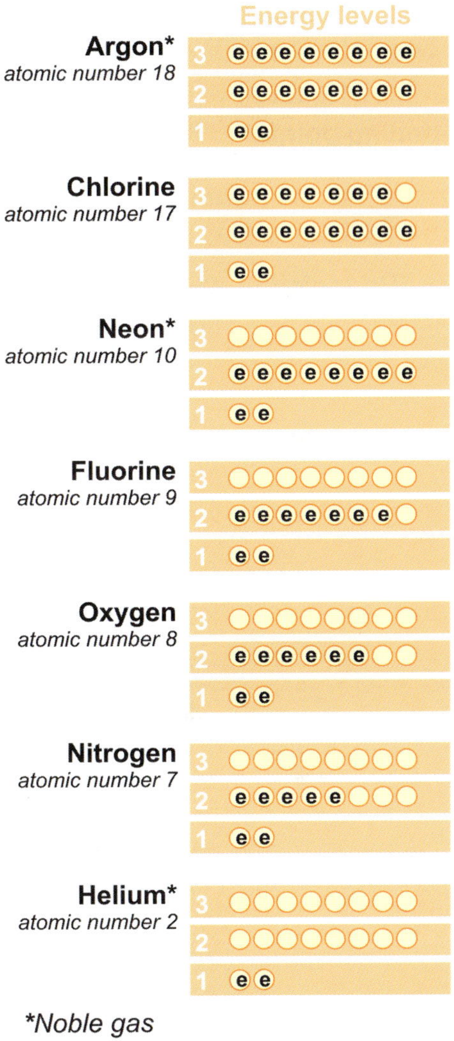

Figure 7.7: *The noble gases have completely filled energy levels. All of the elements which are gas at room temperature have filled or nearly filled energy levels.*

Periodic properties of the elements

The pattern in melting and boiling points

We said earlier that the periodic table arranges elements with common properties in groups (columns). The diagram below shows the melting and boiling points for the first 36 elements. The first element in each row always has a low melting point (Li, Na, K). The melting (and boiling) points rise toward the center of each row and then decrease again.

Periodicity

The pattern of melting and boiling points is an example of **periodicity**. Periodicity means properties repeat each period (row) of the periodic table. Periodicity indicates that a property is strongly related to the filling of electron energy levels. Melting points reflect the strength of intermolecular forces. The diagram shows that intermolecular forces are strongest when energy levels are about half full (or half empty). Elements with half filled energy levels have the greatest number of electrons that can participate in bonding.

Melting and Boiling Points for Elements 1 - 36

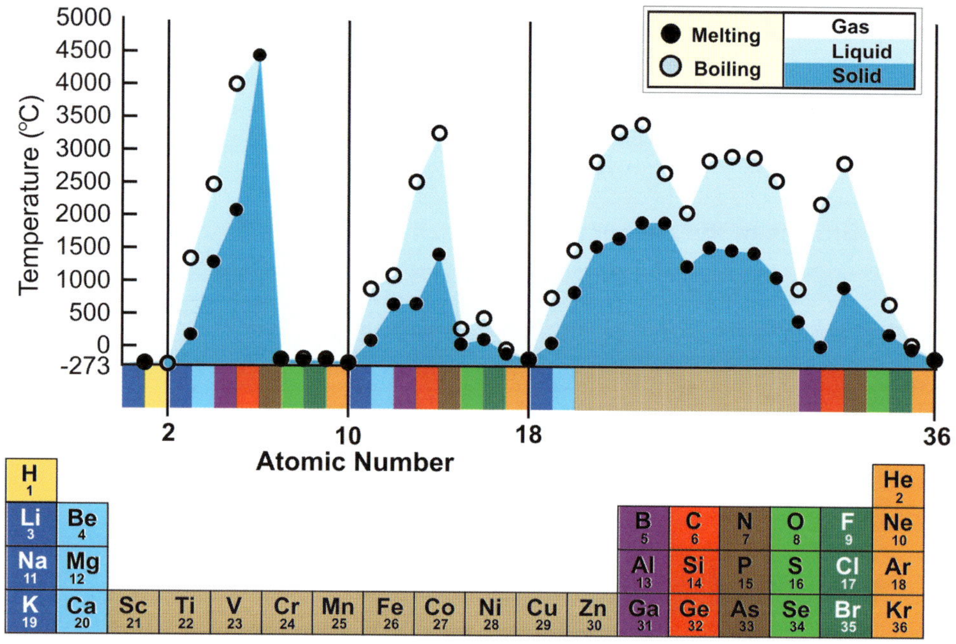

VOCABULARY

periodicity - the repeating pattern of chemical and physical properties of the elements.

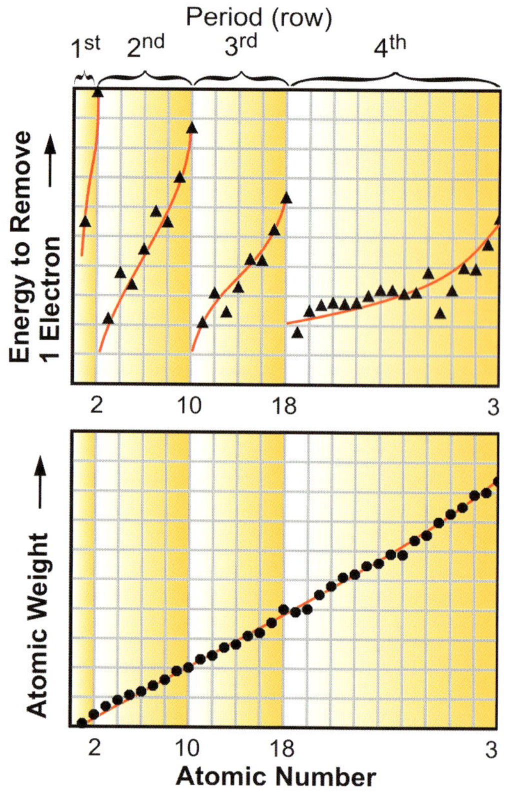

Figure 7.8: *One of these graphs shows periodicity and the other does not. Can you tell which one is periodic? The top graph shows the energy it takes to remove an electron. The bottom graph shows the atomic weight.*

Thermal and electrical conductivity

Metals are good electrical conductors

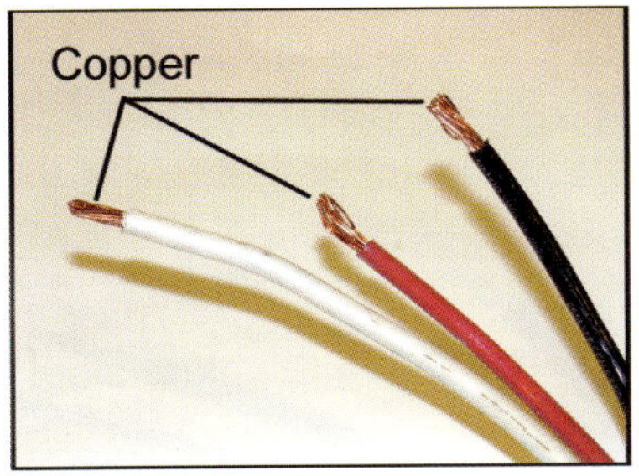

Electricity is something we often take for granted because we use it everyday. Fundamentally, electricity is the movement of electric charge, usually electrons. Some materials allow electrons to flow easily through them. If you connected a battery and a bulb through one of these materials the bulb would light. We call these materials **electrical conductors**. Copper and aluminum are excellent electrical conductors. Both belong to the family of metals, which are elements in the center and left-hand side of the periodic table (Figure 7.9). Copper and aluminum are used for almost all electrical wiring.

Metals are good conductors of heat

If you hold one end of a copper pipe with your hand and heat the other end with the torch, your hand will quickly get hot. That is because copper is a good conductor of heat as well as electricity. Like copper, most metals are good **thermal conductors**. That is one reason pots and pans are made of metal. Heat from a stove can pass easily through the metal walls of a pot to transfer energy to the food inside.

Nonmetals are typically insulators

Elements to the far right of the periodic table are not good conductors of electricity or heatn especially as many are a gas. Because they are so different from metals, these elements are called non-metals. Nonmetals make good **insulators**. An insulator is a material which slows down or stops the flow of either heat or electricity. Air is a good insulator. Air is oxygen, nitrogen and argon.

VOCABULARY

electrical conductor - a material that allows electricity to flow through easily.

thermal conductor - a material that allows heat to flow easily.

insulator - a material that slows down or stops the flow of either heat or electricity.

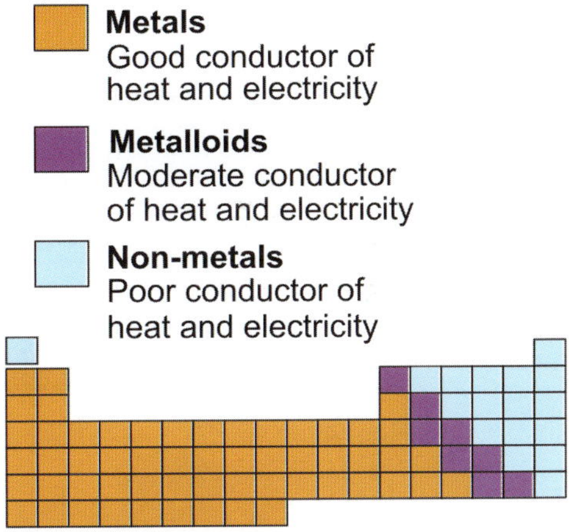

Figure 7.9: *Dividing the periodic table up into metals, metalloids, and non-metals.*

Metals and metal alloys

Steel is an alloy of iron and carbon

When asked for an example of a metal, many people immediately think of **steel**. Steel is made from iron which is the fourth most abundant element in the Earth's crust. However, steel is not pure iron. Steel is an *alloy*. An alloy is a solid mixture of one or more elements. Most metals are used as alloys and not in their pure elemental form. Common steel contains mostly iron with a few percent of carbon. Stainless steel and high strength steel alloys also contain small percentages of other elements such as chromium, manganese, and vanadium. More that 500 different types of steel are in everyday use (Figure 7.10).

Aluminum is light

Aluminum is a metal widely used for structural applications. Aluminum alloys are not quite as strong as steel however aluminum has one third the density of steel. Aluminum alloys are used when weight is a factor, such as for airplane construction. The frames and skins of airplanes are built of aluminum alloys (Figure 7.11).

Titanium is both strong and light

Titanium combines the strength and hardness of steel with the light weight of aluminum. Titanium alloys are used for military aircraft, racing bicycles, and other high performance machines. Titanium is expensive because it is somewhat rare and difficult to work with.

Brass

Brass is a hard, gold-colored metal alloy. Ordinary (yellow) brass is an alloy of 72% copper, 24% zinc, 3% lead, and 1% tin. Hinges, door knobs, keys and decorative objects are made of brass because brass is easy to work with. Because it contains lead however, you should never eat or drink from anything made of ordinary (yellow) brass.

VOCABULARY

steel - an alloy of iron and carbon.

Stainless steel kitchen knife (does not rust)

Ordinary steel nails (will rust)

Figure 7.10: *Nails are made of steel which contains 95% iron and 5% carbon. Kitchen knives are made of stainless steel which is an alloy containing vanadium and other metals.*

Figure 7.11: *This aircraft is made mostly from aluminum alloys. Aluminum combines high strength and light weight.*

Carbon and carbon-like elements

Carbon is an important element for life

Carbon represents less than 100th of a percent of the earth's crust by mass, yet it is the element most essential for life on our planet. Virtually all the molecules that make up plants and animals are constructed around carbon. The chemistry of carbon is so important it has its own name, organic chemistry, which is the subject of Chapter 11 (Figure 7.12).

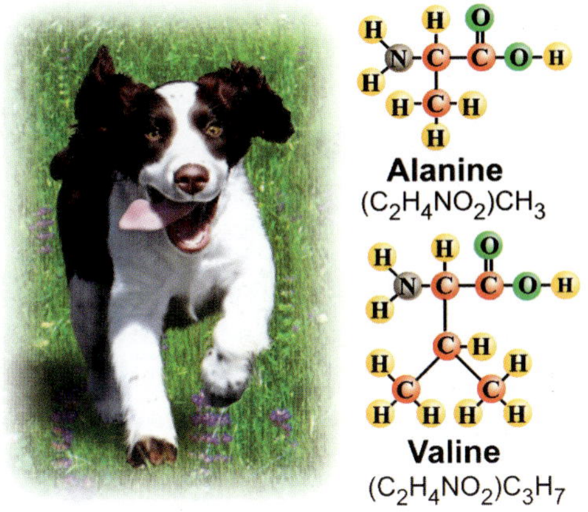

Figure 7.12: *Organic chemistry is the chemistry of living organisms and is based on the element carbon.*

Diamond and graphite

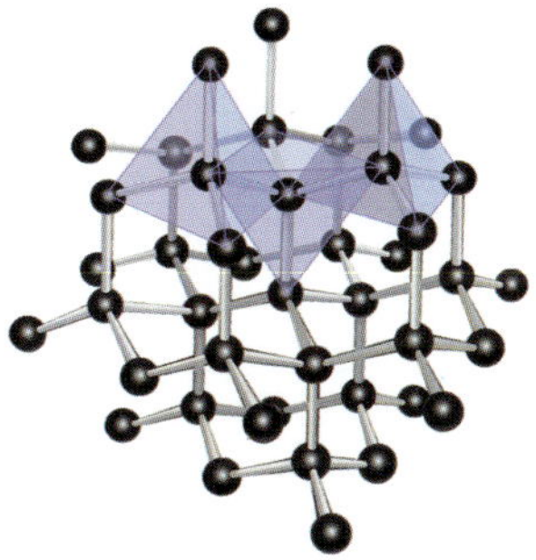

Pure carbon is found in nature in two very different forms. Graphite is a black solid made of carbon that becomes a slippery powder when ground up. Graphite is used for lubricating locks and keys. Diamond is also pure carbon. Diamond is the hardest natural substance known and also has the highest thermal conductivity of any material. Diamond is so strong because every carbon atom in diamond is bonded to four neighboring atoms in a tetrahedral crystal.

Silicon

Directly under carbon on the periodic table is the element silicon. Silicon is the second most abundant element in the Earth's crust second only to oxygen. Like carbon, silicon has four electrons in its outermost energy level. This means silicon can also make bonds with four other atoms. Sand, rocks, and minerals are predominantly made from silicon and oxygen (Figure 7.13). Most gemstones, such as rubies and emeralds, are compounds of silicon and oxygen with traces of other elements. In fact, if you can see a glass window you are looking at (or through) pure silica (SiO_2).

Figure 7.13: *Sand and glass are two common materials based on silicon.*

Silicon and semiconductors

Perhaps silicon's most famous application today is for making semiconductors. Virtually every computer chip or electronic device uses crystals of very pure silicon (Figure 7.14). The area around San Jose, California is known as Silicon Valley because of the electronics companies located there. Germanium, the element just below silicon on the periodic table is also used for semiconductors.

Figure 7.14: *Microelectronics (the small black squares) are constructed on crystals of pure silicon.*

Nitrogen, Oxygen and Phosphorus

Nitrogen and oxygen make up most of the atmosphere

Nitrogen is a colorless tasteless and odorless gas that makes up 78 percent of Earth's atmosphere. Oxygen makes up another 21 percent of the atmosphere. Both oxygen and nitrogen gas consist of molecules with two atoms (N_2, O_2).

Oxygen in rocks and minerals

Oxygen is only 21 percent of the atmosphere however, Oxygen is by far the most abundant element in Earth's crust. Almost 46 percent of the Earth's crust is oxygen. Because it is so reactive, all of this oxygen is bonded to other elements in rocks and minerals in the form of oxides. Silicon dioxide (SiO_2), calcium oxide (CaO), aluminum oxide (Al_2O_3), and magnesium oxide (MgO) are common mineral compounds. Hematite, an oxide of iron (Fe_2O_3) is a common ore from which iron is extracted.

Liquid nitrogen

With a boiling point of -196°C, liquid nitrogen is used for rapid freezing in medical and industrial applications. A common treatment for skin warts is to freeze them with liquid nitrogen.

Oxygen and nitrogen in living organisms

Oxygen and nitrogen are crucial to living animals and plants. For example, proteins and DNA both contain nitrogen. Nitrogen is part of a key ecological cycle. Bacteria in soil convert nitrogen dioxide (NO_2) in the soil to complex proteins and amino acids. These nutrients are taken up by the roots of plants, and later eaten by animals. Waste and dead tissue from animals is recycled by the soil bacteria which return the nitrogen to begin a new cycle.

Phosphorus

Directly below nitrogen in the periodic table is phosphorus. Phosphorus is a key ingredient of DNA, the molecule responsible for carrying the genetic code in all living creatures.One of phosphorus' unusual applications is in "glow-in-the-dark plastic". When phosphorus atoms absorb light, they store energy and give off a greenish glow as they slowly re-emit the energy.

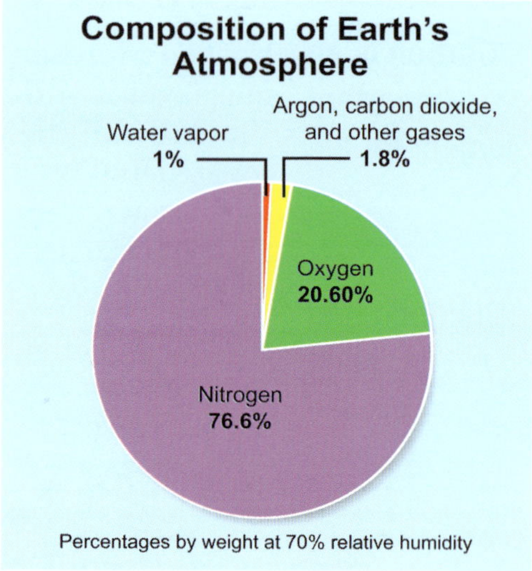

Figure 7.15: *The Earth's atmosphere is predominantly nitrogen and oxygen.*

Figure 7.16: *Oxygen makes up 46% of the mass of Earth's crust. This enormous quantity of oxygen is bound up in rocks and minerals.*

7.2 Section Review

1. Name two elements that are liquid at room temperature.
2. Which of the following is NOT true about the noble gases?
   a. they have completely filled energy levels,
   b. they have weak intermolecular forces
   c. they do not bond with other elements in nature
   d. they have boiling points above room temperature
3. Describe what it means if a chemical or physical property is periodic.
4. Name three elements which are good conductors of electricity.
5. Name three elements which are good conductors of heat.
6. A metalloid is an element which
   a. has properties between those of a metal and a nonmetal
   b. is a good thermal conductor but a poor electrical conductor
   c. is a good electrical conductor but a poor thermal conductor
   d. belongs to the same group as carbon in the periodic table
7. Steel is a metallic-like material but is not a pure element. What is steel?
8. Almost all of the oxygen on the planet Earth is found in the atmosphere. Is this statement true or false?
9. This element is abundant in Earth's crust and combines with oxygen to form rocks and minerals. Which element is it?
10. An element which has strong intermolecular forces is most likely to have
    a. a boiling point below room temperature
    b. a melting point below room temperature
    c. a boiling point very close to its melting point
    d. a very high melting point
11. Which element in Figure 7.17 is likely to be a good conductor of electricity?
12. Which element in Figure 7.17 in his likely to be an insulator?

CHALLENGE

One of the elements with atomic number less than 54 does not exist in nature. It was created in the laboratory.Which element is this and how was it discovered?

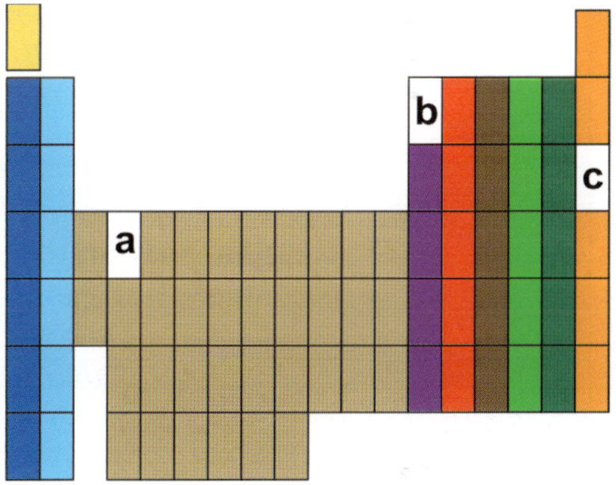

Figure 7.17: *What are the properties of these three elements? (questions 11 and 12)*

HISTORY CONNECTION

Naming the Elements

Spend more than a few minutes looking at the periodic table and you'll probably start to wonder who on Earth came up with some of the element names. Yttrium, for example. How did anyone think that up?

The elements' chemical symbols are just as confusing. Sure, it's easy to see why O stands for oxygen, but why did they pick W for tungsten and K for potassium?

Origin of unusual element symbols

Element	Symbol	Reason
Sodium	Na 11	*Natrium* (Latin). The Romans used *natrum*, from an Egyptian term.
Potassium	K 19	*Kalium* (Latin). From an Arabic term meaning "to roast," because potassium is found in plant ashes.
Iron	Fe 26	*Ferrum* (Latin). From an ancient Semitic word.
Silver	Ag 47	*Argentum* (Latin). Related to the Sanskrit word for "shining brightly."
Tin	Sn 50	*Stannum* (Latin). From Indo-European term meaning "to drip" (tin has a low melting point).
Antimony	Sb 51	*Stibium* (Latin). From an Ancient Egyptian word for the eyebrow make-up they made from Sb_2S_3.
Tungsten	W 74	*Wolfram* (German). The word means "wolf's foam." The mineral wolframite "eats" tin during extraction like a wolf eats sheep.
Gold	Au 79	*Aurum* (Latin). From the word for yellow.
Mercury	Hg 80	*Hydrargyrum* (Latin). From the Greek words for water and silver.
Lead	Pb 82	*Plumbum* (Latin). Probably from a term used in the Aegean area before the time of the Greeks.

Ancient names

It turns out the history of element names is messy and complicated. On the one hand, you have elements whose existence was well-known in many cultures long before the periodic table was developed. Gold is a good example. There are hundreds of names, both ancient and modern, for this element. The English word *gold* comes from the German word for yellow. The element's symbol, Au, comes from the Latin word for yellow, *aurum*.

Modern names

On the other hand, you have elements that have yet to be discovered (or synthesized in a lab). The naming system for these is based on their number in the periodic table. Take element number 118, for example. It's called ununoctium. In English, that translates to "one-one-eight-ium." Using the table below, can you name element 114?

Table 7.1: *IUPAC naming system for new elements*

digit	syllable	digit	syllable
0	nil	6	hex
1	un	7	sept
2	bi	8	oct
3	tri	9	en
4	quad	suffix	-ium (place at end of name)
5	pent		

The names in between

In between the ancient and the undiscovered elements is a long list of elements that were identified, for the most part, between 1600 and 1970. Some of these elements are named for places. Yttrium, for example, is named for Ytterby, a feldspar quarry near Stockholm, Sweden. Yttrium was first isolated from Ytterby rocks.

Some of the elements' names are descriptive. Chlorine (from *khloros*, for yellow-green), iodine (from *ioeides*, for violet) and rhodium (from *rhodon*, for rose) are names derived from the Greek word for the element's color.

The glow-in-the-dark element phosphorus' name means "light bearing" in Greek. The name hydrogen means "water producing." It was suggested by Antoine Lavoisier because when hydrogen burns, water is produced.

One descriptive name reflects an unpleasant characteristic: The element name osmium is taken from the Greek word *osme*, which means "odor." The element got its name because one of its common compounds, OsO_4, smells terrible!

Official naming rights

Since 1949, the International Union for Pure and Applied Chemistry (IUPAC), based in Oxford, England, has been responsible for the international names for the elements. These names are used when chemicals are sold from one country to another. However, countries can use their own names within their borders—and many countries do, especially those that do not use the English alphabet in their native language.

When a new element is identified, the discoverers are awarded the privilege of proposing a name. This is not always a straightforward process, since some of the heaviest elements exist for only fractions of a second. When several labs in different parts of the world are working on similar projects, it can be difficult to determine who should get credit for being first.

However, the elements up through element 109, meitnerium, now have official IUPAC names. Some of these names you will recognize as familiar figures from your study of modern physics. Others, like hassium and dubnium, are named after places where the elements were synthesized. Perhaps in your lifetime you will have a hand in naming a new element!

Questions:

1. Take a look at the periodic table. How many elements can you find that are named after scientists you have studied?
2. The names for cerium and palladium have something in common. Use a library or the internet to find out the origin of these names and explain their relationship
3. Three competing groups proposed names for elements 104 to 108. Find out who they were and how the IUPAC finally resolved the controversy in 1997.

CHAPTER ACTIVITY Name That Element

Each element on the periodic table has a chemical symbol that is an abbreviation of the element's name. Unlike the abbreviations for a U.S. state, these symbol-abbreviations are not always obvious. Many are derived from the element's name in a language such as Latin or German. The chemical symbol for silver is "Ag". Note that the first letter in the symbol is upper case and the second is lower case. Writing symbols this way allows us to represent all of the elements without getting confused. There is a big difference between the element cobalt, with its symbol Co, and the compound carbon monoxide, written as CO. In this activity, you'll make a set of flashcards for 30 elements and then play a game to see who in your class knows their elements.

Materials:

30 blank 8 × 10 cards and markers

What you will do

1. Each person in the class writes the symbol of one of the elements from the list on one of the large cards. Make sure you write the chemical symbol large enough so you can see it all the way across the classroom. The elements suggested below are some of the most common elements.

C	Cu	O	N	He
H	Cl	Mg	Na	K
S	Ca	Mn	Fe	Br
B	Cs	Ag	Au	Pb
I	Si	Al	F	Ne
Ba	Be	Cr	Ni	Hg

2. The teacher collects all the cards and stands in front of the class.
3. The first two players stand next to each other. The teacher holds up a chemical symbol card and the first player to correctly give the name of the element moves on to the next player. The player who didn't answer sits down.
4. The game goes all the way around the classroom, with the player who names the element moving on and the other player sitting down.
5. The player who is left standing at the end of the game is the winner.

Applying your knowledge

a. Find the element whose chemical symbol comes from the Latin word *aurum* which means "shining dawn".
b. What word does the chemical symbol for lead, Pb, come from?
c. Find the element whose chemical symbol comes from the Latin word *natrium*?
d. Which element comes from the Latin word for coal?
e. Another game to play is to see who can come up with the longest word spelled completely with chemical symbols. Some examples are *life*, from lithium (Li) and iron (Fe), and *brook*, from bromine (Br), oxygen (O), and potassium (K).

Chapter 7 Assessment

Vocabulary

Select the correct term to complete the sentences.

physical properties	periodic table	metal
period	group	alloy
nonmetal	atomic weight	noble gases
alkali	halogens	insulators
periodicity	atomic mass unit	thermal conductor
steel	electrical conductor	

Section 7.1

1. Elements that are poor conductors of heat and electricity are classified as ____.
2. A mass equal to $^1/_{12}$ the mass of a carbon-12 atom is the ____.
3. Characteristics of matter that can be seen through direct observation such as density, melting point, and boiling point are called ____.
4. Elements in the first group of the periodic table are the ____.
5. Elements grouped as a column on the periodic table belong to a(n) ____.
6. Elements that are typically shiny and good conductors of heat and electricity are classified as ____.
7. The group containing fluorine, chlorine, and bromine are called the ____.
8. A chart which organizes the elements by chemical properties is known as the ____.
9. The average mass, measured in amu, of all the isotopes of an element is called the ____.
10. Elements grouped as a row on the periodic table belong to a(n) ____.
11. Elements in the group containing helium, sometimes called the inert gases, are known as ____.

Section 7.2

12. A material that allows heat to flow through easily is known as a(n)____.
13. An alloy made by combining iron and carbon is ____.
14. A material that slows or stops the flow of heat or electricity is a(n) ____.
15. The repeating pattern of physical and chemical properties displayed by elements on different periods of the periodic table is known as ____.
16. A solid mixture of metallic elements is known as a(n) ____.
17. A material that allows electricity to flow through easily is known as a(n) ____.

Concepts

Section 7.1

1. List five physical properties of an element.
2. State one important difference between a physical change and a chemical change.
3. What property of elements was used to organize the periodic table?
4. How are the terms group and period used on the periodic table?
5. What is the general location of metals on the periodic table?
6. The energy level of the outermost electrons in an element is the same as the ____ number for that element.

7. Identify each group of elements from the description of the group's general properties:
   a. In the middle of the table, generally good conductors of heat and electricity.
   b. Form toxic gases or liquids in pure form; very active; rarely occur in pure form.
   c. Highly active metals; combine in a ratio of two to one with oxygen.
   d. Do not form chemical bonds with other atoms; commonly occur in pure form.

Section 7.2

8. At room temperature, of the 92 naturally occurring elements, state the number that are:
   a. solid
   b. liquid
   c. gas
9. What is the general location of most of the *gases* on the periodic table?
10. What do elements with the highest melting points have in common?
11. Identify the following elements by their importance to mankind:
    a. The element most essential to life on the planet.
    b. Two elements useful in making semiconductors.
    c. The most abundant element in the Earth's crust.
    d. The most abundant element in the atmosphere.
    e. An element found in proteins and DNA.
    f. An element able to store energy and "glow-in-the-dark".
12. Where do the elements that are good thermal and electrical conductors appear on the periodic table?
13. Where do the elements that are good insulators appear on the periodic table?
14. Why are alloys such as steel and brass commonly made?

Problems

Section 7.1

1. Referring to the periodic table, determine whether each list of elements represents part of a period or a group.
   a. oxygen, sulfur and selenium
   b. nickel, copper and zinc
   c. hydrogen, sodium and potassium
   d. copper, silver and gold
   e. sodium, magnesium and aluminum
2. Magnesium (Mg) has three stable isotopes whose atomic masses are 24 amu, 25 amu, and 26 amu. If the atomic weight of magnesium is given as 24.305 amu, what is the most common isotope of magnesium?
3. Explain why copper is placed in the fourth period on the periodic table.

Section 7.2

4. Using the diagram of the periodic table below, state one property for each element indicated on the table.

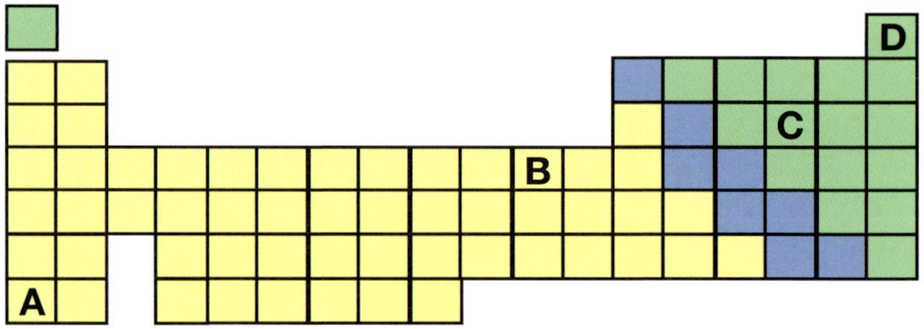

Chapter 8
Molecules and Compounds

What do aspirin, plastic wrap, and vinegar have in common? Give up? They are all compounds made from different combinations of the same three atoms: carbon, hydrogen, and oxygen. By themselves, these atoms cannot reduce pain, keep food fresh, or season food. But when they are chemically combined in certain ways to form compounds, they can be useful in many ways. Study this chapter to learn how millions upon millions of compounds and molecules can form from combinations of less than 100 basic elements

.

Key Questions

1. ***What does the chemical formula H_2O mean?***
2. ***What are chemical bonds, and how do they form?***
3. ***How do scientists show the shape of a molecule?***

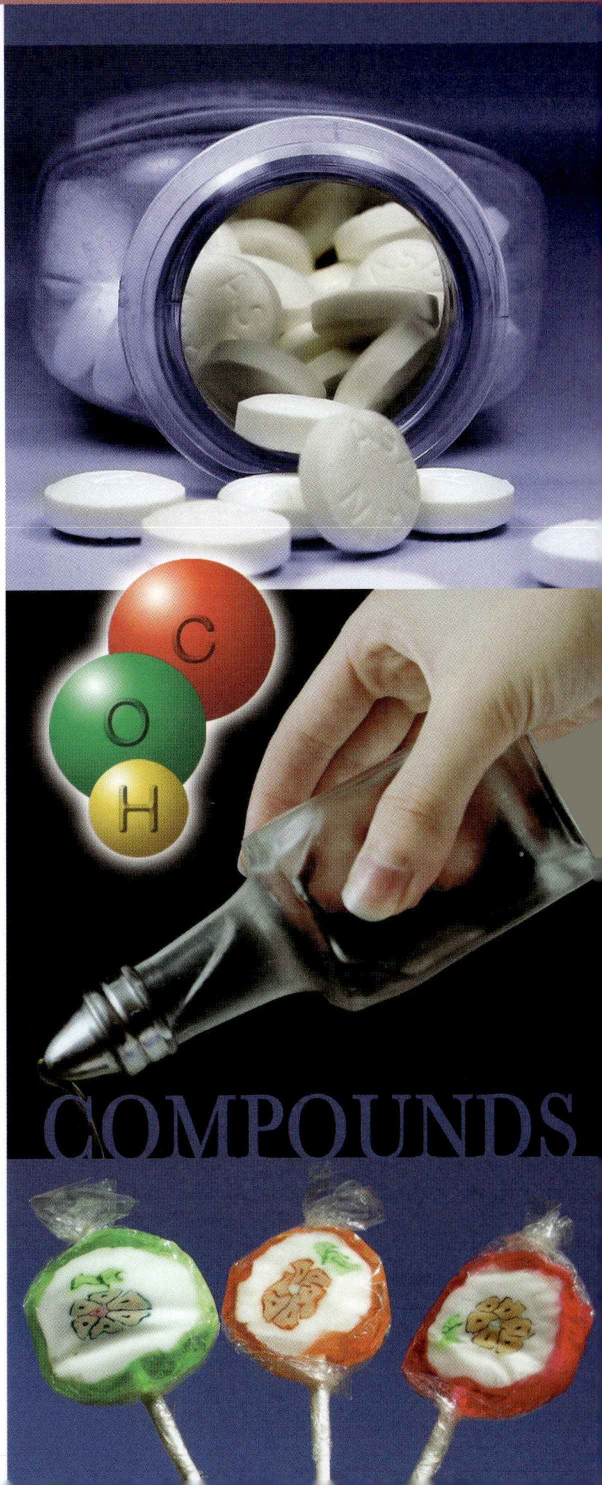

8.1 Compounds and Chemical Bonds

Most matter is in the form of compounds. If a substance is made of a pure element, chances are it will eventually combine with other elements to make a compound. For example, water, H_2O, is a compound made of hydrogen and oxygen atoms. The iron in a nail combines with oxygen in water or air to make a compound called iron oxide, better known as rust. This chapter is about how and why atoms combine into compounds.

Most matter is in the form of compounds and mixtures

Compounds and mixtures A **compound** contains two or more different elements that are chemically bonded together. For example, water (H_2O) is a compound of hydrogen and oxygen atoms bonded together. A **mixture** contains two or more elements and/or compounds that are not chemically bonded together. The atmosphere is a mixture of oxygen and nitrogen but *not* a compound. Nitrous oxide (N_2O), is a compound because oxygen and nitrogen atoms are bonded together in a molecule.

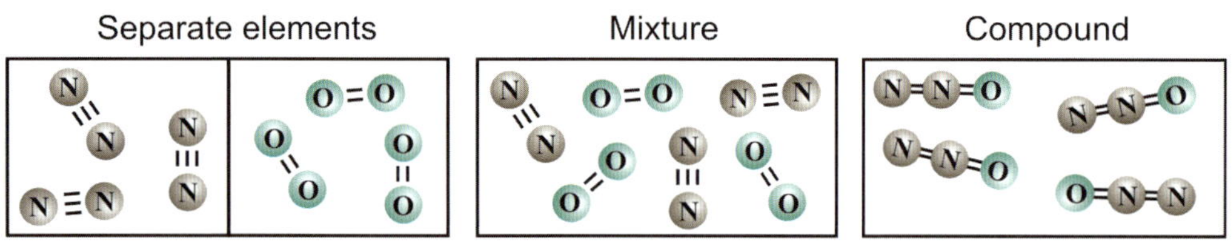

Salt and sugar are compounds Virtually everything you eat and everything in your kitchen is a compound. Salt is a compound of sodium and chlorine. Sugar is a compound of carbon, hydrogen, and oxygen. Animal and vegetable material are made of even more complex compounds, such as proteins and fats which may consist of hundreds or thousands of atoms.

Most matter is mixtures of compounds Most matter is in the form of mixtures of compounds."Sugar" is actually a compound made up of a chemical bond between two simple sugars: glucose and fructose (Figure 8.1). All sugars taste sweet but fructose tastes much sweeter than glucose. There are thousands of different compounds in a single sample of animal or vegetable tissue.

ă VOCABULARY

compound - a substance whose smallest particles include more than one element chemically bonded together. For example, water (H_2O) is a compound of hydrogen and oxygen atoms bonded together.

mixture - a substance that includes more than one type of element and/or compound.

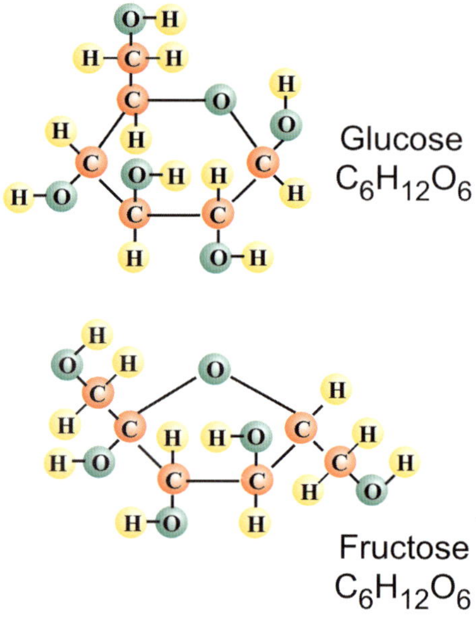

Figure 8.1: *"Pure" sugar is a compound made up of two chemically bonded simple sugars: glucose and fructose.*

Molecules and covalent bonds

Electrons form chemical bonds

A **chemical bond** forms when atoms transfer or share electrons. Two atoms that are sharing one or more electrons are chemically bonded and move together. In a water molecule, each hydrogen atom shares its single electron with the oxygen atom at the center (Figure 8.2). Almost all the elements form chemical bonds easily. This is why most of the matter you experience is in the form of compounds.

A chemical bond forms when atoms transfer or share electrons.

Covalent bonds

A **covalent bond** is formed when atoms share electrons. The bonds between oxygen and hydrogen in a water molecule are covalent bonds (Figure 8.2). There are two covalent bonds in a water molecule, between the oxygen and each of the hydrogen atoms. Each bond represents one electron. In a covalent bond, electrons are *shared* between atoms, not transferred.

Four examples of molecules held together by covalent bonds

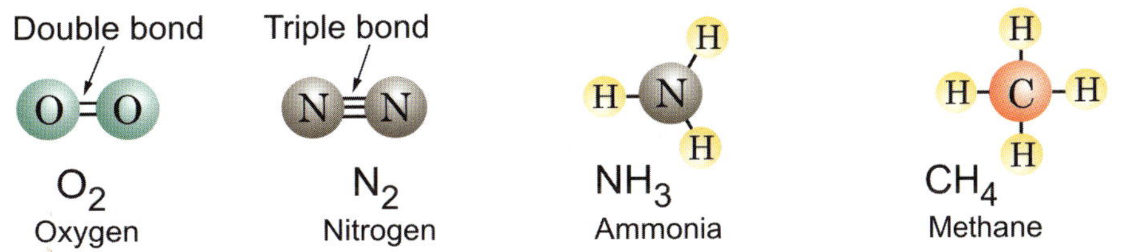

Molecules

A group of atoms held together by covalent bonds is called a **molecule**. Water is a molecule, and so is each of the different sugar molecules on the previous page. Other examples of molecules are methane (CH_4), ammonia (NH_3), oxygen (O_2) and nitrogen (N_2). In the case of oxygen and nitrogen, the bond between atoms is a double or triple covalent bond. A double bond involves two electrons and a triple bond involves three electrons per atom.

VOCABULARY

chemical bond - a bond formed between atoms through the sharing or transferring of electrons.

covalent bond - a type of chemical bond formed by shared electrons.

molecule - a group of atoms held together by covalent bonds in a specific ratio and shape.

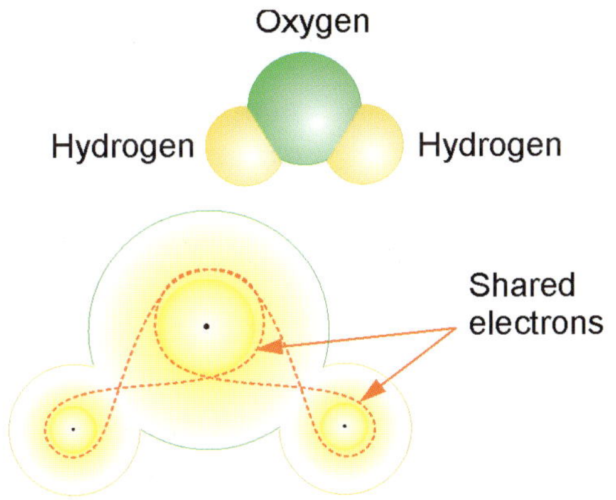

Figure 8.2: *In a covalent bond the shared electrons act like ties that hold a molecule together.*

Chemical formulas and diagrams

The chemical formula

Molecules are represented by a **chemical formula**. The chemical formula tells you the precise number of each kind of atom in the molecule. For example, the chemical formula for water is H_2O. The subscript 2 indicates there are two hydrogen atoms in the molecule. The chemical formula also tells you that water always contains twice as many hydrogen atoms as oxygen atoms. This is important to know if you wish to make water from elemental oxygen and hydrogen.

Reading a chemical formula

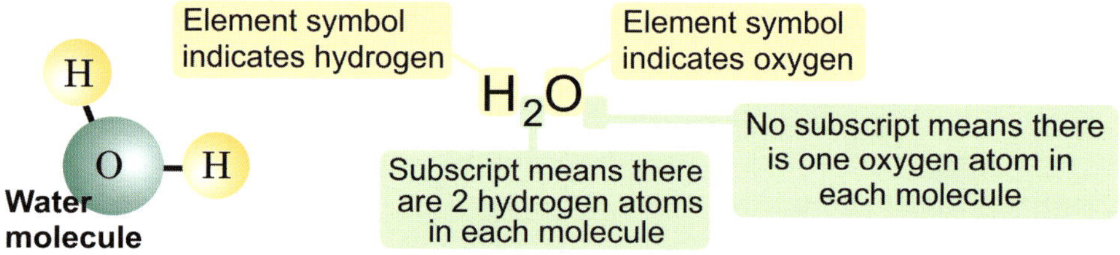

The shape of a molecule is also important to its function and properties. For this reason, molecules are represented by structural diagrams which show the shape and arrangement of atoms. Single bonds between atoms are indicated by solid lines connecting the element symbols. Double and triple bonds are indicated by double and triple lines. Both the chemical formula and structural diagrams are shown in Figure 8.3.

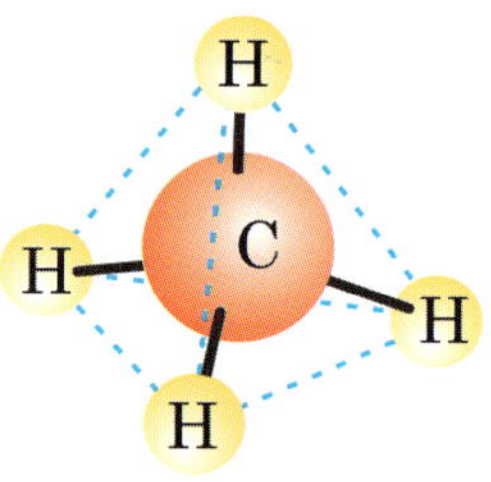

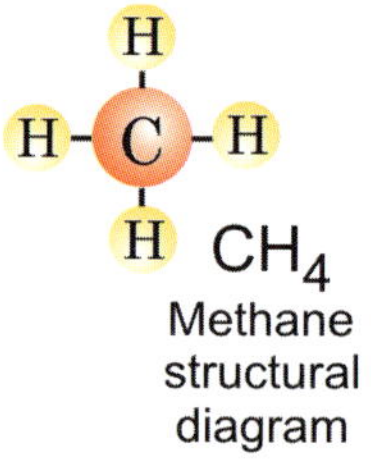

Of course, real molecules are three-dimensional, not flat as shown in the structural diagram. For example a methane molecule has the shape of a 4-sided pyramid called a tetrahedron. Each hydrogen atom is at a corner of the tetrahedron and the carbon atom is at the center.

VOCABULARY

chemical formula - identifies the number and element of each type of atom in a compound. For example, the chemical formula Fe_2O_3 is for a compound with iron (Fe) and oxygen (O) in a ratio of 2 iron atoms for every 3 oxygen atoms.

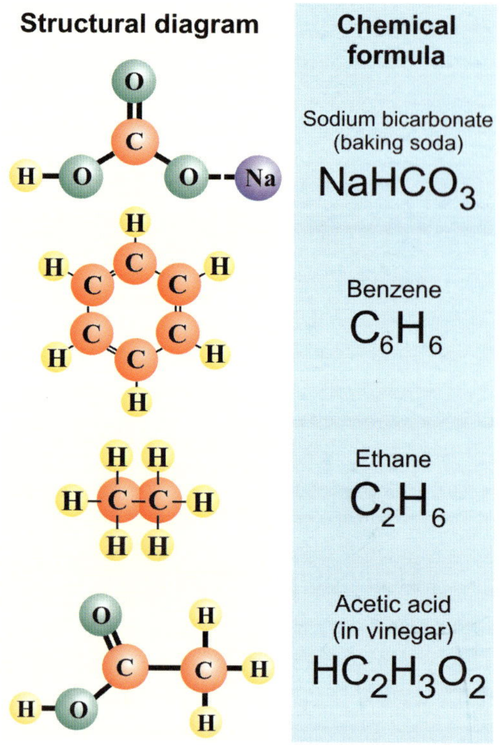

Figure 8.3: *Chemical formulas and structural diagrams.*

Structure and function

Properties come from the molecule

The properties of a compound depend *much* more on the exact composition and structure (shape) of its molecule than on the elements of which it is made. As a good example, aspirin (acetylsalicylic acid) is made from carbon, hydrogen, and oxygen according to the chemical formula $H_8C_9O_4$ (Figure 8.4). This compound has the property of relieving swelling and reducing pain in humans.

Properties depend on the exact chemical formula

By themselves, the elements (H, C, O) do not have the property of reducing pain. Other molecules formed from the same elements have very different properties than aspirin. For example, polyethylene plastic wrap and formaldehyde (a toxic preservative) are also made from carbon, oxygen, and hydrogen. The beneficial properties of aspirin come from the specific combination of exactly 8 hydrogen, 9 carbon, and 4 oxygen atoms. If the ratio of elements was changed, for example removing even one hydrogen, the resulting molecule would not have the properties of aspirin.

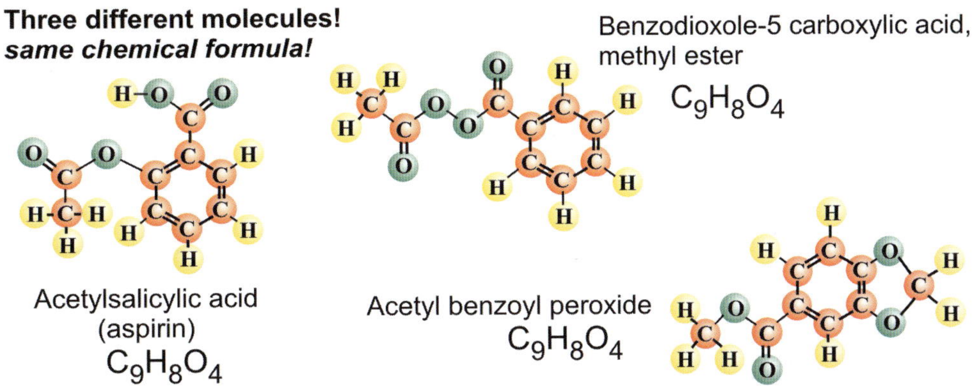

Properties also depend on molecular structure

The structure of a molecule is also important to the properties of a compound. The same 21 atoms in aspirin can be combined in other structures with the same chemical formula! The resulting molecules are something completely different (diagram above) and do not have the beneficial properties of aspirin. *Both chemical formula and structure determine the properties of a compound.*

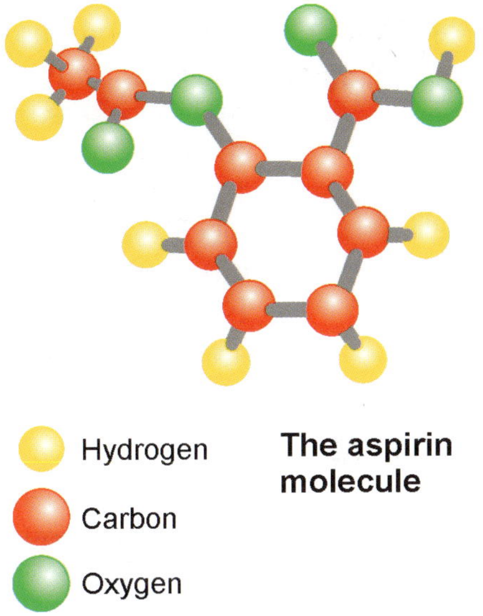

Figure 8.4: *An aspirin molecule.*

CHALLENGE

Like many modern medicines, the active ingredient in aspirin was first discovered in nature. In fact, aspirin's pain-relieving properties were known and used in its natural form long before scientists learned of it. Research the discovery of aspirin to find out the intriguing story of this widely-used medicine.

Ionic compounds

An ion is a charged atom

Not all compounds are made of molecules. For example, sodium chloride (NaCl) is a compound of sodium (Na) and chlorine (Cl) in a ratio of one sodium atom per chlorine atom. The difference is that in sodium chloride, the electron is essentially transferred from the sodium atom to the chlorine atom. When atoms gain or lose an electron they become **ions**. An ion is a charged atom. By losing an electron, the sodium atom becomes a sodium ion with a charge of +1. By gaining an electron, the chlorine atom becomes a chloride ion with a charge of -1 (when chlorine becomes an ion, the name changes to chlor*ide*).

Ionic bonds

Sodium and chlorine form an **ionic bond** because the positive sodium ion is attracted to the negative chloride ion. Ionic bonds are bonds in which electrons are transferred from one atom to another.

Ionic compounds do not form molecules

Ionic bonds are not limited to a single pair of atoms like covalent bonds. In sodium chloride each positive sodium ion is attracted to all of the neighboring chloride ions (Figure 8.5). Likewise, each chloride ion is attracted to all the neighboring sodium atoms. Because the bonds are not just between pairs of atoms, *ionic compounds do not form molecules!* In an ionic compound, each atom bonds with *all* of its neighbors through attraction between positive and negative charge.

The chemical formula for ionic compounds

Like molecular compounds, ionic compounds also have fixed ratios of elements. For example, there is one sodium ion per chloride ion in sodium chloride. This means we can use the same type of chemical formula for ionic compounds and molecular compounds.

Ions may be multiply charged

Sodium chloride involves the transfer of one electron however, ionic compounds may also be formed by the transfer of two or more electrons. A good example is magnesium chloride ($MgCl_2$). The magnesium atom gives up two electrons to become a magnesium ion with a charge of +2 (Mg^{2+}). Each chlorine atom gains one electron to become a chloride ion with a charge of -1 (Cl^-). The ion charge is written as a superscript after the element (Mg^{2+}, Fe^{3+}, Cl^-, etc.).

VOCABULARY

ion - an atom that has an electric charge different from zero. Ions are created when atoms gain or lose electrons.

ionic bond - a bond that transfers an electron from one atom to another resulting in attraction between oppositely charged ions.

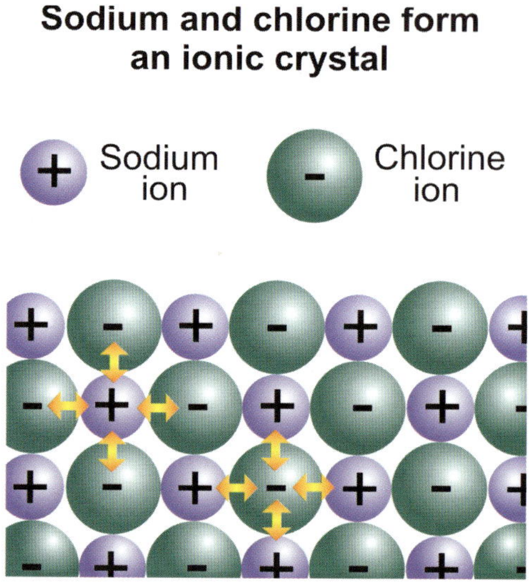

Figure 8.5: *Sodium chloride is an ionic compound in which each positive sodium ion is attracted to all of its negative chloride neighbors and vice versa.*

Why chemical bonds form

Atoms form bonds to reach a lower energy state

Imagine pulling tape off a surface. It takes energy to separate atoms that are bonded together just like it takes energy to pull tape off a surface. If it takes energy to separate bonded atoms, then the same energy must be released when the bond is formed. *Energy is released when chemical bonds form*. Energy is released because chemically bonded atoms have less total energy than free atoms. Like a ball rolling downhill, atoms form compounds because the atoms have lower energy when they are together in compounds. For example, one carbon atom and four hydrogen atoms have more total energy apart than they do when combined in a methane molecule (Figure 8.6).

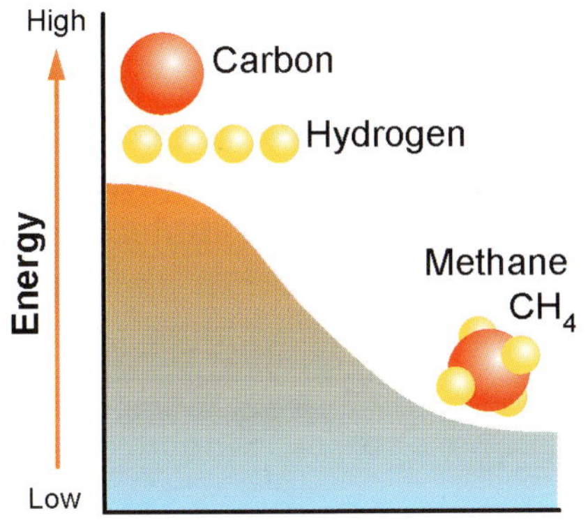

Figure 8.6: *The methane (CH_4) molecule has lower total energy than four separate hydrogen and one separate carbon atom.*

Chemical reactivity

All elements except the noble gases form chemical bonds. However, some elements are much more reactive than others. In chemistry, "reactive" means an element readily forms chemical bonds, often releasing energy. For example, sodium is a highly reactive metal. Chlorine is a highly reactive gas. If pure sodium and pure chlorine are placed together, a violent explosion occurs as the sodium and chlorine combine and form ionic bonds. The energy of the explosion is the energy given off by the formation of the chemical bonds.

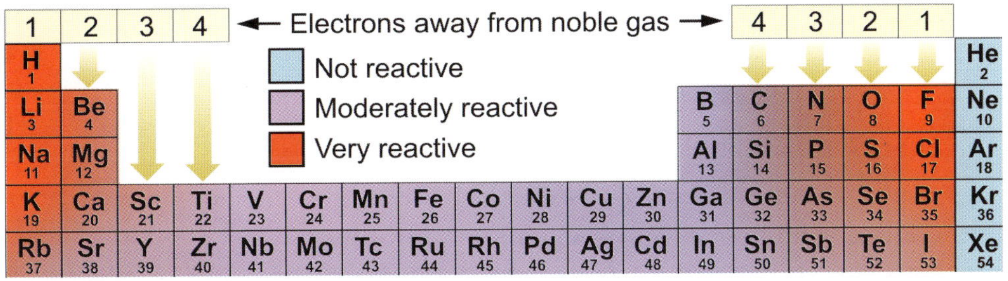

Some elements are more reactive than others

The closer an element is to having the same number of electrons as a noble gas, the more reactive the element is. The alkali metals are very reactive because they are just one electron away from the noble gasses. The halogens are also very reactive because they are also one electron away from the noble gases. The beryllium group and the oxygen group are less reactive because each element in these groups is two electrons away from a noble gas.

CHALLENGE

The noble gases (He, Ne, Ar, ...) are called inert because they do not ordinarily react with anything. You can put sodium in an atmosphere of pure helium and nothing will happen. However, scientists have found that a few noble gases DO form compounds in very special circumstances. Research the topic and see if you can find a compound involving a noble gas.

8.1 Section Review

1. What is the difference between a compound and a mixture?
2. Give an example of a compound, and an example of a mixture.
3. How many atoms of chlorine (Cl) are in the carbon tetrachloride molecule (CCl_4)?
4. Which of the diagrams in Figure 8.7 is the correct structural diagram for carbon tetrachloride (CCl_4)?
5. Write a chemical formula for a compound which has two atoms of oxygen (O) and three atoms of iron (Fe).
6. What is the chemical formula for the molecule in Figure 8.8?
7. How many atoms of hydrogen are in a molecule of acetic acid ($HC_2H_3O_2$)?

Which is the correct CCl_4?

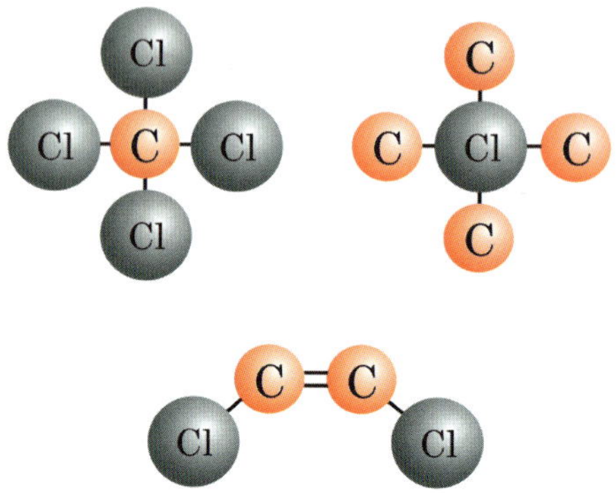

Figure 8.7: *Question 4*

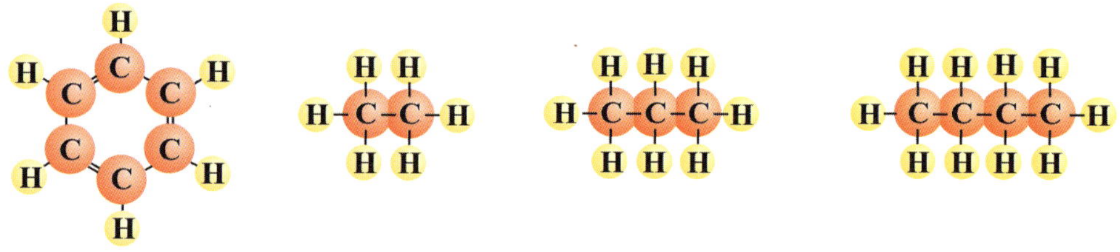

8. Which of the molecules above has the chemical formula C_3H_8?
9. Which of the following statements is FALSE?
   a. The properties of a compound depend more on which elements are present and less on the structure of the molecule.
   b. The properties of a compound depend more on the structure of the molecule and less on which elements are present.
10. Chemical bonds form because
    a. The atoms have more energy bonded together than separated.
    b. The atoms have less energy bonded together than separated.

What is the chemical formula for this molecule?

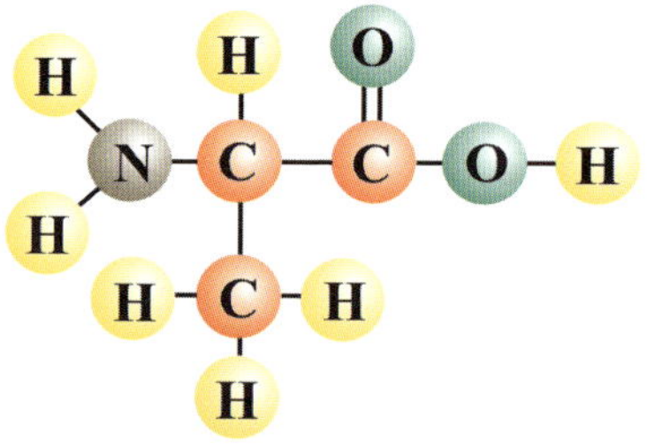

Figure 8.8: *Question 6*

8.2 Electrons and Chemical Bonds

The discovery of energy levels in the atom solved a 2000-year-old mystery. The mystery was why elements combined with other elements only in particular ratios (or not at all). For example, why do two hydrogen atoms bond with one oxygen to make water? Why isn't there a molecule with three (H_3O) or even four (H_4O) hydrogen atoms? Why does sodium chloride have a precise ratio of one sodium ion to one chloride ion? Why do helium, neon, and argon form no compounds with any other element? The answer has to do with energy levels and electrons.

Valence electrons

Valence electrons

Chemical bonds are formed only between the electrons in the highest unfilled energy level. These electrons are called **valence electrons**. You can think of valence electrons as the outer "skin" of an atom. Electrons in the inner (filled) energy levels do not "see" other atoms because they are shielded by the valence electrons. For example, chlorine has 7 valence electrons. The first 10 of chlorine's 17 electrons are in the inner (filled) energy levels.

Most elements bond to reach 8 valence electrons

It turns out that *8 is a magic number for chemical bonding*. All the elements heavier than boron form chemical bonds to try and get to a configuration with eight valence electrons (Figure 8.10). Eight is a preferred number because 8 electrons are a complete (filled) energy level. The noble gases already have a magic number of 8 valence electrons. They don't form chemical bonds because they don't need to!

Light elements bond to reach 2 valance electrons

For elements with atomic number 5 (boron) or less, the magic number is 2 instead of 8. For these light elements, 2 valence electrons completely fills the *first* energy level. The elements H, He, Li, Be,and B, form bonds to reach the magic number of 2.

Hydrogen is special

Because of its single electron, hydrogen can also have 0 valence electrons! Zero is a magic number for hydrogen, as well as 2. This flexibility makes hydrogen a very "friendly" element; hydrogen can bond with almost any other element.

VOCABULARY

valence electrons - electrons in the highest unfilled energy level of an atom. These electrons participate in chemical bonds.

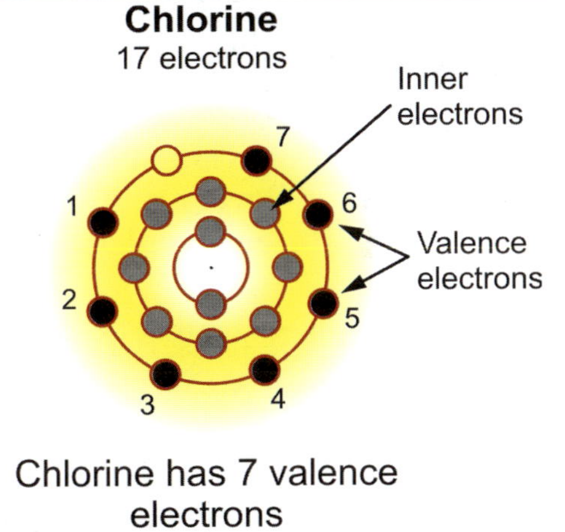

Figure 8.9: *Chlorine has 7 valence electrons. The other 10 electrons are in filled (inner) energy levels.*

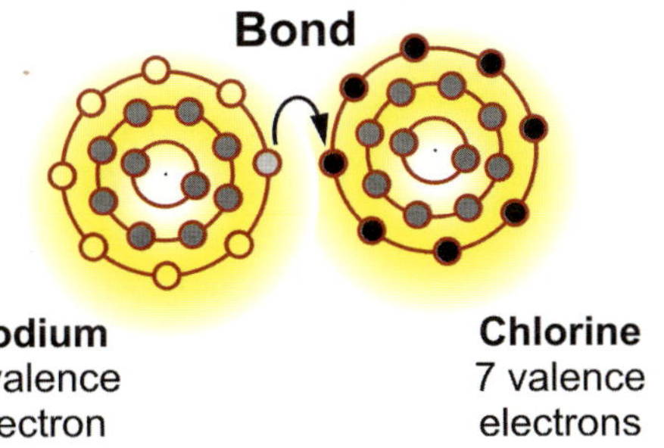

Figure 8.10: *Chlorine and sodium bond so each can reach a configuration with 8 valence electrons.*

Valence electrons and the periodic table

Period 2 elements

The picture below shows how the electrons in the elements in the second period (lithium to neon) fill the energy levels. Two of lithium's three electrons go in the first energy level. Lithium has one valence electron because it's third electron is the only one in the second energy level.

Each successive element has one more valence electron

Going from left to right across a period each successive element has one more valence electron. Beryllium has two valence electrons. Boron has three and carbon has four. Each element in the second period adds one more electron until all 8 spots in the second energy level are full at atomic number 10, which is neon, a noble gas. Neon has 8 valence electrons

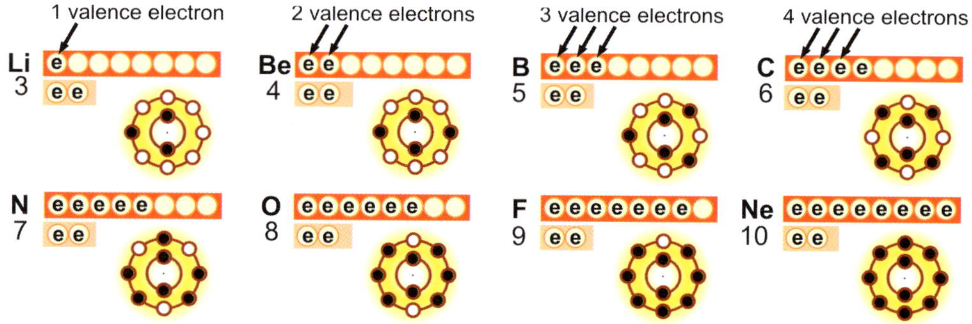

Bonding

Oxygen has 6 valence electrons. To get to the magic number of 8, oxygen needs to add two electrons. *Oxygen forms chemical bonds that provide these two extra electrons*. For example, a single oxygen atom combines with two hydrogen atoms because each hydrogen can supply only one electron. Oxygen combines with one beryllium atom because beryllium can supply two valence electrons to give oxygen its required number of 8 (Figure 8.11).

Double bonds share 2 electrons

Carbon has four valence electrons. That means two oxygen atoms can bond with a single carbon atom, each oxygen sharing two of carbon's four valence electrons. The bonds in carbon dioxide (CO_2) are double bonds because each bond involves 2 electrons (Figure 8.12).

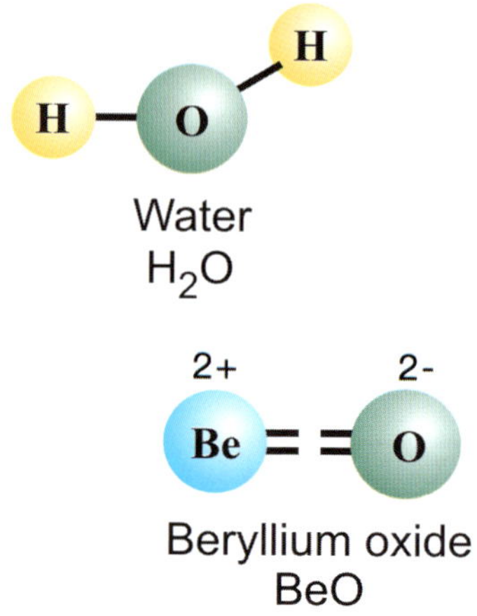

Figure 8.11: *Water (H_2O) and beryllium oxide (BeO).*

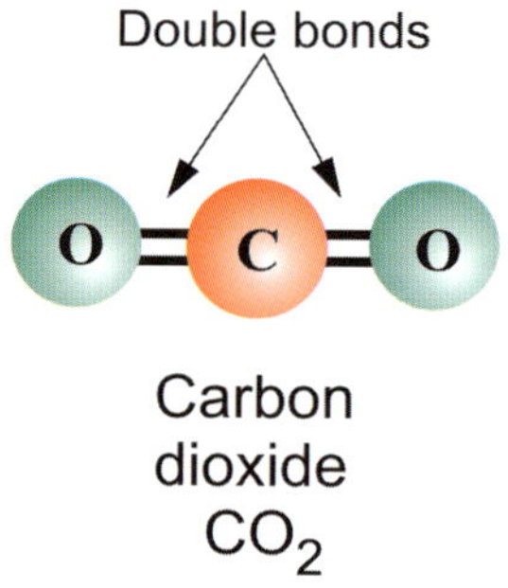

Figure 8.12: *Carbon forms two double bonds with oxygen to make carbon dioxide.*

Lewis dot diagrams

Dot diagrams of the elements

A clever way to keep track of valence electrons is to draw Lewis dot diagrams. A dot diagram shows the element symbol surrounded by one to eight dots representing the valence electrons. Each dot represents one electron. Lithium has one dot, beryllium has two, nitrogen has three, etc. Figure 8.13 shows the dot diagrams for the first 10 elements.

Dot diagrams of molecules

Each element forms bonds to reach one of the magic numbers of valence electrons: 2 or 8. In dot diagrams of a complete molecule each element symbol has either 2 or 8 dots around it. Both configurations correspond to completely filled (or empty) energy levels.

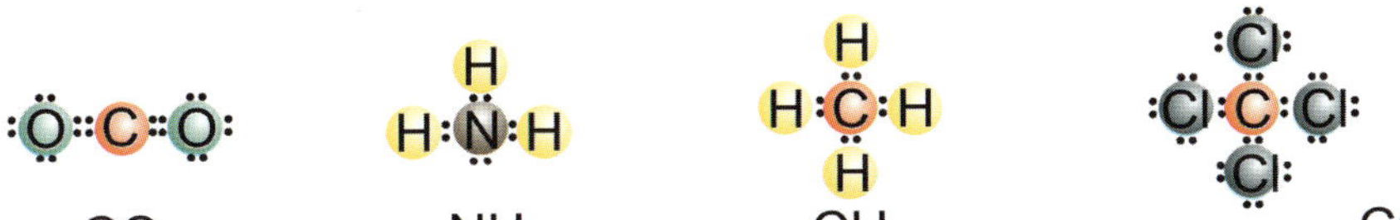

CO_2 Carbon dioxide | NH_3 Ammonia | CH_4 Methane | CCl_4 Carbon tetrachloride

Example dot diagrams

Carbon has four dots and hydrogen has one. One carbon atom bonds with four hydrogen atoms because this allows the carbon atom to have eight valence electrons (8 dots) — four of its own and four shared from the hydrogen atoms. The picture above shows dot diagrams for carbon dioxide (CO_2), methane (CH_4), ammonia (NH_3), and carbon tetrachloride (CCl_4), a flammable solvent.

Dot diagrams

Draw the dot diagram for nitric acid, HNO_3.

1. Looking for: Dot diagram
2. Given: chemical formula HNO_3
3. Relationships:

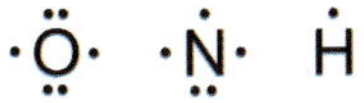

4. Solution:

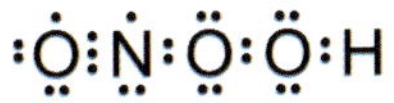

This one is tricky, each element symbol has 2 or 8 dots but one of the bonds (O⁝N) is a double bond.

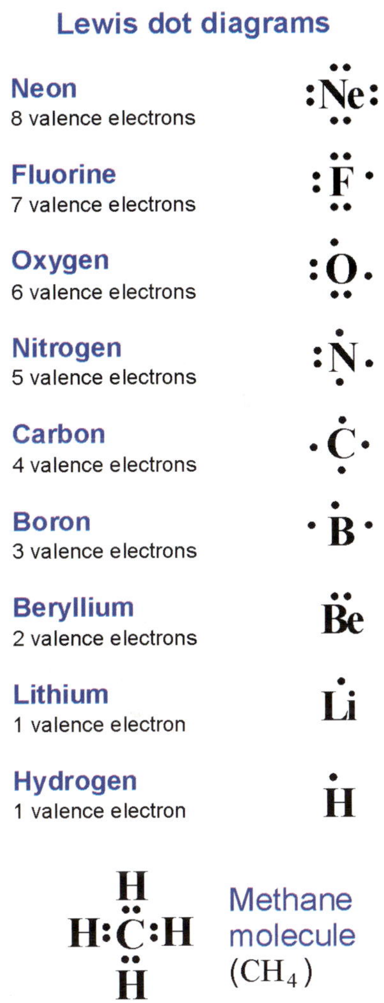

Figure 8.13: *Lewis dot diagrams show valence electrons as dots around the element symbol. Atoms form bonds to get eight valence electrons by sharing with other atoms.*

Oxidation numbers

Oxidation numbers

A sodium atom always ionizes to become Na+ (a charge of +1) when it combines with other atoms to make a compound. Therefore, we say that sodium has an **oxidation number** of 1+. An oxidation number indicates the charge on the remaining atom (ion) when electrons are lost, gained, or shared in chemical bonds. Table 8.1 shows the oxidation numbers for some elements. Notice that the convention for writing oxidation numbers is the opposite of the convention for writing the charge. When writing the oxidation number, the positive (or negative) symbol is written after the number, not before it.

Oxidation numbers and the periodic table

Oxidation numbers correspond closely to an element's group on the periodic table. All of the alkali metals have oxidation numbers of 1+ since these elements all prefer to lose one electron in chemical bonds. All of the halogens have an oxidation number of 1- because these elements prefer to gain an electron in chemical bonds. The diagram below shows the trend in oxidation numbers across the periodic table. Most transition metals have complicated oxidation numbers because they have many more electrons.

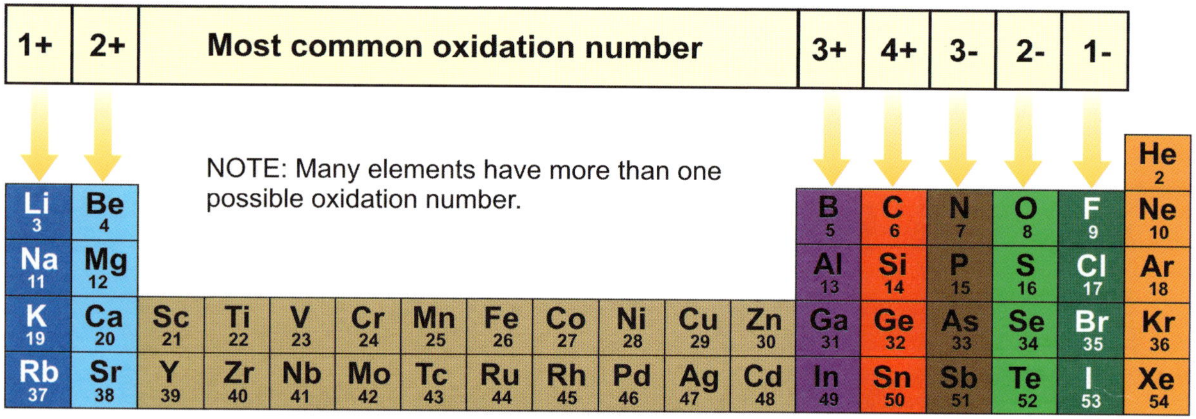

VOCABULARY

oxidation number - indicates the charge of an atom when an electron is lost, gained, or shared in a chemical bond. An oxidation number of +1 means an electron is lost, -1 means an electron is gained.

Table 8.1: Some oxidation numbers

atom	electrons gained or lost	oxidation number
K	loses 1	1+
Mg	loses 2	2+
Al	loses 3	3+
P	gains 3	3-
Se	gains 2	2-
Br	gains 1	1-
Ar	loses 0	0

SOLVE IT!

What is fluorine's oxidation number? If you think it is 1-, you are right. Like the other halogens, fluorine gains one electron, one negative charge, when it bonds with other atoms.

Predicting a chemical formula

Oxidation numbers in a compound add up to zero

When elements combine in molecules and ionic compounds, the total electric charge is always zero. This is because any electron donated by one atom is accepted by another. The rule of zero charge is easiest to apply using oxidation numbers. The total of all the oxidation numbers for all the atoms in a compound must be zero. This important rule allows you to predict many chemical formulas.

The oxidation numbers for all the atoms in a compound must add up to zero

Example, carbon tetrachloride

To see how this works, consider the compound, carbon tetrachloride (CCl_4). Carbon has an oxidation number of 4+. Chlorine has an oxidation number of 1-. It takes four chlorine atoms to cancel with carbon's 4+ oxidation number.

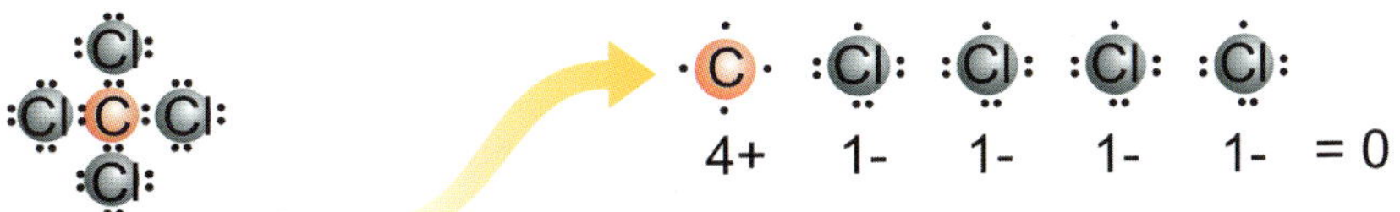

CCl_4
Carbon tetrachloride

The total of the oxidation numbers for each atom in the molecule must be zero.

STUDY SKILLS

Multiple oxidation numbers

Many periodic tables list multiple oxidation numbers for most elements. This is because more complex bonding is possible. This course gives you the fundamental ideas but there is much more!

When multiple oxidation numbers are shown, the most common one is usually in bold type. For example, nitrogen has possible oxidation numbers of 5+, 4+, 3+, 2+ and 3- even though 3- is the most common.

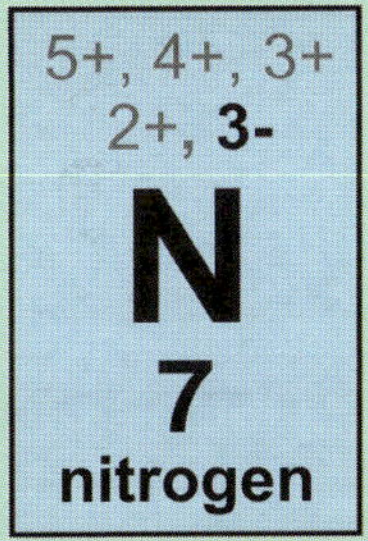

Predict a chemical formula

Iron and oxygen combine to form a compound. Iron (Fe) has an oxidation number of 3+. Oxygen (O) has an oxidation number of 2-. Predict the chemical formula of this compound.

1. Looking for: Chemical formula
2. Given: oxidation numbers Fe 3+ and O 2-
3. Relationships: The oxidation numbers for all the atoms in a compound must add up to zero.
4. Solution: Three oxygen atoms contribute the total oxidation number of 6-. It takes only two iron atoms to get a total oxidation number of 6+. Therefore, the chemical formula is Fe_2O_3.

Your turn...

a. Predict the chemical formula of the compound containing beryllium (2+) and fluorine (1-). **Answer:** BeF_2

Ionic and covalent bonds

Why bonds are ionic or covalent

Whether or not a compound is ionic or covalently bonded depends on how much each element "needs" an electron to get to a magic number (2 or 8). Elements which are very close to the noble gases tend to give or take electrons rather than share them. These elements often form ionic bonds rather than covalent bonds.

Sodium chloride is ionic

As an example, sodium has one electron more than the noble gas, neon. Sodium has a very strong tendency to give up that electron and become a positive ion. Chlorine has one electron less than argon. Therefore, chlorine has a very strong tendency to accept an electron and become a negative ion. Sodium chloride is an ionic compound because sodium has a strong tendency to give up an electron and chlorine has a strong tendency to accept an electron.

Widely separated elements form ionic compounds

On the periodic table, strong electron donors are the left side (alkali metals). Strong electron acceptors are on the right side (halogens). The farther separated two elements are on the periodic table, the more likely they are to form an ionic compound.

Nearby elements form covalent compounds

Covalent compounds form when elements have roughly equal tendency to accept electrons. Elements that are nonmetals and therefore close together on the periodic table tend to form covalent compounds with each other because they have approximately equal tendency to accept electrons. Compounds involving carbon, silicon, nitrogen, and oxygen are often covalent.

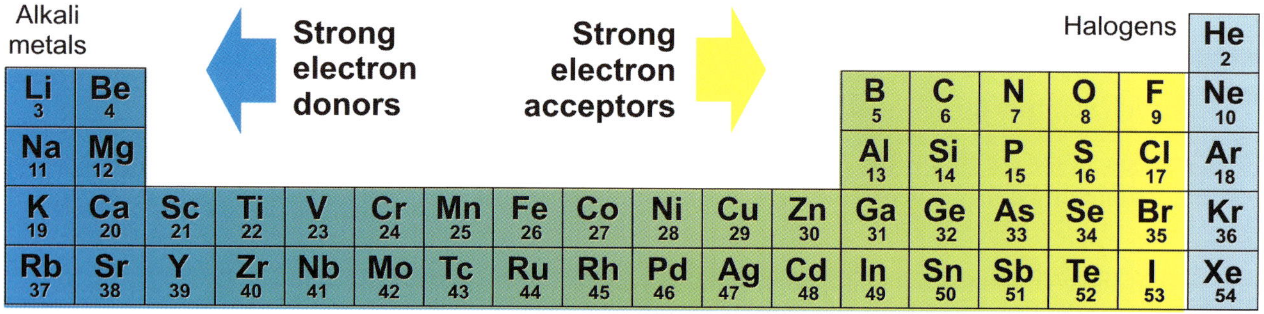

Predicting ionic or covalent bonds

Potassium (K) combines with bromine (Br) to make the salt, potassium bromide (KBr). Is this likely to be an ionic or covalently bonded compound?

1. Looking for: ionic or covalent bond
2. Given: K and Br
3. Relationships: K is a strong electron donor. Br is a strong electron acceptor
4. Solution: KBr is an ionic compound because K and Br are from opposite sides of the periodic table.

Your turn...

a. Is silica (SiO_2) likely to be an ionic or covalently bonded compound?
Answer: covalent

b. Is calcium fluoride (CaF_2) likely to be an ionic or covalently bonded compound?
Answer: ionic

8.2 Section Review

1. Atoms form chemical bonds using
   a. electrons in the innermost energy level,
   b. electrons in the outermost energy level,
   c. protons and electrons.
2. Which of the diagrams in Figure 8.14 shows an element with three valence electrons? What is the name of this element?
3. Which of the following elements will form a double bond with oxygen making a molecule with one atom of the element and one atom of oxygen.
   a. lithium
   b. boron
   c. beryllium
   d. nitrogen
4. Name two elements that have the Lewis dot diagram shown in Figure 8.15.
5. The oxidation number is
   a. the number of oxygen atoms and element bonds with,
   b. the positive or negative charge acquired by an atom in a chemical bond,
   c. the number of electrons involved in a chemical bond.
6. Name three elements that have an oxidation number of 3+.
7. What is the oxidation number for the elements shown in Figure 8.16?
8. When elements form a molecule, what is TRUE about the oxidation numbers of the atoms in the molecule.
   a. The sum of the oxidation numbers must equal zero.
   b. All oxidation numbers from the same molecule must be positive.
   c. All oxidation numbers from the same molecule must be negative

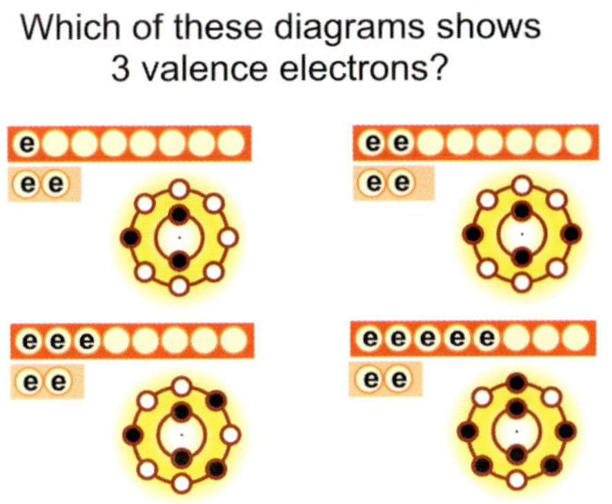

Figure 8.14: *Question 2*

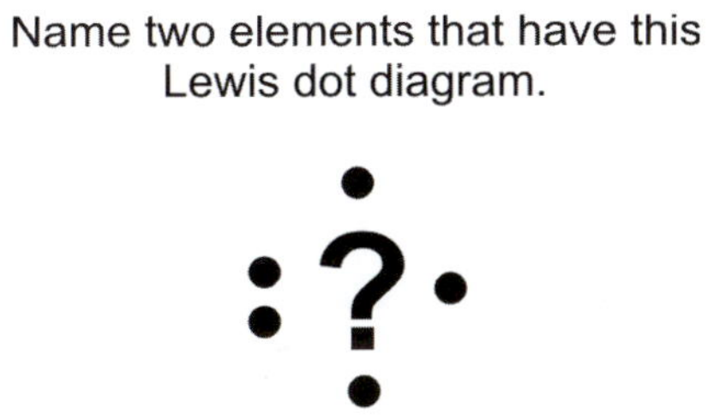

Figure 8.15: *Question 4*

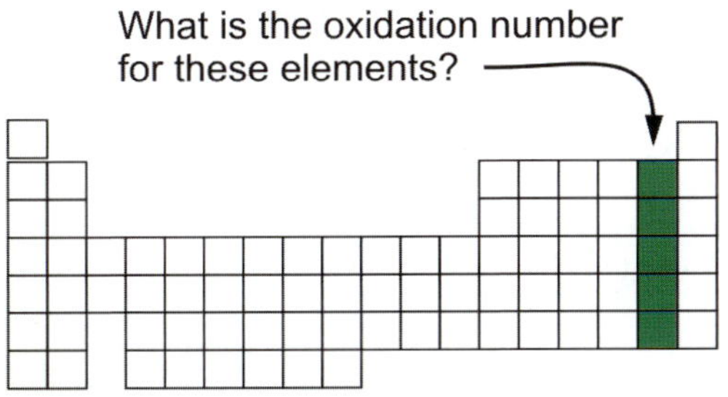

Figure 8.16: *Question 7*

GEOLOGY CONNECTION Salt: Common, But Uncommonly Important

Timbuktu sounds like a faraway place, isolated, inaccessible - almost imaginary. In fact, it is in a remote location: at the edge of the Sahara in the West African nation of Mali. And today, Timbuktu is a ruined city, a harsh place under constant threat from desert sand and winds. It is hard to picture it as it once was - a thriving center of trade and culture. What could this old city have in common with salt?

Beginning more than 900 years ago, Timbuktu became an important crossroads of African trade. Caravans of thousands of camels crossed hundreds of miles of desert to and from the city. Two commodities, which often traded at equal value and were used as money, drove the city's fortunes: gold and salt.

Today salt is among the most common substances we know. Why was it so valuable back then, literally worth its weight in gold? There were two big reasons. First, salt was much harder to find than it is today. Second, it was very important for preserving foods. Before refrigeration, salt was used to keep meat edible for long periods of time. The use of salt as a preservative contributed to the survival and expansion of human civilization.

Salt in the modern world

Today, salt is still important, but not as a food preservative. Of course, it is commonly used to season food. It is also used as a supplement in raising livestock and poultry. Salt has many industrial uses, too, in making paper, soap, detergent, and a variety of chemicals. In many parts of the United States, huge amounts of salt are used to melt ice on highways in winter.

One of the chief ways to get salt is to mine it. Rock salt, also called halite, is mined from salt deposits which are usually deep in the ground. A mammoth salt deposit lies beneath portions of Ohio, Michigan, Pennsylvania, New York, West Virginia, and Ontario, Canada. This Great Eastern Salt Basin is one of the largest salt beds in the world. The salt was evaporated from ancient seas and deposited more than 300 million years ago. Today, in Cleveland, Ohio, this salt is mined from beneath Lake Erie. The salt mine is immense; its two main shafts are nearly 2,000 feet deep.

Photo - courtesy of Ohio Department of Natural Resources

The other common way to get salt is through evaporation. In the evaporation process, salt is gathered from seawater, salt marshes, or salt lakes. San Francisco Bay in California has one of the largest salt operations in the country; it is scooped from the bottom of coastal ponds. The Great Salt Lake in Utah is another huge evaporation site.

The chemistry of salt

Salt is a chemical compound composed of two elements, sodium and chlorine. It is called sodium chloride. The chemical symbol for salt is NaCl. A salt crystal contains one atom of sodium for each atom of chlorine.

Let's take a closer look at these elements. Sodium is on the left side of the periodic table of elements, a metal. Its atomic number is 11. It is soft, light, and silvery-white. Sodium is highly reactive and never found in pure form in nature. It oxidizes in air and reacts violently with water.

Chlorine is on the right side of the periodic table, a halogen. Its atomic number is 17. It is yellow-green in color and poisonous. Chlorine is commonly found in nature as a gas. In liquid form it is a powerful oxidizing and bleaching agent; it is a component of chlorine bleach. Chlorine combines easily with nearly all other elements.

Sodium chloride is a crystal. The particular crystal structure of sodium chloride is cubic. Each sodium ion is surrounded by six chlorine ions. Each chlorine ion is surrounded by six sodium ions. This is known as the halite structure and is common to many minerals.

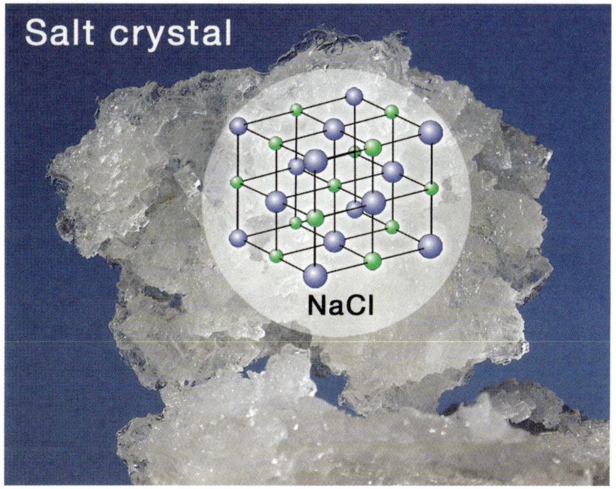

"The salt of the Earth" is precious

Sodium chloride is one of Earth's most common compounds. It makes the oceans salty; it is found in most human tissue; it is plainly essential to life.

Yet long ago, salt was one of the most valuable substances in the world. Having it was as good as money in the bank. Wars were fought over salt. We take common table salt for granted now, but salt has been uncommonly important to human civilization and to life here on Earth.

Questions:

1. Compare the two chief methods of gathering salt.
2. How did salt affect the expansion of human civilization?
3. What are the differences between chlorine and sodium?
4. Describe the crystal structure of sodium chloride.

CHAPTER ACTIVITY Molecular Gumdrop Models

Molecules are the structures that result when two or more atoms bond by sharing electrons. In living things, almost all molecules are made from hydrogen, carbon, nitrogen, oxygen, phosphorus, and sulfur. Molecules are described using formulas. For example, H_2 is a molecule consisting of two hydrogen atoms. CH_4 is a molecule consisting of one carbon atom and four hydrogen atoms. In this activity, you will build some simple molecules out of gumdrops and toothpicks.

Materials:

Toothpicks
White, red, yellow and green gumdrops

What you will do

1. Colored gumdrops will represent the atoms you use to build your molecules: white (or brown) gumdrops represent nitrogen, red gumdrops represent carbon, yellow gumdrops represent hydrogen, and green gumdrops represent oxygen. Toothpicks will represent each bond in your model. Remember that hydrogen can have one bond, oxygen can have two bonds, carbon can make four bonds and nitrogen can have three bonds. The bonding between two atoms can be single, double or triple. So, carbon could make its four bonds with four single bonds, or two double bonds, or one double bond and two single bonds, or one single bond and one triple bond.
2. Build the following molecules. As you build your molecules, try to make sure each atom is as far from other atoms as possible. If a molecule has three hydrogen atoms on it, don't put all three next to each other. Space them evenly around the central atom.
3. Build the following molecules.

H_2	O_2
N_2	H_2O
CH_4	HCN
CO_2	C_2H_6O (two different versions)

Applying your knowledge

a. Once you've built your models, sketch them on paper.
b. If a bond is a pair of electrons shared between two atoms, how many electrons are being shared in H_2?
c. If a bond is a pair of electrons shared between two atoms, how many electrons are being shared in O_2?
d. Electrons are negatively charged. Charges of the same kind repel. Bonds are shared pairs of electrons. How does this explain that atoms are evenly spaced around each other in molecules?
e. There were two different ways (at least) to draw the molecule C_2H_6O. When the same formula can produce different molecules, those molecules are called isomers. Do you think the isomers of C_2H_6O have the same characteristics or different characteristics? Explain.
f. Write the formula for the molecule in the diagram:

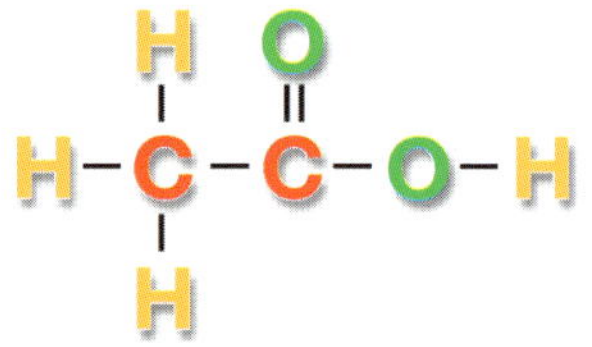

How many toothpicks and gumdrops (and what color!) would you need to build this molecule?

Chapter 8 Assessment

Vocabulary

Select the correct term to complete the sentences.

compound	covalent	ion
mixture	chemical bond	molecule
chemical formula	ionic bond	oxidation number
valence electrons		

Section 8.1

1. When two atoms share or trade electrons a(n) ____ is formed.
2. An atom that has acquired a positive or negative charge is called a(n) ____.
3. To represent the number and type of each element in a molecule, chemists write a ____.
4. A substance whose smallest particles include more than one element chemically bonded together is a ____.
5. The type of chemical bond formed when atoms share electrons is the ____ bond.
6. A substance made of two or more elements or compounds not chemically bonded to each other is a(n) ____.
7. A bond formed when an electron is transferred from one atom to another is a(n) ____.
8. A group of atoms held together by covalent bonds in a specific ratio form a(n) ____.

Section 8.2

9. Electrons in the highest unfilled energy level of an atom, that may participate in chemical bonds are called ____.
10. The number which indicates the charge on an atom when an electron is lost, gained or, shared is called the ____.

Concepts

Section 8.1

1. What is the chemical formula for water? What atoms make up this compound?
2. What is the difference between a compound and a mixture?
3. List 3 examples of a mixture and 3 examples of a compound.
4. Why do atoms form compounds instead of existing as single atoms?
5. What type of bond holds a water molecule together?
6. What do we call the particle that is a group of atoms held together by covalent bonds?
7. List 4 examples of a molecule.
8. What does the subscript "2" in H_20 mean?
9. What do the subscripts in the formula for ethane represent?

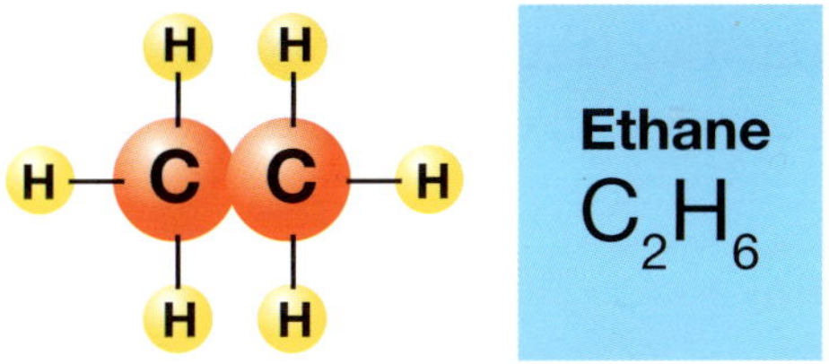

10. Name the two most important factors in determining the properties of a compound.
11. Summarize the differences between a covalent compound and an ionic compound.
12. What happens when chemical bonds form? Why?
13. Which group of elements usually don't form chemical bonds?
14. Name a very reactive group of metals and a very reactive group of nonmetals. Why do they behave this way?

Section 8.2

15. When atoms form chemical bonds, which of their electrons are involved in the bonds?
16. How many electrons represent a complete (filled) outermost energy level for elements heavier than Boron (atomic number greater than 5)?
17. Noble gases usually don't form chemical bonds. Why?
18. What is so special about hydrogen when it comes to forming bonds?
19. Each successive element on a period table going from left to right across a period has what?
20. In a Lewis dot diagram, what is represented by the dots surrounding the element symbol?
21. How many valence electrons does oxygen have? How many more electrons are needed to fill the outermost energy level?
22. How does the oxidation number indicate if an electron will be lost or gained by the bonding atom?
23. Using the periodic table, what is the oxidation number of:
    a. calcium
    b. aluminum
    c. fluoride
24. What is the total electric charge on molecules and compounds?
25. Elements close to the noble gases tend to form what type of bond?
26. Elements that are widely separated on the periodic table tend to form____ compounds.
27. Elements that are close together on the periodic table tend to form ____ compounds.
28. Strong electron donors are on the ____ side of the periodic table, while strong electron acceptors are on the ____ side.

Problems

Section 8.1

1. Label each of the diagrams below as a mixture, compound, or separate elements.

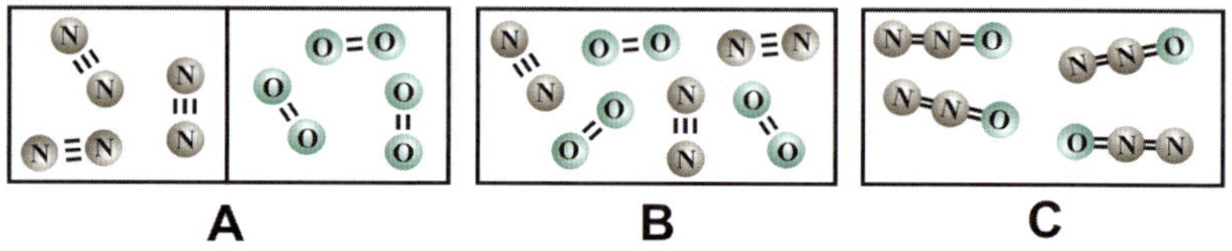

2. For each of the formulas for molecules listed below, name each element and how many of atoms of each element are in that molecule.
    a. $C_6H_{12}O_6$
    b. $CaCO_3$
    c. Al_2O_3
3. Predict the formula for a molecule containing carbon (C) with an oxidation number of 4+ and oxygen (O) with an oxidation number of 2-.
4. Which of the following would be a correct chemical formula for a molecule of N^{3-} and H^+?
    a. HNO_3
    b. H_3N_6
    c. NH_3

5. Referring to the diagram of the periodic table in chapter 7, determine which element in each pair is more active:
   a. Li or Be
   b. Ca or Sc
   c. P or S
   d. O or Ne

Section 8.2

6. In order for nitrogen to form a compound with other elements, how many additional electrons are required to give nitrogen the required number of electrons in its outermost energy level?

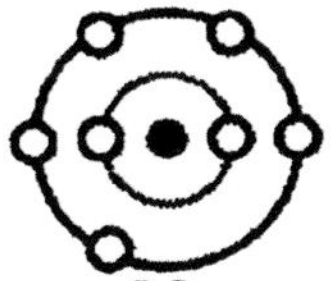

7. Draw Lewis dot diagrams for the following:
   a. An atom of hydrogen (one valence electron)
   b. An atom of oxygen (6 valence electrons).
   c. A molecule of water, H_2O.
   d. A molecule of carbon dioxide, CO_2.
8. Using the periodic table:
   a. Determine the oxidation number of Ca and Cl.
   b. Write the chemical formula for calcium chloride.
9. Give the most common oxidation number and how many electrons are gained or lost for the following elements:
   a. oxygen (O)
   b. boron (B)
   c. lithium (Li)
   d. potassium (Na)
   e. magnesium (Mg)
   f. aluminum (Al)
   g. carbon (C)
   h. iodine (I)
10. Use the periodic table to determine the type of bond most likely formed between the elements:
    a. carbon and oxygen
    b. lithium and fluorine
    c. carbon and carbon
    d. carbon and nitrogen
11. Carbon and oxygen combine to form a gas called carbon dioxide. Carbon (C) has an oxidation number of 4+ and oxygen (O) has an oxidation number of 2-.
    a. What is the total of the oxidation numbers for all the atoms in carbon dioxide?
    b. Predict the formula for carbon dioxide.
    c. Is carbon dioxide an ionic or covalently bonded compound?
12. Carbon and hydrogen combine to form a gas called methane. Carbon (C) has an oxidation number of 4+ and hydrogen has an oxidation number of 1-.
    a. What is the total of all the oxidation numbers for all the atoms in methane?
    b. Predict the formula for methane.
    c. Is methane an ionic or covalently bonded compound?
13. The chemical formula for a molecule of glucose is $C_6H_{12}O_6$.
    a. How many atoms of carbon are in a molecule of glucose?
    b. How many atoms of hydrogen are in a molecule of glucose?
    c. How many atoms of oxygen are in a molecule of glucose?

UNIT 4

Matter and Change

TRY THIS AT HOME

Why do pennies get dull over time instead of staying shiny? Put 1/4 cup white vinegar or lemon juice and 1 teaspoon of salt into a bowl and stir until the salt dissolves. Dip a dull penny halfway into the liquid and hold it there for ten to twenty seconds, then remove the penny.

What do you see? Pennies don't stay shiny because the copper slowly reacts with air to form a dull coating. What do you think happened to the dull coating when you dipped the penny in the vinegar and salt solution?

WARNING — This lab contains chemicals that may be harmful if misused. Read cautions on individual containers carefully. Not to be used by children except under adult supervision.

Chapter 9
Acids, Bases, and Solutions

Water is essential to all living things on Earth. Consider your body for example - it is about 65% water by weight. For every hour of vigorous exercise, you may lose as much as a half-gallon of your body's water supply through sweating and exhaling! You also lose small amounts of salts when you sweat. If the lost water and dissolved salts are not replaced, eventually your body will stop functioning. You can replace lost fluid by drinking water. To quickly replace salts, many athletes consume sports drinks. Why is water such an important substance for living creatures?

Sweat and sports drinks are both examples of solutions – they are mostly water with dissolved substances. In this chapter, you will learn about solutions. You will also learn about some special solutions called acids and bases. Among other things, acids create the bitter taste in food and can dissolve some rocks. Bases are slippery, like soap. Both acids and bases play a key role in maintaining your body's internal chemical balance.

Key Questions

1. ***What is the difference between 10, 14, and 24-karat gold?***

2. ***Why does salt dissolve in water but substances like chalk and sand do not?***

3. ***What are acids and bases?***

9.1 Water and solutions

Water is one substance that makes our planet unique. All life on Earth depends on this useful combination of hydrogen and oxygen atoms. In our solar system, only Earth has liquid water in such great abundance. Because we seem to have so much water, it is easy to take it for granted. Think about what you did yesterday. How often did you use water and how much? Now think about how yesterday would have been different if you didn't have fresh water!

Examples of solutions

A solution is homogeneous at the molecular level

A **solution** is a mixture of two or more substances that is uniform at the molecular level. Uniform means there are no clumps bigger than a molecule and the solution has the same ingredients everywhere. Grape soda is a solution you have probably consumed. All the particles in grape soda, from the flavor molecules to the color molecules, are evenly dispersed throughout the bottle (Figure 9.1).

An alloy is a solution of two or more solids

Although we often think of solutions as mixtures of solids in liquids, solutions exist in every phase; solid, liquid, or gas. Solutions of two or more solids are called **alloys**. Steel is an alloy (solution) of iron and carbon. Fourteen-karat gold is an alloy of silver and gold. "Fourteen-karat" means that 14 out of every 24 atoms in the alloy are gold atoms and the rest are silver atoms.

Muddy water is not a solution. Particles of soil are small, but still contain thousands of atoms and molecules. A true solution contains only individual molecules which are not clumped together into larger particles.

VOCABULARY

solution - a mixture of two or more substances that is uniform at the molecular level

alloy - a solution of two or more soilds.

Figure 9.1: *Examples of solutions.*

Solvents and solutes

Solvent and solute

A solution contains at least two components: a **solvent**, and a **solute**. The solvent is the part of a mixture that is present in the greatest amount. For example, the solvent in grape soda is water. The remaining parts of a solution (other than the solvent) are called solutes. Sugar, coloring dyes, flavoring chemicals, and carbon dioxide gas are solutes in grape soda.

Dissolving

When the solute particles are evenly distributed throughout the solvent, we say that the solute has **dissolved**. The picture shows a sugar and water solution being prepared. The solute (sugar) starts as a solid in the graduated cylinder on the left. Water is added and the mixture is carefully stirred until all the solid sugar has dissolved. Once the sugar has dissolved the solution is clear again.

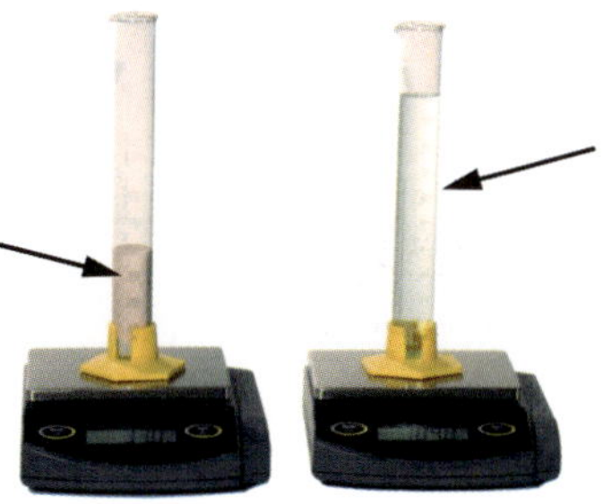

The molecular explanation for dissolving

On the molecular level, dissolving of a solid (like sugar) occurs when molecules of solvent interact with and separate molecules of solute (Figure 9.2). Most of the time, substances dissolve faster at higher temperatures. This is because higher temperature molecules have more energy and are more effective at knocking off molecules of solute. You may have noticed that sugar dissolves much faster in hot water than in cold water.

Why solutes are ground up into powder

Dissolving can only occur where the solvent can touch the solute. Most things that are meant to be dissolved, like salt and sugar, are ground up to a powder to increase the surface area. Increased surface area speeds dissolving because more solute is exposed to the solvent.

VOCABULARY

solvent - the component of a solution that is present in the greatest amount

solute - any component of a solution other than the solvent

dissolve - to separate and disperse a solid into individual molecules or ions in the presence of a solvent.

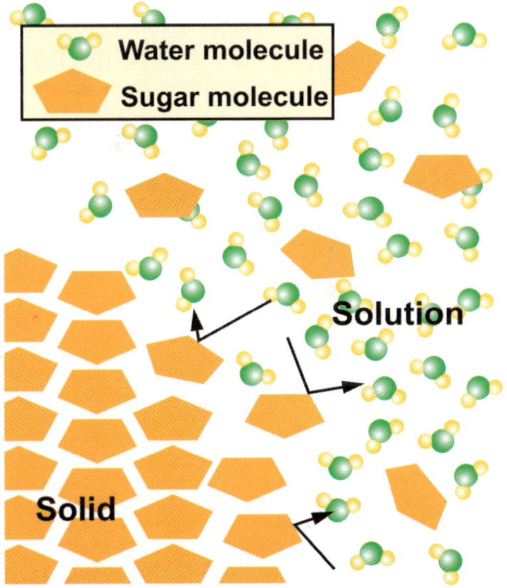

Figure 9.2: *The molecular explanation for a solid dissolving in a liquid. Molecules of solvent interact with and carry away molecules of solute.*

Solubility

What is solubility? The term **solubility** means the amount of solute (if any) that can be dissolved in a volume of solvent. Solubility is often listed in grams per 100 milliliters of solvent. Solubility is always given at a specific temperature since temperature strongly affects solubility. For example, Table 9.1 tells you that 200 grams of sugar can be dissolved in 100 milliliters of water at 25°C.

Insoluble substances do not dissolve Notice in Table 9.1 that chalk and talc do not have solubility values. These substances are **insoluble** in water because they do not dissolve in water. You can mix chalk dust and water and stir them all you want but you will still just have a mixture of chalk dust and water. The water will not separate the chalk dust into individual molecules because chalk does not dissolve in water.

Saturation Suppose you add 300 grams of sugar to 100 milliliters of water at 25°C? What happens? According to Table 9.1, 200 grams will dissolve in the water. *The rest will remain solid.* That means you will be left with 100 grams of solid sugar at the bottom of your solution. Any solute added in excess of the solubility does not dissolve. A solution is **saturated** if it contains as much solute as the solvent can dissolve. Dissolving 200 grams of sugar in 100 milliliters of water creates a saturated solution because no more sugar will dissolve.

How much salt will dissolve in water?

Seawater is a solution of water, salt and other minerals. How much salt can dissolve in 200 milliliters of water at 25°C?

1. Looking for: Grams of solute (salt)
2. Given: Volume (200 ml) and temperature of solvent
3. Relationships: 37.7 grams of salt dissolves in 100 milliliters of water at 25°C (Table 9.1)
4. Solution: If 37.7 grams dissolves in 100 milliliters then twice as much, or 75.4 grams will dissolve in 200 milliliters.

VOCABULARY

solubility - the amount of solute that can be dissolved in a specific volume of solvent under certain conditions. For example, 200 grams of sugar can be dissolved in 100 milliliters of water at 25°C.

insoluble - a substance is insoluble in a particular solvent if it does not dissolve in that solvent.

saturated - a solution is saturated if it contains as much solute as the solvent can dissolve.

Table 9.1: Solubility of some materials in water

Common name	Solubility at 25 °C (grams per 100 mL H_2O)
table salt ($NaCl$)	37.7
sugar ($C_{12}H_{22}O_{11}$)	200
baking soda ($NaHCO_3$)	approx. 10
chalk ($CaCO_3$)	insoluble
talc (Mg silicates)	insoluble

Concentration

How do you express solution concentration? In chemistry, it is important to know the exact **concentration** of a solution—that is the exact amount of solute dissolved in a given amount of solvent. The mass-percent is an accurate way to describe concentration. The concentration of a solvent in mass-percent is the mass of the solute divided by the total mass of the solution.

$$\text{Concentration} = \frac{\text{mass of solute}}{\text{total mass of solution}} \times 100\%$$

Mass percent example Suppose you dissolve 10.0 grams of sugar in 90.0 grams of water. What is the mass percent of sugar in the solution (Figure 9.3)?

$$\text{Concentration} = \frac{10\text{g sugar}}{(10\text{g} + 90\text{g})\ \text{solution}} \times (100\%) = 10\%$$

Describing very low concentrations Parts per million (ppm), parts per billion (ppb), and parts per trillion (ppt) are commonly used to describe very small concentrations of pollutants in the environment. These terms are measures of the ratio (by mass) of one material in a much larger amount of another material. For example, a pinch (gram) of salt in 10 tons of potato chips is about 1 g salt per billion g chips, or a concentration of 1 ppb.

Concentration

How many grams of salt do you need to make 500 grams of a solution with a concentration of 5% salt?

1. Looking for:	mass of salt (solute)
2. Given:	concentration (5%) and total mass of solution (500 g)
3. Relationships:	concentration = mass of solute ÷ total mass of solution
4. Solution:	0.05 = mass of salt ÷ 500g mass of salt = 0.05 × 500g = 25 grams

VOCABULARY

concentration - the ratio of solute to solvent in a solution. For example, a 10% sugar solution contains 10 g of sugar for every 90 g of water.

Figure 9.3: *Preparing a sugar solution with a concentration of 10%.*

CHALLENGE

Lead is toxic to humans and therefore there are limits on the allowable concentration of lead in drinking water. What is the maximum concentration of lead in drinking water allowed by the Environmental Protection Agency (EPA)?

Equilibrium and supersaturation

Dissolving and un-dissolving

When a solute like sugar is mixed with a solvent like water, *two* processes are actually going on continuously.

- Molecules of solute dissolve and go into solution
- Molecules of solute come out of solution and become "un-dissolved"

When the concentration is lower than the solubility, the dissolving process puts molecules into solution faster than they come out. The concentration increases and the mass of un-dissolved solute decreases. However, dissolving and un-dissolving are still going on!

Equilibrium concentration

The more molecules that are in solution (higher concentration) the faster molecules come out of solution. As the concentration increases, the un-dissolving process also gets faster until the dissolving and un-dissolving rates are exactly equal. When the rate of dissolving equals the rate of coming out of solution, we say **equilibrium** has been reached. At equilibrium, a solution is *saturated* because the concentration is as high as it can go.

Supersaturation

According to the solubility table in Figure 9.4, at 80°C, 100 g of water reaches equilibrium with 360 grams of dissolved sugar. At lower temperatures, less sugar can dissolve. What happens if we cool the saturated solution? As the temperature goes down, sugar's solubility also goes down and the solution becomes **supersaturated**. A supersaturated solution means there is more dissolved solute than the maximum solubility.

Growing crystals

A supersaturated solution is unstable. The excess solute comes out of solution and returns to its un-dissolved state. This is how the large sugar crystals of rock candy are made. Sugar is added to boiling water until the solution is saturated. As the solution cools, it becomes supersaturated. Solid sugar crystals form as the sugar comes out of the supersaturated solution.

ă VOCABULARY

equilibrium - occurs when a solution has the maximum concentration of dissolved solute; the dissolving rate equals the rate at which molecules come out of solution (un-dissolve).

supersaturated - a concentration greater than the maximum solubility.

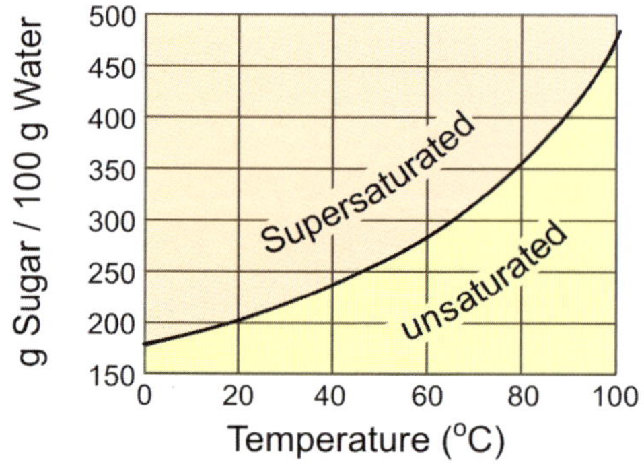

Temp (°C)	g Sugar / 100 g H_2O	Temp (°C)	g Sugar / 100 g H_2O
0	177	50	259
10	189	60	284
20	204	70	318
30	219	80	360
40	238	90	410

Figure 9.4: *The process for making rock candy uses a supersaturated solution of sugar in water.*

The solubility of gases and liquids

Gas dissolves in water

Gases can also dissolve in liquids. When you drink carbonated soda, the fizz comes from dissolved carbon dioxide gas (CO_2). Table 9.2 lists the solubility of CO_2 as 1.74 grams per kilogram of water at room temperature and atmospheric pressure (1 atm).

Solubility of gas increases with pressure

The solubility of gases in liquids increases with pressure. Soda is fizzy because the carbon dioxide was dissolved in the liquid at high pressure. When you pop the tab on a can of soda, you release the pressure. The solution immediately becomes supersaturated, causing the CO_2 to bubble out of the water and fizz.

Dissolved oxygen

Table 9.2 also shows that 0.04 grams of oxygen dissolves in a kilogram of water. Dissolved oxygen keeps fish and other underwater animals alive (Figure 9.5). Just like on land, oxygen is produced by underwater plants as a by-product of photosynthesis.

Solubility of gas decreases with temperature

When temperature goes up, the solubility of gases in liquid goes down (Figure 9.6). When the water temperature rises, the amount of dissolved oxygen decreases. Less dissolved oxygen means less oxygen for fish. When the weather is warm, fish stay near the bottom of ponds and rivers where there is cooler, more oxygenated water.

Solubility of liquids

Some liquids, such as alcohol, are soluble in water. Other liquids, such as corn oil, are not soluble in water. Oil and vinegar (water solution) salad dressing separates because oil is not soluble in water. Liquids that are not soluble in water may be soluble in other solvents. For example, vegetable oil is soluble in mineral spirits, a petrolem-based solvent used to thin paints.

Table 9.2: Solubility of gases in water at 21°C and 1 atm.

Gas	Solubility
Oxygen (O_2)	0.04 g/kg water
Nitrogen (N_2)	0.02 g/kg water
Carbon dioxide (CO_2)	1.74 g/kg water

Figure 9.5: *Fish and other aquatic life are sustained by dissolved oxygen in water.*

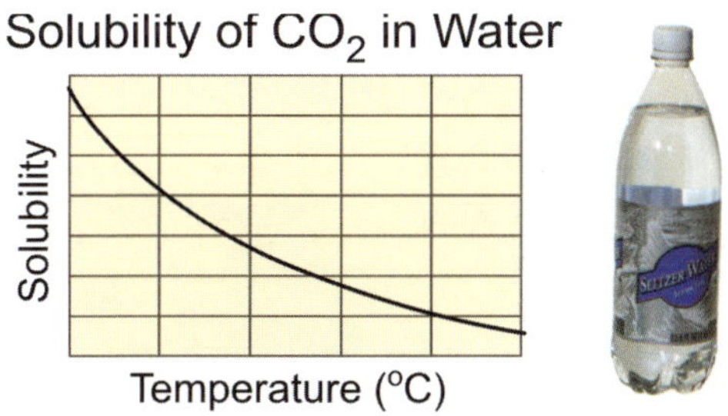

Figure 9.6: *The solubility of gases in water decreases as temperature increases.*

Water as a solvent

The universal solvent Water is often called the "universal solvent." While water doesn't dissolve everything, it does dissolve many different types of substances such as salts and sugars. Water is a good solvent because of the way the H_2O molecule is shaped (Figure 9.7).

Water is a polar molecule A water molecule has a negative end (pole) and a positive end. This is because electrons are shared unequally; pulled toward the oxygen atom and away from the two hydrogen atoms. The oxygen side of the molecule has a partially negative charge and the hydrogen side of the molecule has a partially positive charge. A molecule (like water) with a charge separation is called a **polar** molecule.

How water dissolves salt The polar molecules of water dissolve many ionic compounds. Suppose a sodium chloride (table salt) crystal is mixed with water. The polar water molecules surround the sodium and chlorine atoms in the crystal. This causes the ions in the crystal to separate. Because opposites attract, the negative ends of the water molecules are attracted to the Na+ ions and the positive ends are attracted to the Cl- ions. Water molecules surround the Na+ and Cl- ions and make a solution (Figure 9.8).

Water dissolves many molecular compounds When sucrose is mixed with water, the individual molecules of sucrose become separated from each other and are attracted to the opposite poles of the water molecules. Because sucrose is a covalent compound, the sucrose molecules do not dissociate into ions but remain as neutral molecules in the solution.

Water does not dissolve oils

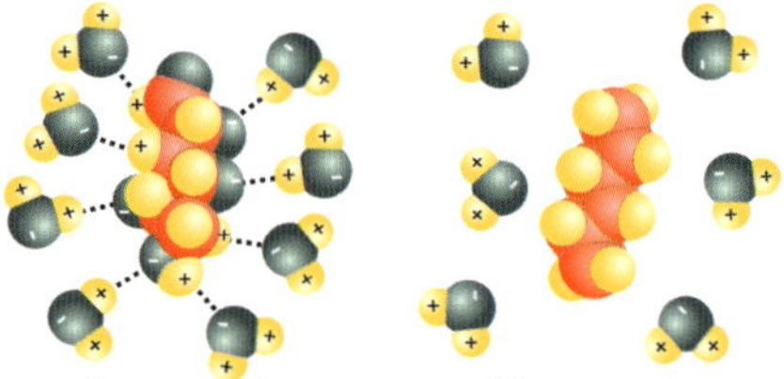

Oil does not dissolve in water because water is a polar molecule and oil molecules are nonpolar. In general, like dissolves like: water dissolves polar substances and non-polar solvents (like mineral spirits) dissolve non-polar substances.

VOCABULARY

polar - describes a molecule that has charge separation, like water.

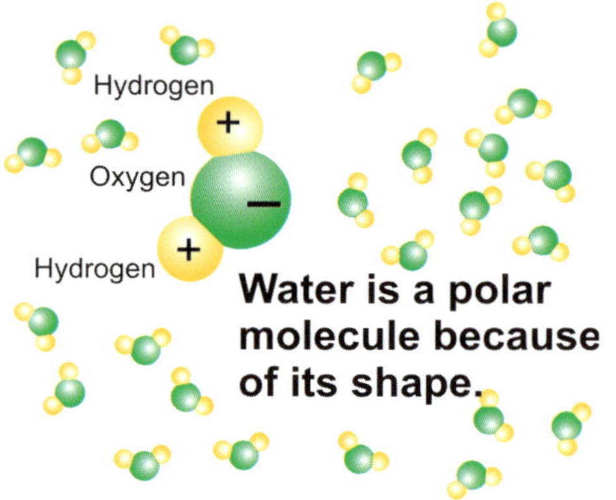

Figure 9.7: *Water is a polar molecule because it has a negative pole and a positive pole.*

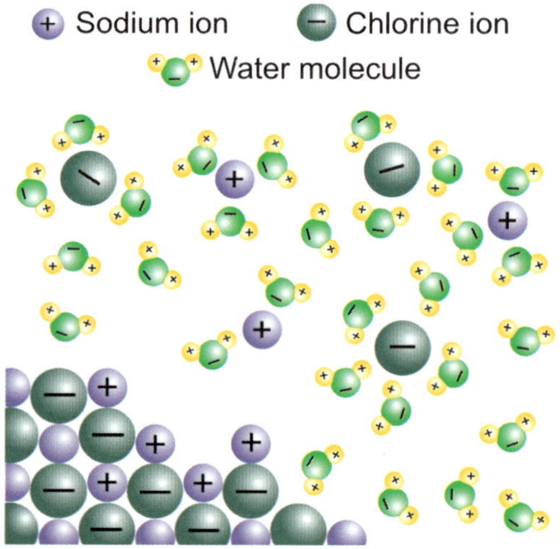

Figure 9.8: *Water dissolves sodium chloride to form a solution of sodium (+) and chlorine (-) ions.*

9.1 Section Review

1. One of the following is NOT a solution. Choose the one that is *not* a solution and explain why.
   a. steel
   b. ocean water
   c. 24-karat gold
   d. muddy water
   e. orange soda
2. For each of the following solutions, name the solvent and the solute.
   a. saltwater
   b. seltzer water (hint: what causes the fizz?)
   c. lemonade made from powdered drink mix
3. Give an example of a solution in which the solute is *not* a solid and the solvent is *not* a liquid.
4. When can you say that a solute has dissolved?
5. Does sugar dissolve faster in cold water or hot water? Explain your answer.
6. Jackie likes to put sugar on her breakfast cereal. When she has eaten all of the cereal, there is some cold milk left in the bottom of the bowl. When she dips her spoon into the milk, she notices a lot of sugar is sitting at the bottom of the bowl. Explain what happened in terms of saturation.
7. Describe exactly how you would make 100 grams of a saltwater solution that is 20% salt. In your description, tell how many grams of salt and how many grams of water you would need.
8. Why is water often called the "universal solvent"?

CHALLENGE

Larry opens a new bottle of soda. He quickly stretches a balloon over the opening of the bottle. As he gently shakes the bottle, the balloon expands! Explain what is happening to cause the balloon to expand. Use at least three vocabulary words from this section. Draw a diagram to illustrate your explanation.

9.2 Acids, Bases, and pH

Acids and bases are among the most familar of all chemical compounds. Some of the acids you may have encountered include acetic acid (found in vinegar), citric acid (found in orange juice), and malic acid (found in apples). You may be familiar with some bases including ammonia in cleaning solutions and magnesium hydroxide found in some antacids. The pH scale is used to describe whether a substance is an acid or a base. This section is about properties of acids and bases, and how the pH scale works.

VOCABULARY

acid - a substance that produces hydronium ions (H_3O^+) when dissolved in water. Acids have a pH less than 7.

What are acids?

Properties of acids

An **acid** is a compound that dissolves in water to make a particular kind of solution. Some properties of acids are listed below and some common acids are shown in Figure 9.9. Note: you should NEVER taste a laboratory chemical!

- Acids create the sour taste in food, like lemons.
- Acids react with metals to produce hydrogen gas (H_2).
- Acids change the color of blue litmus paper to red.
- Acids can be very corrosive, destroying metals and burning skin through chemical action.

Acids make hydronium ions

Chemically, an acid is any substance that produces *hydronium ions* (H_3O^+) when dissolved in water. When hydrochloric acid (HCl) dissolves in water it ionizes, splitting up into hydrogen (H+) and chlorine (Cl-) ions. Hydrogen ions (H+) are attracted to the negative oxygen end of a water molecule, combining to form hydronium ions.

What an acid does in water

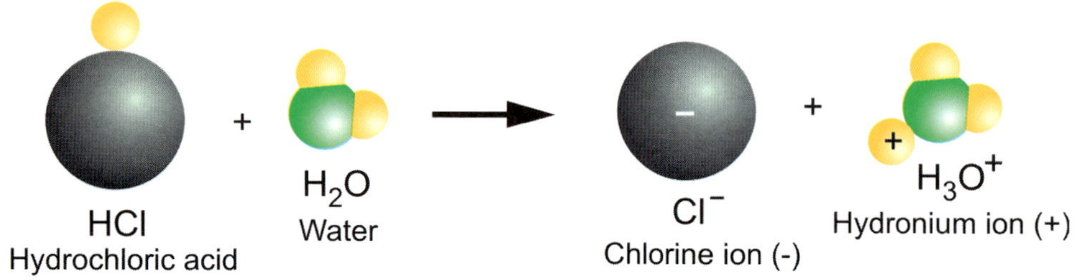

Figure 9.9: *Some weak acids you may have around your home.*

Bases

Properties of bases

A **base** is a compound that dissolves in water to make a different kind of solution, opposite in some ways to an acid. Some properties of bases are listed below and shown in Figure 9.10.

- Bases create a bitter taste.
- Bases have a slippery feel, like soap
- Bases change the color of red litmus paper to blue.
- Bases can be very corrosive, destroying metals and burning skin through chemical action.

Bases produce hydroxide ions

A base is any substance that dissolves in water and produces *hydroxide ions* (OH^-). A good example of a base is sodium hydroxide ($NaOH$), found in many commercial drain cleaners. This compound dissociates in water to form sodium (Na^+) and hydroxide ions:

What a base does in water

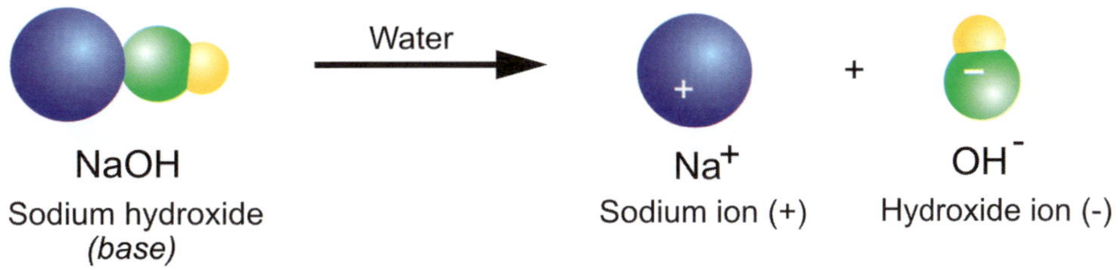

Ammonia is a base

Ammonia (NH_3), found in cleaning solutions, is a base because it dissociates in water to form hydroxide ions. Notice that a hydroxide ion is formed when ammonia *accepts* H^+ ions from water molecules in solution as shown below. How is this different than $NaOH$?

What ammonia (base) does in water

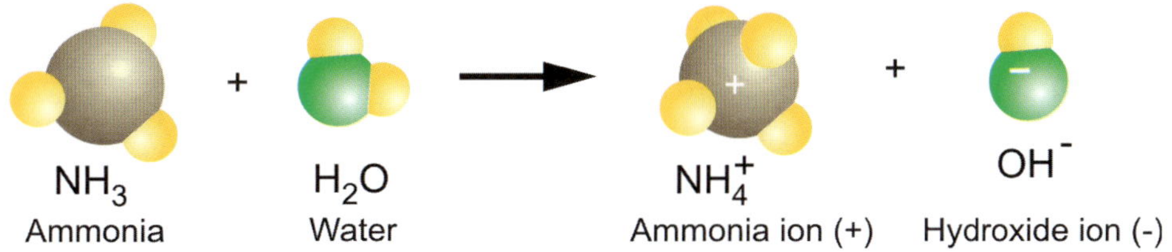

VOCABULARY

base - a substance that produces hydroxide ions (OH -) when dissolved in water. Bases have a pH greater than 7.

Figure 9.10: *Some common bases.*

Strength of acids and bases

The strength of acids The strength of an acid depends on the concentration of hydronium ions the acid produces when dissolved in water. Hydrochloric acid (HCl) is a *strong acid* because HCl completely dissolves into H^+ and Cl^- ions in water. This means that every molecule of HCl that dissolves produces one hydronium ion.

Acetic acid is a weak acid Acetic acid ($HC_2H_3O_2$), (in vinegar), is an a *weak acid*. When dissolved in water, only a small percentage of acetic acid molecules ionize (break apart) and become H^+ and $C_2H_3O_2^-$ ions. This means that a small number of hydronium ions are produced compared to the number of acetic acid molecules dissolved (Figure 9.11).

The strength of bases The strength of a base depends on the relative amount of hydroxide ions (OH^-) produced when the base is mixed with water. Sodium hydroxide (NaOH) is considered a strong base because it dissociates completely in water to form Na^+ and OH^- ions. Every unit of NaOH that dissolves creates one OH^- ion (Figure 9.12). Ammonia (NH_3) on the other hand, is a weak base because only a few molecules react with water to form NH_4^+ and OH^- ions.

Water can be a weak acid or a weak base One of the most important properties of water is its ability to act as both an acid and as a base. In the presence of an acid, water acts as a base. In the presence of a base, water acts as an acid. In pure water, the H_2O molecule ionizes to produce both hydronium and hydroxide ions. This reaction is called the *dissociation of water*.

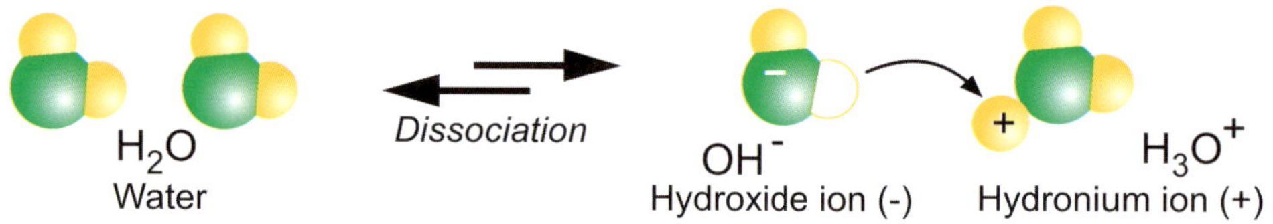

What does the double arrow mean? The double arrow in the equation means that the dissociation of water can occur in *both* directions. This means that water molecules can ionize and ions can also form water molecules. However, water ionizes so slightly that most water molecules exist whole, not as ions.

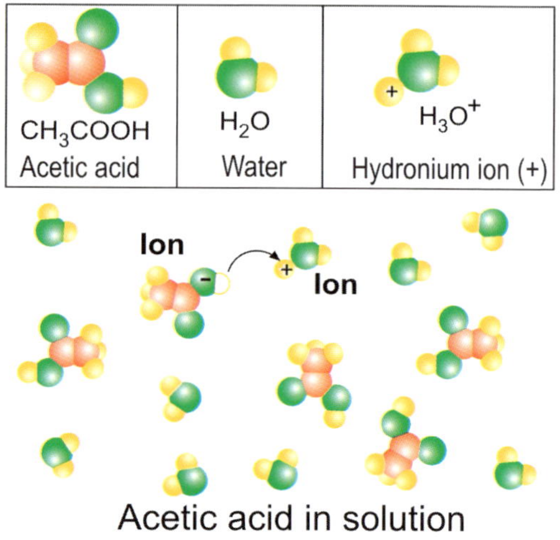

Figure 9.11: *Acetic acid dissolves in water, but only a few molecules ionize (break apart) to create hydronium ions.*

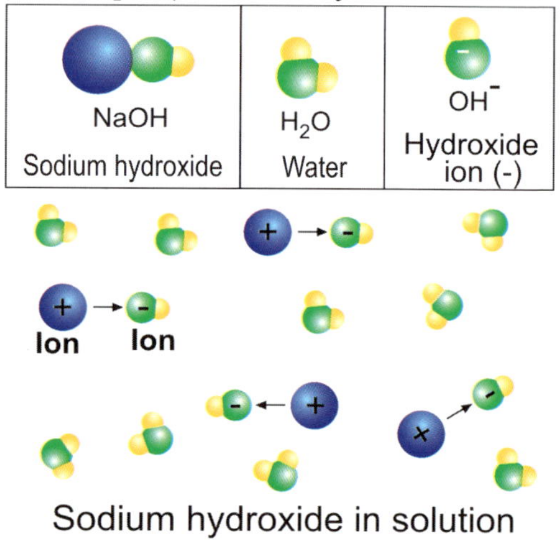

Figure 9.12: *Sodium hydroxide (NaOH) is a strong base because every NaOH unit contributes one hydroxide (OH^-) ion.*

pH and the pH scale

What is pH? The pH scale describes the concentration of hydronium ions in a solution. The **pH scale** ranges from 0 to 14. A pH of 7 is neutral, neither acidic nor basic. Distilled water has a pH of 7.

Acids have a pH less than 7. A concentrated solution of a *strong acid* has the *lowest* pH. Strong hydrochloric acid has a pH of 1. Seltzer water is a weak acid at a pH of 4. Weaker acids have a pH nearer to 7.

A base has a pH greater than 7. A concentrated solution of a *strong base* has the *highest* pH. A strong sodium hydroxide solution can have a pH close to 14. Weak bases such as baking soda have pH closer to 7.

The pH scale

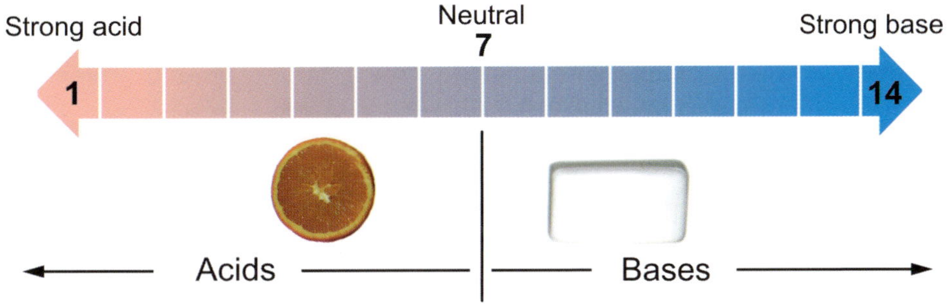

The pH of common substances Table 9.3 lists the pH of some common substances. It turns out that many foods we eat or ingredients we use for cooking are acidic. On the other hand, many of our household cleaning products are bases.

pH indicators Certain chemicals turn different colors at different pH. These chemicals are called pH indicators and they are used to determine pH. The juice of boiled red cabbage is a pH indicator that is easy to prepare. Red cabbage juice is deep purple and turns various shades ranging to yellow at different values of pH. Litmus paper is another pH indicator that changes color (Figure 9.13).

VOCABULARY

pH - pH measures the acidity of a solution

pH scale - the pH scale goes from 1 to 14 with 1 being very acidic and 14 being very basic. Pure water is neutral with a pH of 7.

Table 9.3: The pH of some common chemicals.

Household chemical	Acid or base	pH
lemon juice	acid	2
vinegar	acid	3
soda water	acid	4
baking soda	base	8.5
bar soap	base	10
ammonia	base	11

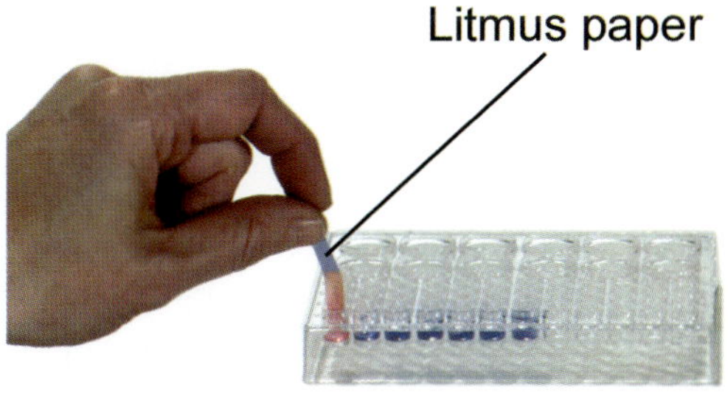

Figure 9.13: *Red and blue litmus paper are pH indicators that test for acid or base.*

pH in the environment

The best pH for plants

The pH of soil directly affects nutrient availability for plants. Most plants prefer a slightly acidic soil with a pH between 6.5 and 7.0. Azaleas, blueberries, and conifers grow best in more acid soils with a pH of 4.5 to 5.5 (Figure 9.14). Vegetables, grasses and most other shrubs do best in less acidic soils with a pH range of 6.5 to 7.0.

Effects of pH too high or low

In highly acid soils (pH below 4.5) too much aluminum, manganese and other elements may leach out of soil minerals and reach concentrations that are toxic to plants. Also, at these low pH values, calcium, phosphorus and magnesium are less available to plant roots. At pH values of 6.5 and above, (more basic), iron and manganese become less available.

pH and fish

The pH of water directly affects aquatic life. Most freshwater lakes, streams, and ponds have a natural pH in the range of 6 to 8. Most freshwater fish can tolerate pH between 5 and 9 although some negative effects appear below pH of 6. Trout (like the california golden) are among the most pH tolerant fish and can live in water with a pH from 4 to 9.5.

pH and amphibians

Frogs and other amphibians are even more sensitive to pH than fish. The California tree frog, and other frogs prefer pH close to neutral and don't survive below pH of 5.0. Frogs eggs develop and hatch in water with no protection from environmental factors. Research shows that even pH below 6 has a negative effect on frog hatching rates.

pH ranges in nature

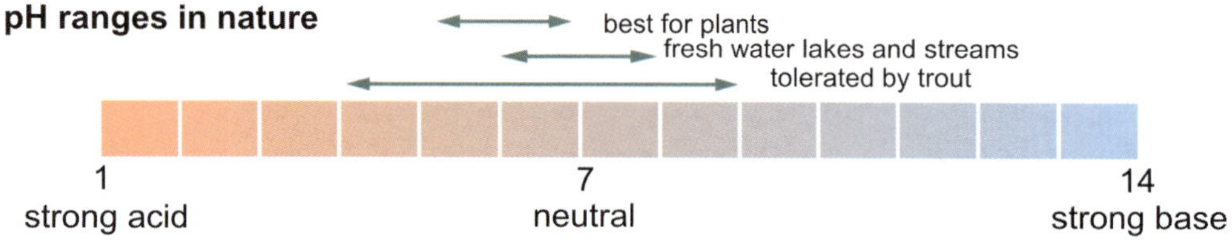

CHALLENGE

Test your soil pH

Many garden centers offer soil testing kits. These kits are specially designed to measure the pH of soil. Measure the soil around your house or school. What sorts of plants would thrive in the pH range is that you measure?

Figure 9.14: *Blueberries grow best in acid soils that have a pH between 5.0 and 5.5.*

Acids and bases in your body

Acids and bases play a role in digestion

Many reactions, such as the ones that occur in your body, work best at specific pH values. For example, acids and bases are very important in the reactions involved in digesting food. As you may know, the stomach secretes hydrochloric acid (HCl), a strong acid (pH 1.4). The level of acidity in our stomachs is necessary to break down the protein molecules in food so they can be absorbed. A mucus lining in the stomach protects it from the acid produced (Figure 9.15).

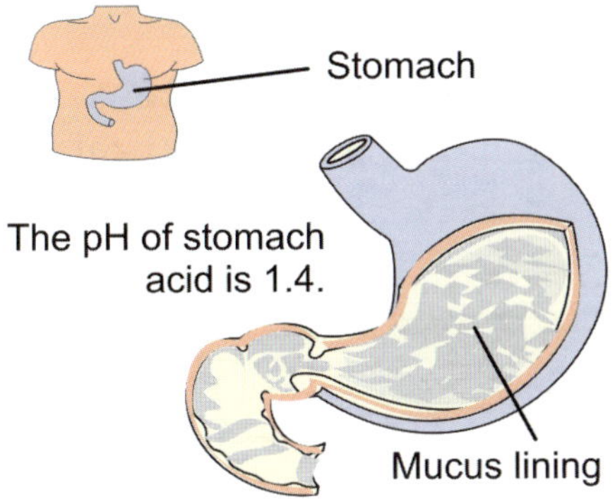

Figure 9.15: *The stomach secretes a strong acid (HCl) to aid with food digestion. A mucus lining protects the stomach tissue from the acid.*

Ulcers and heartburn

Very spicy foods, stress, or poor diet can cause the stomach to produce too much acid, or allow stomach acid to escape from the stomach. An *ulcer* may occur when the mucus lining of the stomach is damaged. Stomach acid can then attack the more sensitive tissues of the stomach itself. Infections by the bacteria *h. pylori* can also damage the mucus lining of the stomach, leading to ulcers. The uncomfortable condition called *heartburn* is caused by excessive stomach acid backing up into the esophagus. The esophagus is the tube that carries food from your mouth to your stomach. The esophagus lacks the mucus lining of the stomach and is sensitive to acid. Eating very large meals can lead to heartburn because an overflowing stomach pushes acid up into the esophagus.

pH and your blood

Under normal conditions the pH of your blood is within the range of 7.3 - 7.5, close to neutral but slightly basic. Blood is a watery solution that contains many solutes including the dissolved gases carbon dioxide (CO_2) and oxygen. Dissolved CO_2 in blood produces a weak acid. The higher the concentration of dissolved CO_2, the more acidic your blood becomes.

Blood pH is controlled through breathing

Your body regulates the dissolved CO_2 level by breathing. For example, if you hold your breath, more carbon dioxide enters your blood and the pH falls as your blood becomes more acid. If you hyperventilate, less carbon dioxide enters your blood and the opposite happens. Blood pH starts to rise, becoming more basic. Your breathing rate regulates blood pH through these chemical reactions (Figure 9.16).

Figure 9.16: *Under normal conditions, your blood pH ranges between 7.35 and 7.45. Holding your breath causes blood pH to drop. High blood pH can be caused by hyperventilating.*

Neutralization reactions

Neutralization

When an acid and a base are combined, they neutralize each other. Neutralization occurs when the positive ions from the base combine with the negative ions from the acid. This process also goes on in your body. As food and digestive fluids leave the stomach, the pancreas and liver produce bicarbonate (a base) to neutralize the stomach acid. Antacids such as sodium bicarbonate have the same effect.

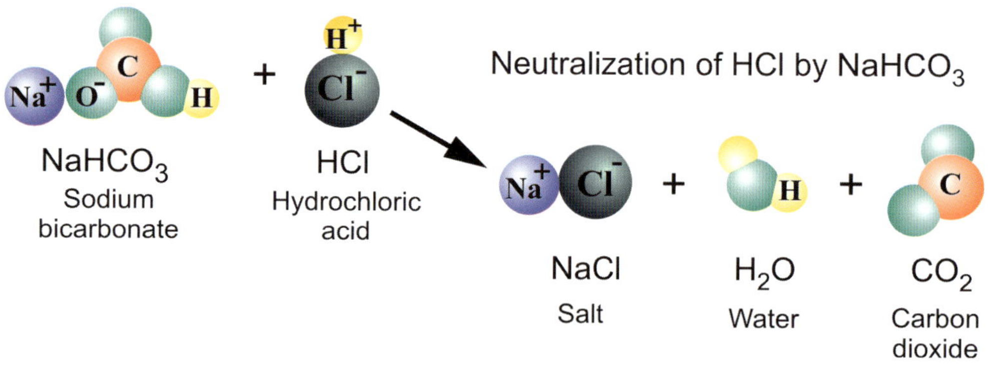

Adjusting soil pH

Neutralization reactions are important in gardening and farming. For example about 1/4 of the yards in the US have soil which is too acidic (pH less than 5.5) to grow grass very well. For this reason, many people add *lime* to their yard every spring. A common form of lime is ground-up calcium carbonate ($CaCO_3$) made from natural crushed limestone. Lime is a weak base and undergoes a neutralization reaction with acids in the soil to raise the pH.

Neutralization of acid in soil

For example, sulfuric acid (H_2S) in soil reacts with the calcium carbonate to form the salt calcium sulfate ($CaSO_4$) also known as gypsum. Sulfuric acid is in acid rain and is created in the atmosphere from pollutants in the air. Many of the walls of buildings and homes are made with "plaster board" which is a sheet of gypsum (plaster) covered with paper on both sides.

CHALLENGE

Test your soil

Almost any garden center carries soil test kits. These kits have pH test papers inside and are designed to help gardeners measure the pH of their soil.

Get a soil test kit and test samples of soil from around your home or school. Repeat the test taking new soil samples after a rainfall to see if the pH changes. See if you can answer these questions:

- What kinds of plants thrive in the pH of the soil samples you tested?
- Is the soil the proper pH for the plants you found where you took your soil samples?
- What kinds of treatments are available at your local garden center for correcting soil pH?

9.2 Section Review

1. List three ways that acids and bases are different.
2. Many foods are acidic. List four examples.
3. Answer these questions about water:
    a. Is water an acid, a base, neither, or both?
    b. What is the pH of water?
4. Nadine tests an unknown solution and discovers that it turns blue litmus paper red, and it has a pH of 3.0. Which of the following could be the unknown solution?
    a. sodium hydroxide
    b. vinegar
    c. ammonia
    d. soap
5. What makes a strong acid strong?
6. What makes a strong base strong?
7. Give two examples of a pH indicator.
8. Describe in your own words how the amount of carbon dioxide dissolved in your blood affects your blood pH.
9. Two years ago, you joined a project to study the water quality of a local pond. During the second spring, you notice that there are not as many tadpoles (first stage in frog development) as there were last year (Figure 9.17). You want to know if the number of tadpoles in the pond is related to the pH of the pond. The records that document the water quality and wildlife started ten years ago. Describe the steps you would take to determine whether a change in the pH of the pond water is affecting the population of frogs and their ability to reproduce.

Figure 9.17: *Because their young develop underwater, outside the female's body, frogs and other amphibians are very sensitive to the pH of their environment.*

HEALTH CONNECTION

Keeping Your (pH) Balance

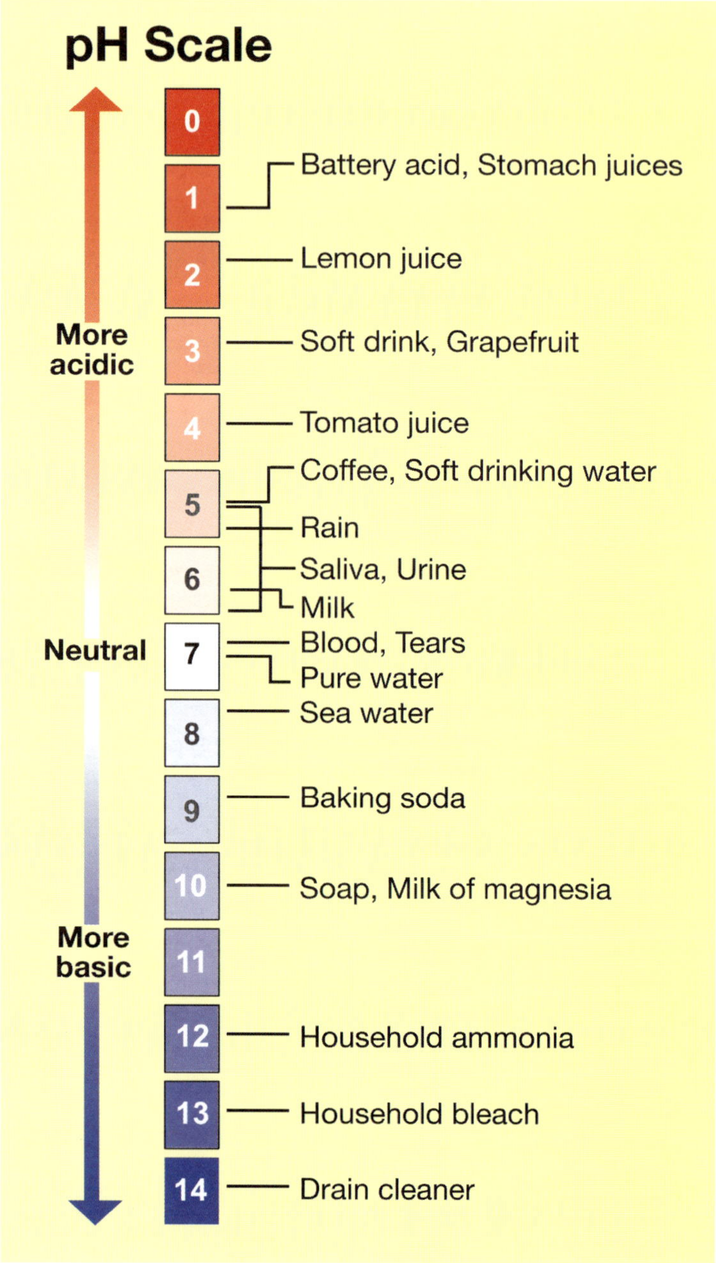

You know that solutions have a pH. A test paper dipped in a solution will show whether it is an acid (acidic) or a base (alkaline). Did you know the same is true of the blood that courses through your body? Our bodies are constantly adjusting to keep our blood pH in a normal range between 7.35 and 7.45. Yes, we are slightly alkaline.

Acids and bases are everywhere. Many of our favorite foods are acidic: Lemons and oranges, for instance, contain citric acid. We depend on gastric acid in our stomachs to digest our food - and if we eat or drink unwisely, we suffer stomach pains caused by that same gastric acid. Then to counter the acid, we have antacids and baking soda.

Imbalances

The human body's many different processes produce a great deal of acid, which then must be removed. For example, our lungs can dispose of excess acid; carbon dioxide can form carbonic acid by causing us to breathe faster. Our kidneys remove excess acid from the blood and dispose of it in urine. But disease or extreme conditions can interfere with the body's self-adjusting system.

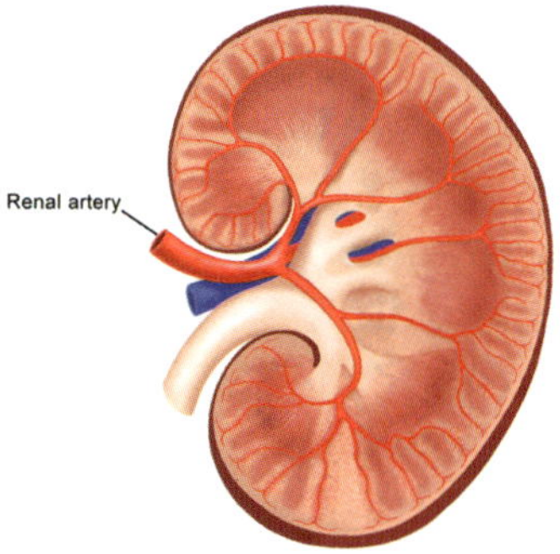

There are two types of imbalance. We can have too much acid in our body fluids, or those fluids can be too alkaline. Too much acid is called acidosis and too much alkali is called alkalosis.

These imbalances are either respiratory or metabolic. When the lungs are not functioning properly, the imbalance is respiratory. When the body's physical and chemical processing of substances is not functioning properly, the imbalance is metabolic.

Acidosis and its causes

Respiratory acidosis occurs when the lungs cannot remove all of the carbon dioxide produced by the body. As a result, body fluids become too acidic. This can be caused by almost any lung disease, such as asthma, or by a deadly habit like cigarette smoking.

Treatment may include drugs that expand the air passages in the lungs. Inhaled oxygen may be used to raise the oxygen level in the blood. Stopping smoking is a given among these attempts to restore the body's pH balance.

Metabolic acidosis is a pH imbalance in which the body has too much acid. The body does not have enough bicarbonate needed to neutralize the excess acid. This can be caused by a disease like diabetes, or by severe diarrhea, heart or liver failure, kidney disease, or even prolonged exercise.

A result of prolonged exercise is a buildup of lactic acid, which causes the blood to become acidic. Fluids can restore the body's pH balance, which is why various sports drinks are popular among athletes. Those drinks are formulated to help the body maintain its pH balance under stress.

Alkalosis and its causes

The opposite of acidosis, alkalosis is the result of too much base in the body's fluids. Respiratory alkalosis is caused by hyperventilation, that is, extremely rapid or deep breathing that makes the body lose too much carbon dioxide. It can be provoked by anxiety. In such a case, the person may breathe (or be helped to breathe) into a paper bag. Why? Because the bag retains the exhaled carbon dioxide and it can be taken back in. Altitude or any disease that causes the body to lose carbon dioxide may also cause hyperventilation. Metabolic alkalosis is a result of too much bicarbonate in the blood. Other types of alkalosis are caused by too little chloride or potassium. Alkalosis symptoms include confusion, muscle twitching or spasms, hand tremors, nausea, and lightheadedness.

Balancing act: food, drink, exercise, calm, acid, base ...

By nature our slightly alkaline pH needs to remain balanced there. Yet, what we eat and drink changes our pH. If you eat a lot of meat such as hamburgers, steak, and chicken, your body produces more acid than someone who eats a lot of vegetables and fruits. If we don't balance what we eat, the body has to rely on reserves. For example, if you eat a lot of meat and no vegetables, your pH becomes acidic. Your kidneys can handle only so much acid and, in this case, the reserve the body would use is bicarbonate from your bones to help neutralize the acid level.

This is just one more instance in which the food we eat can affect our bodies in many ways. Maintaining a balanced diet is the first step toward good health—a little on the alkaline side.

Questions:

1. What two organs regulate the acid-base balance?
2. What is a common cause of hyperventilation?
3. How is the alkalosis caused by hyperventilation treated?
4. Name a leading cause of respiratory acidosis.

CHAPTER ACTIVITY Acid Rain and Stone Structures

Acid rain resulting from air pollution is a growing problem in the industrialized world. It can have devastating effects on the pH of our lakes. Interestingly, while the pH of some lakes has dropped dramatically in recent years, the pH of some nearby lakes during the same time period has changed very little. The type of rock that is found beneath and around the body of water is is what makes the difference. Calcium carbonate, which is found in marble and limestone, has the ability to neutralize acid rain while other types of rocks and minerals have no effect. In this activity, we will make solutions of water and soak chips of different types of rock to see their effect on a dilute acid.

We will measure the pH of each solution using an indicator solution. The indicator solution appears red at a pH of 4, orange to yellow at a pH of 6.5, green to blue at a pH of 9, and violet to red-violet at pH of 10.

Materials:

Chalk and/or marble chips, beakers, granite chips, white vinegar, and Universal Indicator solution

What you will do

1. Place a small sample of marble chips in two 50 mL beakers.
2. Place a small sample of granite chips in one other beaker.
3. Be sure to keep some of the original indicator solution as a control.
4. Add 10 mL of vinegar-Universal Indicator solution to each.
5. Record the time and color of each solution according to the data table below.

Applying your knowledge

a. Compare the pH changes of the solutions containing granite and limestone.

b. Does the solution become more or less acidic as time passes?

c. What affect do you think acid rain has on marble statues?

d. Spelunking (cave exploration) is very popular in Ireland due to the large deposits of limestone. Explain how these caves could have formed.

Time	Color of solution (marble)	Approx. pH (marble)	Color of Solution (granite)	Approx. pH (granite)
0 min.				
5 min.				
10 min.				
20 min.				
30 min.				
40 min.				
overnight				

Chapter 9 Assessment

Vocabulary

Select the correct term to complete the sentences.

solvent	acid	solute
equilibrium	pH	concentration
base	solubility	polar
alloy	solution	supersaturated

Section 9.1

1. The substance that dissolves particles in a solution is called the ____.
2. The substance that is dissolved in a solution is called the ____.
3. A mixture of two or more substances that is uniform at the molecular level is called a(n) ____.
4. A solution of two or more metals is known as a(n) ____.
5. A water molecule is an example of a(n) ____ molecule.
6. When the dissolving rate equals the rate at which molecules come out of solution, the solution is in ____.
7. The exact amount of solute dissolved in a given amount of solvent is the ____ of a solution.
8. A(n) ____ solution has a concentration greater than the maximum solubility.

Section 9.2

9. A substance that produces hydronium ions (H_3O^+) in solution is called a(n) ____.
10. A substance that produces hydroxide ions (OH^-) in solution is called a(n) ____.
11. ____ measures the acidity of a solution.

Concepts

Section 9.1

1. What would happen to the solubility of potassium chloride in water as the water temperature increased from 25°C to 75°C?
2. What are two ways to increase the dissolving rate of sugar in water?
3. Water is described as a polar molecule because it has:
   a. a positive and a negative pole.
   b. two positive poles.
   c. two negative poles.
   d. no charge.
4. Water is a solvent in which of the following solutions?
   a. Air
   b. Liquid sterling silver
   c. Saline (salt) solution
5. Very small concentrations are often reported in ppm. What does "ppm" stand for? Give three examples of concentrations that are described in "ppm".
6. How would the fish in a lake be affected if large amounts of hot water from a power plant or factory were released into the lake?
7. When you open a can of room-temperature soda, why is it more likely to fizz an spill over than a can that has been refrigerated?
8. What happens to a supersaturated solution when more solute is added?

Section 9.2

9. What determines the strength of an acid?

10. What determines the strength of a base?
11. What is the pH of a neutral solution?
12. Indicate whether the following properties belong to an acid (**A**), a base (**B**), or both (**AB**):
    a. ____ Creates a sour taste in food.
    b. ____ Creates a bitter taste in food.
    c. ____ Changes the color of red litmus paper to blue.
    d. ____ Changes the color of blue litmus paper to red.
    e. ____ Can be very corrosive.
13. When hydroxide ions are added to a solution, does the pH increase or decrease?
14. Your stomach uses hydrochloric acid to break down the protein molecules in the food you eat. Give two reasons why this acid doesn't destroy your stomach and intestines during digestion.
15. Are hydronium ions contributed to a solution by an acid or a base?
16. If you add water to a strong acid, how will the pH of the diluted acid compare to the pH of the original acid?
    a. lower
    b. higher
    c. the same
17. How can ammonia (NH_3) be a base if it doesn't contain any hydroxide ions?

Problems

Section 9.1

1. What is the mass percent of table salt in a solution of 25 grams of salt dissolved in 75 g of water?
2. You add 20 grams of baking soda (NaHCO3) to 100 mL of water at 25°C.
    a. Approximately how much of the baking soda will dissolve in the water?
    b. What happens to the rest of the baking soda?
    c. How could you increase the amount of baking soda that will dissolve in 100 mL of water?
3. How much of the following materials will dissolve in 300 mL of water at 25°C?
    a. table salt
    b. sugar
    c. chalk
4. How many grams of sugar do you need to make a 20% solution by mass in 500 g of water?

Section 9.2

5. Solution A has a pH of 3 and solution B has a pH of 10.
    a. Which solution is a base?
    b. Which solution is an acid?
    c. What would happen if you combined both solutions?
6. Which of the following pH values is the most acidic?
    a. 1
    b. 3
    c. 7
    d. 8
7. Luke and Sian want to plant a vegetable garden in their yard. A soil testing kit measures the soil pH at 5.0, but the lettuce they want to plant in their garden does best at a pH of 6.5. Should they add an acid or a base to the soil to make it the optimum pH for growing lettuce?

Chapter 10
Chemical Reactions

When some people think of *chemistry*, they often think of things that suddenly give off smelly odors, or explode. The process that makes the "stink" in a "stink bomb" and an explosion are examples of *chemical reactions*. Chemical reactions occur everywhere around you every day. Chemical reactions happen in the engines of automobiles to make them go. Chemical reactions happen in plants that release oxygen into the atmosphere.

Would you be surprised to learn that chemical strong enough to dissolve metal is in your stomach all the time? The chemical is hydrochloric acid and it is a necessary part of the chemical process of breaking food down into nutrients that can be used by your body. In fact, chemical reactions are responsible for most of the food digestion that occurs in your body each day, as well as many other body processes. In this chapter you will learn what chemical reactions are and how they occur.

Key Questions

1. ***Why are baking soda and vinegar such good ingredients for making a volcano model?***

2. ***What does it mean to balance a chemical reaction equation?***

3. ***When something burns, is there a chemical reaction taking place?***

10.1 Understanding Chemical Reactions

If you leave a tarnished copper penny in acid for a few minutes the penny becomes shiny again. The copper oxide that tarnished the penny was removed by a chemical reaction with the acid. Chemical reactions are the process through which chemical changes happen.

Chemical changes rearrange chemical bonds

Chemical change Ice melting is an example of a *physical change*. During a physical change, a substance changes its form but remains the same substance. A *chemical change* turns one or more substances into different substances that usually have different properties. An example of chemical change is burning wood into ashes.

Using chemical changes We use chemical changes to create useful materials. The rubber in car tires is an example of a material that has been modified by chemical changes. A chemical change called *vulcanization* inserts pairs of sulfur atoms into the long chain molecules of natural rubber. The sulfur ties adjacent molecules together like rungs on a ladder and makes vulcanized rubber much harder and more durable.

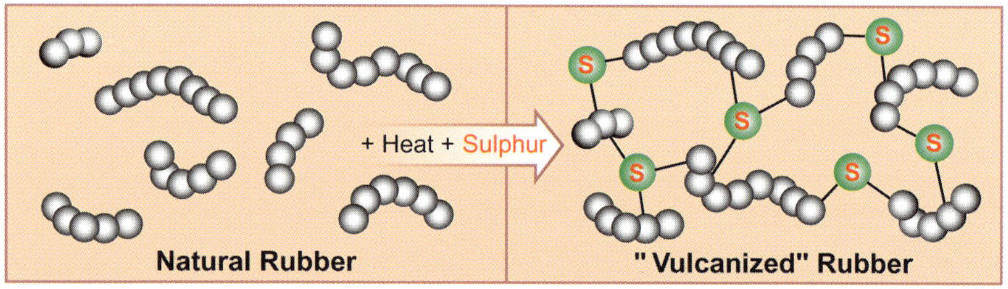

Recognizing chemical change A **chemical reaction** is a system of chemical changes that involves the breaking and reforming of chemical bonds to create new substances. A chemical reaction occurs when you mix baking soda with vinegar. The mixture bubbles violently as carbon dioxide gas, a new substance, is formed. The temperature of the mixture also gets noticeably colder. Bubbling, new substances, and temperature change can all be evidence of chemical change (Figure 10.1).

ă VOCABULARY

chemical reaction - a process that rearranges chemical bonds to create new substances.

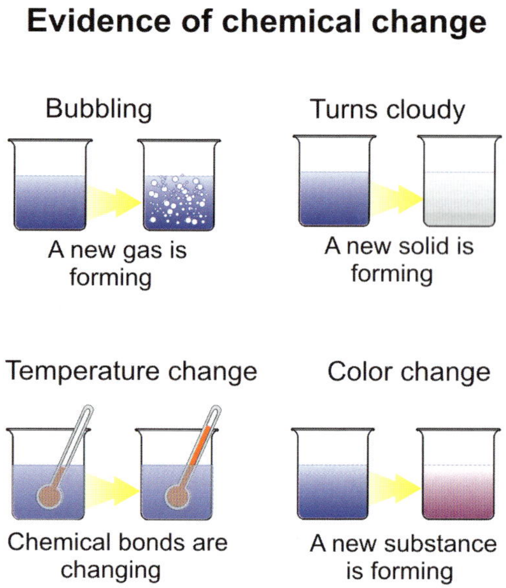

Figure 10.1: *Four observations that may be evidence of chemical change.*

Products and reactants

Products and reactants

How do we show the chemical reaction between baking soda and vinegar (Figure 10.2)? In cooking, you start with *ingredients* that are combined to make different *foods*. In chemical reactions, you start with **reactants** that are combined to make **products**. The reactants are substances which are combined and changed in the chemical reaction (baking soda and vinegar). The products are the new substances which result from the chemical reaction. The reactants and products may include atoms, compounds, and energy.

Products may change phase in a reaction

The substances in a chemical reaction may be in different phases. In this reaction the reactants are a solid (baking soda) and a liquid solution (vinegar). What are the products? The bubbling that goes on is a clue that at least one of the products is a gas.

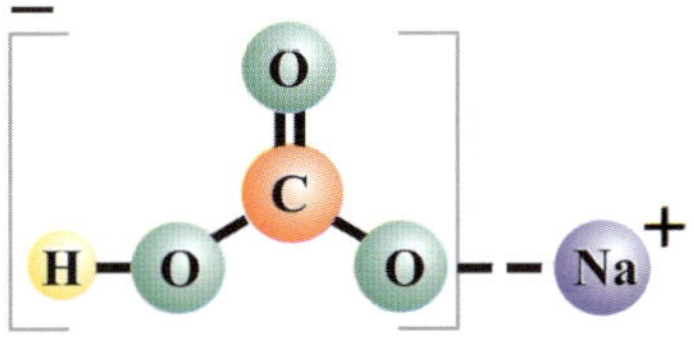

Sodium bicarbonate
(baking soda)
$NaHCO_3$

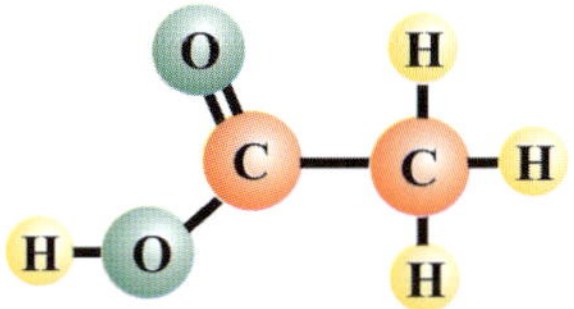

Acetic acid
(dissolved in vinegar)
$HC_2H_3O_2$

Start by counting the atoms in the products

Chemically, the components that react are sodium bicarbonate ($NaHCO_3$) in the baking soda and acetic acid ($HC_2H_3O_2$) in the vinegar. The first step in understanding the reaction is to see what atoms are in the reactants. If you count them you see there are four elements: carbon, oxygen, sodium, and hydrogen. *The reaction must rearrange these same atoms into the products.*

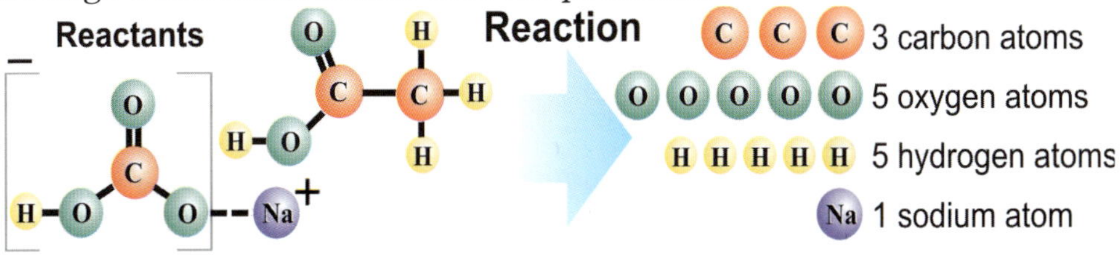

VOCABULARY

reactants - the substances which are combined and changed in the chemical reaction.

products - the new substances which result from a chemical reaction.

Figure 10.2: *The chemical reaction between baking soda and and vinegar.*

The products of the reaction

Figuring out the products

The chemical reaction rearranges the same atoms in the reactants to become new compounds in the products. In this case, the 3 carbon, 5 oxygen, 5 hydrogen, and 1 sodium atoms are rearranged to become sodium acetate ($NaC_2H_3O_2$), water (H_2O) and carbon dioxide (CO_2). Note that *the same exact atoms in the reactants are rearranged to make the products*. No new atoms are created!

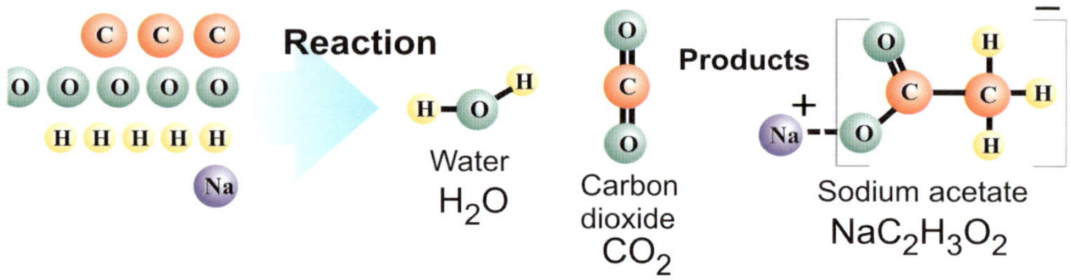

The whole reaction

We can now see the whole reaction clearly. Only three chemical bonds actually change. The sodium ion jumps to the acetic acid to make sodium acetate. The rest of the bicarbonate breaks up into water and carbon dioxide. Since carbon dioxide is a gas, that explains the bubbles observed during the reaction.

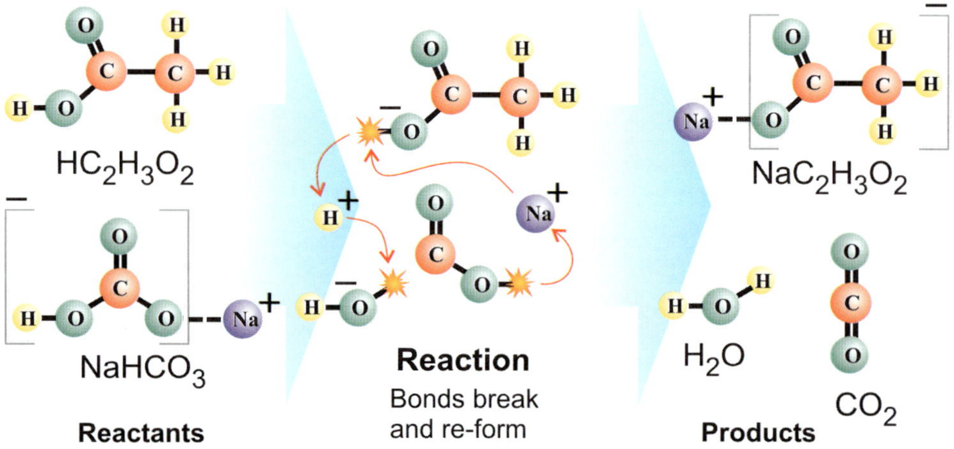

$$HC_2H_3O_2 + NaHCO_3 \longrightarrow NaC_2H_3O_2 + H_2O + CO_2$$

A chemical reaction rearranges the atoms of the reactants to form the new compounds of the products

No new atoms are created!

CHALLENGE

A similar reaction occurs when an effervescent tablet is dropped into water. The tablet contains sodium bicarbonate and citric acid. The reactants are:

$NaHCO_3 + H^+$ (from citric acid) + H_2O

Can you figure out what the products are? Hint: This reaction creates a bubbling gas, too.

Chemical equations

Understanding a chemical equation

A **chemical equation** is an abbreviated way to show the exact numbers of atoms and compounds in a chemical reaction. Without drawing elaborate diagrams, we can write the baking soda and vinegar reaction as a chemical equation. The arrow shows the direction the reaction goes, from reactants to products.

Chemical equation

$$HC_2H_3O_2 + NaHCO_3 \longrightarrow NaC_2H_3O_2 + H_2O + CO_2$$

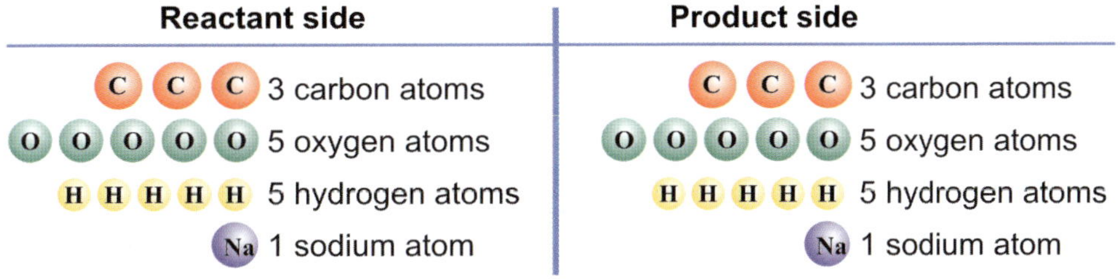

Conservation of mass

Notice that there are the exact same number of each type of atom on the reactant side of the equation as there are on the product side. There are three carbon atoms in the reactants and three carbon atoms in the products. This demonstrates that *chemical reactions conserve mass*. The total mass of the reactants is equal to the total mass of the products because they are the same atoms! They have just been rearranged into new compounds.

Demonstrating conservation of mass

Once you understand atoms and reactions, conservation of mass is a perfectly obvious result. The mass is the same because *the atoms are the same atoms*. Of course, demonstrating the conservation of mass in this reaction is tricky because one of the products is a gas! It took a long time before people realized that mass is conserved because they were fooled by their own measurements. If you compared the mass of the reactants and products as shown in Figure 10.3, what do you think you will find? Can you think of a way to do the experiment so that no mass escapes being measured?

VOCABULARY

chemical equation - an equation of chemical formulas that shows the exact numbers of atoms and compounds in a chemical reaction. For example:

$$2H_2 + O_2 \rightarrow 2H_2O$$

Conservation of mass

The mass of the reactants equals the mass of the products

Would doing the reaction this way demonstrate the conservation of mass?

Figure 10.3: *If you tried to demonstrate conservation of mass this way, what would you find? Would the mass of the reactants equal the mass of the products?*

Balancing chemical equations

Multiple compound reactions

The baking soda and vinegar reaction was a good one to learn about chemical equations because only one kind of each compound appeared. However, many reactions involve more than one compound of each type. A good example is the reaction that combines hydrogen and oxygen to produce water.

Start with the un-balanced reaction

The reaction combines hydrogen and oxygen molecules as shown.

Oxygen | Hydrogen | Water

$$O_2 + H_2 \longrightarrow H_2O$$

Count the atoms on each side

Next we count the atoms to see if there are the same number of each type of atom on the reactant and product sides of the equation.

Reactant side	Product side
2 oxygen atoms	1 oxygen atom
2 hydrogen atoms	2 hydrogen atoms

There is one more oxygen atom on the reactant side than there is on the product side. This means the reaction equation is not *balanced*. A balanced chemical equation has the same number of each type of atom on the product side and the reactant side.

The balanced chemical equation

To balance the equation, we add another water molecule to the product side. Now there are equal numbers of oxygen atoms on both sides. However, there aren't equal numbers of hydrogen atoms, so another hydrogen molecule is added to the reactant side.

Oxygen | Hydrogen | Water

$$O_2 + 2H_2 \longrightarrow 2H_2O$$

Reactant side	Product side
2 oxygen atoms	2 oxygen atoms
4 hydrogen atoms	4 hydrogen atoms

Understanding the numbers in a chemical equation

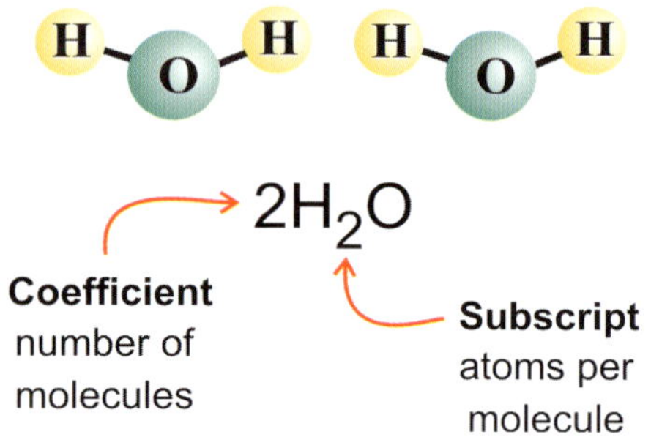

The large number 2 in "$2H_2O$" tells you there are *two molecules* of H_2O in the reaction. The large number is called a *coefficient*. If a coefficient is not written it is understood to be "1."

The little numbers (subscripts) tell you how many atoms of each element there are in *one molecule*. For example, the subscript 2 in H_2O means there are two hydrogen atoms in a single water molecule.

CHALLENGE

The hydrogen-oxygen reaction releases a lot of energy and many people hope hydrogen will replace gasoline as a fuel for cars.

10.1 Section Review

1. What is the difference between a physical change in matter and a chemical change in matter?
2. In the following list, decide whether each item is a physical or a chemical change:
   a. liquid water freezes into solid ice
   b. wood burns to ashes
   c. a window shatters when hit with a rock
   d. an old car sits in a junkyard and rusts
   e. a cup of hot chocolate gives off steam
3. Answer the following questions about this chemical equation:

$$CH_4 + 2O_2 \rightarrow CO_2 + 2H_2O$$

   a. How many carbon atoms are on the reactant side of the equation? How many hydrogen? How many oxygen?
   b. How many carbon atoms are on the product side of the equation? How many hydrogen? How many oxygen?
   c. Is this equation balanced? How do you know?
   d. What does the coefficent "2" in front of the H_20 mean?
4. Which of the following reactions is balanced?
   a. $CS_2 + 3O_2 \rightarrow CO_2 + SO_2$
   b. $2N_2O_5 + NO \rightarrow 4NO_2$
   c. $P_4 + 5O_2 \rightarrow P_2O_5$
   d. $Cl_2 + 2Br \rightarrow 2Cl + Br_2$
5. What does "mass is conserved in a chemical reaction" mean?
6. It has been said that chemical equations are sentences in the language of chemistry. List three things that chemical equations can tell you about a chemical reaction.

Balancing chemical equations

(1) Start with the correct chemical formula for each compound that appears as a reactant or product.

(2) Write down the equation for the reaction (un-balanced).

(3) Count the number of atoms of each element in the reactants and the products.

(4) Adjust the coefficient of each reactant or product until the total number of each type of atom is the same on both sides of the equation. This is done by trial and error.

Important reminder: You can NOT change subscripts in order to balance an equation. For example, calcium chloride has the chemical formula, $CaCl_2$. You can NOT change the subscript on Cl from 2 to 3 and make $CaCl_3$ to get an extra chlorine atom. $CaCl_3$ is a totally different compound than $CaCl_2$. *You can only change coefficients to balance equations.*

10.2 Energy and chemical reactions

If you have ever felt the heat from a campfire or fireplace, you have experienced the energy from a chemical reaction. *Burning* is a chemical reaction that *gives off* energy in the form of heat and light. In plants, photosynthesis is a reaction that *uses* energy from sunlight (Figure 10.4). In fact, *all chemical reactions involve energy.*

The two types of reactions

Energy is involved in two ways

Energy is involved in chemical reactions in two ways. (1) At the start of a chemical reaction, some (or all) bonds between atoms in the reactants must be broken so that the atoms are available to form new bonds. (2) Energy is released when new bonds form as the atoms recombine into the new compounds of the products. We classify chemical reactions based on how energy used in (1) compares to energy released in (2).

Exothermic reactions

If forming new bonds releases *more* energy than it takes to break the old bonds, the reaction is **exothermic.** Once started, exothermic reactions tend to keep going because each reaction releases enough energy to start the reaction in neighboring molecules. A good example is the burning of hydrogen in oxygen. If we include energy, the balanced reaction looks like this.

$$\underset{\text{Oxygen}}{O_2} + \underset{\text{Hydrogen}}{2H_2} \longrightarrow \underset{\text{Water}}{2H_2O} + \textbf{Energy}$$

Endothermic reactions

If forming new bonds in the products releases *less* energy than it took to break the original bonds in the reactants, the reaction is **endothermic**. *Endothermic reactions absorb energy.* These reactions need energy to keep going. An example of an important endothermic reaction is *photosynthesis.* In photosynthesis, plants use energy in sunlight to make glucose and oxygen from carbon dioxide and water.

$$\underset{\text{Carbon dioxide}}{6\ CO_2} + \underset{\text{Water}}{6\ H_2O} + \textbf{Energy} \longrightarrow \underset{\text{Glucose}}{C_6H_{12}O_6} + \underset{\text{Oxygen}}{6\ O_2}$$

ă VOCABULARY

exothermic - a reaction is exothermic if it releases more energy than it uses.

endothermic - A reaction is endothermic if it uses more energy than it releases.

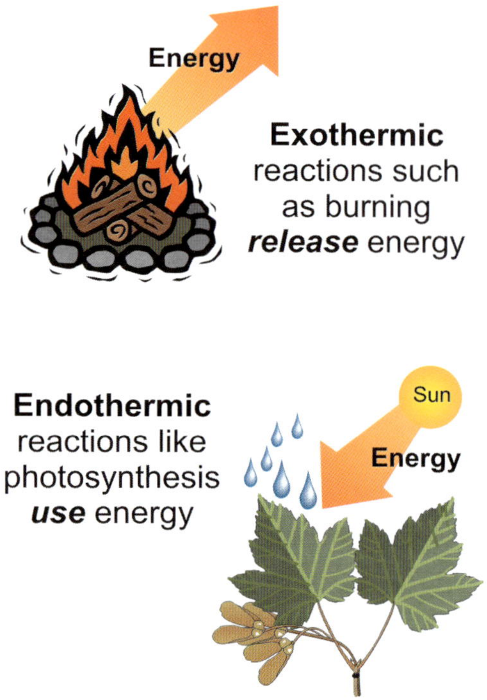

Figure 10.4: *Exothermic and endothermic reactions.*

Activation energy

An interesting question Exothermic reactions occur because the atoms arranged as compounds of the products have lower energy then they had when arranged as compounds of the reactants. Since this is true, why don't all of the elements immediately combine into the molecules that have the lowest possible energy?

Activation energy The answer has to do with **activation energy**. Activation energy is the energy needed to start a reaction and break chemical bonds in the reactants. Without enough activation energy, a reaction will not happen even if it releases energy when it does happen. That is why a flammable material, like gasoline, does not burn without a spark or flame. The spark supplies the activation energy to start the reaction.

Reaction: $2H_2 + O_2 \rightarrow 2H_2O + energy$

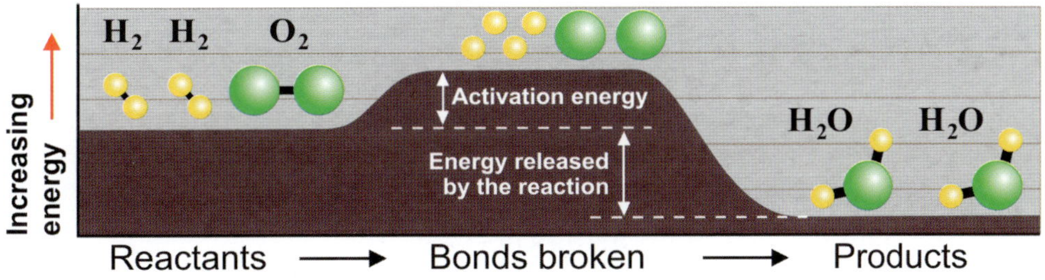

An example reaction The diagram above shows how the energy flows in the reaction of hydrogen and oxygen. The activation energy must be supplied to break the molecules of hydrogen and oxygen apart. Combining 4 free hydrogen and oxygen atoms into 2 water molecules releases energy. The reaction is exothermic because the energy released by forming water is greater than the activation energy. Once the reaction starts, it supplies its own activation energy and quickly grows (Figure 10.5).

Thermal energy A reaction starts by itself when thermal energy is greater than the activation energy. Any reaction which could start by itself, probably already has! The compounds and molecules that we see around us are ones that need more activation energy to change into anything else.

VOCABULARY

activation energy - energy needed to break chemical bonds in the reactants to start a reaction.

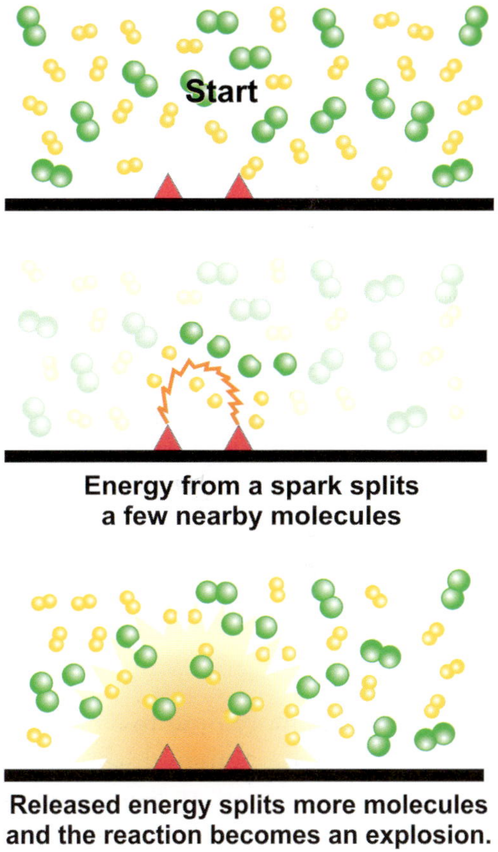

Figure 10.5: *Because energy released by one reaction supplies activation energy for new reactions, exothermic reactions can grow quickly once activation energy has been supplied.*

Addition reactions

Compounds are made in addition reactions

In an **addition reaction**, two or more substances combine to form a new compound. A good example of an addition reaction is the formation of rust (iron oxide, Fe_2O_3) from pure iron and dissolved oxygen in water. Iron oxide is an ionic crystal because the bonds between iron and oxygen are ionic bonds. Ionic crystals do not have structure diagrams like molecules however, they do have the specific ratio of atoms given by the chemical formula (2 Fe to every 3 O).

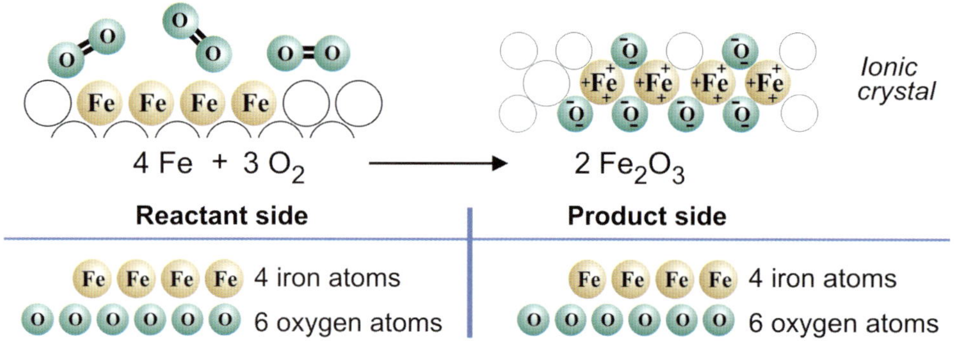

Polymerization is an addition reaction

As you saw earlier in this chapter, polymers are large molecules made up of repeating segments. The process of creating these molecules is called **polymerization**. Polymerization is a series of addition reactions that join small molecules into very long chain molecules. Polymers are made by joining smaller molecules called monomers.

Acid rain

Some fossil fuels, like coal, contain sulfur. When these fuels are burned, the sulfur reacts with oxygen in the air to form sulfur dioxide in an addition reaction. In air polluted with sulfur dioxide, sulfur trioxide is create when sulfur dioxide reacts with oxygen. Finally, *sulfuric acid* is produced by a third addtion reaction of sulfur trioxide with water.

VOCABULARY

addition reaction - two or more substances chemically combine to form a new compound.

polymerization - a series of addition reactions that join small molecules into large chain molecules.

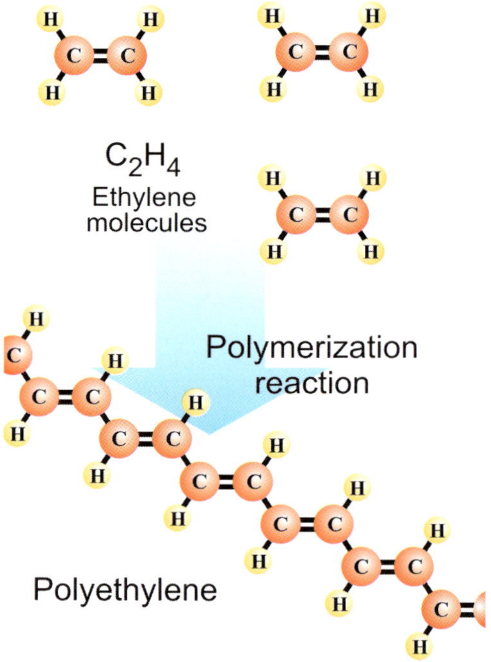

Figure 10.6: *Polymerization is a series of successive addition reactions that combine small molecules into large chain molecules.*

Decomposition reactions

Describing the phase of a product or reactant

In many cases, you wish to know the form of the products and reactant in a reaction. Are they solid, liquid or gas? Are they dissolved in water? The small symbols in the parentheses (*s, l, g, aq*) next to each chemical formula indicate the phase of each component in the reaction (Figure 10.7).

Decomposition reactions

A chemical reaction in which a single compound is broken down to produce two or more smaller compounds is called a **decomposition reaction**. The simplest kind of decomposition is the breakdown of a binary compound into its elements, as in the decomposition of water into hydrogen and oxygen with electricity:

Decompositon of Water

$$2\,H_2O_{(l)} \longrightarrow 2\,H_{2\,(g)} + O_{2\,(g)}$$

Larger compounds can also decompose to produce other compounds, as in the decomposition of baking soda with heat:

$$2\,NaHCO_{3\,(s)} \xrightarrow{heat} CO_{2\,(g)} + Na_2CO_{3\,(s)} + H_2O_{(l)}$$

Baking soda — Carbon dioxide — Soda ash — Water

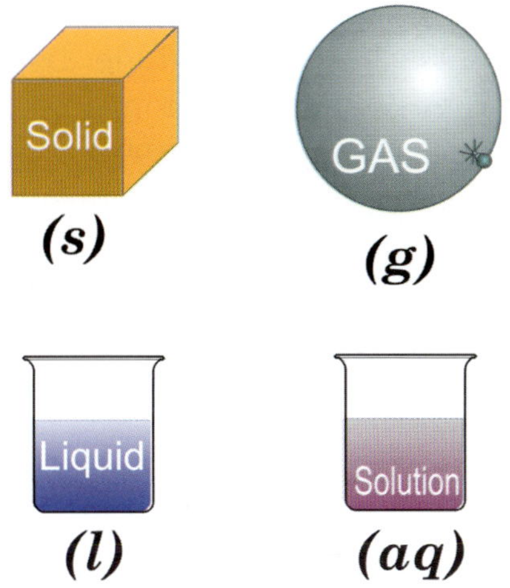

symbol	meaning
(s)	substance is a solid
(l)	substance is a liquid
(g)	substance is a gas
(aq)	substance is dissolved in solution (aqueous)

Figure 10.7: *What do the symbols shown in parentheses mean?*

Write a balanced decomposition reaction

Write a balanced chemical equation for the decomposition reaction of potassium chlorate ($KClO_3$) into potassium chloride (KCl) a salt substitute in food, and oxygen (O_2).

1. Looking for: The balanced chemical equation
2. Given: Products ($KClO_3$) and reactants (KCl, O_2)
3. Relationships: The total number of each type of atom must be the same on both sides of the equation.
4. Solution: The unbalanced equation is: $KClO_3 \rightarrow KCl + O_2$
   The oxygen atoms are not balanced so we need to add molecules of $KClO_3$ and O_2
   The balanced equation is: $2KClO_3 \rightarrow 2KCl + 3O_2$

Displacement and precipitation reactions

In single displacement, one element replaces another

In single-displacement reactions, one element replaces a similar element in a compound. For example, if you place an iron nail into a beaker of copper (II) chloride solution, you will begin to see reddish copper forming on the iron nail. In this reaction, iron *replaces* copper in the solution and the copper *falls out* of the solution as a metal:

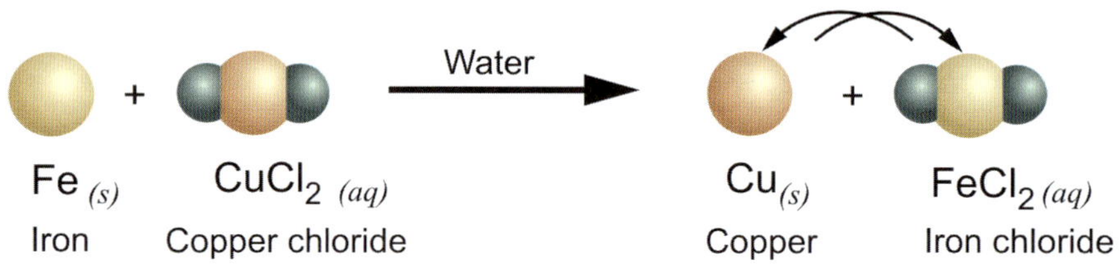

Precipitation occurs when one product is insoluble

In many reactions dissolved substances react to form substances that are no longer soluble. The insoluble product drops out of solution, forming a **precipitate**. A precipitate is a solid product that comes out of solution in a chemical reaction. Precipitates usually form many small particles which cause a cloudy appearance in a solution (Figure 10.8).

The limewater test for carbon dioxide is a precipitation reaction. In this test, a gas suspected of containing carbon dioxide is bubbled through a solution of CaO_2H_2 (limewater). Any carbon dioxide in the gas reacts to form a precipitate, turning the solution milky-white.

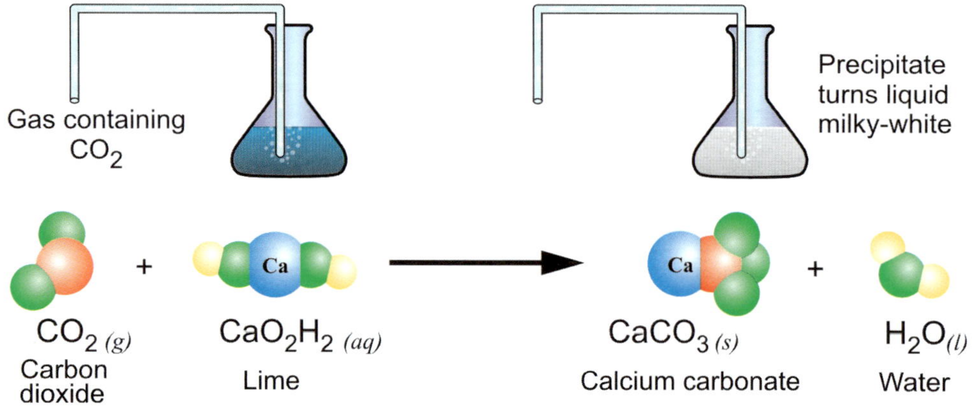

VOCABULARY

precipitate - a solid product that comes out of solution in a chemical reaction.

Figure 10.8: *The formation of a cloudy precipitate is evidence that a double-displacement reaction has occurred.*

Combustion reactions

Petroleum is a mixture of hydrocarbons

Almost 40 percent of all the energy we use comes from petroleum (oil) and 2/3 of that is gasoline and diesel fuel. Petroleum is not a single substance but a complex mixture of many substances created over millions of years by the decay of plants and animals. The major elements in petroleum are hydrogen and carbon, with smaller amounts of oxygen, nitrogen, and sulfur.

Refining

The *refining* process separates petroleum into molecules with different numbers of carbon atoms. The smaller molecules are used in gasoline. Heavier molecules become kerosene and heating oil. The heaviest molecules become tar and asphalt used for roads (Figure 10.9).

Range of molecule sizes	End use
C_1 - C_{12}	Gasoline and light fuels, such as aviation fuel
C_{12} - C_{18}	Kerosene and heating oil
C_{19} - C_{30}	Grease, motor oil, wax
C_{31} - C_{36+}	Tar and asphalt

The reactions of burning gasoline

In a perfect reaction, all the hydrocarbon molecules are completely burned to into carbon dioxide and water. Unfortunately, in an engine not all the fuel burns completely and pollutants such as carbon monoxide are also formed. Impurities in fuel, such as sulfur and nitrogen in the air, also have reactions that form pollutants such as oxides of nitrogen and sulfuric acid.

Perfect combustion reaction

$$2\,C_8H_{18} + 25\,O_2 \xrightarrow{\text{spark}} 16\,CO_2 + 18\,H_2O$$

Iso-octane, Oxygen, Carbon dioxide, Water

Actual combustion reaction

$$2\,C_8H_{18} + 27\,O_2 + N_2 + S \longrightarrow 15\,CO_2 + CO + 17\,H_2O + 2\,NO + H_2SO_4$$

Iso-octane, Oxygen, Nitrogen, Sulfur, Carbon dioxide, Carbon monoxide, Water, Nitrogen oxides, Sulfuric acid

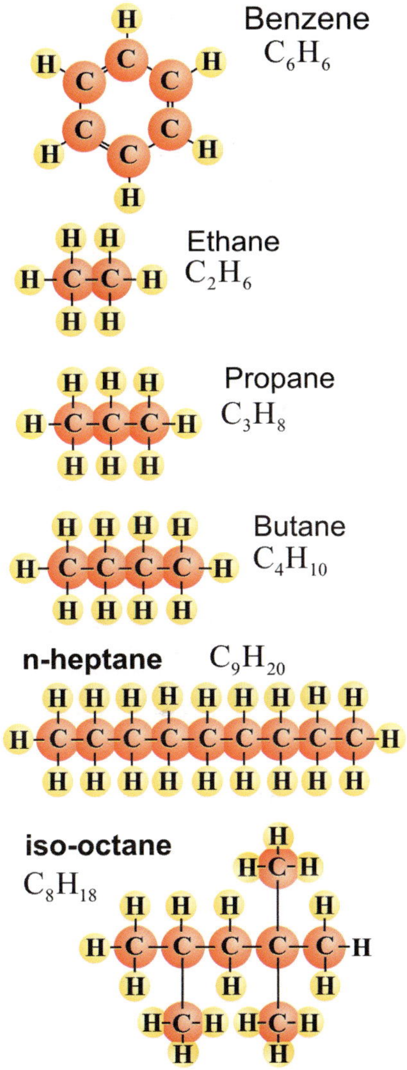

Figure 10.9: *Some of the many molecules found in gasoline. These are examples of hydrocarbons, molecules made with only hydrogen and carbon.*

Nuclear reactions

What is a nuclear reaction? Nuclear reactions change the nucleus of an atom. Until just 100 years ago people looked for a way to turn lead into gold. With today's understanding of nuclear reactions, it is now possible. However, we don't do it very often because the process is much more expensive than gold itself!

Nuclear versus chemical Because they affect the nucleus itself, nuclear reactions can change one element into a different element. Nuclear reactions can also change an isotope into a different isotope of the same element. Remember, isotopes of the same element have the same number of protons but different numbers of neutrons in the nucleus. By comparison, chemical reactions do *not* change the types of atoms. Chemical reactions only rearrange atoms into different compounds.

Nuclear reactions involve more energy than chemical reactions *Nuclear reactions involve much more energy than chemical reactions.* The energy in a nuclear reaction is much greater because nuclear reactions involve the strong nuclear force, the strongest force in the universe. Chemical reactions involve electrical forces. The electrical force acting on an electron far from the nucleus is much smaller than the strong force acting on a proton or neutron *inside* the nucleus. The difference in strength between the forces involved is the reason nuclear reactions are so much more energetic than chemical reactions (Figure 10.10).

Mass and energy in nuclear reactions Mass and energy are conserved together but *not* separately in nuclear reactions. This is because nuclear reactions can convert mass into energy. If you could take apart a nucleus and separate all of its protons and neutrons, the separated protons and neutrons would have more mass than the nucleus does all together. This bizarre fact is explained by Einstein's formula ($E = mc^2$), which tells us that mass (m) can be converted to energy (E), when multiplied by the speed of light (c) squared. The mass of a nucleus is reduced by the energy that is released when the nucleus comes together.

VOCABULARY

nuclear reaction - a process that changes the nucleus of an atom and may turn one element into a completely different element.

Chemical reactions	Nuclear reactions
What part of the atom is involved?	
Outer electrons	Nucleus (protons and neutrons)
What changes?	
Atoms are rearranged into new molecules but the atoms stay the same	Atoms may change into atoms of a different element
How much energy is involved?	
A small amount	A huge amount

Figure 10.10: *Comparing nuclear and chemical reactions.*

10.2 Section Review

1. What two ways is energy involved in chemical reactions?
2. Explain the difference between an exothermic and an endothermic reaction.
3. This is the chemical equation for the formation of rust:

$$4Fe_{(s)} + 3O_{2(g)} \rightarrow 2Fe_2O_{3(s)} + \text{energy}$$

   a. What do the symbols "(s)" and "(g)" mean?
   b. Is this reaction endothermic or exothermic?
   c. Is this a decomposition, addition, or displacement reaction?
4. Explain what *activation energy* is, and give an example that proves you understand its meaning.
5. What is a *polymer*, and what type of chemical reaction produces a polymer?
6. Identify the following reactions as: addition, decomposition, displacement, or combustion reactions.
   a. $2KClO_3 \rightarrow 2KCl + 3O_2$
   b. $Mg + 2AgNO_3 \rightarrow Mg(NO_3)_2 + 2Ag$
   c. $6Li + N_2 \rightarrow 2Li_3N$
   d. $2C_3H_7OH + 9O_2 \rightarrow 6CO_2 + 8H_2O$
7. List three ways that nuclear reactions are different from chemical reactions.
8. Petroleum refineries can be found throughout the United States. What does it mean to *refine* petroleum, and why must this process be performed?

CHALLENGE

Propane - a common fuel

Propane, C_3H_8, is a fuel that is used by cooks and campers every day. It is burned in oxygen to make a flame that can cook food, provide heat and light, and even run refrigerators. Write the complete, balanced chemical equation for the combustion of propane.

CHEMISTRY CONNECTION

The Science of Hot and Cold Packs

Have you ever had an injury, such as to you knee or back? If so, chances are you have used a cold or hot pack to treat the injury and provide pain relief. Cold and hot packs can be made ready in an instant without refrigeration or a microwave. Although it seems that these packs magically work, it's really a mini chemistry lab that lies within the plastic wrap.

Endothermic and exothermic reactions

Thermochemistry is the study of the heat given off or absorbed during a chemical reaction. A chemical reaction refers to when substances are mixed and new substances are formed. In chemical reactions, bonds between atoms are broken or formed, and heat may be given off or taken up during the process.

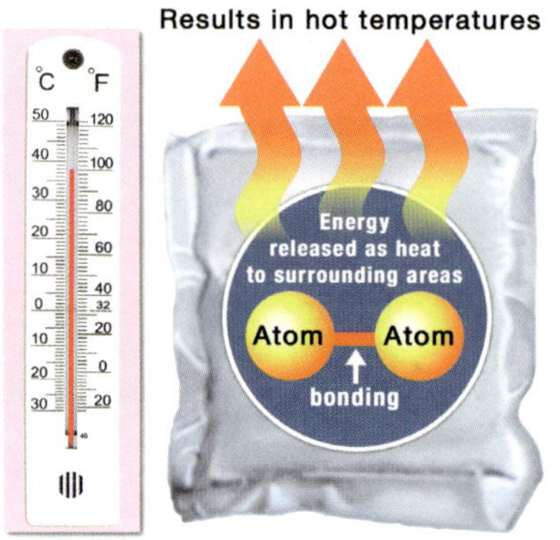

With many chemical reactions, energy may be released in the form of heat. This is called an exothermic reaction. When an exothermic reaction occurs, bonds are made between atoms. During this bond making, energy is lost to the surroundings. As a result, exothermic reactions produce hot temperatures or may even be explosive.

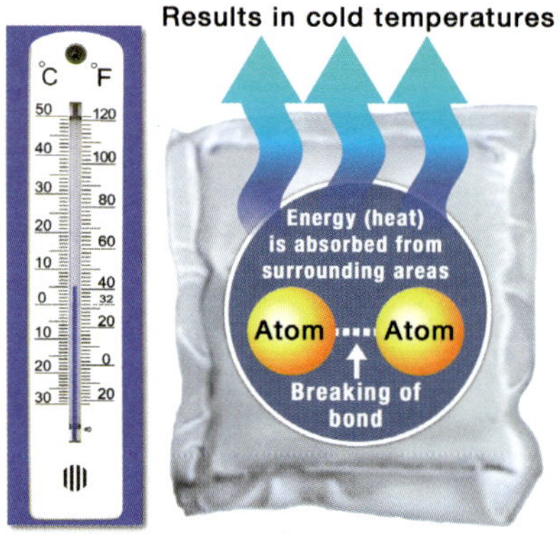

Some chemical reactions can only be completed if they absorb energy. When a chemical reaction absorbs heat from the environment, it is called an endothermic reaction. In endothermic reactions, bonds are broken between atoms. Here, the energy required for bond breaking is absorbed from the surroundings. As a result, endothermic reactions produce cold temperatures.

How hot and cold packs work

Let's look at the so-called "magic" inside a common hot and cold packs. Usually these packs are pouches made of thick, strong plastic. The pouches contain a dry chemical in the form of a powder. Within the large pouch of hot and cold packs is an inner pouch made of thin, weak plastic. This inner pouch contains water. The packs are make active when the seal of the inner water pouch is broken and the contents are vigorously shaken. Water is released and mixes with the chemical in the outer pouch to create either an exothermic or endothermic reaction.

Sounds simple, but let's look at the chemicals involved. Hot packs are made of chemicals that produce an exothermic reaction when mixed with water. Usually, hot packs contain calcium chloride or magnesium sulfate. These chemicals release heat energy when mixed with water and raise the temperature of the pack.

Cold packs work much the same way as hot packs, except different chemicals are used. Cold packs are made of chemicals that produce an endothermic reaction when mixed with water. Often, the chemical used in cold packs is ammonium nitrate. It absorbs heat energy and lowers the temperature of the pack.

The concentrations of the water and chemicals determine how hot or cold a pack gets. Commercial hot and cold packs typically last for about twenty minutes.

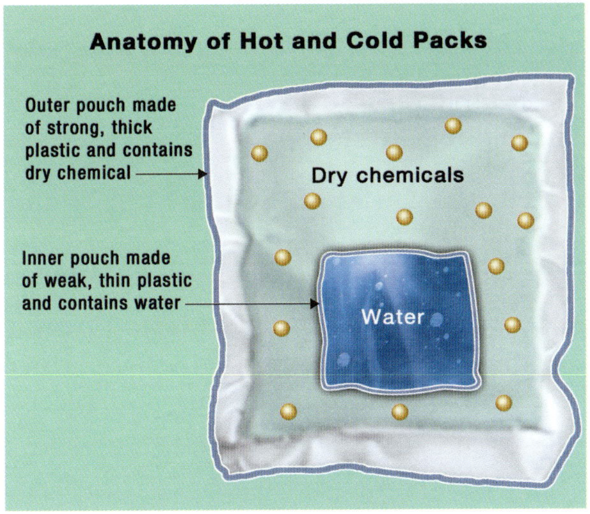

Hot and cold therapy

Physical therapists and sports trainers often use hot packs and cold packs to treat patients and athletes with injuries. However, the type of pack that should be used depends on the nature of injury.

Applying heat to your body can improve the flexibility of your tendons and ligaments. Tendons are bands of cordlike tissue that connect bone to muscle, while ligaments are cordlike tissues that connect bone to bone. Heat therapy can also reduce muscle spasms, reduce pain, and increase blood flow. The exact way in which heat relieves pain is not known. However, researchers think that heat inactivates nerves fibers that can force muscles to spasm. Heat may also induce the release of endorphins. Endorphins are chemicals in our body that block the transmission of pain by our nerves. Heat applied to body parts also relaxes the walls of blood vessels, resulting in increased blood flow. Health care professionals recommend using heat to untighten muscles and increase overall flexibility. However, it is best to avoid heating up already inflamed joints.

Like heat therapy, cold therapy may also be used to reduce muscle spasms. Muscle spasms are reduced with cold therapy because muscles fibers become less sensitive to being stretched. Cold is also useful for reducing pain and swelling. Cold therapy slows pain by reducing the speed of nerve impulses. Most tissue swelling is drastically decreased when cold and compression are applied to an injured area. The cold temperature constricts the walls of blood vessels, while the compression reduces the blood flow to the injured body part. Cold therapy is best used to reduce inflammation and swelling caused by sprains, strains, and bruises.

A variety of inexpensive, disposable hot and cold packs can be purchased at pharmacy stores. So next time you experience a minor injury, just reach for the magic pack that fits your needs.

Questions:

1. What are the differences between endothermic and exothermic reactions?
2. What are the structural components of hot packs and cold packs?
3. What are the differences between hot packs and cold packs?
4. How do physical therapists and sports trainers use hot packs and cold packs to treat patients and athletes with injuries?

CHAPTER ACTIVITY

Explore Hot and Cold Packs

All chemical reactions are either exothermic (release energy) or endothermic (absorb energy). In addition, some physical processes such as dissolution (dissolving) can also release or absorb energy. This is the basis for commercially available hot packs and cold packs. Most hot and cold packs work by breaking a membrane that separates a solid and water. Once the membrane is broken the solid dissolves in the water. Depending on the nature of the compound, heat is either released (hot pack) or removed from the environment (cold pack) during the process.

Materials:

Thermometer, Styrofoam cups, hot pack, cold pack, safety goggles, scissors, goggles, apron

What you will do

For the Hot Pack

1. Cut apart the outer pouch of hot pack and pour the solid into a styrofoam cup.

SAFETY NOTE: Do not touch the chemicals from the hot pack with your hands. Wear goggles and an apron!

1. Carefully cut the corner of the inner pouch and pour the water into another cup.
2. Meaure and record the temperature of the water in the cup.
3. Pour the water into the cup containing the solid and quickly transfer the thermometer to the mixture.
4. Stir the mixture until it dissolves and record its final temperature.

For the Cold Pack

1. Repeat the above procedure with the cold pack.

SAFETY NOTE: Do not touch the chemicals from the cold pack with your hands. Wear goggles and an apron!

Experiment	Starting Temperature (°C)	Final Temperature (°C)
Part A: Hot Pack		
Part B: Cold Pack		

Questions:

a. What is the change in temperature for for the hot pack? What is the change in temperature for the cold pack?
b. What compounds are used in commercial hot packs and cold packs?
c. Why does your skin feel cool when a cold pack is applied?

Chapter 10 Assessment

Vocabulary

Select the correct term to complete the sentences.

reactant	chemical equation	endothermic
exothermic	addition reaction	chemical reaction
products	polymerization	activation energy
nuclear reaction		decomposition reaction

Section 10.1

1. A(n) ____ occurs when you mix baking soda and vinegar.
2. The new substances that are created in chemical reactions are called ____.
3. A substance that changes during a chemical reaction is a(n) ____.
4. A(n) ____ is a short hand description of a chemical reaction using chemical formulas and symbols.

Section 10.2

5. A reaction is ____ if it releases more energy than it uses.
6. ____ is needed to start a reaction and break chemical bonds in the reactants.
7. In a(n) ____, the nucleus of an atom is changed, and one element may become a completely different element.
8. A series of addition reactions that join small molecules into large chain molecules is known as ____.
9. A reaction is ____ if it uses more energy than it releases.
10. Combining iron and oxygen to form rust is an example of a chemical reaction called a(n) ____.
11. The reaction that breaks down water into hydrogen and oxygen using electricity is known as a(n) ____.

Concepts

Section 10.1

1. Is tearing a piece of paper a physical change or a chemical change?
2. What happens to chemical bonds during chemical reactions?
3. The substance produced when iron is oxidized is:
   a. water.
   b. oxygen.
   c. iron precipitate.
   d. rust.
4. The reactants in the equation $2H_2 + O_2 \rightarrow 2H_2O$ + energy are:
   a. hydrogen and energy.
   b. hydrogen and oxygen.
   c. water and energy.
   d. oxygen and water.
5. The number of atoms of each element on both sides of a chemical equation must always be:
   a. greater than one.
   b. less than two.
   c. different.
   d. equal.
6. The chemical formula $3H_2O$ means
   a. three atoms of hydrogen and three atoms of oxygen.
   b. six atoms of hydrogen and three atoms of oxygen.
   c. three atoms of water.
   d. three atoms of hydrogen and two atoms of oxygen.
7. How do balanced chemical equations illustrate the law of conservation of mass?

8. Which is an example of the use of activation energy?
   a. Plugging in an iron
   b. Playing basketball
   c. Holding a match to paper
   d. Eating
9. What physical and chemical changes occur when a wax candle burns?

Section 10.2

10. What conditions must be met in order for a reaction to be considered exothermic?
11. A "instant cold pack" is a plastic bag with a packet of water surrounded by crystals of ammonium nitrate. To activate the cold pack, you squeeze the plastic bag to release the water. When the water contacts the ammonium nuitrate crystals, a reaction occurs and the pack becomes icy cold. Is the reaction inside the cold pack an endothermic or an exothermic reaction?
12. List two or more combustion reactions that are a part of your everyday life.
13. Calcium chloride and silver nitrate react to form a *precipitate* of silver chloride in a solution of calcium nitrate. This is an example of:
    a. a combustion reaction.
    b. a dispacement reaction.
    c. polymerization.
14. Explain why *mass* is not necessarily conserved in a nuclear reaction.
15. Write the balanced chemical equation for the decomposition of lithium carbonate (Li_2CO_3) into lithium oxide (Li_2O) and carbon dioxide (CO_2).

Problems

Section 10.1

1. Calculate the number of atoms of each element shown in each of the following
   a. $CaSO_4$
   b. $4NaOCl$
   c. $Fe(NO_3)_2$
   d. $2Al_2(CO_3)_3$
2. Is this chemical equation balanced?
   $2C_4H_{10}\ (g) + 13O_2\ (g) = 8CO_2\ (g) \rightarrow 10H_2O\ (l)$
3. The mass of an iron bolt was 5.4 grams when it was manufactured. After being bolted to an outdoor structure for several months, the mass of the bolt was found to have increased by 0.2 grams. Given the following balanced equation for the reaction, does this example support the law of conservation of mass? Why or why not?
   $4Fe\ (s) + 3O_2\ (g) \rightarrow 2Fe_2O_3\ (s)$

Section 10.2

4. Many drain cleaners are a mixture of sodium hydroxide and aluminum filings. When these two substances mix in water, they react to produce enough heat to melt the fat in a clogged drain. The bubbles produced are hydrogen gas. The complete reaction occurs in two steps:
   step 1: $Al(s) + NaOH(aq) \rightarrow Al(OH)_3(s) + Na^+(aq)$
   step 2: $Na^+(aq) + H_2O \rightarrow Na_2O(s) + H_2(g)$
   a. Classify step 1 of the reaction as: addition, displacement, or decomposition.
   b. Is this an endothermic or an exothermic reaction?
   c. Balance each equation for each step of the reaction.

Chapter 11

The Chemistry of Living Systems

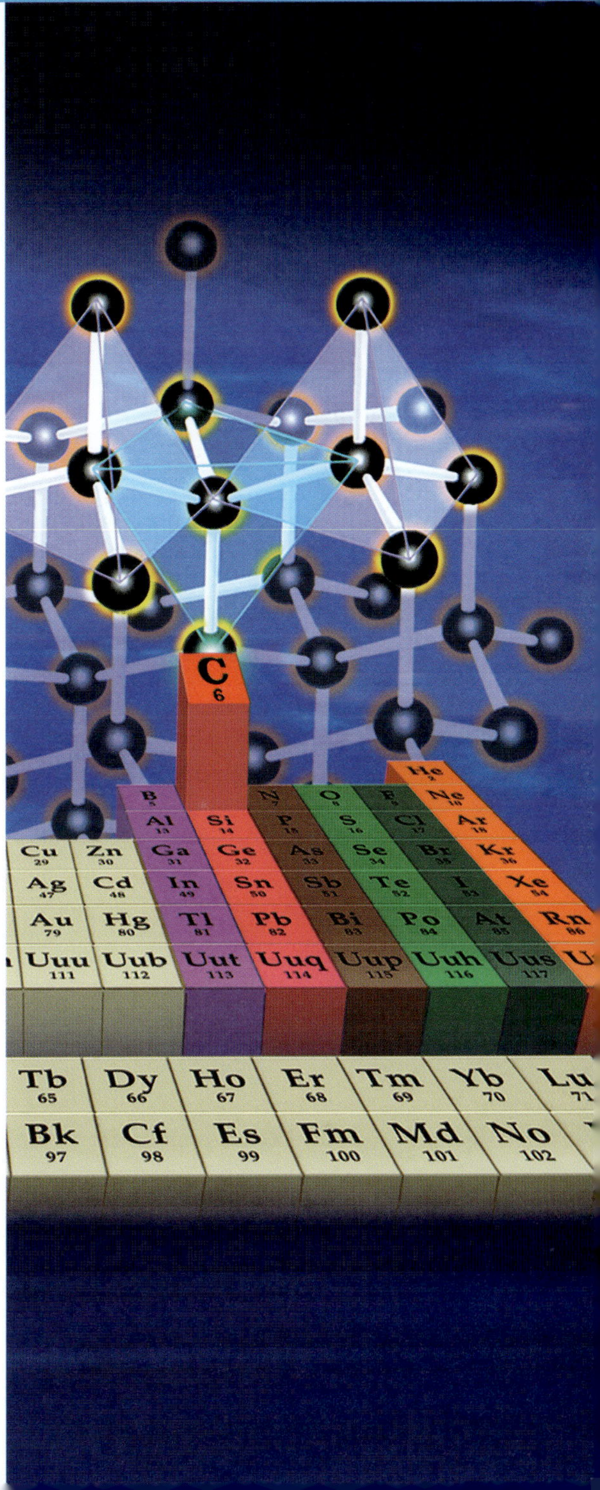

What is chemistry? If the image that comes to your mind is of a complicated array of tubes and smoking beakers, factories and stinky fumes, this chapter will paint quite a different picture. Chemistry is *you*. Like all living organisms, your body is an incredibly complex chemical machine taking in chemicals from food and causing countless chemical reactions to occur every second. In so many ways, the fundamental processes of life are chemistry.

If you take away the water, the rest of the human body is 53 percent carbon by weight. The chemistry of living things is the chemistry of carbon and its compounds. Carbon is the basic building block in the complex molecules that make up all living things. This chapter is your introduction to a branch of chemistry — organic chemistry — that is devoted solely to carbon and carbon compounds.

Key Questions

1. ***Why is carbon so important to living things?***

2. ***What are carbohydrates, fats, proteins, and DNA?***

3. ***How does a living creature control its chemistry?***

11.1 The Chemistry of Carbon

The chemistry of life is largely the chemistry of the element carbon. About 65 percent of the human body is water. Of the remainder, 91 percent is only four elements: carbon, oxygen, nitrogen, and hydrogen (Figure 11.1). Of those four, carbon is the largest fraction at 53 percent. Carbon atoms often serve as the backbone to which oxygen, nitrogen and hydrogen are connected. Carbon is so important because it is the lightest element that can make up to four bonds at the same time, including bonds with itself.

VOCABULARY

organic chemistry - the chemistry of carbon and carbon compounds.

Carbon molecules

Carbon forms ring and chain molecules

Carbon has four electrons in its outer energy level, and therefore four valence electrons. Carbon can share one or more electrons to make covalent bonds with itself or as many as four other elements. Carbon molecules come in three basic forms: straight chains, branching chains, and rings. All three forms are found in important biological molecules. For example, glucose is a sugar made by plants and valine is an amino acid found in proteins — and both are built on carbon.

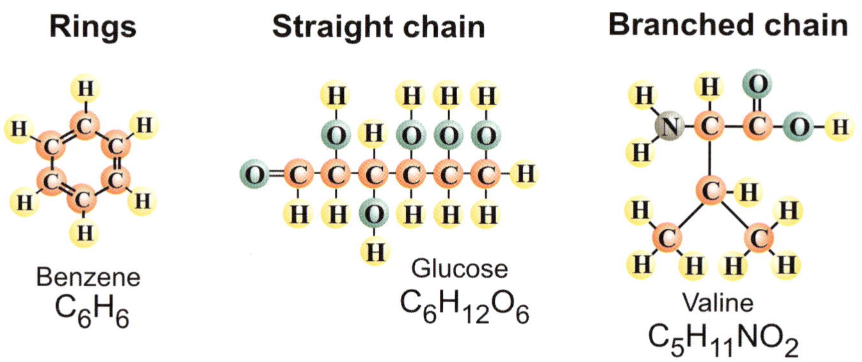

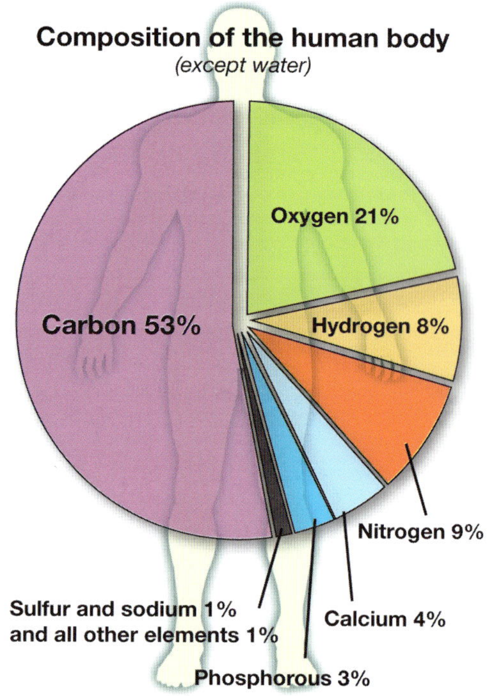

Figure 11.1: *After water, carbon is the most abundant element in the human body.*

Organic chemistry is the chemistry of carbon

The three basic shapes are often combined (chains and rings) in the same molecule. **Organic chemistry** is the branch of chemistry that specializes in carbon and carbon compounds. Organic molecules are found in all living things and also in many nonliving substances such as candle wax and polyethylene plastic.

Molecules in plants and animals

The four types of biological molecules

Living creatures are such complex organisms that even today we have much to learn about the chemical reactions that take place inside us. Scientists classify the organic molecules in living things into four basic groups: carbohydrates, fats, proteins, and nucleic acids. All living things contain *all four types* of molecules. And each type of molecule includes thousands of different chemicals, some specific to plants, some to animals. It is only in the past few decades that biotechnology has been able to reveal the rich chemistry of living things.

Carbohydrates

Carbohydrates are mainly composed of carbon, hydrogen, and oxygen in a ratio of about 1:2:1. Carbohydrates exist as small molecules, like glucose, and long-chain molecules, like starches. Table sugar is a carbohydrate called sucrose. Sucrose is composed of two simple sugars, glucose and fructose, which are chained together (Figure 11.2).

Proteins

Proteins are large molecules composed of carbon, hydrogen, oxygen, nitrogen, and trace elements. Skin and muscle tissue are composed primarily of protein. A single protein may contain several thousand atoms in a complex structure.

Fats

Fats are medium-to-large nonpolar molecules that do not dissolve in water. Structurally, fats are long chains of carbon and hydrogen with different elements added every so often. Cholesterol is a fat that makes up part of the outer membrane of cells. Cholesterol is naturally essential to cells, but unnaturally high cholesterol may lead to heart disease.

Nucleic acids

Nucleic acids such as DNA store the genetic code that allows organisms to reproduce. DNA is a huge molecule with millions of individual atoms. All the information that makes you a human is stored as a coded sequence of component molecules within DNA.

VOCABULARY

carbohydrates - energy-rich sugars and starches.

proteins - large molecules found in animal and plant tissue.

fats - energy-rich hydrocarbon chain molecules.

nucleic acids - biological molecules such as DNA that have the ability to store the genetic code.

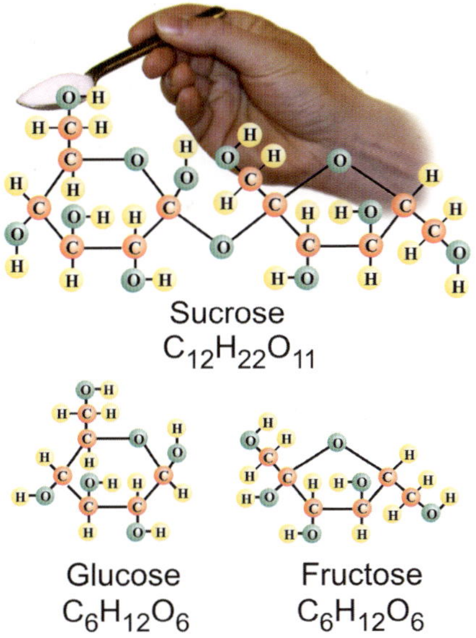

Figure 11.2: *Table sugar is sucrose, a carbohydrate made of two simple sugars: glucose and fructose.*

Carbohydrates

How carbohydrates are used

Carbohydrates are relatively small molecules used to store and transfer energy in living systems. Plant cells use energy from the sun to build carbohydrates from carbon dioxide and water. Animal cells consume carbohydrates and extract the energy by breaking the carbohydrates into smaller molecules again.

Starches and sugars

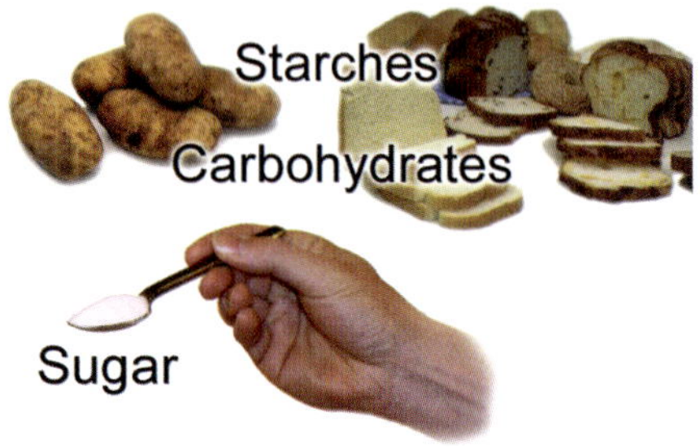

Carbohydrates are classified as either sugars or starches. Sugars are the smaller of the two types. Sugars break down relatively quickly in the body, releasing energy within a short time of being eaten. Glucose is the simplest sugar and is dissolved directly into the bloodstream.

Glucose is the simplest sugar

Glucose is the primary energy source for cells. When dissolved in water, the chain structure of a glucose molecule curls around on itself to become a ring (Figure 11.3). The glucose molecule can have several variations in the order of the OH - H - OH groups on either side of the carbon backbone. Some animals are so specialized that they can only digest one form of glucose and not the others.

Starches are chains of sugar

Starches are long chains of simple sugars joined together to make natural polymers. Because starches are larger molecules, they are slower to break down in the body and therefore can provide energy for a longer period than sugars. Corn, potatoes, and wheat contain substantial amounts of starches (Figure 11.4).

Cellulose

Cellulose is the primary molecule in plant fibers, including wood. The long-chain molecules of cellulose are what give wood its strength. Like starch, cellulose is made from chains of thousands of glucose molecules. However, in starch all the glucose units are the same orientation. In cellulose, alternate glucose units are inverted. This difference makes cellulose difficult for animals to digest. Trees grow so large partly because so few animals can digest wood.

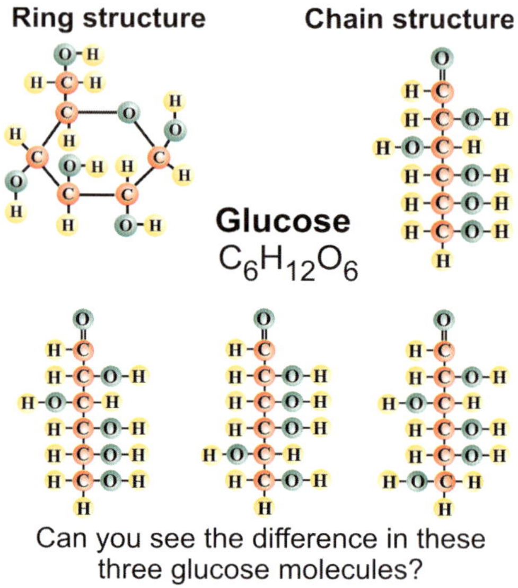

Figure 11.3: *Different structures of the glucose molecule.*

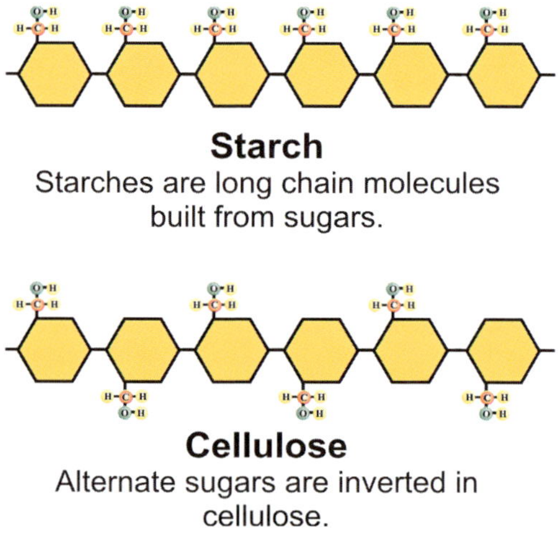

Figure 11.4: *Starch and cellulose.*

Photosynthesis

The importance of photosynthesis The energy that supports life on Earth starts with a reaction that takes energy from sunlight and stores it as chemical bonds in molecules of glucose and other simple sugars. This reaction is called **photosynthesis**. Photosynthesis occurs mostly in plants and in some types of bacteria.

The food chain Photosynthesis is the foundation of the food chain on Earth. At the bottom of the food chain are producers, plants that take energy from the sun and convert it to chemical energy in glucose and other organic molecules. Animals (including ourselves) ultimately get energy from photosynthesis, because we eat plants or other animals that eat plants. Nearly all the energy in living things can be traced to this important reaction.

Photosynthesis releases oxygen Photosynthesis also produces the oxygen in our atmosphere. Without photosynthetic organisms, Earth could not support life. Although oxygen is a common element on Earth, it is usually trapped by rocks and minerals in compounds like calcium carbonate ($CaCO_3$).

Photosynthesis removes CO_2 Photosynthesis removes carbon dioxide from the atmosphere. For every glucose molecule produced, six molecules of carbon dioxide are removed from the air, and six molecules of oxygen are produced. Carbon dioxide absorbs infrared radiation and therefore traps heat in the atmosphere. If too much carbon dioxide is present, the planet cannot cool itself by radiating energy into space. Higher levels of carbon dioxide may be responsible for the warming of Earth by several degrees over the past 200 years. Can you think of ways to stabilize carbon dioxide levels?

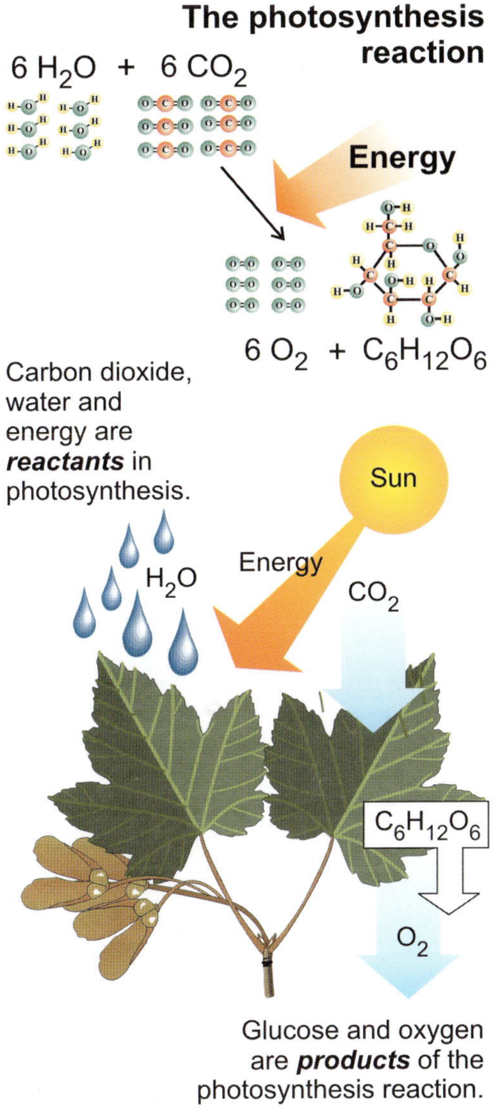

Figure 11.5: *Photosynthesis is a chemical reaction that is the basis for the food chain on Earth.*

Respiration

Digestion Animals get energy and nutrients by breaking up glucose, starch, and other organic molecules. The digestive system breaks food down into molecules the body can use. Proteins are split into amino acids. Carbohydrates are reduced to simple sugars. Fats are split into glycerol and fatty acids. These nutrients are then absorbed into the blood and transported to all the cells of the body.

Cellular respiration On the molecular level, **cellular respiration** breaks down glucose into water and carbon dioxide again, extracting energy in the process. The reactions of respiration proceed in many steps, but the end result is that glucose and oxygen are used up and carbon dioxide and water are produced. Respiration is almost the reverse of photosynthesis, releasing energy that originally came from the sun.

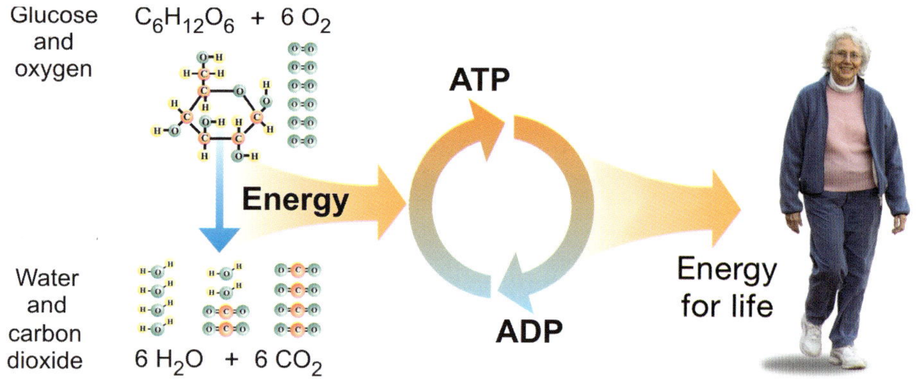

The ATP cycle Each cell converts the energy in glucose into chemical energy stored in molecules of ATP. A cycle between ADP and ATP is the energy source of cells. In a series of complex reactions that also require oxygen, one molecule of glucose is used to convert a maximum of 36 to 38 molecules of ADP to ATP. The ATP molecule is like a battery that distributes energy to where it is needed. Cells use the energy by converting the ATP back into ADP and the cycle starts over. Phosphorus is a critical part of the ADP-ATP cycle and one reason this element is an important nutrient.

VOCABULARY

cellular respiration - the reactions in cells that release energy from glucose.

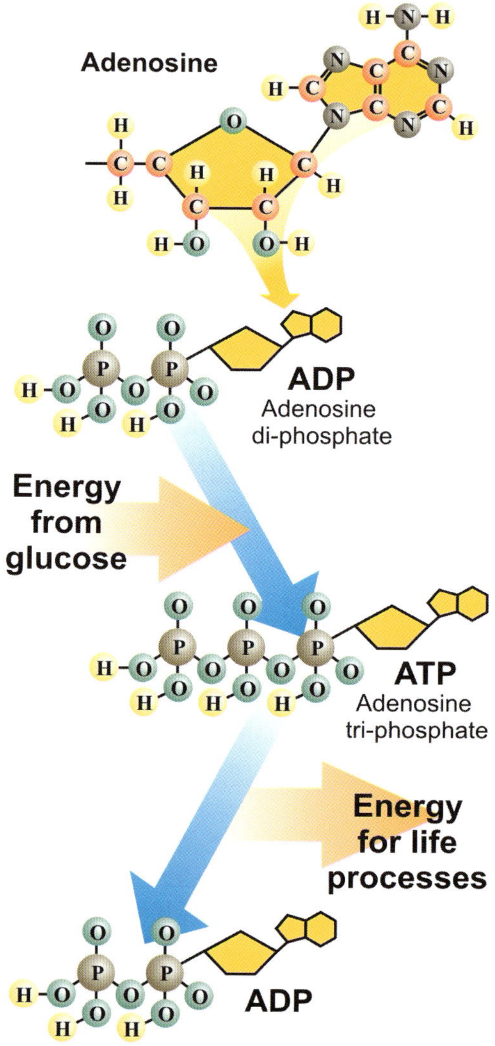

The importance of water

Why water is necessary

Liquid water is essential to life as we know it. The human body is typically between 60 and 65 percent water by weight. Most of the chemical reactions that sustain life *only work in solution*. Therefore, when scientists look for life on other planets, the first thing they look for is water. We believe Mars either had or has water on its surface or beneath its surface. That raises the tantalizing possibility that life may exist there.

There are three important characteristics of water that make it essential for life.

Water is a good solvent

Water is a good solvent. In order to have a chemical reaction, molecules must be able to move around and contact each other. In a solid, this is just not possible. However, in a solution, molecules can move relatively large distances carrying energy and nutrients throughout a cell. Water also allows transport through the body on a larger scale. For example, oxygen is required by cells throughout the body, but it comes into the body in a centralized place: your lungs. Red blood cells absorb oxygen in the lungs and are carried throughout the body so they can distribute the oxygen.

Liquid over a wide temperature range

Water exists as a liquid over a large range of temperatures. In fact, virtually all living organisms on Earth are most active between the freezing and boiling point of water. The wide range over which water remains liquid allows most of Earth to be habitable most of the time. Very few biological processes can proceed when completely frozen because molecules: (a) cannot reach each other, and (b) have less thermal energy for activating reactions.

High specific heat

Water has a high specific heat — one of the highest of any substance known. Water's high specific heat means it takes a lot of energy to raise the temperature a small amount. This property of water helps living organisms maintain a stable body temperature even though outside temperatures may fluctuate a great deal.

Figure 11.6: *Water is essential to our planet and to living things, including you.*

11.1 Section Review

1. About 80 percent of all chemical compounds on Earth contain carbon. Why is carbon found in so many compounds?
2. Complete the table by filling in the missing information:

Biological molecule	Composed of what atoms?	Example	Importance
carbohydrate		sugar	store and transfer energy
fat			
protein	C, H, N, O and trace elements	found in skin and muscle tissue	
nucleic acid		makes up DNA	protein synthesis and heredity

3. Photosynthesis is a critically important process. Why?
4. Does photosynthesis involve a physical change or a chemical change?
5. List the reactants and products for:
   a. Photosynthesis.
   b. Cellular respiration.
6. Why is water essential to life?
7. What are the characteristics of water that make it life-sustaining?

CHALLENGE

Termites are insects that eat wood. Do a little research on termites to answer the following:

a. What biological molecule can termites digest that most animals have great difficulty digesting?
b. What are some signs that a building may be infested with termites?
c. What parts of the world have the greatest problems with termites?

11.2 Proteins, Fats, and Nucleic Acids

Carbohydrates are the simplest of the important biological molecules. Proteins, fats, and nucleic acids are more complex molecules, including thousands of individual atoms in a single molecule. Nucleic acids found in DNA are at the core of genetics, an active area of scientific research. The creation and functions of proteins are another area of active research. While we know a tremendous amount, we still have much to learn.

Fats

Fats provide long-term energy storage Fats are high-energy molecules that plants and animals use to store energy in reserve for longer periods. Sugars break down too quickly to store energy reserves in a body. Fats are more complex molecules that take much longer to break down. Chemically, fats and oils are similar. Oils are fats that are liquid at room temperature.

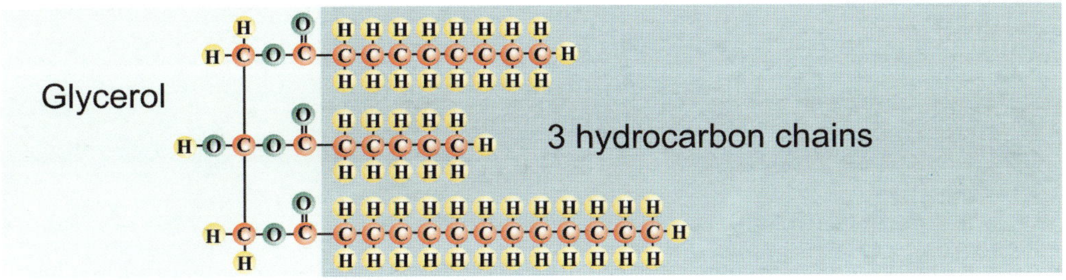

Saturated fats A fat molecule has a two-part structure. The first part is called *glycerol*. Attached to the glycerol are 3 hydrocarbon chains. In a **saturated fat** the carbon atoms are surrounded by as many hydrogens as possible (Figure 11.8).

Unsaturated fats An **unsaturated fat** has fewer hydrogen atoms than it could have, meaning some of the carbon molecules have double bonds with each other instead of with hydrogen. Chemical processing of food adds some hydrogen to unsaturated fats in a process called *hydrogenation*. Because they are harder to digest, *partially hydrogenated* fats have have a longer shelf life however, research is showing that partially hydrogenated fats may be unhealthy.

VOCABULARY

saturated fat - a fat molecule in which each carbon is bonded with two hydrogen atoms.

unsaturated fat - a fat molecule that has less hydrogen atoms than a saturated fat.

Figure 11.7: *Fats and oils are high-energy molecules organisms use to store energy reserves.*

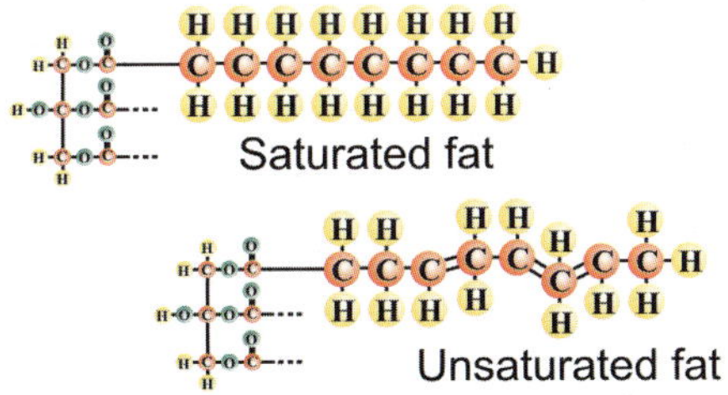

Figure 11.8: *The hydrocarbon chains in saturated and unsaturated fats.*

Proteins

Proteins are large molecules

Proteins are basic molecular building blocks of cells and all parts of animals. Muscle, skin, blood, and internal organs contain proteins. Second only to DNA, proteins are among the largest organic molecules. A relatively small protein is shown in Figure 11.9.

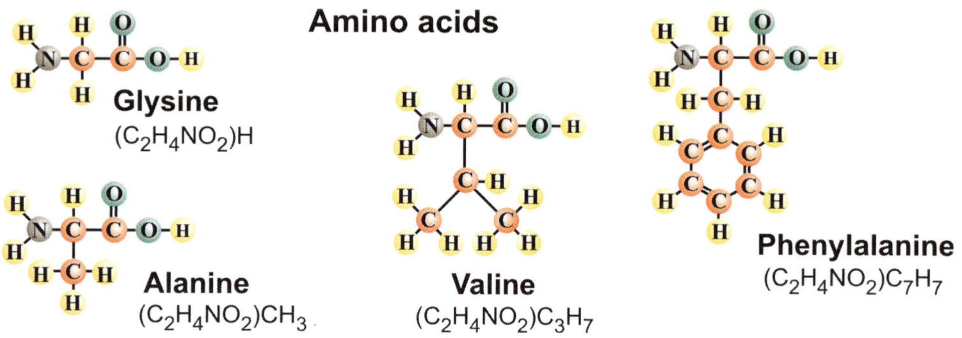

Proteins are made of amino acids

Amino acids are the building blocks of proteins. Virtually all proteins found in animals are made from only 20 different **amino acids**. The amino acids in a protein form multiple chains that fold around each other in complex structures (Figure 11.10).

Shape and function

Only certain parts of a protein are chemically active. The shape of a protein determines which active sites are exposed. Many proteins work together by fitting into each other like a lock and key. This is one reason proteins that do the same function in one organism do not work in another organism. For example, a skin protein from an animal cannot replace a skin protein from a human.

Amino acids from food are used to build proteins

Food supplies new proteins that a body needs to live and grow. However, proteins from one organism cannot be used by another. Fortunately, the same 20 amino acids are found in proteins from almost all living things. In your body, digestion breaks down food protein into its component amino acids. Cells reassemble the amino acids into new proteins suitable for your body's needs.

VOCABULARY

amino acids - organic molecules that are the building blocks of proteins.

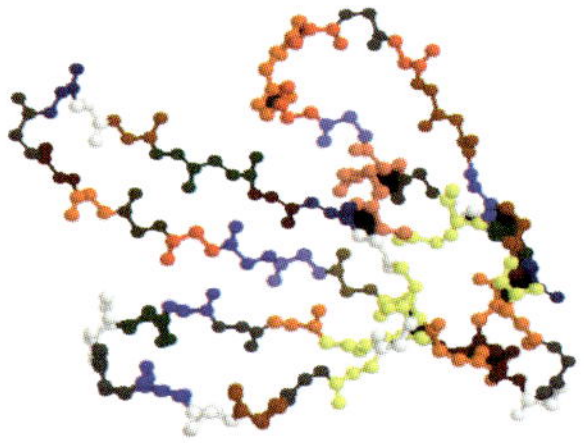

Figure 11.9: *This small protein called erabutoxin B is the active ingredient in sea snake venon.*

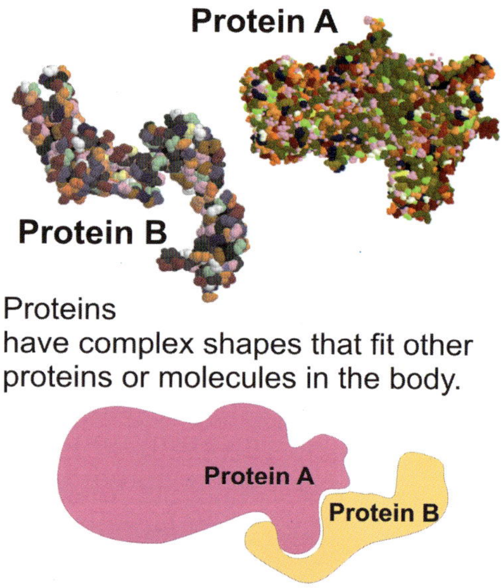

Figure 11.10: *The shape of a protein determines how it functions.*

Enzymes

The control problem

Thousands of chemical reactions are going on in your body each *second,* involving thousands of chemicals. The reactions proceed at just the right rate to produce energy as it is needed. When you exercise, the reaction rate increases because your body needs more energy. How does your body control its chemical reactions?

The temperature problem

Sugar (glucose) does not turn into water and carbon dioxide by itself. Outside the body, this reaction needs the intense heat of a flame. Yet your body causes this reaction to occur at only 37°C. How does the body cause reactions like this to occur at low temperature?

The answer is that *enzymes* allow your body to initiate chemical reactions at low temperature and to control the rate of reactions.

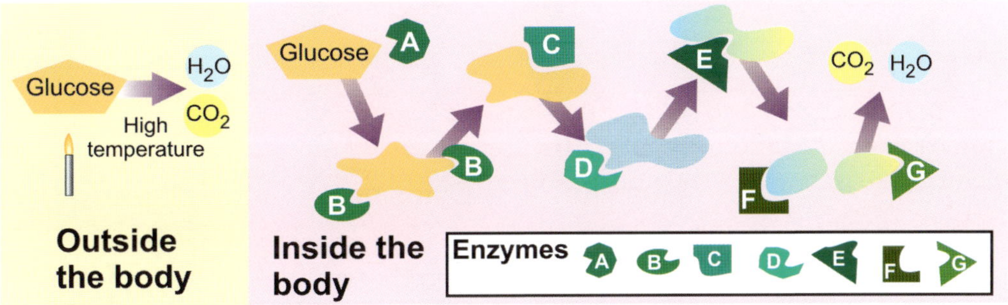

How enzymes solve the temperature problem

Enzymes are special proteins that are **catalysts** for chemical reactions. A *catalyst* is a chemical that allows a reaction to have a much lower activation energy than it normally would. You can think of catalysts like *helper molecules* that allow a reaction to proceed in many small steps instead of all at once. Each step uses only the thermal energy provided by ordinary body temperature.

How enzymes solve the control problem

The body controls the rate of reactions by regulating the amount of enzymes produced. For example, when a cell needs more energy, it produces more enzymes to break down glucose. Without those enzymes, glucose molecules stay together and store their energy for when it is needed.

VOCABULARY

catalyst - a chemical that allows a reaction to have a much lower activation energy than it normally would have.

enzymes - special proteins that are catalysts for chemical reactions in living things.

How do enzymes work?

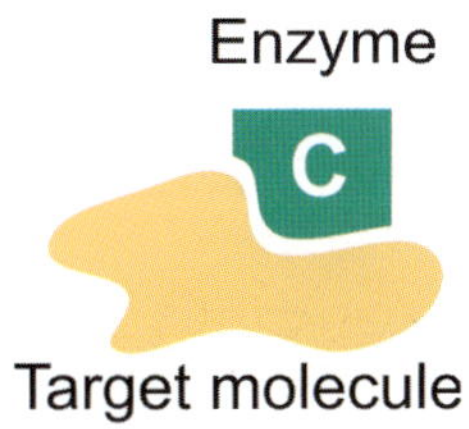

Enzyme molecules have special shapes that allow them to bind to their target molecule. The body has thousands of different enzymes because each one is highly specific and only works on its target molecule.

Enzymes are quite sensitive to temperature and pH. Most will not work outside a narrow range of temperature and pH.

DNA and nucleic acids

Protein synthesis Cells must continually create the proteins they need from amino acids. This process is called **protein synthesis** and it occurs inside every cell of your body. How does protein synthesis work? How are the instructions for building proteins remembered and carried out?

DNA The answer involves DNA, a nucleic acid. A DNA molecule is put together like a twisted ladder, or *double helix* (Figure 11.11). Each side of the ladder is made of 5-carbon sugars called deoxyribose and phosphate groups. The nitrogen bases are paired in the center of the ladder. DNA is among the largest molecules known. A single DNA molecule contains more than one million atoms.

The 4 nitrogen bases

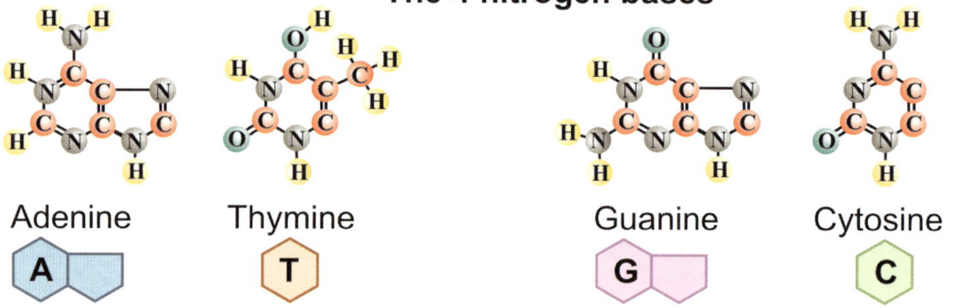

The four nitrogen bases There are four nitrogen bases in two matched pairs, adenine (A) with cytosine (C) and thymine (T) with guanine (G). The assembly instructions for building a protein are coded in the sequence of nitrogen bases on one side of the ladder. For example, TAA-GCT-AGG-GCT-GGC-GGC-TAA tells the cell: start-alanine-argenine-alanine-glycine-glycine-stop. This code would result in a protein with that sequence of five amino acids..

TTT	Phenylalanine	GCT	Alanine	GGT	Glycine
TTA	Leucine	CCC	Proline	GGC	Glycine
ATG	Methionine	GTT	Valine	AGG	Argenine
TAA	stop	ACA	Threonine	AGT	Serine

VOCABULARY

protein synthesis - using the information in DNA to assemble proteins from amino acids.

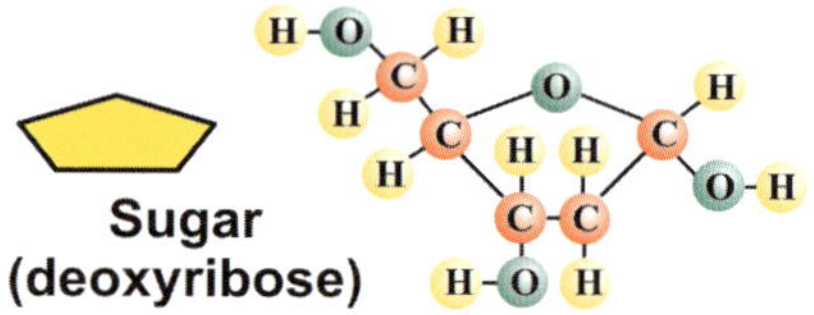

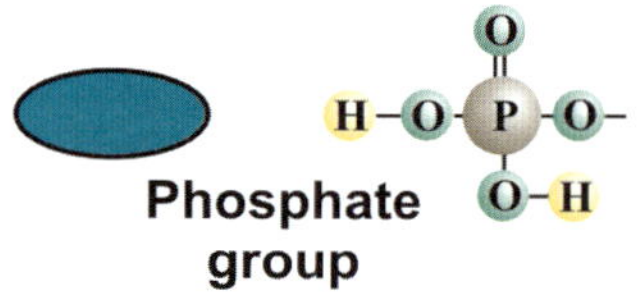

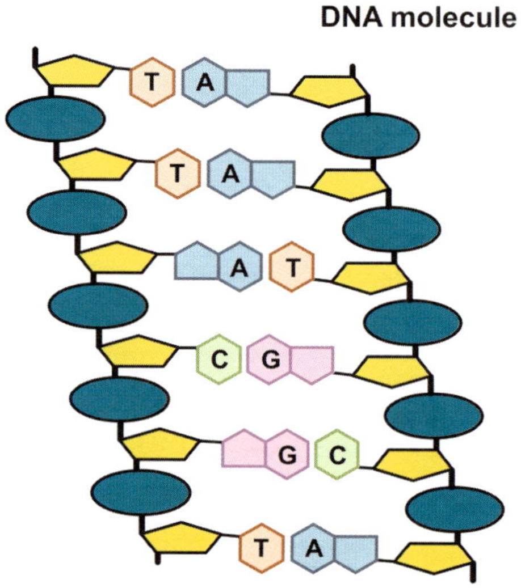

Figure 11.11: *The DNA molecule.*

DNA and reproduction

What reproduction does When an organism reproduces, it must pass on the chemical information for how to create every single protein in the organism. This is an incredible amount of information considering how many proteins there are and how complex a protein is.

Splitting the DNA molecule Fortunately, the DNA molecule is able to make exact replicas of itself. When a cell reproduces, enzymes split the DNA molecule down the center. Each half of the molecule contains a complementary code of nitrogen bases. Since guanine only pairs with cytosine and adenine only pairs with thymine, each half of the molecule contains the complete genetic information for how to make proteins.

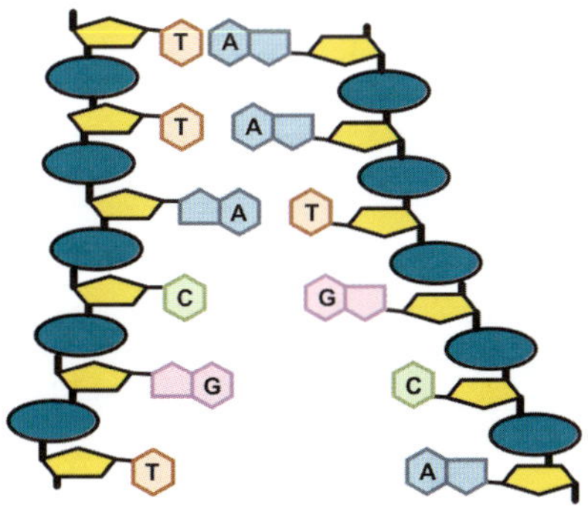

A DNA molecule "unzips" down the center

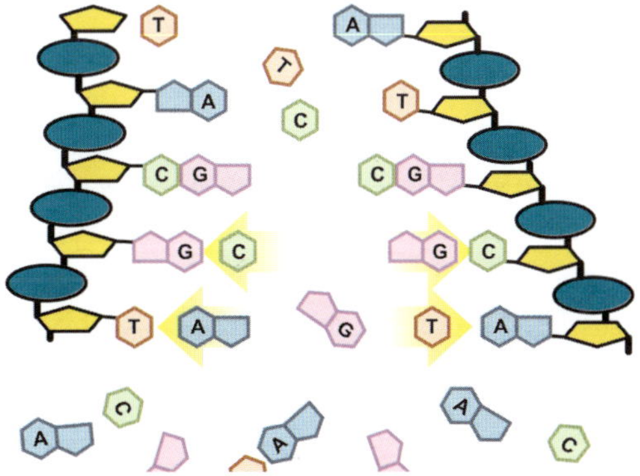

Each half is used as a template for making a new, identical DNA molecule

Rebuilding identical DNA molecules Other enzymes called *polymerases* move along the unzipped DNA molecule rebuilding the nitrogen bases on each side. Still more enzymes rebuild the sugar and phosphate backbone on top of the completed nitrogen base pairs. At the end of the process, there are two identical DNA molecules.

Error checking Another set of enzymes compares the old and new DNA strands for errors and corrects them by replacing nitrogen bases where necessary. We believe DNA replication occurs with less than one error out of every *billion* base pairs.

VOCABULARY

mutation - change in the sequence of base pairs in DNA that may be passed on to successive generations.

Mutations and evolution

Even with odds of 1 in a billion, over time, the sequence of bases in a DNA molecule does change through random replication errors. Radiation from the environment and other processes also change DNA. Changes in DNA are called **mutations**. Changes in DNA lead to new proteins, and changes in living organisms that are passed on in successive generations. This is the chemical basis for evolution.

Vitamins

Vitamins Most of the chemicals required for life can be synthesized by your own body, like proteins. However, there are certain chemicals necessary for the chemistry of life that the human body does not make. Collectively, these are called vitamins and minerals. In addition to carbohydrates, fats, and proteins, your body must get vitamins and minerals from food.

Vitamin C Ascorbic acid (Figure 11.12), also known as vitamin C, is required for synthesis of several important chemicals in your brain and nervous system. Vitamin C is also needed to synthesise compounds used in the transfer of energy within cells (ADP/ATP). Vitamin C must be supplied daily through food.

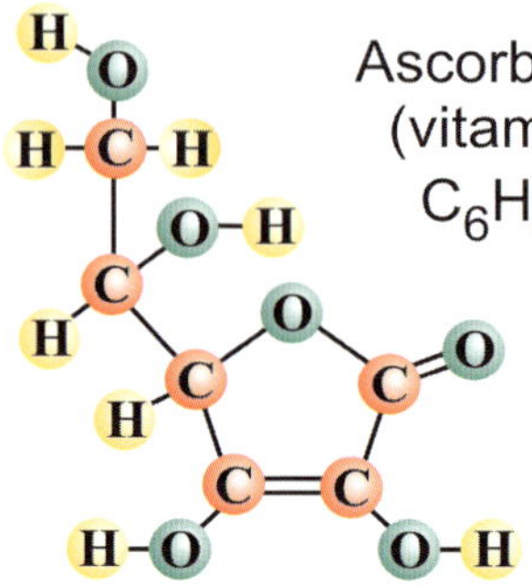

Figure 11.12: Ascorbic acid, also known as vitamin C.

Vitamin D Vitamin D includes several fat-soluble compounds known chemically as *calciferols*. Vitamin D is not a true vitamin since it can be synthesized by your skin when cholesterol reacts with ultraviolet light. However, sunscreens and clothes block UV rays from reaching the skin and can result in vitamin D deficiency. To help prevent this possibility, foods such as milk are being fortified with vitamin D2 or vitamin D3. A severe deficiency of vitamin D leads to softening of the bones called *rickets* in children and *osteomalacia* in adults.

The B vitamins The B vitamins include several compounds that must be obtained from food. The B vitamins often work together to bolster metabolism, maintain healthy skin and muscle tone, enhance immune and nervous system function, and promote cell growth and division, including that of the red blood cells that help prevent anemia. All B vitamins are water soluble, and are dispersed throughout the body and must be replenished daily, with any excess excreted in the urine.

B vitamins
Vitamin B-1 (thiamine)
Vitamin B-2, also vitamin G (riboflavin)
Vitamin B-3, also vitamin P or vitamin PP (niacin)
Vitamin B-5 (pantothenic acid)
Vitamin B-6 (pyridoxine and pyridoxamine)
Vitamin B-7, also vitamin H (biotin)
Vitamin B-9, also vitamin M (folic acid)
Vitamin B-12 (cyanocobalamin)

Folate Folate is another vitamin especially important during periods of rapid cell division and growth such as infancy and pregnancy. Folate is needed to make DNA and RNA. Both adults and children need folate to make normal red blood cells and prevent anemia.

11.2 Section Review

1. What is the difference between a saturated and unsaturated fat? Why are *partially hydrogenated fats* useful when making potato chips — and not particularly healthy for humans to eat?
2. Simple sugars are the building blocks of carbohydrates. What are the simple units that make up proteins?
3. All of the amino acids share something in common. What is it?
4. What type of biological molecule is an *enzyme*, and why are enzymes so important to living things?
5. Why is DNA important to the process of protein synthesis?
6. How does an organism pass on the chemical information for making proteins to the next generation?
7. Complete the table by filling in the missing information. You will have to do some research to fill in the last column.

Vitamin	Examples of why it is important	Foods that supply this vitamin
C		
D	strong bones	
B		
folate	needed to make RNA, DNA, and red blood cells	

8. One of the DNA sequences in Figure 11.13 is impossible. Which one is wrong and why is it wrong?
9. Which of the DNA sequences in Figure 11.14 contains a mutation?
10. A vitamin is a chemical
    a. in food that is needed but not produced in the body.
    b. that is produced in the body and not needed from food.
    c. that is found in food but not used by the body.
    d. that is found in the body but not found in food.

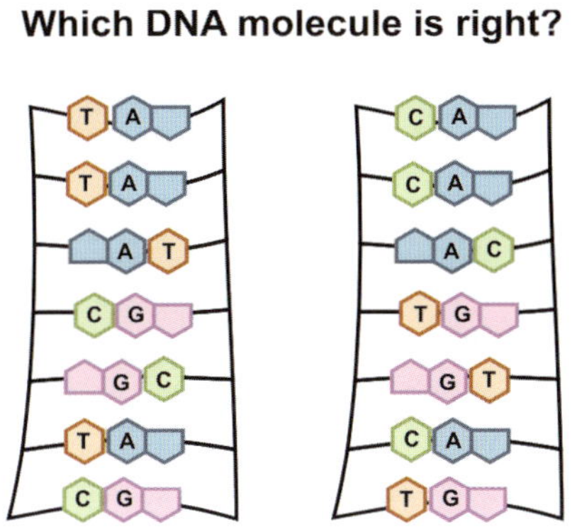

Figure 11.13: *Only one of the DNA molecules shown is correct. Which one is it (question 8)?*

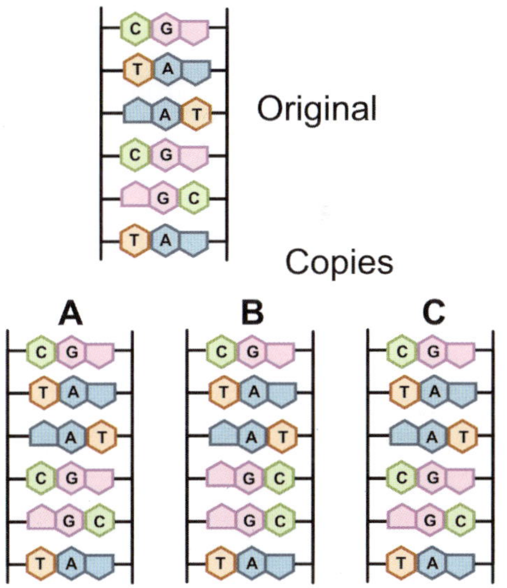

Figure 11.14: *A DNA molecule and three copies of the same molecule.*

HEALTH CONNECTION Good Fats vs. Bad Fats

What is your favorite food? French fries, ice cream, cookies, maybe, or candy? Does just reading that make your mouth water? Many of us eat these tasty treats every now and then, but eating too much of high-fat foods can be bad for our health. What does "high-fat" actually mean, and are all fats bad for us?

The lowdown on fats

We need a reasonable amount of fat in our diets. Fat helps support cell function and helps our bodies absorb vitamins. But a diet too high in certain fats can lead to many health problems, including obesity, diabetes, and heart disease.

High cholesterol is a leading risk factor for heart disease. It can cause the deposit of fatty buildups in our arteries called plaques. Plaques narrow the arteries and so reduce blood flow, a condition called atherosclerosis.

Cholesterol-carrying compounds called lipoproteins play a key role in the development of atherosclerosis and heart disease. Low-density lipoproteins (LDL) transport cholesterol from the liver to the rest of our body. When there is too much LDL cholesterol in the blood, it can begin to build up on the walls of the arteries. LDL cholesterol is called "bad cholesterol" because of this buildup. High-density lipoproteins (HDL) transport cholesterol from the blood back to the liver for removal. HDL cholesterol is less likely to be deposited in the arteries, and so is referred to as the "good cholesterol."

A low-fat diet will help to reduce the risk for heart disease. But what truly affects our health is the type of fat and the total amount we eat.

Unsaturated, saturated, and trans fats

Based on their chemical structure, fats are either unsaturated or saturated. An unsaturated fat has two or more carbon atoms that are not bonded to hydrogen. In a saturated fat, all the carbon atoms have the maximum number of hydrogen atoms attached.

Unsaturated fats, known as "good fats," lower LDL and reduce the risk for heart disease. Unsaturated fats are liquid at room temperature and are found in olive and canola oils, avocados, and some nuts. Fatty fishes, such as salmon and tuna, are good sources of unsaturated fat.

Saturated fats are referred to as "bad fats." Our livers easily convert them to LDL cholesterol, which increases the risk for heart disease.

Saturated fats are solid at room temperature and are found in whole-milk dairy products like cheese, cream, and butter, and also in meat and poultry. Coconut and palm oils are among plant foods that are high in saturated fats.

Trans fatty acids, or trans fats, are included in the "bad fats" that raise LDL cholesterol levels. Food manufacturers produce trans fats by adding hydrogen to liquid vegetable oils, a process known as hydrogenation. The more hydrogenated the oil, the harder it is at room temperature. For example, a tub of spreadable margarine is less hydrogenated than a stick of margarine. Trans fats are found in most fried foods and in many processed foods, such as cakes and cookies.

The new food pyramid

For over 100 years, the U.S. Department of Agriculture (USDA) has made recommendations on the types and amounts of food we should eat. Recently, the USDA developed "MyPyramid," an interactive food guidance system. MyPyramid incorporates recommendations from the USDA's 2005 Dietary Guidelines for Americans. Using the USDA website (www.MyPyramid.gov), and entering their age, gender, and activity level, consumers can design their own healthy food plans.

The MyPyramid symbol encourages people to personalize their food choices and to incorporate daily physical activity in their lives. In terms of fat intake, the USDA recommends low-fat or fat-free milk products, that less than 10 percent of calories come from saturated fats, and that trans fats be kept to a minimum. Most fats we eat should come from unsaturated sources, such as fish, nuts, and vegetable oils.

Changes on food labels

Nutrition Facts

Serving Size 28g **Tortilla chips**

Amount Per Serving	
Calories 138	Calories from Fat 58
	% Daily Value*
Total Fat 7g	10%
Saturated Fat 1g	3%
Trans Fat	
Cholesterol 0mg	0%
Sodium 119mg	5%
Total Carbohydrate 18g	6%
Dietary Fiber 1g	6%
Sugars 0g	
Protein 2g	
Vitamin A 0% • Vitamin C	0%
Calcium 5% • Iron	4%

*Percent Daily Values are based on a 2,000 calorie diet. Your daily values may be higher or lower depending on your calorie needs.

NutritionData.com

By Jan. 1, 2006, the Food and Drug Administration (FDA) is requiring that food labels list trans fats. This will be in addition to the listing of overall fat, saturated fat, and cholesterol. The FDA hopes that the more detailed label will give consumers a complete picture of the foods they buy - and help them make healthier food choices and ultimately live healthier lives.

Questions:

1. Why are low-density lipoproteins (LDL) called "bad cholesterol" and high-density lipoproteins (HDL) called "good cholesterol"?
2. How do unsaturated, saturated, and trans fats affect LDL and HDL levels?
3. What changes are being made to food labels by Jan. 1, 2006, and why?

CHAPTER ACTIVITY The Scoop on Nutrition Labels

All packaged foods are required to contain nutrition labels to help consumers choose healthy foods. A nutrition label shows the amount of calories, fat, cholesterol, carbohydrates, protein, and several vitamins and minerals in one serving of the food.

The exact amount of each nutrient a person needs depends on gender, age, activity level, and weight. An average female teenager should consume approximately 2200 calories and not more than 73 grams of fat per day. A male should consume 2800 calories and not more than 93 grams of fat. Protein requirements depend on body mass and amount of physical activity. Teenagers need between 1 and 1.5 grams of protein per kilogram of body mass.

Materials:

Nutrition labels from food packages or from the internet
Poster board

What you will do

1. Look at a variety of nutrition labels, including those for foods considered to be healthy and those that are unhealthy. You can use the internet to find information on foods that don't come in packages, such as fruits and vegetables.
2. Suppose your doctor recommends a diet of 2400-2600 calories, fewer than 80 grams of fat, and at least 65 grams of protein. Use nutrition labels to select an assortment of food to eat in one day that meets these requirements and includes 100% of the daily requirement of vitamins A, vitamin C, vitamin D, calcium, and iron.
3. Make a poster describing your menu for the day. List the number of calories, grams of fat and protein, and percentage of each nutrient.

Applying your Knowledge

a. The term "empty calories" is used to describe foods that have a high number of calories but little nutritional value. List three foods that contain empty calories.

b. Which of the foods you selected for your menu contains the greatest amount of protein?

c. Which of the foods contains the greatest amount of fat?

d. Which three of your selected foods would you consider to be the most healthy? Explain why you chose these foods and what you think it means to say that a food is healthy.

Nutrition Facts

Serving Size 1 fruit 27/8 (140g)

Amount Per Serving	
Calories 69	Calories from Fat 2
	% Daily Value*
Total Fat 0g	0%
Saturated Fat 0g	0%
Trans Fat	
Cholesterol 0mg	0%
Sodium 1mg	0%
Total Carbohydrate 18g	6%
Dietary Fiber 3g	12%
Sugars 12g	
Protein 1g	

Vitamin A	7%	• Vitamin C	138%
Calcium	6%	• Iron	1%

*Percent Daily Values are based on a 2,000 calorie diet. Your daily values may be higher or lower depending on your calorie needs:

	Calories	2,000	2,500
Total Fat	Less than	65g	80g
Sat Fat	Less than	20g	25g
Cholesterol	Less than	300mg	300mg
Sodium	Less than	2,400mg	2,400mg
Total Carbohydrate		300g	375g
Fiber		25g	30g

Calories per gram:
Fat 9 • Carbohydrate 4 • Protein 4

NutritionData.com

Chapter 11 Assessment

Vocabulary

Select the correct term to complete the sentences.

nucleic acid	fat	carbohydrates
photosynthesis	unsaturated	proteins
cellular respiration	organic chemistry	partially hydrogenated
catalyst	protein synthesis	nitrogen bases
amino acids	mutations	enzymes

Section 11.1

1. The branch of chemistry that specializes in carbon and carbon compounds is called ____.
2. The chemical energy that supports the food chain on Earth comes from a reaction called ____.
3. The reaction that breaks down glucose and releases its stored energy is called ____.
4. Sugars and starches are classified as ____.
5. DNA is an example of a(n) ____.

Section 11.2

6. High-energy ____ molecules are used to store energy in reserve.
7. ____ are made up of amino acids.
8. When a fat molecule has two hydrogen atoms bonded to each carbon atom, it is called a ____ fat.
9. When a fat molecule has some carbon atoms double bonded to each other, along with hydrogen atoms, it is called a(n) ____ fat.
10. ____ are organic molecules that are the building blocks of proteins.
11. ____ allow your body to initiate chemical reactions and control the reaction rates.
12. Changes in DNA are called ____.
13. Enzymes are a type of ____ for chemical reactions.
14. The process the cells in your body use to build proteins from amino acids is called ____.
15. The molecular components within DNA that contain the code for building proteins from amino acids are ____.

Concepts

Section 11.1

1. What do all organic molecules have in common?
2. What makes carbon uniquely suited to being the basis for biological molecules?
3. Describe the four types of biological molecules. Give an example for each type:
   a. carbohydrate
   b. fat
   c. protein
   d. nucleic acid
4. What elements are carbohydrates made of?
5. Why do sugars break down so quickly in your body?
6. What is the difference between a starch molecule and a cellulose molecule?
7. Why does the high specific heat of water make it essential to life?
8. Explain how photosynthesis and respiration are related carbon reactions.

9. Which process adds oxygen (O_2) to Earth's atmosphere?
   a. photosynthesis
   b. cellular respiration
   c. protein synthesis
10. What process removes carbon dioxide (CO_2) from the atmosphere?
    a. photosynthesis
    b. cellular respiration
    c. protein synthesis
11. What process does the diagram illustrate?

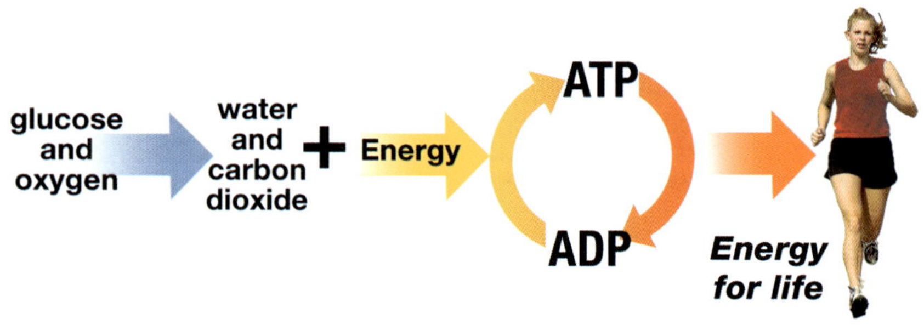

12. Digestion breaks down food into molecules the body can use. What type of molecules are each of the following broken down into?
    a. proteins
    b. carbohydrates
    c. fats

Section 11.2

13. Describe how your body allows chemical reactions to occur at low temperature?
14. What is the function of fats in the human body?
15. What is the role of a catalyst in a chemical reaction? Describe how enzymes act as catalysts.
16. What is the structure of DNA called?
    a. Nitrogen bases
    b. Protein synthesis
    c. Double helix
17. How are mutations the chemical basis for evolution?
18. In the process of DNA reproduction, how are errors fixed?
19. Nitrogen bases:
    a. are amino acids.
    b. hold the codes for building proteins.
    c. initiate chemical reactions.
20. Which function does DNA perform?
    a. It reproduces itself exactly.
    b. It controls chemical reactions in the body.
    c. It provides energy for cells.

Problems

Section 11.1

1. Classify each of the following carbohydrates as containing mostly sugar, starch, or cellulose:
   a. a stack of firewood
   b. rice
   c. jelly beans
   d. a shirt made of cotton
   e. an apple
2. The human body is made mostly of:
   a. carbon, oxygen, nitrogen, and hydrogen.
   b. oxygen, calcium, carbon, and hydrogen.
   c. hydrogen, iron, nitrogen, and oxygen.

3. All plants use the process of photosynthesis. However, this process wasn't always understood. In one classic experiment, a small plant and its soil were weighed. The plant was given only water for a solid year. At the end of the year, the plant weighed much more than it did at the end of the first of the year. The soil weighed the same amount. Where did the extra weight of the plant come from?
4. A product of cellular respiration is energy. What is this energy used for?
5. Which of the following compounds are organic?
   a. nucleic acid
   b. CH_4
   c. H_2O
   d. hydrochloric acid
   e. table salt
   f. sugar

Section 11.2

6. Identify each of the following as a carbohydrate, fat, protein, or nucleic acid.
   a. glucose
   b. DNA
   c. cholesterol
   d. cellulose
   e. olive oil
7. An organic compound contains carbon, hydrogen, oxygen, and nitrogen. Could this compound be a fat? Could it be a nucleic acid? Explain.
8. What is the relationship between proteins and nucleic acids?
9. About how many different amino acids are found in animal proteins?
   a. 2
   b. 4
   c. 20
10. Which of the following is NOT part of the process for the body to get the essential proteins it needs?
    a. protein synthesis
    b. digestion of food protein into amino acids
    c. the manufacturing of amino acids from fats
11. Of the four nitrogen base pairs, adenine always pairs with:
    a. adenine
    b. guanine
    c. thymine
    d. cytosine
12. Your body produces proteins it needs through the process of protein synthesis. How does your body obtain the vitamins it needs?
13. The diagram shows an enzyme and three different molecules. Which of the three molecules would this enzyme target for a reaction?

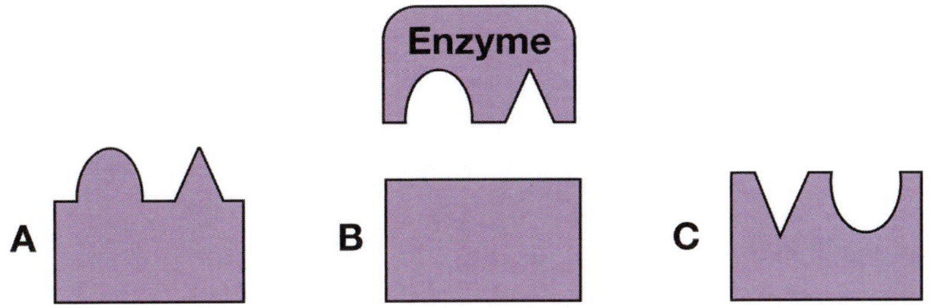

UNIT 5

Motion and Force

TRY THIS AT HOME

Experiment with falling objects. Drop a sheet of notebook paper and a paperback book from the same height. Which one hits the ground first? Why? Next, crumple the sheet of paper into a ball and drop it from the same height as the book. Does one hit the ground before the other, or do they hit at about the same time? Why? Draw and label diagrams to explain your results.

Chapter 12

Distance, Time, and Speed

Do you know which animal is the fastest? You may have heard that the cheetah is the fastest land animal. A cheetah is able to run at a top speed of 70 mph (113 kph). But the cheetah is not the fastest animal on Earth. Scientists have measured the speeds of peregrine falcons to be over 200 mph (332 kph)! Falcons reach this top speed when diving through the air toward prey on the ground. In comparison, the fastest human can reach a speed of only 23 mph (37 kph) while running over a short distance. People are able to use their own power to move much more quickly with a bicycle. The world record speed of a cyclist is 81 mph (130 kph). This record was reached by a person on a recumbent bicycle in which the rider sits with his or her legs stretched in front of the body rather than below. Recumbent bicycles are twice as fast as upright bicycles. One reason is that they are more aerodynamic. Another reason is pedaling a recumbent bicycle requires different leg muscles than pedaling an upright bicycle. These muscles are stronger and allow the rider to pedal with more force.

Key Questions

1. *How can an object's exact location be described?*
2. *What is the relationship between distance, time, and speed?*
3. *How can graphs be used to represent motion?*

12.1 Distance, Direction, and Position

If asked how far you live from your school, you may say the distance is 5 miles or 8 kilometers. But if you asked where your house is located compared to the school, you would probably give a different answer. To describe the exact location or position, you must specify a *direction* as well as a distance.

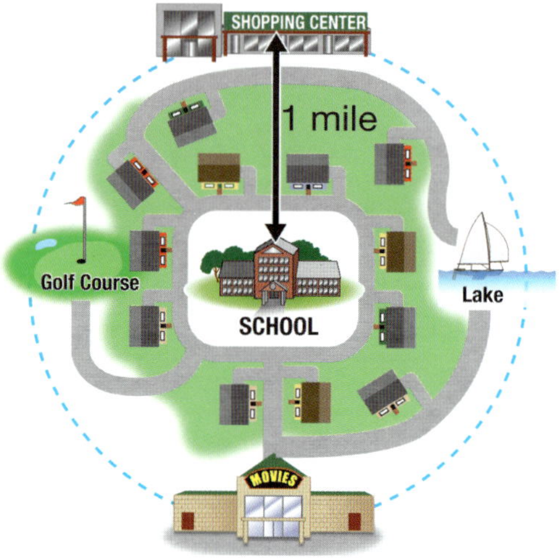

Figure 12.1: *All of the points on the circle are two miles from the school.*

Direction

Distance Distance is the amount of space between two points. A distance measurement does not include information about the direction you would have to travel to get from one point to the next. If you say you live a distance of one mile from school, you could mean anywhere on the circle of points shown in Figure 12.1.

Different ways to give directions To find the exact location of your house, you must give both the distance and the *direction* from school. There are several ways of specifying direction. You might tell someone to turn right in front of the school, go 4 miles on Main Street, turn left on 7th Avenue, go 3 miles, and stop at 623 7th Avenue, and then walk left to your front door. The direction words left and right, along with up and down, are used in everyday life because they are simple to follow. You could also give directions to your house using the compass directions north, south, east, and west.

Navigation Compass directions are printed on maps and used by ships and planes for navigation. The four perpindicular directions are north, south, east, and west. Directions in between these four are given with angles relative to one of the four principle directions. For example, a pilot flying from New York to Boston must direct the plane at an angle of 36 degrees north of east (Figure 12.2). This angle is measured counter-clockwise from east, which corresponds to the way angles are given in math. This same direction could also be given as 54 degrees east of north. Most compasses are labeled in degrees clockwise from north.

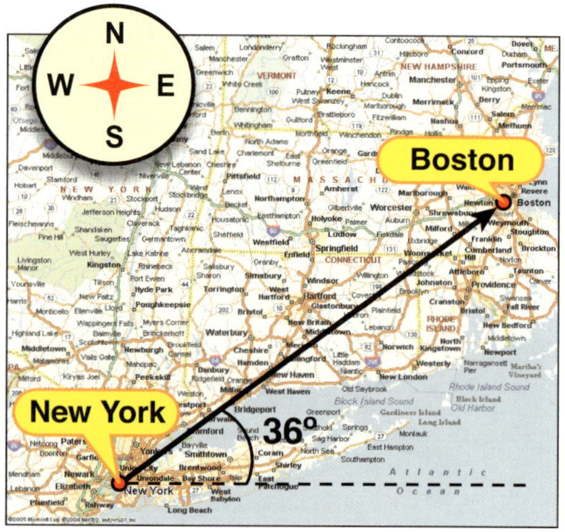

Figure 12.2: *Boston is a distance of 188 miles (303 km) at a direction of 36 degrees north of east from New York.*

Position

Location compared to an origin

An object's **position** is its location at one instant in time. A position is always given compared to an **origin**. The origin is a fixed reference point. When giving directions to your house from school, the school is the origin. The position of your house might be 1 mile north of school.

Positive and negative numbers

Both positive and negative numbers are used to describe position. A positive position usually means in front of, to the right of, or above the origin. A negative position usually means behind, to the left of, or below the origin. It is important to define the positive and negative directions if you are doing an experiment.

Doing experiments

If an object is moving in a line, its position is usually given in meters away from its starting point. For example, suppose you are doing an experiment with a car on a track. It is easiest to define the origin at the position where the car starts (Figure 12.3).

Using a number line

A number line can be used to think about position in straight-line motion. The origin is at zero. If you start at the origin and move 3 meters to the right, your position is +3 meters. If you then move 5 meters to the left, your position is -2 meters. You can add each change in position to calculate your final position.

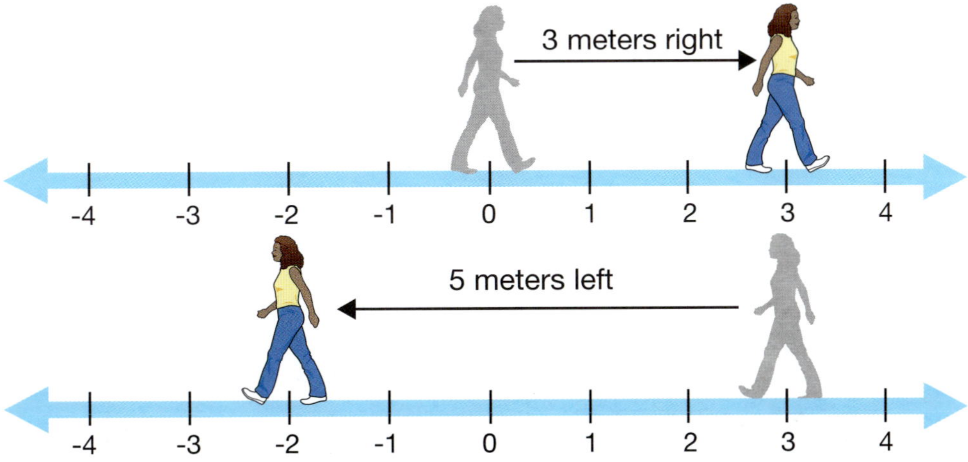

VOCABULARY

position - the location of an object compared to a reference point.

origin - a fixed reference point.

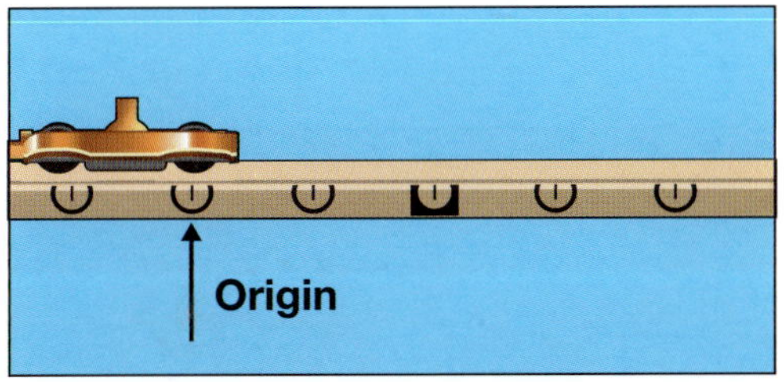

Figure 12.3: *It is convenient to define the origin at the end of the track where the motion begins.*

Position in two dimensions

Motion in a line, on a plane, and in space

If an object can only move in a line, its motion is called one-dimensional. Motion on a flat surface (a plane) is two-dimensional. How you describe an object's position depends on whether you are studying its motion in one or two dimensions. A single number fully describes position in one dimension. Two numbers are needed to describe position in two dimensions (Figure 12.4).

Coordinates describe position

Two number lines at right angles to each other are used to show position in two dimensions. The two number lines make up what is called an *x-y plane*. Two numbers, called *coordinates*, are used to describe the position. The x-coordinate describes the position to the left/right or east/west. The y-coordinate describes the position up/down or north/south.

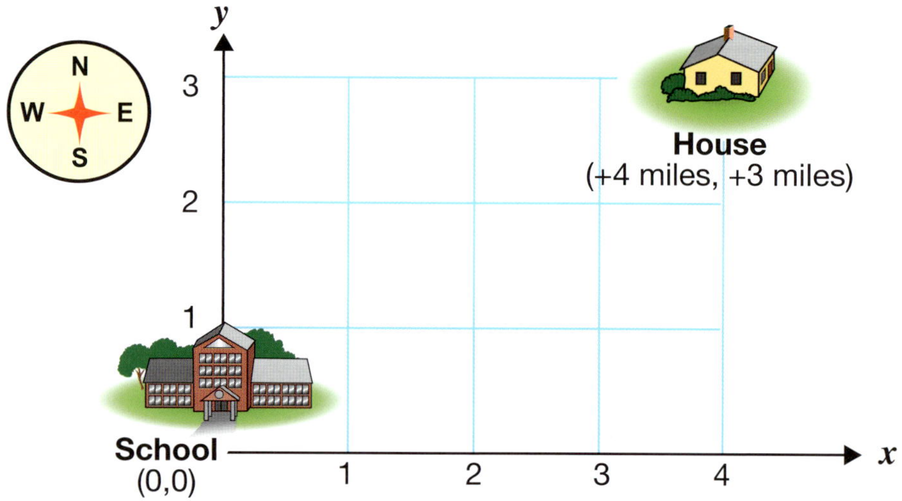

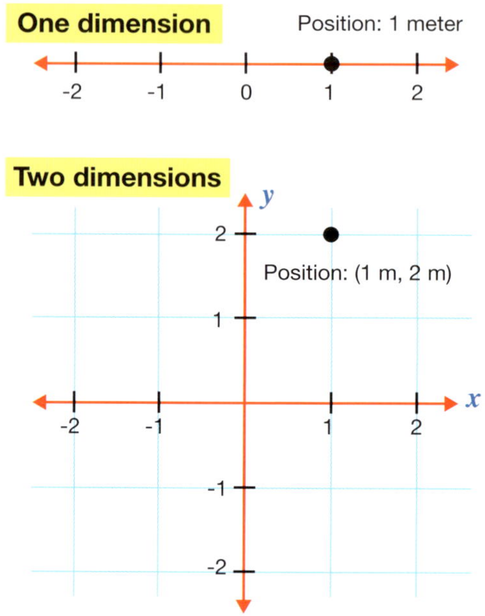

Figure 12.4: *The number of numbers needed to describe an object's position is the same as the number of dimensions.*

You must define an origin

You can describe the position of your house compared to your school on an *x-y* plane. Let your school be the origin. It is convenient to define the positive *x*-direction as east and the positive *y*-direction as north.Your school is 4 miles east and 3 miles north. The coordinates for this position are (+4 miles, +3 miles).

12.1 Section Review

1. What two pieces of information do you need to know to get from one location to another?
2. What is the difference between *distance* and *position*?
3. You start at the origin and walk 3 meters east, 7 meters west, and 6 meters east. Where are you now?
4. Give an example of a situation in which you would describe an object's position in:
   a. one dimension
   b. two dimensions
   c. three dimensions

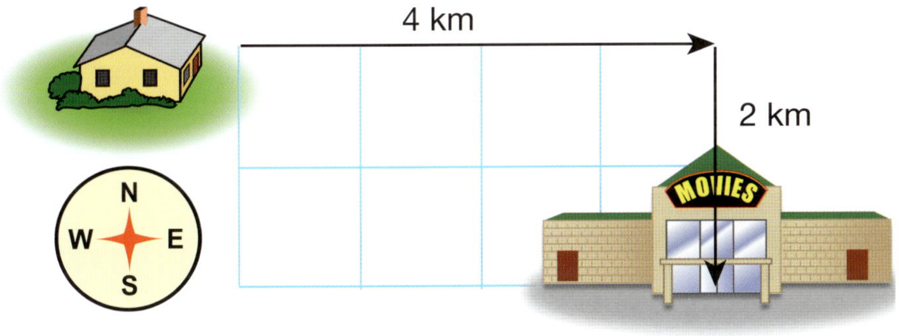

5. A movie theater is 4 kilometers east and 2 kilometers south of your house.
   d. Using your house as the origin, give the coordinates of the movie theater.
   e. After leaving the movie theater, you drive 5 kilometers west and 3 kilometers north to a restaurant. What are the coordinates of the restaurant? Use your house as the origin.
6. Does the origin of an object always have to be at zero on a number line or x-y plane? Why or why not?

CHALLENGE

A plane takes off in San Francisco at noon and flies toward the southeast. An hour later, it is 400 kilometers east and 300 kilometers south of its starting location.

Assuming the plane flew in a straight line, how far did it travel?

How many degrees south of east did the plane fly?

12.2 Speed

Speed is the most common measurement used to describe the motion of objects. Saying a race car, runner, or plane is "fast" is not enough to accurately describe its speed. In this section, you will learn a precise definition of speed. Once you know an object's speed, you can figure out how far it can go in a certain amount of time. Or you can predict how long it will take to get somewhere.

Speed

An example of speed

Imagine two bicycles moving along the road at different speeds. The picture below shows the position of each at one second intervals. The fast bicycle (the bottom one) moves three times the speed of the slow one. The fast bicycle moves 3 meters each second, while the slow bicycle moves only 1 meter each second.

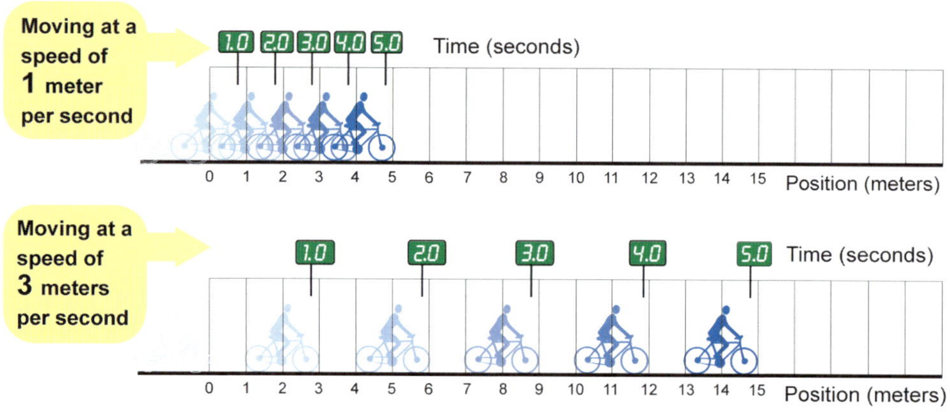

Speed is distance divided by time

The **speed** of a bicycle is the distance it travels divided by the time it takes. At 1 m/s, a bicycle travels one meter each second. At 3 m/s, it travels three meters each second. Both bicycles in the diagram are moving at **constant speed**. Constant speed means the same distance is traveled every second. The snapshots are evenly spaced, so you know the distance traveled by each bicycle is the same each second.

VOCABULARY

speed - the distance an object travels divided by the time it takes.

constant speed - speed of an object that travels the same distance each second.

MY JOURNAL

We depend on cars, buses, trains, and planes to get us from place to place at high speeds. But people living 200 years ago could not travel so easily. Five miles per hour is about what a horse can comfortably go for a long distance. Compare the lives of people today with those of people living 200 years ago. Disucss the ways being able to travel at speeds greater than ten miles per hour has changed our everyday lives.

Velocity

Direction Speed tells you *how fast* something is moving. Sometimes it is also important to know the *direction* an object is moving. An air traffic controller must keep track of all the planes flying in an area. Information about the direction each is headed is as important as knowing the speed.

Velocity is speed with direction Speed with direction is called **velocity**. Like position, the compass headings north, south, east, and west are often used for the velocity of planes and ships. The words up, down, left, and right can also be used to describe velocity.

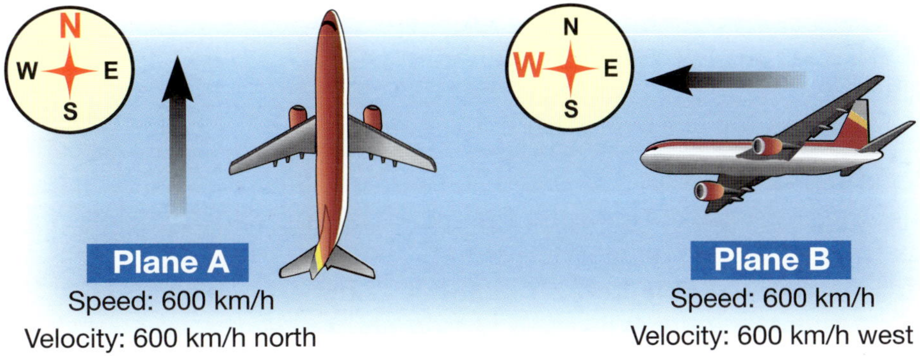

Comparing velocities Two objects with the same velocity must have the same speed and must be moving in the same direction. Both planes in the diagram above have a speed of 600 km/h. Plane A has a velocity of 600 km/h north, and plane B has a velocity of 600 km/h west. The planes are moving at the same speed, but their velocities are different.

Constant velocity An object with constant velocity has both constant speed and constant direction. A plane flying with a constant velocity of 600 km/h north keeps moving toward the north. The plane can only do this if it moves in a straight line. Any object with constant velocity must move in a straight line (Figure 12.5). A car going around a bend at a constant 30 mph does not have a constant velocity because its direction is changing (Figure 12.6).

VOCABULARY

velocity - speed with direction.

Figure 12.5: *A plane with a constant velocity of 600 km/h west moves in a straight line.Its speed and direction remain the same.*

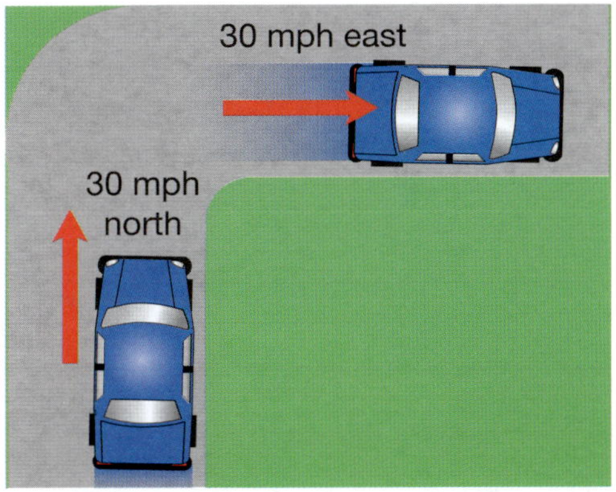

Figure 12.6: *A car going around a bend at a constant 30 mph has a constant speed, but its velocity is not constant.*

Calculating speed

Speed is distance divided by time

Speed is a measure of the *distance* traveled in a given amount of *time*. You calculate speed by dividing the distance traveled by the time taken. For example, if you drive 150 kilometers in 1.5 hours, then the speed of the car is 150 kilometers divided by 1.5 hours or 100 kilometers per hour (Figure 12.7). The speed found by dividing the total distance by the total time is the **average speed**. A car speeds up and slows down during a trip as it moves up and down hills, gets stuck in traffic, and reaches intersections. When talking about speed, we usually mean the average speed over a certain time period.

What does "per" mean?

The word "per" means "for every" or "for each." The speed of 100 kilometers per hour is short for saying 100 kilometers *for each* hour. You can also think of "per" as meaning "divided by." The quantity before the word per is divided by the quantity after it. For example, 150 kilometers divided by 1.5 hours (or per every 1.5 hours) equals 100 miles per hour.

Units for speed

Since speed is a ratio of distance over time, the units for speed are a ratio of distance units over time units. In the metric system, distance is measured in centimeters, meters, or kilometers. If distance is in kilometers and time in hours, then speed is expressed in kilometers per hour (km/h). Other metric units for speed are centimeters per second (cm/s) and meters per second (m/s). Speed is also commonly expressed in miles per hour (mph). Table 12.1 shows different units commonly used for speed.

Table 12.1: Common units for speed

Distance	Time	Speed	Abbreviation
meters	seconds	meters per second	m/s
kilometers	hours	kilometers per hour	km/h
centimeters	seconds	centimeters per second	cm/s
miles	hours	miles per hour	mph
feet	minutes	feet per minute	ft/min, fpm

VOCABULARY

average speed - the total distance divided by the total time for a trip.

$$\frac{\text{150 kilometers}}{\text{1.5 hours}} = \text{100 kilometers (km/h)}$$

Figure 12.7: *A driving trip with an average speed of 100 km/h*

Relationships between distance, speed, and time

Mixing up distance, speed, and time

A common type of question in physics is: "How far do you go if you drive for two hours at a speed of 100 km/h?" You know how to get speed from time and distance. How do you get distance from speed and time? The answer is the reason mathematics is the language of physics. An equation can be used to calculate speed, distance, or time if two of the three values are known.

Calculating speed

Let the letter the letter d stand for "distance traveled" and the letter t stand for "time taken." The letter v is used to represent "speed" because it refers to the word *velocity*. You follow the same steps whether you are calculating speed or velocity. The only difference is that when giving an object's velocity, you must include the direction of its motion in addition to the speed.

SPEED

Speed (m/s) → $v = \dfrac{d}{t}$ ← *Distance traveled* (meters) (numerator); *Time taken* (seconds) (denominator)

There are three ways to arrange the variables that relate distance, time, and speed. You should be able to work out how to get any one of the three variables if you know the other two (Figure 12.8).

Using formulas

Remember that the words or letters stand for the values that the variables have. You can think about each letter as a box that will eventually hold a number. Maybe you do not know yet what the number will be. Once we get everything arranged according to the rules, we can fill the boxes with the numbers that belong in each one. The last box left will be our answer.

Equation	gives you	if you know
$v = d \div t$	speed	distance and time
$d = vt$	distance	speed and time
$t = d \div v$	time	distance and speed

Figure 12.8: *Different forms of the speed equation.*

STUDY SKILLS

Solved example problems can be very helpful when you are studying for a test. Cover up the solution to a problem, and then try to solve it on your own. Fully write out the steps you follow.

Look at the answer to see if you are correct. If your solution is wrong, compare your steps to those in the example problem. Figure out where you made a mistake. Then try to solve the additional problems below the example problem.

12.2 Section Review

1. If something moves at a constant speed, what do you know about the distance it moves each second?
2. What is the difference between speed and velocity?
3. A ball rolls along the ground at a constant velocity. Describe the path it follows.
4. What type of speed does the v in the formula $v = d/t$ represent?
5. Calculate the average speed (in km/h) of a car that drives 280 kilometers in 4 hours.
6. You ride your bicycle at an average speed of 15 km/h for 2 hours. How far did you go?
7. How long (in seconds) will it take you to run 100 meters if you run at 5 m/s?
8. A boat sails at an average speed of 20 km/h for two days. How far does the boat go?

CHALLENGE

You leave your house at 7:30 AM to go to school. You arrive at 7:50 AM, and the school is 3 miles away.

What was your average speed in mph?

If the speed limit on the road was 40 mph, how long should it have taken you?

Explain why your calculated time does not match the actual time it took you to get to school.

12.3 Graphs of Motion

Motion graphs are an important tool used to show the relationships between position, speed, and time. For example, meteorologists use graphs to show the motion of hurricanes and other storms. A graph can show the location and speed of a storm at different points in time. The graph can be used to help predict the path of a storm and the time when it will reach a certain location. In this section, you will learn how to make and use graphs of position versus time and speed versus time to describe the motion of different objects.

Position vs. time graphs

Recording data Suppose you are helping a friend who is training for a track meet. She wants to know if she is running at constant speed during a section of the race. You mark the track in 50-meter increments and measure her time at each position during a practice run. The data for your experiment is shown in Figure 12.9.

Calculating speed The data shows that your friend took 10 seconds to run each 50-meter segment. Because the time was the same for each segment, you know her speed was the same for each segment. You can use the formula $v=d/t$ to calculate the speed. Dividing 50 meters by 10 seconds results in a speed of 5 meters per second.

Graphing the data You make a graph of the data by plotting the four points on graph paper and connecting them with a smooth line. Notice that when moving from each data point to the next, the graph goes over 10 seconds and up 50 meters. This causes the points to fall exactly in a straight line. A position vs. time graph that is a straight line always means the object moves the same distance during each time period. An object moving at a constant speed always creates a position vs. time graph that is a straight line.

Position and Time Data for a Runner

Time (s)	Position (m)
0	0
10	50
20	100
30	150

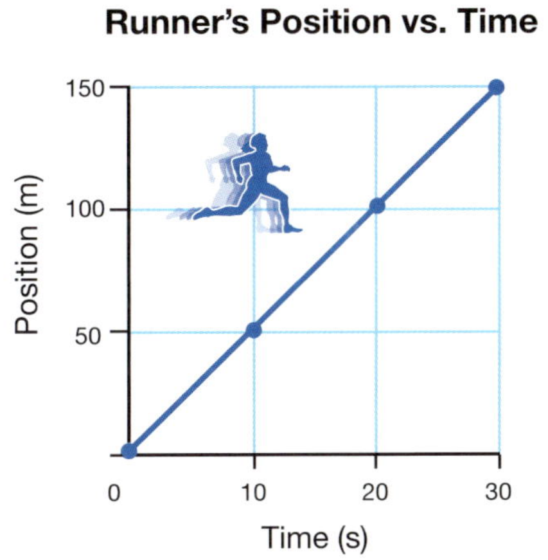

Figure 12.9: *The data table and position vs. time graph for a runner.*

Slope

Comparing graphs You can use position vs. time graphs to compare the motion of different objects. Figure 12.10 shows the position vs. time graph for two people running along a 600-meter section of a jogging path. Both runners start at the beginning of the path (the origin) at the same time. Runner A takes 100 seconds to cover 600 meters, and runner B takes 150 seconds. Using $v=d/t$, runner A's speed is 6 m/s and runner B's speed is 4 m/s. You can see that runner A's speed is faster by looking at the two lines on the graph. Runner A's line is steeper. A steeper line on a position vs. time graph means a faster speed.

A steeper line on a position vs. time graph means a faster speed.

Calculating slope The steepness of a line is measured by finding its slope. The **slope** of a line is the ratio of the "rise" (vertical change) to the "run" (horizontal change). The diagram below shows you how to calculate the slope of a line. The rise is equal to the height of the triangle. The run is equal to the length along the base of the triangle. Here, the x-values represent time and the y-values represent position. The slope of a position versus time graph is therefore a distance divided by a time, which equals speed. The units for the speed are the units for the rise (meters) divided by the units for the run (seconds) or meters per second.

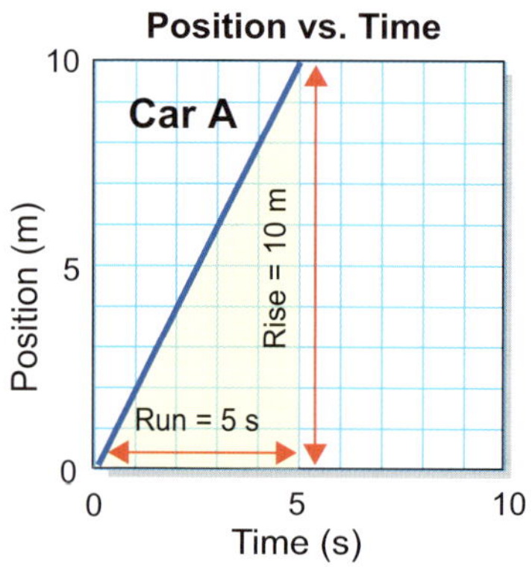

$$\text{Slope} = \frac{\text{rise}}{\text{run}} = \frac{10\text{ m}}{5\text{ s}} = 2\text{ m/s}$$

The slope of position vs.time is the **speed.**

VOCABULARY

slope - the ratio of the rise (vertical change) to the run (horizontal change) of a line on a graph.

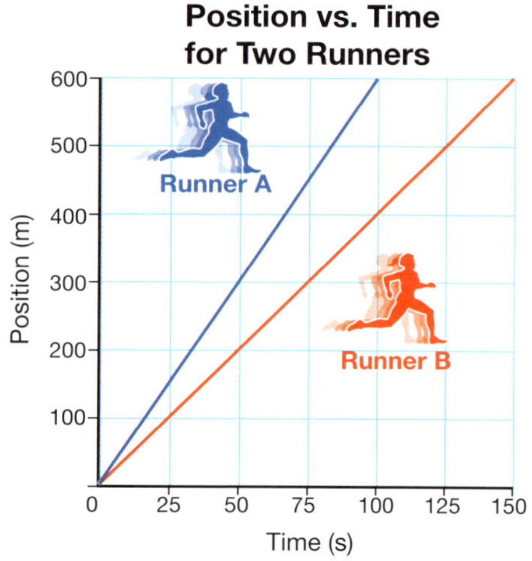

Figure 12.10: *A position vs. time graph for two runners.*

Speed vs. time graphs

Constant speed on a speed vs. time graph

The speed versus time graph has speed on the *y*-axis and time on the *x*-axis. The graph in Figure 12.11 shows the speed versus time for a ball rolling at constant speed on a level floor. On a speed vs. time graph, constant speed is shown with a straight horizontal line. If you look at the speed on the *y*-axis, you see that the ball is moving at 1 m/s for the entire 10 seconds. Figure 12.12 is the position versus time graph for the ball. Both of the graphs show the exact same motion. If you calculate the slope of the lower graph, you will find that it is 1 m/s, the same as the speed in Figure 12.11.

Calculating distance

A speed versus time graph also can be used to find the *distance* the object has traveled. Remember, distance is equal to the speed multiplied by the time. Suppose we draw a rectangle on the speed versus time graph between the *x* - axis and the line showing the speed. The area of the rectangle (shown below) is equal to its length times its height. On the graph, the length is equal to the time and the height is equal to the speed. Therefore, the area of the graph is the speed multiplied by the time. This is the distance the ball traveled.

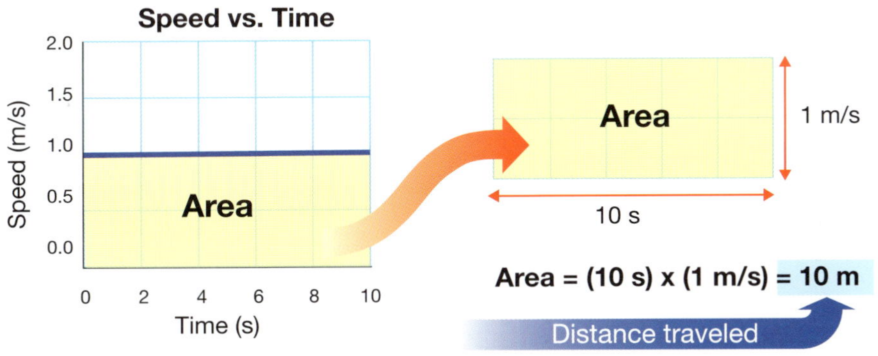

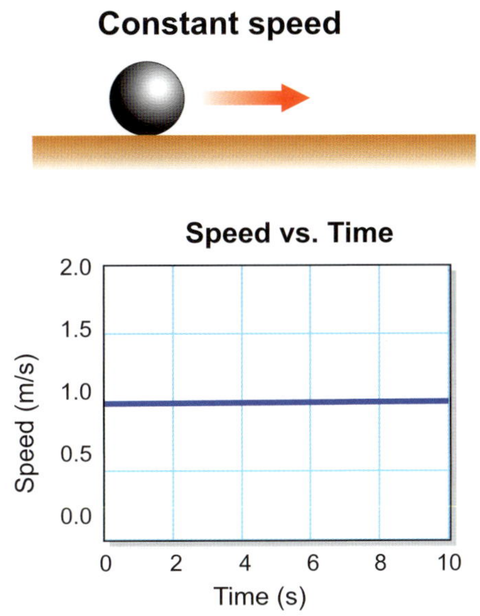

Figure 12.11: *The speed versus time graph for a ball rolling on a level floor at a constant speed of 1 m/s.*

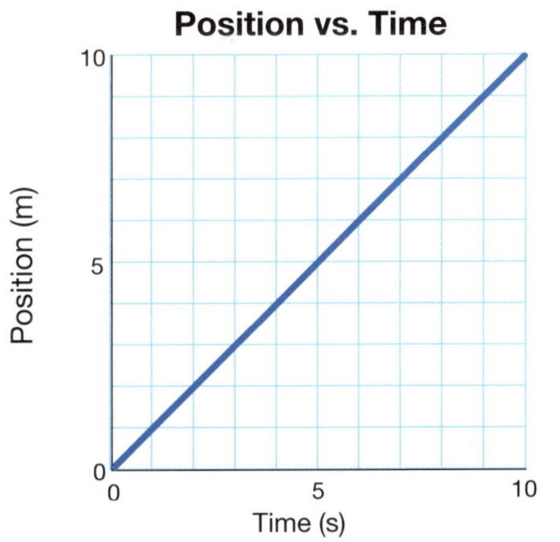

Figure 12.12: *The position versus time graph that shows the exact same motion as the speed versus time graph above.*

The speed vs. time graph for changing motion

Speeding up and slowing down

The graphs you have been learning about until now have all shown motion at a constant speed. But objects rarely move at the same speed for a long period of time. A speed vs. time graph is useful for showing the motion of an object that is speeding up or slowing down. If a speed vs. time graph slopes up, then the speed is increasing. If it slopes down, then the speed is decreasing. If the graph is horizontal, then the object is moving at a constant speed.

The graphs in Figure 12.13 show the motion of three different cars for 10 seconds. Each begins at a speed of 20 m/s. The first car moves at a constant 20 m/s for the entire 5 seconds. The second car speeds up from 20 m/s to 30 m/s. The third car slows down from 20 m/s to 10 m/s.

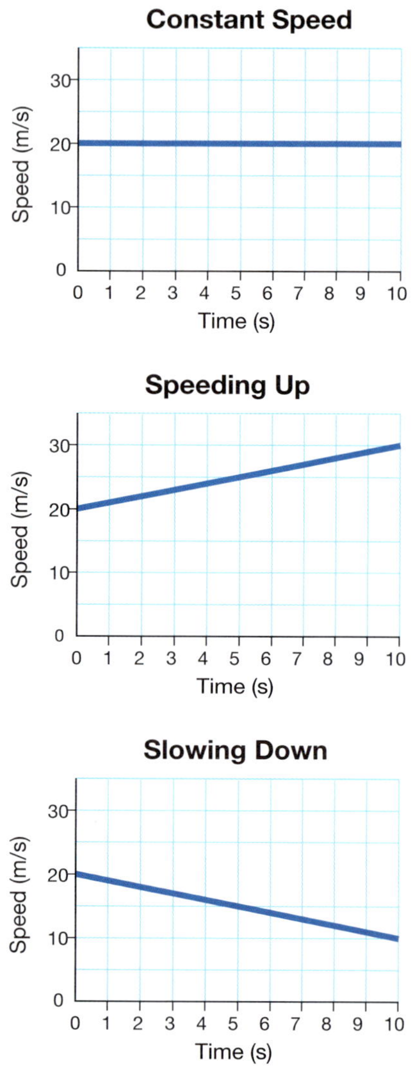

Figure 12.13: *A speed vs. time graph can show you whether an object is speeding up, slowing down, or moving at a constant speed.*

Drawing a speed vs. time graph

Draw a speed vs. time graph that shows the same motion as the position vs. time graph to the right.

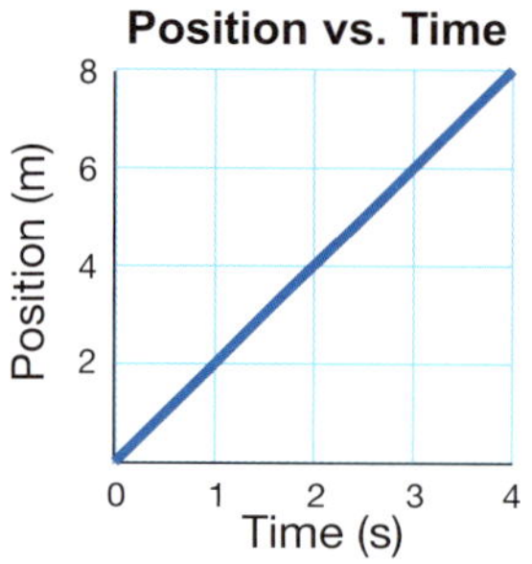

1. Looking for: You are asked to draw a speed vs. time graph.
2. Given: You are given the position vs. time graph.
3. Relationships: Speed equals the slope of the position vs. time graph.
4. Solution: The object is moving at a constant speed. The slope equals the rise divided by the run.

$$\text{slope} = \frac{\text{rise}}{\text{run}} = \frac{12\text{ m}}{4\text{ s}} = 12\text{ m/s}$$

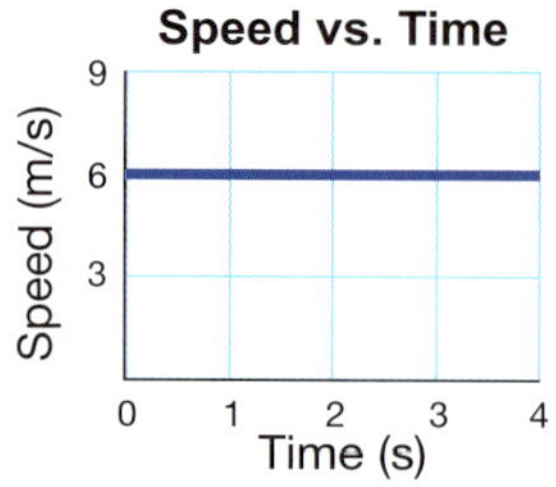

12.3 Section Review

1. What does the slope of the line on a position versus time graph tell you about the object's speed?
2. On a graph of position versus time, what do the x-values represent? What do the y-values represent?
3. The data table below shows the position and time for kayaker paddling in a river . Make a graph of the data, and use it to calculate the kayaker's speed.

Time (s)	Position (m)
0	0
2	6
4	9
6	12
8	15
10	18

4. The graph to the right shows the position and time for two runners in a race. Who has the faster speed? Explain how to answer this question without doing calculations.
5. Calculate the speed of each runner in the graph to the right.
6. Maria walks at a constant speed of 2 m/s for 8 seconds.
   a. Draw a speed versus time graph for Maria's motion.
   b. How far does she walk?

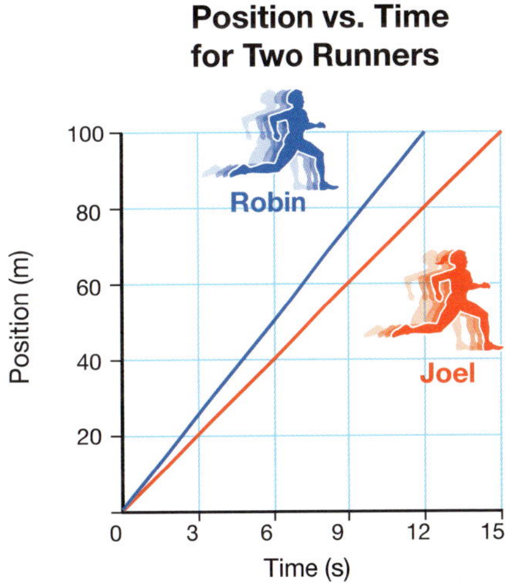

CHALLENGE

Draw the speed versus time graph that goes with the position versus time graph shown below.

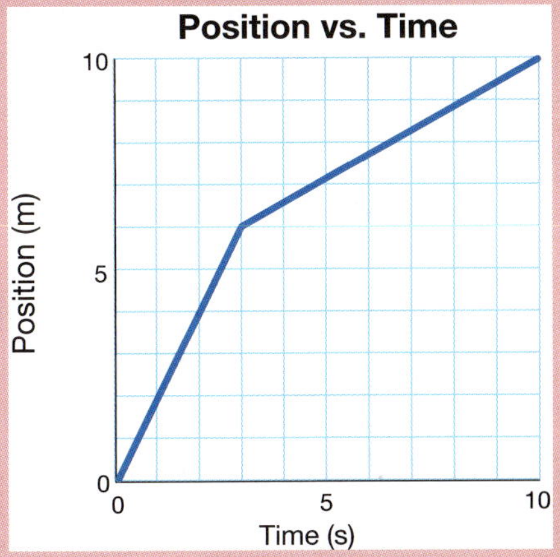

TECHNOLOGY CONNECTION

Roller Coasters: The Physics Behind the Thrills

The first roller coaster in the United States was built in 1884 at Coney Island, in Brooklyn, N.Y. A nickel bought you a one-minute ride and patrons have been paying for that thrill ever since. Are you among the thrill seekers?

That first roller coaster was all wood and ran in a straight line. Even if you have not yet ridden your first roller coaster, you know modern roller coasters have come a long way (and can be found far from Coney Island). Roller coasters are now made of wood or steel, and the tracks include a variety of twists and turns and loops and drops that make the ride scary and thrilling.

The keys to the thrill factor are velocity and acceleration. Velocity is the rate at which the position of an object—say, a roller coaster car—changes. Acceleration is the rate at which the velocity of an object changes. The thrill of the ride comes from the sudden accelerations. The fun begins at the top of the first peak on the track; that is where velocity and acceleration get their start.

Materials make a difference

Not so long ago, almost every roller coaster was still made of wood. But wood has limitations that made those coasters less popular once steel roller coasters made their debut. Steel coasters have replaced most of the wooden ones.

The two offer quite different riding experiences. A wooden coaster is often a rougher and "wilder" ride. This is because wood tends to be less rigid than steel. A wooden coaster's tracks usually move anywhere from a few inches to a few feet in response to the force of the cars rolling on the rails. The track is designed to do this; its swaying makes for a frightening ride, which is what riders want, don't they?

If you are a coaster enthusiast, you know that some wooden frames include inversions. An inversion is when the roller coaster's cars are upside down. And some cars fly through steeply banked (though not inverted) curves called "overbanked" turns. Most wooden tracks cannot support inversions, and overbanked turns don't work as well in wood. On wooden coasters, the motion is mainly up and down. That is more than enough for designers to still create thrilling wood roller coasters for big amusement parks.

How a roller coaster makes its thrills

While they have are clearly different, wooden and steel roller coasters have even more in common. They both rely on rapid changes in velocity to provide excitement. When your velocity changes rapidly your body can feel weightless, like you are falling. You can feel sudden sideways forces, like trying to turn a corner too fast in a car. You can even feel pressed into your seat with twice your normal weight. The challenge to a roller coaster designer is to create rapid changes in velocity that also keep the rider safe!

Gravity applies a constant downward force on a roller coaster car. The anticipation builds on the first big hill as the motor drags the car slowly up to the top. Upon reaching the top, the car picks up speed as it seems to fall down the first big hill. The first hill is usually straight so the car gains velocity by increasing its speed. This is where you feel weightless, like you are falling because, *you are falling*!

The next change in velocity comes in the first turn. The coaster's tracks constrain the car so it can only move along the track. A rapid change in the car's velocity vector is created by forcing the car to make a sharp turn at high speed. This is where you feel thrown to one side of the car or get squashed against your fellow thrill-seekers. The speed of the roller coaster car stays about the same through the turn. The change in velocity is in direction, not in speed.

Some roller coasters feature corkscrew-like tracks that whip the car through a combination of vertical and turning motion that puts everyone upside down. Here is where you feel pressed into your seat with more than your usual weight. This feeling is created by changing the direction of the velocity vector in the vertical direction, when the car curves over as it speeds through the corkscrew. Because of the tight turns and high speeds, corkscrew tracks must be extremely strong. Wooden roller coasters could not have supported a corkscrew at the speeds of modern steel coasters.

Most roller coasters rely on the potential energy from the first big hill. This is converted to kinetic energy in the motion of the car. As the ride progresses, friction slows the car down and the speed gets lower. The second and third hills are lower than the first because the car does not have enough energy to climb a bigger hill without additional power.

If you ever wondered how roller coasters have gotten bigger and bigger as new designs have been introduced, now you know. Simple physics!

Questions:

1. What is velocity?
2. How does a roller coaster create the sensations of falling, and being thrown around?
3. How does a roller coaster car move without an engine powering it?
4. Why do you think the first hill in a roller coaster's design is the highest?

CHAPTER ACTIVITY

How fast are you?

Speed is how fast something moves in relation to a reference point without regard to the direction. Speed is found by dividing the distance traveled by the total time the object has traveled. An object can travel at a constant rate or the speed may vary. When speed varies during a trip, you can find the average speed for the entire trip. In this activity you and a partner will each calculate your average speed in different units.

Materials

Tape measure, meterstick, or ruler
Stopwatch or watch with a second hand
Pieces of tape

What you will do

1. Decide how you and your partner will be moving. You can walk, run, roll, or move in any other way you choose.
2. Find an open area outside, in a hallway, or in another location where you can do this activity.
3. Mark your starting point (origin) with a piece of tape.
4. Measure at least five evenly spaced positions in meters along the path you are going to follow. Mark these positions with tape. For example, if you are running 60 meters, place a piece of tape at the starting point and at every 10 meters. If you are crawling only 5 meters, mark off every 1 meter.
5. Start at the origin and move along the length of your path. Your partner will start the timer once you start moving. Your partner should record the time for each marked position. For example at the origin time is zero. At the 1-meter mark the time might be 2 seconds, at the 2-meter mark the time might be 4 seconds and so on.
6. Record your data in a table like this:

Position (m)	Time (s)

7. Switch roles and repeat the activity with the other person moving. Record your data in the table.
8. Make a position vs. time graph to show each person's motion. Put both sets of data on the same graph. It might be helpful to use two different colors to plot the points.

Applying your knowledge

a. Explain how you can use your graph to figure out who had the faster average speed.
b. Explain how you can use your data table to figure out who had the faster average speed.
c. Look at each person's line on the graph. How can you use the graph to tell whether you moved at a constant speed? Did you move at a constant speed? Did your partner?
d. Calculate each person's average speed in meters per second and in centimeters per second.

Chapter 12 Assessment

Vocabulary

Select the correct term to complete the sentences.

average speed	constant speed	slope
origin	speed	position
velocity		

Section 12.1

1. An ____ is a fixed reference point.
2. The location of an object compared to a reference point is called its ____.

Section 12.2

3. When an object is traveling the same distance every second it has a ____.
4. The ____ of an object is the total distance traveled divided by the total time for the trip.
5. ____ is the distance an object moves divided by the time it takes.

Section 12.3

6. The ____of a line is the ratio of rise to run.

Concepts

Section 12.1

1. If left, south, west, and down are negative, then which of the following directions must be positive?
    a. north
    b. up
    c. east
    d. right
2. Give the coordinates of a pizza shop located 4 miles west and 2 miles north of your house.

Section 12.2

3. Can you use a car's speedometer (only) to tell if the car is moving at a constant velocity? Why or why not?
4. List three common units for measuring speed.
5. Write the speed equation 3 different ways to find time, distance, and speed.

Section 12.3

6. When comparing two different lines on a position vs. time graph, how can you tell which object is faster?
7. Which of the graphs below shows an object that is stopped?

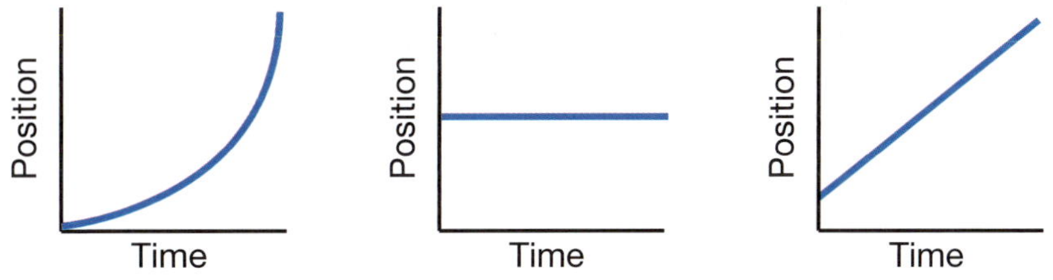

8. Which of the graphs above shows an object moving at a constant speed?

Problems

Section 12.1

1. A number line is defined with positive to the right and negative to the left. Starting at -12 meters, you walk 7 meters to the right. What is your new position? Are you to the right or left of the origin?
2. You define the origin to be the 50 cm mark on a meter stick. What is the coordinate of an object located at the 75 cm mark on the meterstick?

3. You use an x-y plane to represent your position. Starting at (+150 m, -50 m), you walk 20 meters west and 30 meters north. What are your new coordinates?

Section 12.2

4. Use the speed equation to complete the chart:

distance (m)	speed (m/s)	time (s)
	10	6
45	5	
100		2

5. The French high-speed train travels at 300 km/h. How long would it take the train to travel 1500 km at this speed?

6. Lance Armstrong's teammate, George Hincapie, averaged a speed of 33.63 km/h in the 15th stage of the Tour de France, which took 4 hours. How far did he travel in the race?

7. A snail crawls 300 cm in 1 hour. Calculate the snail's speed in each of the following units:

   a. centimeters per hour (cm/h)

   b. centimeters per minute (cm/min)

   c. meters per hour (m/h)

Section 12.3

8. The chart below gives the position and time data for a boat in a crew race. Graph the data and find the boat's speed.

Time (s)	Position (m)
0	0
50	250
100	500
150	750
200	1000
250	1250
300	1500

9. Draw the position vs. time graph for a person walking at a constant speed of 1 m/s for 10 seconds. On the same set of axes, draw the graph for a person running at a constant speed of 4 m/s.

10. Calculate the speed represented by each position time graph below.

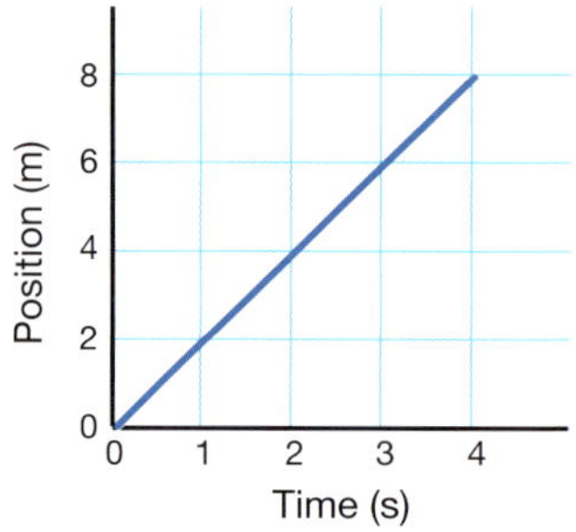

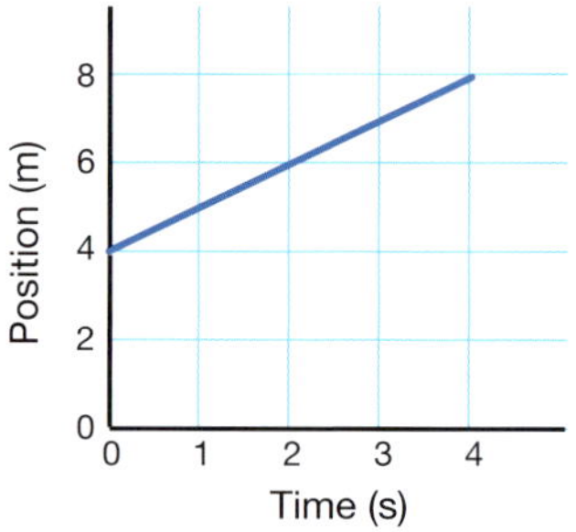

11. Draw the speed vs. time graph that shows the same motion as each position vs. time graph above.

12. The speed vs. time graph to the right shows the motion of a person on inline skates. Calculate the distance the person moves.

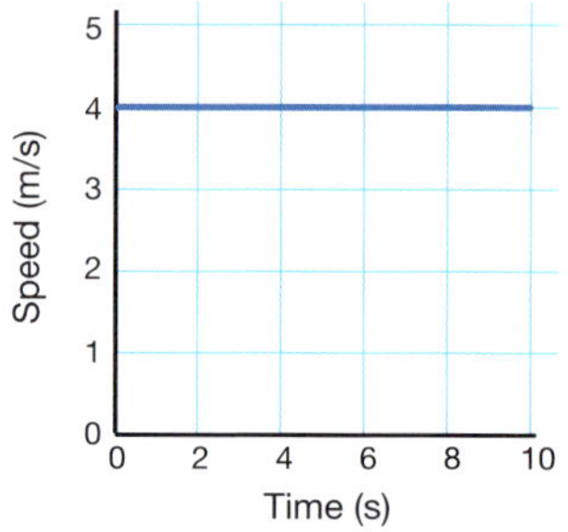

13. Draw a speed versus time graph for each of the following situations:

   a. A person walks along a trail at a constant speed.

   b. A ball is rolling up a hill and gradually slows down.

   c. A car starts out at rest at a red light and gradually speeds up.

Chapter 13

Forces

On May 27, 1931, a train called *Empire Builder* encountered the amazing force of a tornado as it moved across Minnesota. The tornado's force was so great that as the train moved along the track at 60 miles per hour, five of its 60-ton cars where lifted from the rails! One car was lifted and thrown 80 feet away into a ditch. Moving heavy railroad cars requires a tremendous amount of force.

Forces are created and applied every time anything moves. Forces such as weight are even present when things are not moving. Your body uses forces every moment of your life from the beating of your heart to walking up stairs. Understanding how forces are created and described is fundamental to understanding nature. Read this chapter to learn more about how forces are created, measured, described, and used in daily life.

Key Questions

1. *What is a force?*
2. *How are forces measured?.*
3. *What is friction, and is friction ever useful?*

13.1 Forces

We first introduced the idea of force in Chapter 2. Since force is such an important concept, this whole chapter is about forces, where they come from, how they are measured, and how they are added and subtracted.

The cause of forces

What forces are A force is a push or pull, or any action that has the ability to change motion. The key word here is *action*, force is an action. You need force to start things moving and also to make any change to an object's motion once it is moving. Forces can be used to increase or decrease the speed of an object, or to change the direction in which an object is moving.

How are forces created? Forces are created in many different ways. For example, your muscles create force when you swing a tennis racket. Earth's gravity creates a force called weight that pulls on everything around you. On a windy day, the movement of air can create forces. Each of these actions can create force because they all can change an object's motion.

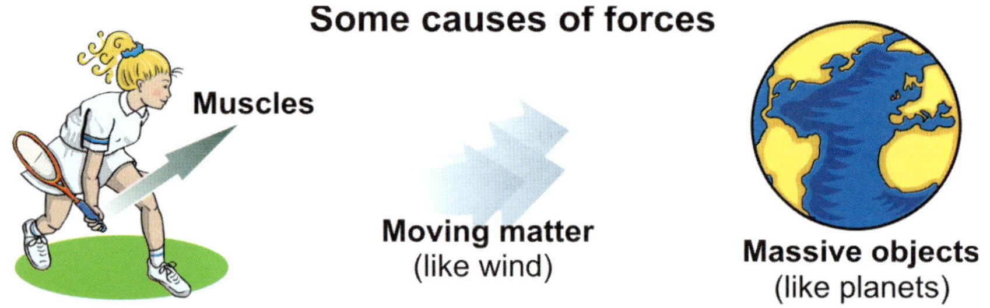

The four elementary forces Fundamentally, forces come from the interaction between atoms and energy. Just as a few elements can make millions of kinds of matter, all of the forces we know of in the universe come from four elementary forces. The in Figure 13.1 summarizes the four elementary forces. You met them before when we learned about the atom and will meet them again when we learn about astronomy.

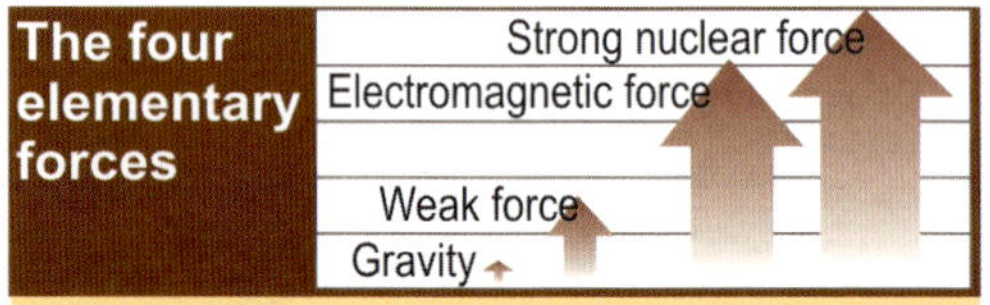

Strong nuclear force	This force holds the nucleus of an atom together. This force is very strong but only reaches a very short distance.
Electromagnetic force	This force acts between positive and negative charges. This force holds atoms together into molecules.
Weak force	This force causes some kinds of radioactivity.
Gravity	This force causes all masses to attract each other. Your weight comes from the mass of the Earth attracting the mass of your body.

Figure 13.1: *All forces in the universe come from only four elementary forces.*

Units of force

Pounds If you are mailing a package at the post office, how does the clerk know how much to charge you? The package is placed on a scale and you are charged based on the package's weight. For example, the scale shows that the package weighs 5 pounds. The pound is a unit of force commonly used in the United States. When you measure weight in pounds on a scale, you are measuring the force of gravity acting on the object (Figure 2.3).

The origin of the pound The pound measurement of force is based on the Roman unit libra, which means "balance" and is the source for pound's abbreviation, "lb." The word "pound" comes from the Latin word pondus, which means "weight." The definition of a pound has varied over time and from country to country.

Newtons Although the pound is commonly used to express force, scientists prefer to use the newton. The newton (N) is the metric unit of force. The newton is defined by how much a force can change the motion of an object. A force of one newton is the exact amount of force needed to cause a mass of one kilogram to speed up by one m/sec each second (Figure 13.2). We call the unit of force the newton because force in the metric system is defined by Newton's laws. The newton is a useful way to measure force because it connects force directly to its effect on mass and speed.

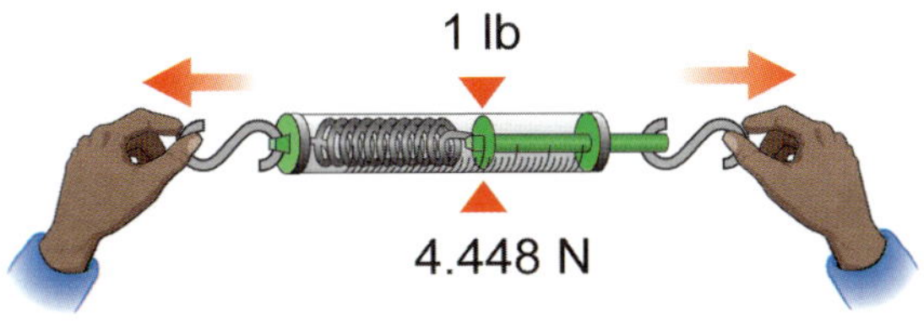

1 pound = 4.448 newtons

Converting newtons and pounds The newton is a smaller unit of force than the pound. One pound of force equals 4.48 newtons. How much would a 100-pound person weigh in newtons? Recall that 1 pound = 4.48 newtons. Therefore, a 100-pound person weighs 448 newtons.

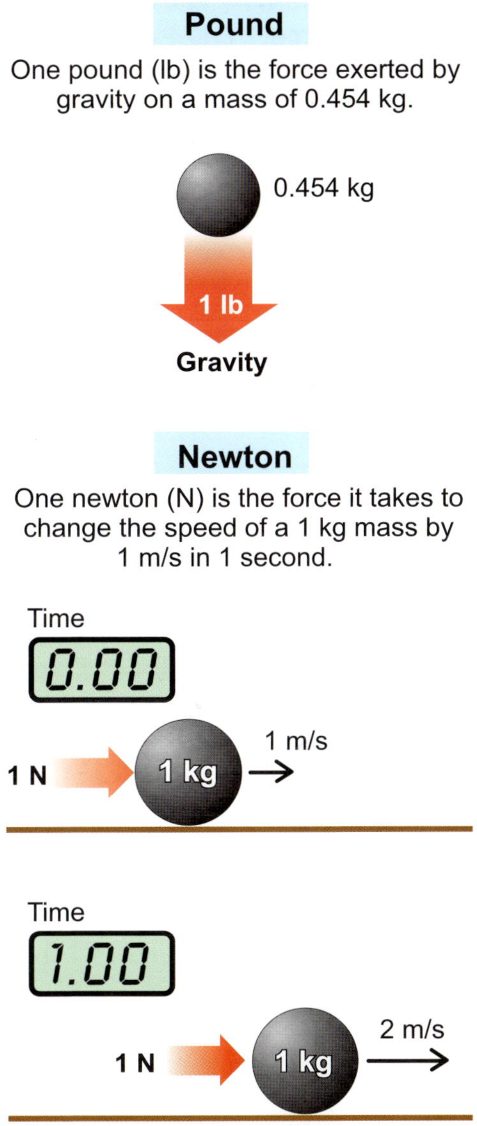

Figure 13.2: *The definitions of the newton and pound.*

Gravity and weight

Gravity's force depends on mass

The force of gravity on an object is called weight. At Earth's surface, gravity exerts a force of 9.8 N on every kilogram of mass. That means a 1-kilogram mass has a weight of 9.8 N, a two-kilogram mass has a weight of 19.6 N, and so on. On Earth's surface, the weight of any object is its mass multiplied by 9.8 N/kg. Because weight is a force, it is measured in units of force such as newtons and pounds.

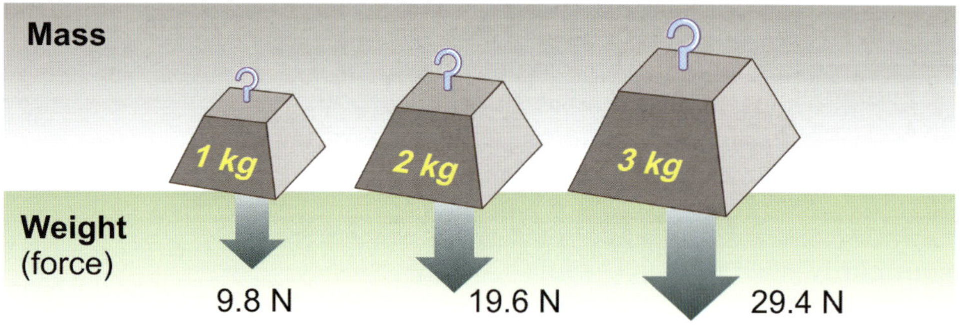

At Earth's surface each kg of mass feels **9.8** N of weight from Earth's gravity.

Weight and mass are not the same thing

We all tend to use the terms *weight* and *mass* interchangeably. However, in science *weight and mass are not the same thing*. In fact, weight is caused by nearby mass. You have weight because the huge mass of Earth is right next to you. It is easy to confuse mass and weight because heavy objects (more weight) have lots of mass and light objects (less weight) have little mass. Always remember the difference when doing physics. Mass is a fundamental property of an object measured in kilograms (kg). Weight is a *force* measured in *newtons* (N) that depends on mass and gravity.

Mass and weight are not the same thing!

Weight is less on the moon

A 10-kilogram rock has a mass of 10 kilograms no matter where it is in the universe. A 10-kilogram rock's weight however, can vary greatly depending on where it is. On Earth, the rock weighs 98 newtons. But on the moon, where $g = 1.6$ N/kg, it only weighs 16 newtons (Figure 13.3)!

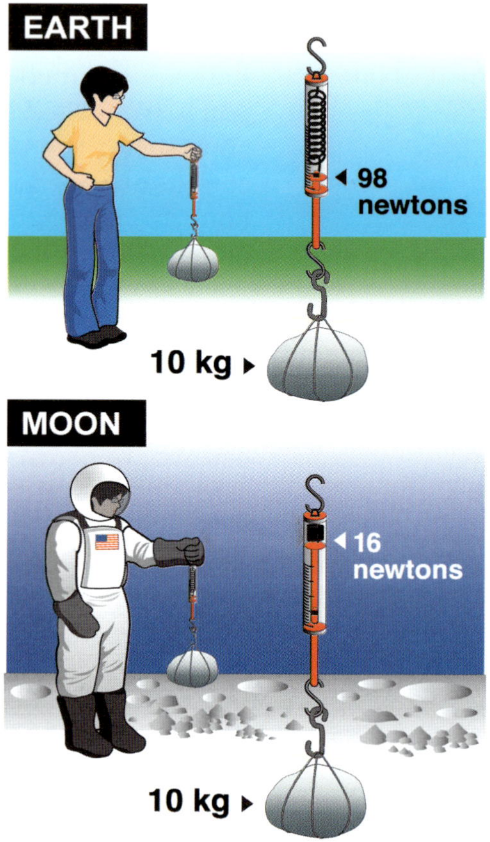

Figure 13.3: *A 10-kilogram rock weighs 98 newtons on Earth but only 16 newtons on the moon.*

Calculating weight

The weight equation

The weight equation (Figure 13.4) can be rearranged into three forms. You can use it to calculate weight, mass, or the strength of gravity by using the correct form of the equation. The three forms are summarized below.

Use . . .	. . . if you want to find . . .	. . . and you know . . .
$W=mg$	weight (W)	mass (m) and strength of gravity (g)
$m=W/g$	mass (m)	weight (W) and strength of gravity (g)
$g=W/m$	strength of gravity (g)	weight (W) and mass (m)

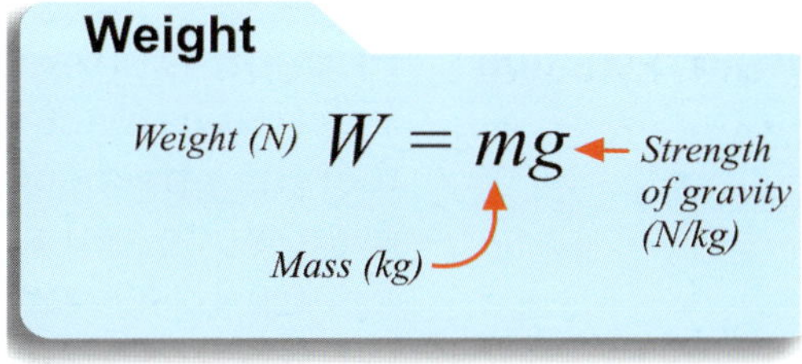

Figure 13.4: *The weight equation.*

Weight and mass

Calculate the weight of a 60-kilogram person (in newtons) on Earth and on Mars ($g = 3.7\ m/sec^2$).

1. Looking for: You are asked for a person's weight on Earth and Mars.
2. Given: You are given the person's mass and the value of g on Mars.
3. *Relationships:* $W=mg$
4. Solution: For the person on Earth:
   $W=mg$
   $W = (60\ kg)(9.8\ N/kg) = 588$ newtons

   For the person on Mars:
   $W=mg$
   $W = (60\ kg)(3.7\ N/kg) = 222$ newtons

   Notice that while the masses are the same, the weight is much less on Mars.

Your turn...

a. Calculate the mass of a car that weighs 19,600 newtons on Earth. **Answer:** 2000 kg

b. A 70-kg person travels to a planet where he weighs 1,750 N. What is the value of g on that planet? **Answer:** 25 N/kg

Friction

What is friction? **Friction** is a force that resists the motion of objects or surfaces. Friction can act when an object is moving or when it is at rest. You feel the effects of friction when you swim, ride in a car, walk, and even when you sit in a chair. Many kinds of friction exist and Figure 13.5 shows some common examples.

The cause of friction If you looked at a piece of wood, plastic, or paper through a powerful microscope, you would see tiny hills and valleys on the surface. As surfaces slide (or try to slide) across each other, the hills and valleys grind against each other and cause friction. Contact between the surfaces can cause the tiny bumps to change shape or wear away. If you rub sandpaper on a piece of wood, friction affects the wood's surface and makes it either smoother (bumps wear away) or rougher (they change shape).

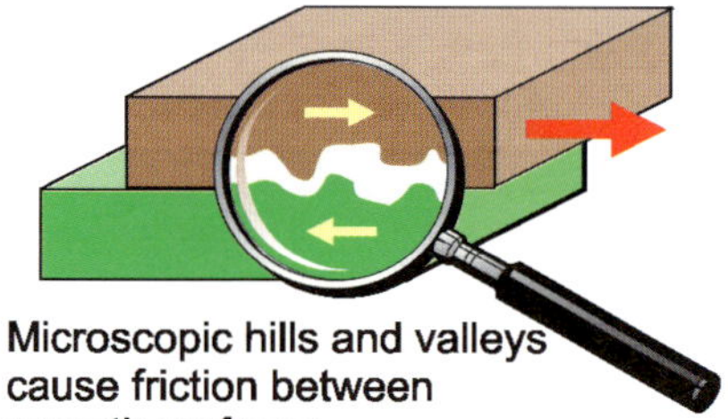

Microscopic hills and valleys cause friction between smooth surfaces

Two surfaces are involved Friction depends on *both* of the surfaces in contact. The force of friction on a rubber hockey puck is very small when it is sliding on ice. But the same hockey puck sliding on a piece of sandpaper feels a large friction force. When the hockey puck slides on ice, a thin layer of water between the rubber and the ice allows the puck to slide easily. Water and other liquids such as oil can greatly reduce the friction between surfaces.

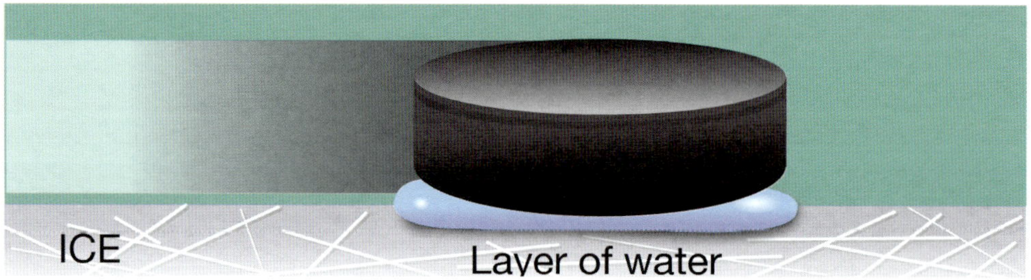

VOCABULARY

friction - a force that resists the motion of objects or surfaces.

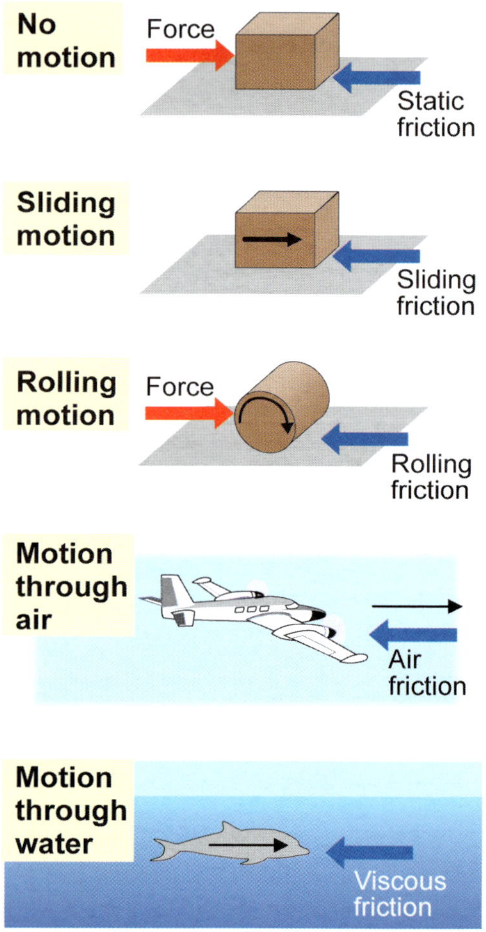

Figure 13.5: *There are many types of friction.*

Identifying friction forces

Direction of the friction force

Friction is a force, measured in newtons just like any other force. You draw the force of friction with a force vector. To figure out the direction of friction, always remember that friction is a *resistive* force. The force of friction acting *on* a surface always points opposite the direction of motion *of that surface*. Imagine pushing a heavybox across the floor (Figure 13.6). If you push to the right, the sliding friction acts to the left on the surface of the box touching the floor. If you push the box to the left, the force of sliding friction acts to the right. This is what we mean by saying friction resists motion.

Static friction

Static friction keeps an object at rest from moving. Imagine trying to push a heavy box with a small force. The box stays at rest because the static friction force acts against your force. If the box does not move, the friction force is the same strength as your force, so the net force is zero. As you increase the strength of your push, the static friction also increases. Eventually your force becomes stronger than the maximum static friction force and the box starts to move (Figure 13.6). The force of static friction balances your force up to a limit. The limit of the friction force depends on the types of surfaces and the weight of the object you are pushing.

Sliding friction

Sliding friction is a force that resists the motion of an object moving across a surface. If you push a box across the floor toward the right, sliding friction acts toward the left. If you stop pushing the box, sliding friction keeps acting as long as the box is moving. Without your force acting, sliding friction slows the box to a stop.

Comparing static and sliding friction

How does sliding friction compare with static friction? If you have ever tried to move a heavy sofa or refrigerator, you probably know the answer. It is harder to get something moving than it is to keep it moving. The reason is that static friction is greater than sliding friction for almost all combinations of surfaces.

VOCABULARY

static friction - the friction force that resists the motion between two surfaces that are not moving.

sliding friction - the friction force that resists the motion of an object moving across a surface.

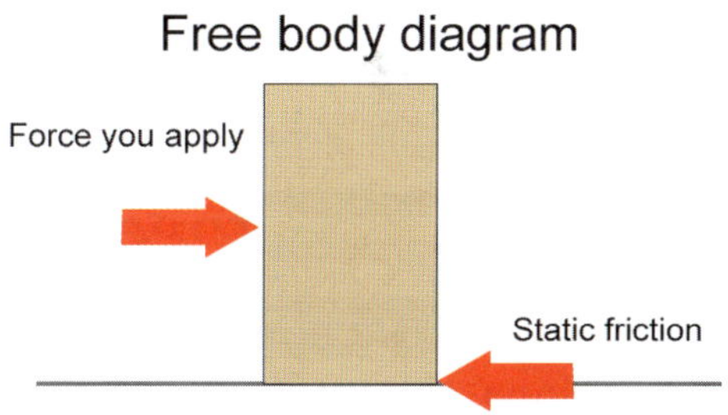

Figure 13.6: *The direction of friction is opposite the direction the box is pushed.*

A model for friction

Different amounts of friction The amount of friction that exists when a box is pushed across a smooth floor is very different from when it is pushed across a carpet. The friction between two surfaces depends on the types of materials, degrees of roughness, how clean they are, and other factors. Even the friction between two identical surfaces changes as the surfaces are polished by sliding across each other. No one model or formula can accurately describe the many processes that create friction. Even so, some simple models are useful.

An example Suppose you use a spring scale pull a piece of paper across a table. It is easy to pull the paper because the friction force is so small. The paper slides smoothly, and the scale measures a very small force. Do you believe the friction force between the paper and the table is a value that cannot be changed? How might you test this question?

Friction and the force between surfaces Suppose you place a brick on the piece of paper (Figure 13.7). The paper becomes much harder to slide. You must pull with a greater force to keep the paper moving. The two surfaces in contact are still the paper and the tabletop, so why does the brick have an effect? The brick causes the paper to press harder into the table's surface. The tiny hills and valleys in the paper and in the tabletop are pressed together with a much greater force, so the friction increases.

The greater the force squeezing two surfaces together, the greater the friction force.

Friction is often proportional to weight The friction force between two smooth, hard surfaces is approximately proportional to the perpindicular (normal) force the surfaces exert on each other. The greater the force squeezing the two surfaces together, the greater the friction force. Consider sliding a heavy box across a floor. The force between the bottom of the box and the floor is the weight of the box. Therefore, the force of friction is proportional to the weight of the box. If the weight doubles, the force of friction also doubles.

Figure 13.7: *Friction increases greatly when a brick is placed on the paper.*

Reducing the force of friction

All surfaces experience some friction Any motion where surfaces move across each other or through air or water always causes some friction. Unless a force is constantly applied, friction will slow all motion to a stop eventually. For example, bicycles have low friction, but even the best bicycle slows down if you coast on a level road. It is impossible to completely get rid of friction, but it can be reduced.

Lubricants reduce friction in machines Putting a liquid such as oil between two sliding surfaces keeps them from touching each other. The tiny hills and valleys don't become locked together, so the friction is less. The liquid also keeps them from wearing away quickly. You add oil to a car's engine so that the pistons will slide back and forth with less friction. Even water can be used as to reduce friction between objects if they are not too hot.

Ball bearings Ball bearings reduce friction on spinning objects (Figure 13.8). Ball bearings change sliding motion into rolling motion, which has much less friction. For example, a metal shaft rotating in a hole rubs and generates a great amount of friction. Ball bearings that go between the shaft and the inside surface of the hole allow it to spin more easily. The shaft rolls on the bearings instead of rubbing against the walls of the hole. Well-oiled bearings rotate easily and greatly reduce friction.

Magnetic levitation Another method of decreasing friction is to separate the two surfaces with a cushion of air. A hovercraft floats on a cushion of air created by a large fan. Magnetic forces can also be used to separate surfaces. A magnetically levitated (or maglev) train uses magnets that run on electricity to float on the track once the train is moving (Figure 13.9). There is no contact between train and track, so there is far less friction than with a standard train on tracks. The ride is smoother, so maglev trains can move at very fast speeds. Maglev trains are not yet in wide use because they are much more expensive to build than regular trains. They may become much more popular in the future.

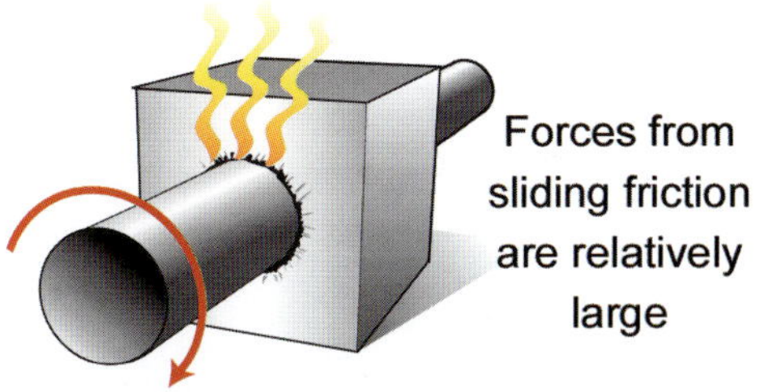

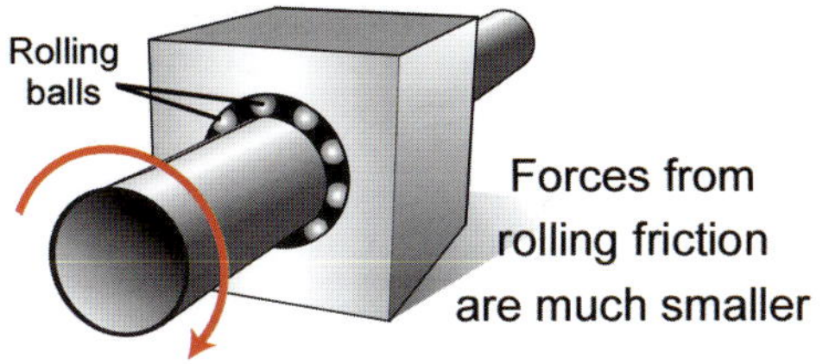

Figure 13.8: *The friction between a shaft (the long pole in the picture) and an outer part of a machine produces a lot of heat. Friction can be reduced by placing ball bearings between the shaft and the outer part.*

Figure 13.9: *With a maglev train, there is no contact between the moving train and the rail — and thus little friction.*

Using friction

Friction is useful for brakes and tires

There are many times when friction is very useful. For example, the brakes on a bicycle create friction between two rubber *brake pads* and the rim of the wheel. Friction between the brake pads and the rim makes the bicycle slow down or stop. Friction is also needed to make a bicycle go. Without friction, the bicycle's tires would not grip the road.

Weather condition tires

Friction is also important to anyone driving a car. Rain and snow act like a cushion that separates a car's tires from the road. As a tire rolls over a wet road, the rubber squeezes the water out of the way. This causes there to be good contact between rubber and road surface. Tire treads have grooves that allow space for water to be channeled away where the tire touches the road (Figure 13.10). Special groove patterns, along with tiny slits, have been used on snow tires to increase traction in snow. These tires keep snow from getting packed into the treads. Tires can also slightly change shape to grip the uneven surface of a snow-covered road.

Nails

Friction is the force that keeps nails in place (Figure 13.11). The material the nail is hammered into, such as wood, pushes against the nail from all sides. Each hit of the hammer pushes the nail deeper into the wood, increasing the length of the nail being compressed. The strong compression force creates a large friction force and holds the nail in place.

Cleated shoes

Shoes are designed to increase the friction between their soles and the ground. Many types of athletes, including football and soccer players, wear shoes with cleats that increase friction. Cleats are like teeth on the bottom of the shoe that dig into the ground. Players wearing cleats can apply much greater forces against the ground to help them move and to keep from slipping.

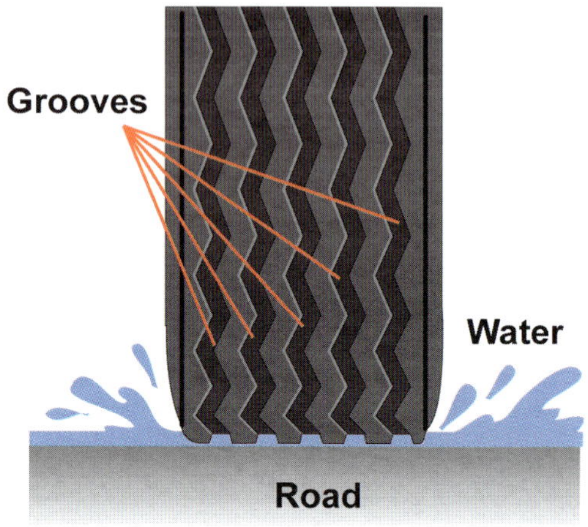

Figure 13.10: *Grooved tire treads allow space for water to be channeled away from the road-tire contact point, allowing for more friction in wet conditions.*

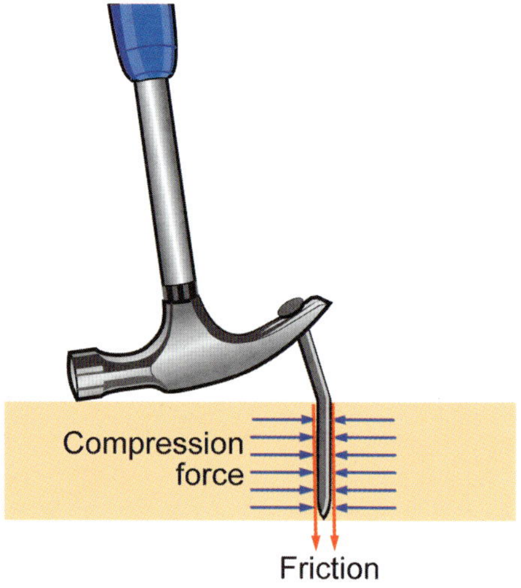

Figure 13.11: *Friction is what makes nails hard to pull out and gives them the strength to hold things together.*

13.1 Section Review

1. Name three situations in which force is created. Describe the cause of the force in each situation.
2. Which of the following are units of force?
   a. kilograms and pounds
   b. newtons and pounds
   c. kilograms and newtons
3. Which is greater: a force of 10 N or a force of 5 lbs.?
4. Does the mass of an object change if the object is moved to another planet? Explain your answer.
5. What is the weight (in newtons) of a bowling ball which has a mass of three kilograms?
6. If the strength of gravity is 9.8 newtons per kilogram, that means:
   a. each newton of force equals 9.8 pounds.
   b. each pound of force equals 9.8 newtons.
   c. each newton of mass weighs 9.8 kilograms
   d. each kilogram of mass weighs 9.8 newtons.
7. Name three devices or inventions which are designed to decrease friction.
8. Name three devices or inventions which are designed to increase friction.
9. If the force squeezing two surfaces together is decreased, the force of friction between the two services will most likely
   a. increase.
   b. decrease.
   c. stay about the same.
10. An astronaut in a space suit has a mass of 100 kilograms. What is the weight of this astronaut on surface of the moon where the strength of gravity is approximately 1/6 that of Earth.

Calculating mass from weight

What is the mass of an object with a weight of 35 newtons? Assume the object is at the Earth's surface.

1. Looking for:	Mass
2. Given:	Weight = 35N
3. Relationships:	The weight equation W = mg
4. Solution:	m = W/g = (35 N)/(9.8 N/kg) = 35.7 kg

Your turn...

a. Which is greater: a force of 100 N or the weight of 50 kilograms at Earth's surface. **Answer**: The weight of 50 kg is greater.

b. The mass of a potato is 0.5 kg. Calculate its weight in newtons. **Answer**: 4.9 N

13.2 Addition of Forces and Equilibrium

50 newtons would be a good description of the strength of a force. But what about the direction? The direction of a force is also important. This is especially true when forces are added together. Most real situations involve more than one force acting at a time, so adding forces is necessary to understand what is going on. In this section you will learn that force is a *vector*. A vector is a quantity that includes information about both size (strength) and direction

Vectors

When a quantity and a unit are sufficient

Some things in science can be completely described by a quantity and a unit. Temperature is a good example of this. If you are sick and use a thermometer to find out your temperature, it might show 101°F. The "quantity" of your temperature is 101, and degrees Fahrenheit is the unit of measurement. The value of 101°F is a complete description of the temperature because you do not need any more information.Length, time, and speed are also quantities that can be completely described with a single number and a unit.

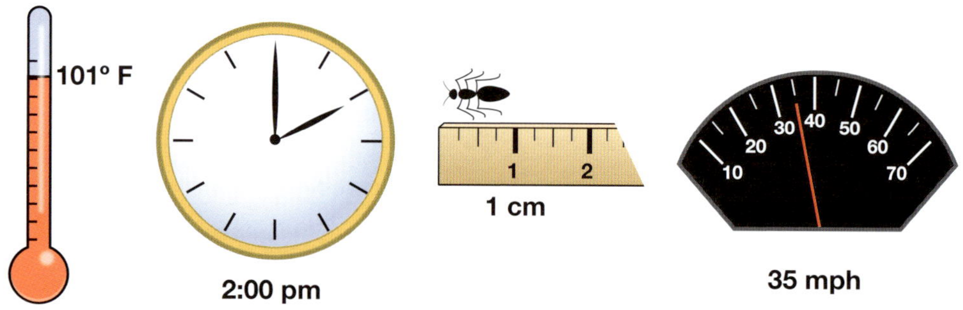

Vectors have direction

Sometimes a single number does not include enough information to describe a measurement. When giving someone directions to your house, you must include both the distance and the direction. The information "two kilometers north" is an example of a **vector**. A vector is a quantity that includes both a **magnitude** and a direction. Other examples of vectors are velocity, acceleration, and force.

ă VOCABULARY

vector - a quantity that includes both a magnitude (size) and a direction.

magnitude - describes the size component of a vector.

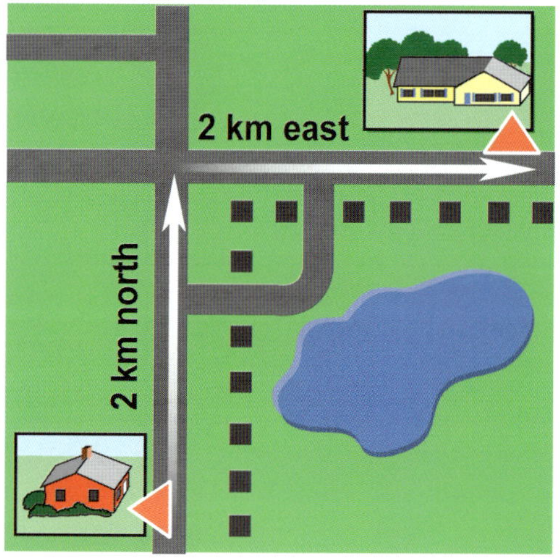

Figure 13.12: *Vectors are useful in giving directions.*

The force vector

What is a force vector? A force vector has units of newtons, just like all forces. In addition, the force vector also includes enough information to tell the direction of the force. You can describe a force vector's direction in words, such as 5 newtons north or 2 newtons down. Or you can use an arrow to draw the force vector in graph form.

Using positive and negative numbers Positive and negative numbers indicate opposite directions. For example, suppose a person pushes with a force of 10 newtons to the right (Figure 13.13). The force vector is + 10 N. A person pushing with the same force to the left would create a force vector of - 10 N. The negative sign indicates the -10 N force is in the opposite direction from the + 10 N force. We usually choose positive values to represent forces directed up, to the right, East or North.

Drawing a force vector It is sometimes helpful to show the strength and direction of a force vector as an arrow on a graph. The length of the arrow represents the strength of the force. The arrow points in the direction of the force. The *x*- and *y*-axes show the strength of the force in the *x* and *y* directions.

Scale When drawing a force vector, you must choose a scale. For example, if you are drawing a vector showing a force of 5 N pointing straight up (y-direction) you might use a scale of 1 cm = 1 N. You would draw the arrow five centimeters long pointing along the y-direction on your paper (Figure 13.14). A 5 N horizontal force would be drawn starting with a 5 cm line, as shown below..

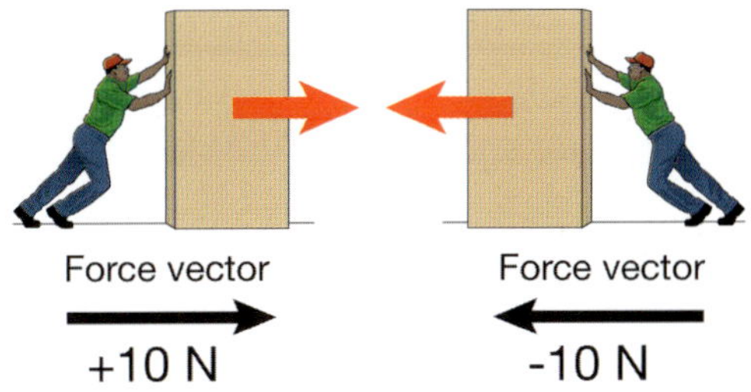

Figure 13.13: *Positive and negative numbers are used to indicate the direction of force vectors.*

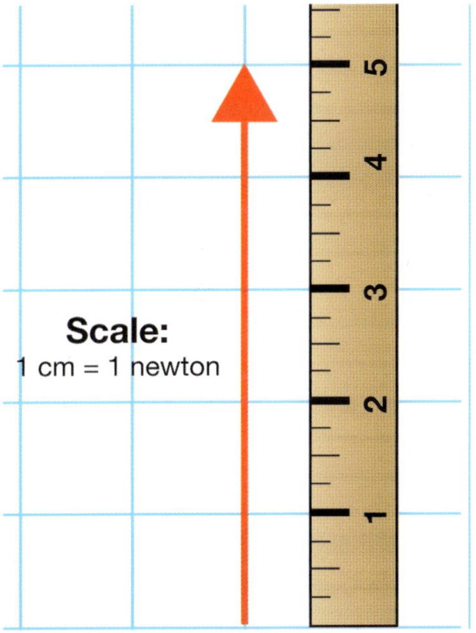

Figure 13.14: *You must use a scale when drawing a vector.*

Adding force vectors, the net force

An example To figure out if or how an object will move, you must look at all of the forces acting on it. The sum of all the forces on an object is called the **net force**.The word *net* means total but also means the direction of the forces has been taken into account. Consider an airplane in flight (Figure 13.15). Four forces act on the plane: weight, drag (air friction), the thrust of the engines, and the lift force caused by the flow of air over the wings. For a plane to fly at a constant speed in a level path, the forces must all balance. When the forces are **balanced**, the net force is zero.

When net force isn't zero A pilot must always be aware of these four forces and know how to change the net force on the plane to speed up, slow down, lift off, and land. For example, to speed up there must be a net force in the forward direction. The thrust must be greater than the drag. To climb, there must be an upward net force. This happens when the lift force is greater than the plane's weight.

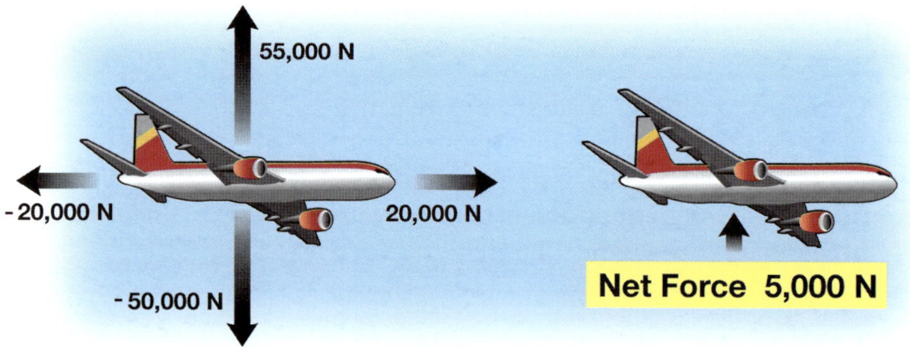

Adding x-y components To calculate the net force on an object, you must add the forces in each direction separately. Remember to define positive and negative directions for both the x-direction and y-direction. In the diagram above, +x is to the right and +y is up. The net force in the x-direction is zero because the +20,000 N and -20,000 N sum to zero. The net force in the y-direction is +5,000 N (+55,000 N - 50,000 N). The plane climbs because there is a positive (upward) net force.

VOCABULARY

net force - the sum of two or more forces on an object.

balanced forces - result in a zero net force on an object.

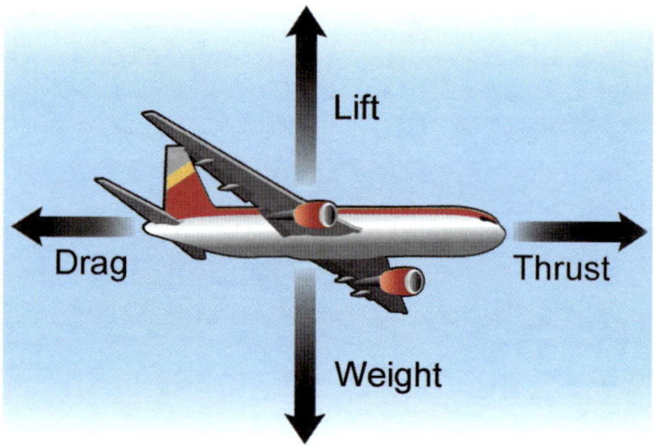

Figure 13.15: *Four forces act on a plane as it flies.*

Equilibrium and normal forces

Definition of equilibrium When the net force on an object is zero, we say the object is in **equilibrium**. An object at rest and in equilibrium will stay at rest. This also means that when an object is at rest, you know the net force on it must be zero.

Normal force Imagine a book sitting on a table (Figure 13.16). Gravity pulls the book downward with a force equal to the book's weight. The book is at rest, so the net force must be zero. But what force balances the weight? The table exerts an upward force on the book called the **normal force**. The word *normal* here has a different meaning from what you might expect. In mathematics, normal means *perpendicular*. The force the table exerts is perpendicular to the table's surface. The normal force is also sometimes called the support force.

When normal force is created A normal force is created whenever an object is in contact with a surface. The normal force has *equal strength* to the force pressing the object into the surface, which is often the object's weight. The normal force has *opposite direction* to the force pressing the object into the surface. For example, the weight of a book presses down on the table surface. The normal force is equal in strength to the book's weight but acts upward on the book, in the opposite direction from the weight.

Strength of the normal force What happens to the normal force if you put a brick on top of the book? The brick makes the book press harder into the table. The book does not move, so the normal force must be the same strength as the total weight of the book and brick (Figure 13.17). The normal force increases to keep the book in balance.

How the normal force is created How does a table "know" how much normal force to supply? The answer is that normal force is very similar to the force exerted by a spring. When a book sits on a table, it squeezes the atoms in the table together by a tiny amount. The atoms resist this squeezing and try to return the table to its natural thickness. The matter in the table acts like a bunch of very stiff springs. The amount of compression is so small you cannot see it, but it can be measured with instruments.

VOCABULARY

equilibrium - state in which the net force on an object is zero.

normal force - the force a surface exerts on an object that is pressing on it.

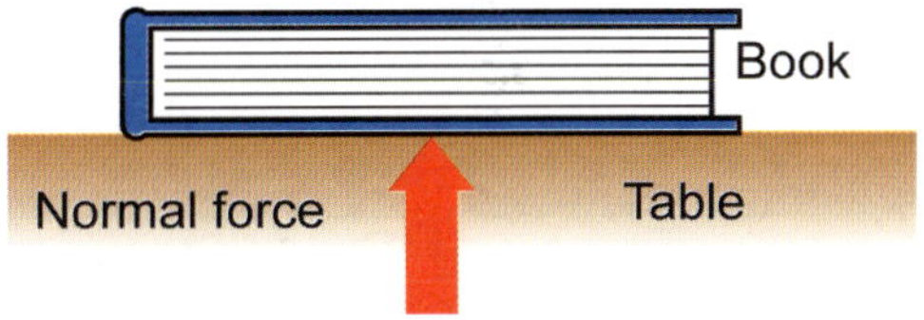

Figure 13.16: *The normal force and the weight are equal in strength and opposite in direction.*

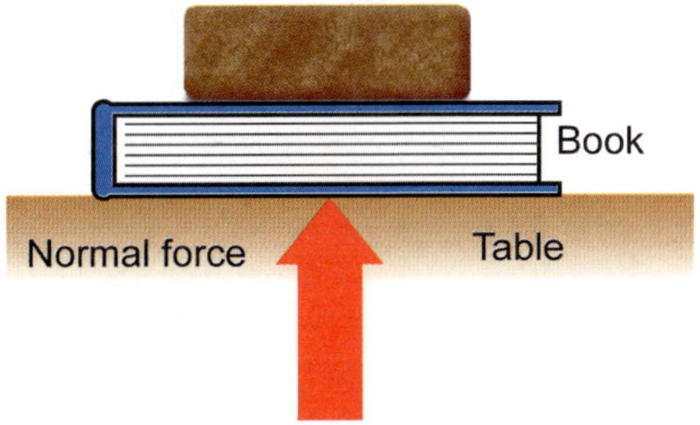

Figure 13.17: *The normal force is greater if a brick is placed on the book*

The free-body diagram

Forces on a free-body diagram

How do you keep track of many forces with different directions? The answer is to draw a **free-body diagram**. A free-body diagram contains only a single object, like a book or a table. All connections or supports are taken away and replaced by the forces they exert on the object. An accurate free-body diagram includes *every* force acting on an object, including weight, friction and normal forces.

An example

As an example of a free-body diagram, consider a 30-newton book resting on a table that weighs 200 newtons. The book is on one corner of the table so that its entire weight is supported by one leg. Figure 13.18 shows a free-body diagram of the forces acting on the table.

Finding the forces

Because the table is in equilibrium, the net force on it must be zero. The weight of the book acts on the table. The weight of the table acts on the floor. At every point where the table touches the floor (each leg) a normal force is created. The correct free-body diagram shows six forces. The normal force at each of three legs is one-quarter the weight of the table (50 newtons). The leg beneath the book also supports the weight of the book (80 N = 50 N + 30 N).

The purpose of a free-body diagram

By separating an object from its physical connections, a free-body diagram helps you identify all forces and where they act. A normal force is usually present at any point an object is in contact with another object or surface. Forces due to weight may be assumed to act directly on an object, often at its center.

Positive and negative forces

There are two ways to handle positive and negative directions in a free-body diagram. One way is to make all upward forces positive and all downward forces negative. The second way is to draw all the forces in the direction you believe they act on the object. When you solve the problem, if you have chosen correctly, all the values for each force are positive. If one comes out negative, it means the force points in the opposite direction from what you guessed.

VOCABULARY

free-body diagram - a diagram showing all the forces acting on an object.

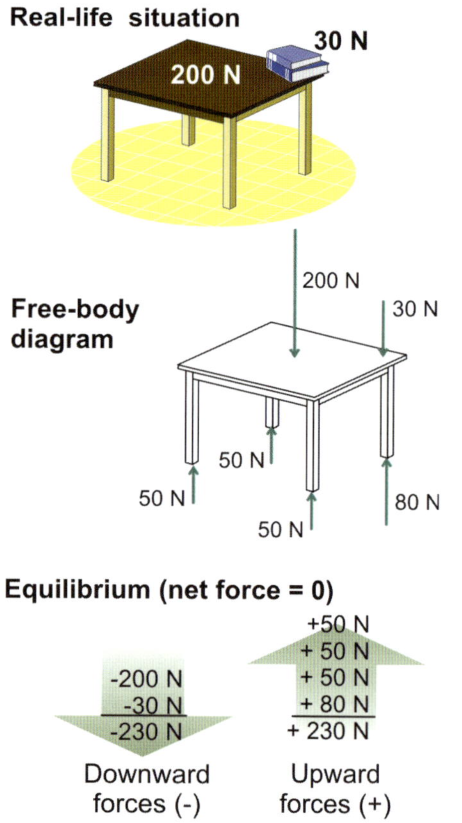

Figure 13.18: *A free body diagram showing the forces acting on a table that has a book resting on one corner.*

Forces from springs and ropes

Types of forces

Examples of the forces you might draw in a free-body diagram are shown in Figure 13.19. You are already familiar with weight, normal force, friction, and contact forces. You learned about electrical and magnetic forces in Chapter 6. Other forces you might encounter are forces from springs or ropes. These forces have special properties.

VOCABULARY

tension - a pulling force that acts in a rope, string, or other object.

compression - a squeezing force that can act on a spring.

Tension forces

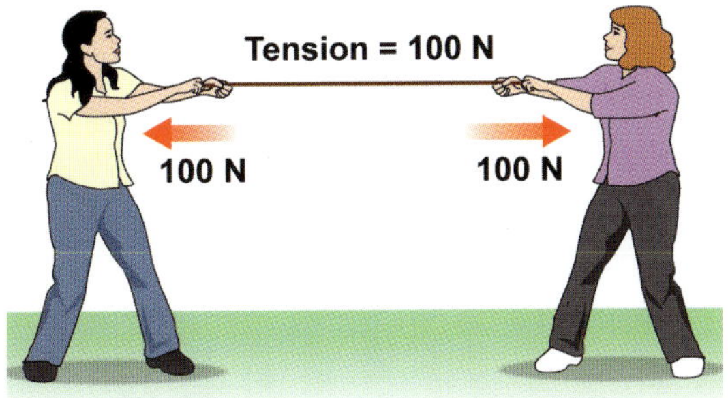

Each person pulls with a force of 100 newtons so the tension is 100 newtons

Tension is a force that acts in a rope, string, or other object that is pulled. *Tension always acts along the direction of the rope.* A rope in tension exerts equal forces on whatever is connected to either end. For example, the two people in the diagram above are each pulling on the rope with a force of 100 newtons. The tension in the rope is 100 newtons. Ropes or strings do *not* carry pushing forces. This is obvious if you have ever tried pushing a rope.

The force from springs

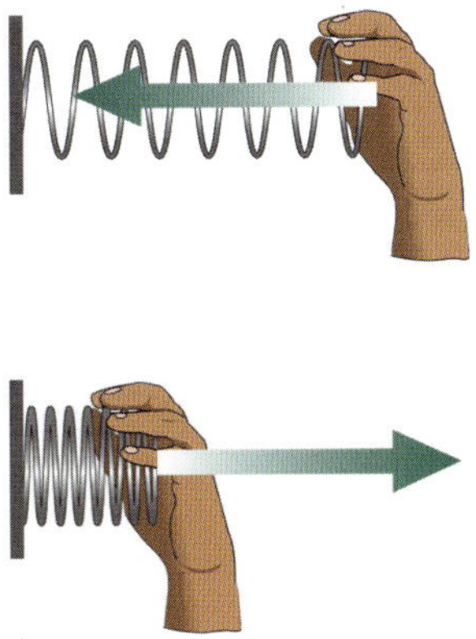

The most common type of spring is a coil of metal or plastic that creates a force when you stretch it or compress it. The force created by stretching or compressing a spring always acts to return the spring to its natural length. When you stretch a spring, it pulls back on your hand as the spring tries to return to its original length. When you apply a **compression** force to a spring and make it shorter, it pushes on your hand as it tries to return to its original length.

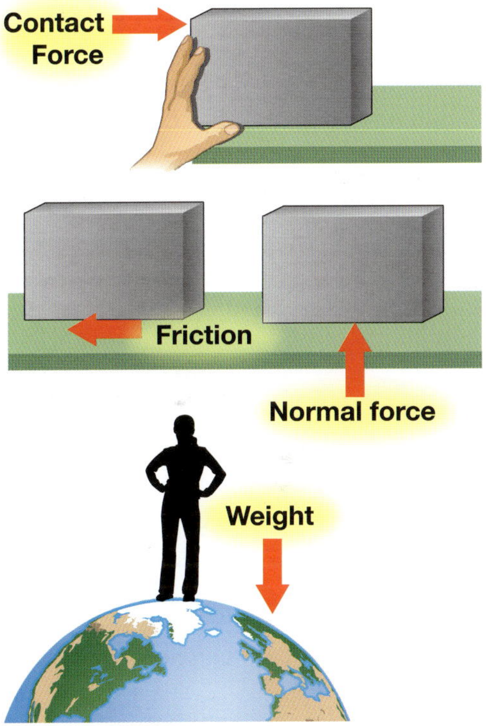

Figure 13.19: *Some examples of forces that can be included in a free-body diagram.*

Solving equilibrium problems

Finding the net force

For an object to be in equilibrium, all the forces acting *on the object* must add to zero (Figure 13.20). The net force *in each direction* must be zero. That means the total force in the horizontal (x) direction must be zero and total force in the vertical (y) direction also must be zero. You cannot mix forces in the horizontal direction with forces in the vertical direction.

Balancing forces

If you are trying to find an unknown force on an object in equilibrium, the first step is always to draw a free-body diagram. Then use the fact that the net force is zero to find the unknown force. To be in equilibrium, forces must balance both horizontally and vertically. Forces to the right must balance forces to the left, and upward forces must balance downward forces.

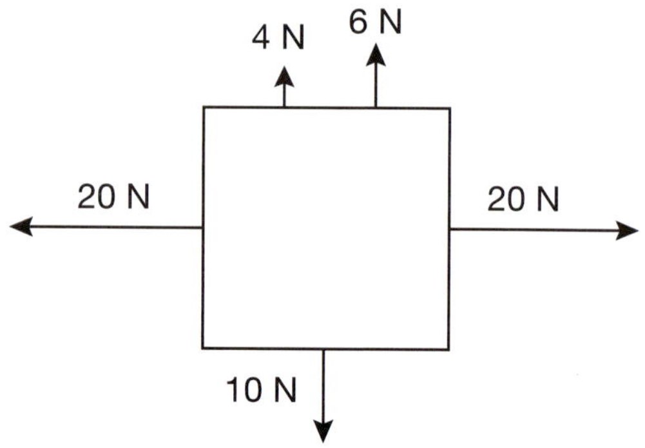

Figure 13.20: *An object is in equilibrium if the vertical forces balance and the horizontal forces balance.*

Equilibrium

Two chains are used to support a small boat weighing 1500 newtons. One chain has a tension of 600 newtons. What is the force exerted by the other chain?

1. Looking for: You are asked for an unknown tension in a chain.
2. Given: You are given the boat's weight in newtons and the tension in one chain in newtons.
3. Relationships: The net force on the boat is zero.
4. Solution: Draw a free-body diagram.
   The force of the two chains must balance the boat's weight.
   $600\text{ N} + F_{chain2} = 1500\text{ N}$ $\quad F_{chain2} = 900\text{ N}$

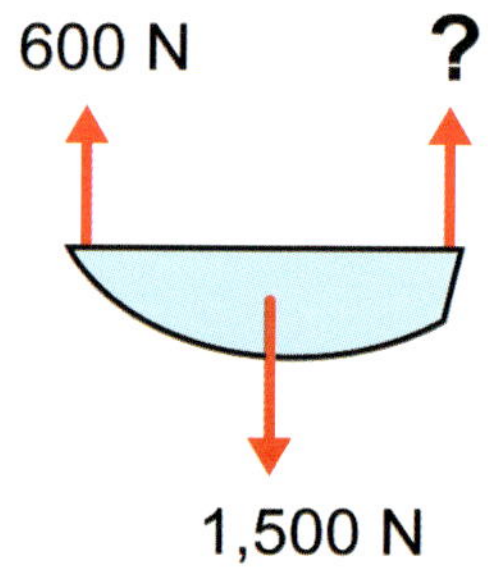

Your turn...

a. A heavy box weighing 1000 newtons sits on the floor. You press down on the box with a force of 450 newtons. What is the normal force on the box? **Answer:** 1450 newtons

b. A 40-newton cat stands on a chair. If the normal force on each of the cat's back feet is 12 newtons, what is the normal force on each front foot? (You can assume it is the same on each.) **Answer:** 8 newtons

13.2 Section Review

1. A vector is an example of a physical quantity that
   a. includes information about force and mass.
   b. includes information about temperature.
   c. includes information about quantity and direction.
   d. includes information about forces only.
2. The loudness of sound can be measured in decibels. Do you think loudness is a vector or not? Explain why or why not.
3. The diagram in shows three forces acting on a pencil. What is the net force acting on the pencil?
4. If an object is an equilibrium, that means
   a. the net force on the object is zero.
   b. the object has zero total mass.
   c. no forces are acting on the object.
   d. only normal forces are acting on the object.
5. A train is climbing a gradual hill. The weight of the train creates a downhill force of 150,000 newtons. Friction creates an additional force of 25,000 newtons acting in the same direction (downhill). How much force does the train's engine need to produce so the train is in equilibrium.
6. Draw a free body diagram of your own body sitting on a chair. Include all forces acting on your body.
7. If a force has a negative value, such as - 100 N, that means the force
   a. is less than 100 N in strength.
   b. acts in the opposite direction from a + 100 N force.
   c. is a normal force.
8. A child weighing 150 newtons is sitting in a swing. The swing is supported by two ropes, one on each side. What is the tension in one of the ropes?

Figure 13.1: *The pencil and forces for question three.*

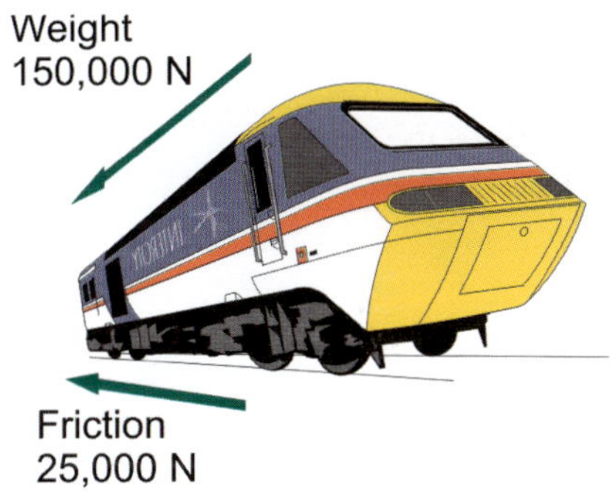

Figure 13.2: *The train and forces for question five.*

TECHNOLOGY CONNECTION

Defy Gravity? It Can Be Done

How can something that weighs almost a million pounds overcome the force of gravity and stay off the ground for hours? It almost sounds impossible. Yet it happens thousands of times every day, as large airplanes like the Boeing 747 fly all over the world.

When you see a 747 parked at an airport gate, the plane looks as big as a building. How does this enormous object manage to fly? Simple physics—really!

Newton's first law of motion states that an object will remain at rest unless external forces affect it. A 747 jet plane parked on the ground will not move until external forces make it move. For that plane to not only move but fly, four basic forces are at work: weight, lift, thrust, and drag.

Weight and lift

Weight is a force caused by gravity. The gravitational pull of Earth causes objects to have weight and generate force. All forces have magnitude (size) and direction. The magnitude of weight varies from object to object, depending on mass. A suitcase has less magnitude than an airplane. The direction of weight in an airplane is down, toward Earth's center.

Lift is a force that goes in the opposite direction of weight. The magnitude of lift depends mainly on the size and shape of an airplane's wing.

When an airplane has the two opposite forces of lift and weight affecting it, the cumulative effect will determine the motion of the airplane. If lift is much weaker than weight, the airplane will remain on the ground. But when the force of lift is stronger than the force of an airplane's weight, it will leave the ground and fly.

Controls allow the pilot to keep the airplane at a constant altitude once it is airborne. The forces of lift and weight are balanced, and the motion of the airplane does not change along its perpendicular axis.

Thrust and drag

Thrust is the force that moves an airplane forward. This is the heavy work. On the ground, airplanes move on wheels, which reduce the force of friction from the ground. But to get that 747 rolling, then airborne, then to keep it flying - this requires a lot of thrust. And it is provided by the jet's engines.

The direction of thrust is forward. It moves horizontally to the airplane. The magnitude of thrust depends on the number of engines and their power. Imagine the power required to get a 747 moving from a standing start to the takeoff speed of 180 miles per hour. (It's quite a moving - and thrilling - experience, that's for sure.)

Drag is a force that goes in the opposite direction of thrust. Drag is horizontal to the airplane, just as thrust is. But the direction of drag is toward the back of the airplane. The magnitude of drag depends mainly on the shape of the airplane and its wing, and the plane's velocity.

When an airplane is airborne, air resistance causes drag. The more air resistance, the greater the magnitude of the drag. When you ride your bike, you feel the force of the wind in your face even on a calm day. This is air resistance, or drag. If you ride into a strong wind, or pedal faster, the magnitude of the drag increases. The same is true of the drag on an airplane flying into a heavy wind.

Once the airplane is airborne, controls allow the pilot to keep the airplane at a constant speed. At this point the forces of thrust and drag will remain balanced until the wind speed changes or the pilot reduces power.

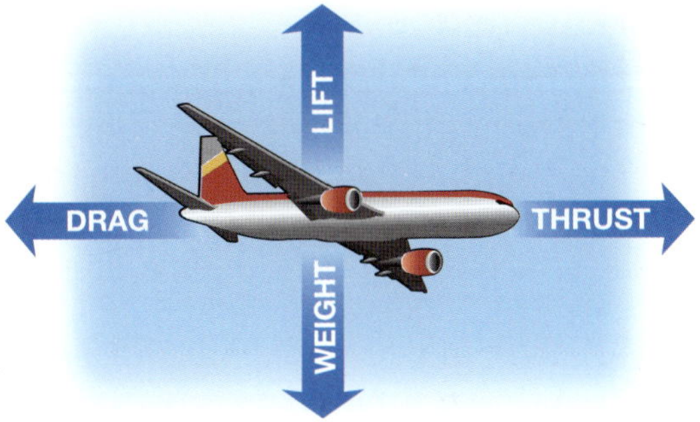

Combining forces

A 747 has a lot of mass and requires an enormous amount of thrust to get it in motion. But how does all that relate to the airplane becoming airborne?

The answer lies in the motion of air on the wing. Airplane wings are shaped to deflect airflow over the top. The air pattern is different on top of the wing from the pattern on the bottom, and so is the air pressure. There is greater air pressure underneath the wing than on top of it, and this, quite logically, produces lift.

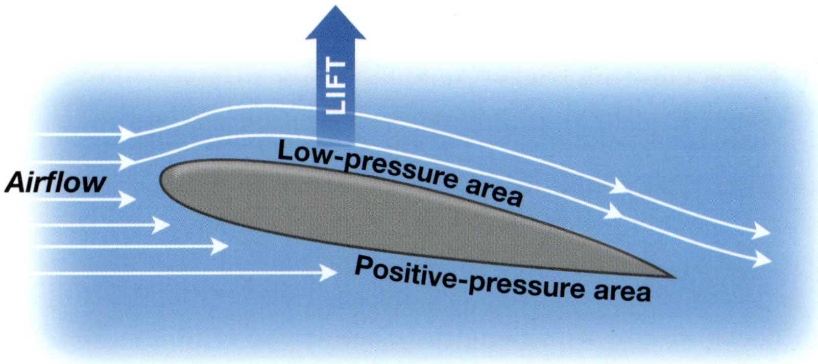

In order to produce enough airflow over and under the wing to maintain lift, a tremendous amount of thrust is needed. That thrust must be powerful enough to overcome the force of drag. When lift has overcome the weight of the airplane, and thrust has overcome drag, the forces on the plane reach a new balance. An airplane can cruise for as long as its fuel lasts or until thrust is reduced in preparation for landing.

Questions:

1. Describe the force of weight and how it affects an airplane in flight.
2. Describe the force of lift and how it affects an airplane in flight.
3. Describe the force of thrust and how it affects an airplane in flight.
4. Describe the force of drag and how it affects an airplane in flight.

CHAPTER ACTIVITY Kilograms to Newtons: Make Your Own Measurements!

The unit for measuring force called the newton is named after Sir Isaac Newton. Newton came up with the idea of gravity as he tried to find an explanation for the orbits of the planets around the Sun and the moon around Earth. He realized that planets could only be held in their orbits if a force of attraction existed between the planets and the Sun. He called this force gravity. The force of gravity is also called an object's weight. In the activity below you will explore the concept of a newton.

Materials:

50, 100, 200, and 500-gram masses
30 x 1/2 inch steel washers
Spring scale(s) calibrated in newtons for measuring above objects

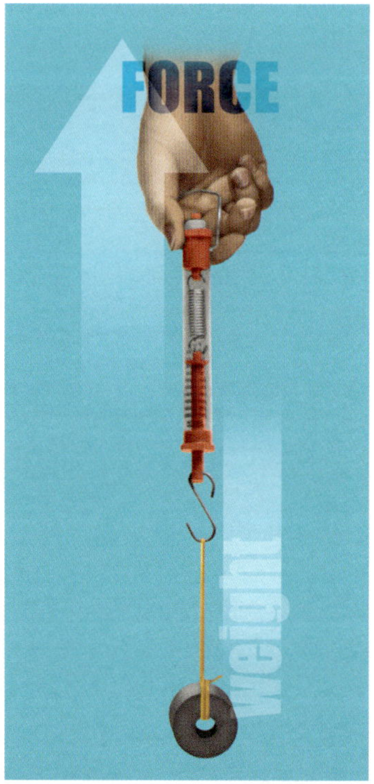

What you will do

1. Copy the data table. Convert each mass in grams to kilograms. There are 1000 grams in a kilogram, so you must divide by 1000.
2. Check to see that your scale is calibrated to read zero when nothing hangs from it.
3. Hang the 50-gram mass on the scale and measure the force of gravity (weight) in newtons. Record your value in the table.
4. Repeat with the 100, 200, and 500-gram masses.
5. Measure the weight of 10 washers, 20 washers, and 30 washers. Use what you learned during this activity to calculate the mass of these groups of washers.

Object	Mass (kilograms)	Force of gravity (newtons)
50-gram mass		
100-gram mass		
200-gram mass		
500-gram mass		
10 washers		
20 washers		
30 washers		

Applying your knowledge

a. If you know the mass of an object in kilograms, how can you calculate its weight in newtons on Earth?
b. If you know the weight of an object in newtons on Earth, how can you calculate its mass in kilograms?
c. The three blocks are the same size, but do they have the same weight? Why?

Chapter 13 Assessment

Vocabulary

Select the correct term to complete the sentences.

vector	static friction	net force
weight	normal force	sliding friction
friction	tension	free-body diagram
equilibrium	newtons	mass

Section 13.1

1. ____ is the force that resists the motion of objects or surfaces.
2. When a box is being pushed across a floor, ____ acts between the box and the floor.
3. If you stand still on the side of a hill, ____ keeps you from sliding down the hill.

Section 13.2

4. The ____ is the combination of all the forces acting on an object.
5. A(n) ____ has both an amount and a direction.
6. The ____ of an object changes due to the gravitational force acting on it.
7. In the metric system, all forces are measured in ____.
8. An object will have the same ____ if it is on Earth, the moon, or any other planet.
9. When a rope is being pulled from both ends, the force acting through it is called ____.
10. The ____ acts perpendicular to the surface of a book resting on a table, and balances the weight of the book.
11. A ____ is a drawing that shows all the forces affecting an object.
12. An object is in ____ when all of the forces on the object balance and there is no net force on the object.

Concepts

Section 13.1

1. Which has more mass, a 10 N object on the moon, or a 10 N object on Earth?
2. Explain why ice generally has very little friction when it is in contact with other materials.
3. When you put your groceries on a conveyer belt at the supermarket, friction plays a part in how the objects move down the belt. Describe how the friction acts on your food, and why this is the case.
4. Describe the difference between sliding friction and static friction.
5. Explain how friction keeps a nail in place in a block of wood. If you try to pull out the nail, which way does the friction act?

Section 13.2

6. For each of the following, tell whether it is a vector.
   a. 20 N west
   b. 42°C
   c. 3:15 pm
   d. 5 km down
   e. 20 m/s
   f. 30 m/s to the right
7. Draw the following vectors on a piece of paper and show the scale you use:
   a. 20 m/s west
   b. 3 miles north
   c. 4 N southeast

8. What four main forces act on an airplane in flight? If the plane accelerates forward, which two forces must be out of balance? To fly on a level path, which two forces must be in balance?

9. Describe the motion of the race car shown in the figure to the right. Is it speeding up or slowing down?

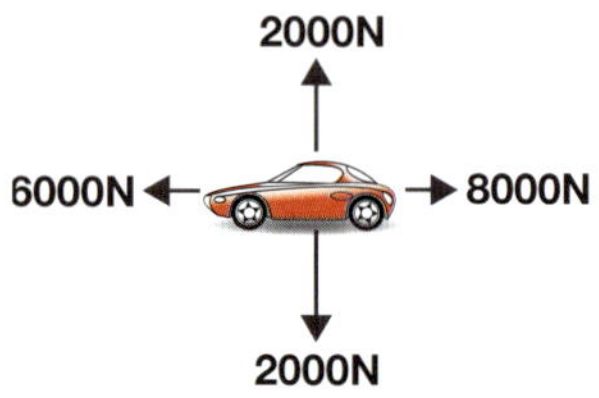

10. Which of the following diagrams correctly shows the normal force on the block of wood sliding down the incline?

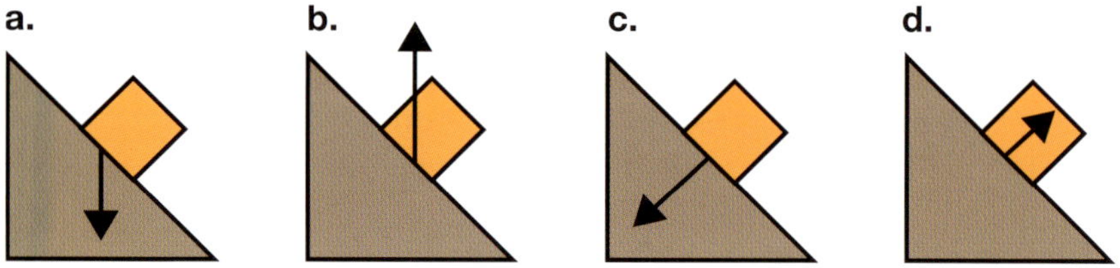

11. Draw a free-body diagram for the forces acting on the parachutist shown below. Don't forget about air friction!

Problems

Section 13.1

1. How much does a 40-kg student weigh on Earth, in newtons?

2. How much mass does a 50000-N truck have?

3. An empty 100-kg refrigerator is pushed across a floor, and a student measures a frictional force of 200 N. When the refrigerator is filled with 25 kg of food, how much friction will be present when the now filled refrigerator is pushed across the floor?

Section 13.2

4. How would you label the vector below? (each space = 1N)

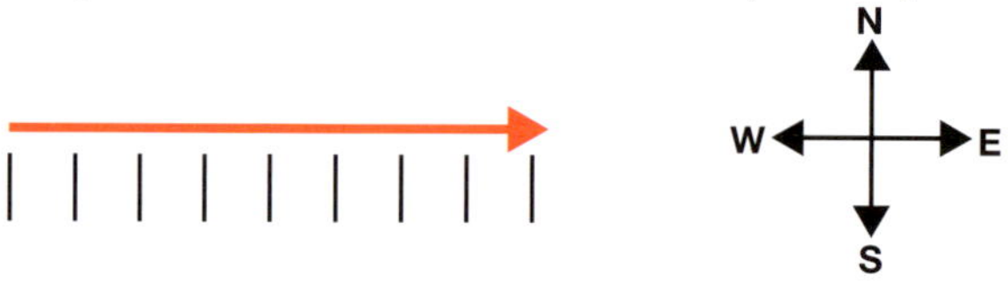

5. A 4-kg book rests on a table. What is the weight of the book? What is the normal force on the book and in which direction does it act?

6. Two friends decide to build their strength by having a tug of war each day. They each pull with a force of 200 N.
   a. How much tension is on the rope?
   b. One day, one of the friends is sick and cannot work out. The other friend decides to build strength by tying a rope around a tree and pulling on the rope. How much must the single friend pull in order to get the same workout as he normally does? What is the tension in the rope in this case? Explain.
   c. In both cases above, what is the net force on the rope if neither person is moving, and the tree stays put?

Chapter 14

Force and Motion

In January 1993, the 53rd space shuttle mission crew brought some toys on board. During the flight, crew members took the toys out and played with them to see how they would work in "microgravity." Could you imagine trying to shoot a foam ball through a hoop while floating around in the space shuttle? How about running a toy car around a loop track – would that work? This chapter will help you use laws of motion to explain the motion of objects on Earth, and then you will be able to predict how toys would work in space. Do an Internet search on "toys in space" to learn more about the interesting space shuttle experiments.

Sir Isaac Newton, who lived from 1642 – 1727, answered many questions about motion. Many historians believe Newton's ideas about motion were the beginning of modern science. Read this chapter and you will know all about motion too!

Key Questions

1. ***Why is a bowling ball harder to move than a golf ball?***

2. ***What would happen if Sir Isaac Newton had a skateboard contest with an elephant?***

3. ***What happens to the speed of an object as it falls freely?***

14.1 Newton's First and Third Laws

Sir Isaac Newton (1642-1727), an English physicist and mathematician, was one of the most brilliant scientists in history. Before age 30, he had made many important discoveries in physics and had even invented a new kind of mathematics called calculus. Newton's three laws of motion are probably the most widely used natural laws in all of science. The laws explain the relationships between the forces acting on an object, the object's mass, and its motion. This section discusses Newton's first and third laws.

> **VOCABULARY**
>
> **force** - a push or pull, or any action that is able to change motion. Force is measured in newtons (N).

Force changes motion

Changing an object's motion Suppose you are playing miniature golf and it is your turn. What must you do to make the golf ball move toward the hole? Would you yell at the ball to make it move? Of course not! You would hit the ball with the golf club to get it rolling. In physics language, "hit the ball" means the golf club applies a *force* to the ball. This force is what changes the ball from being at rest to being in motion (Figure 14.1). *Motion can change only through the action of a force.*

What is force? A **force** is a push or pull, or any action that is able to change motion. The golf ball will stay at rest until you apply force to set it in motion. Once the ball is moving, it will continue to move in a straight line at a constant speed, unless another force changes its motion. You need force to start things moving and also to make any change to their motion once they are moving. Forces can be used to increase or decrease the speed of an object, or to change the direction an object is moving. Recall that velocity is speed with direction. Any change in velocity requires force.

How are forces created? Forces are created in many different ways. For example, your muscles create force when you swing the golf club. Earth's gravity creates forces that pull on everything around you. On a windy day, the movement of air can create forces. Each of these actions can create force and they all can change an object's motion.

Figure 14.1: *Force is the action that has the ability to change motion. Without force, the motion of an object cannot be started or changed.*

The law of inertia

Force is required to change motion There can be no change in motion without force. Anytime there is a change in motion a force must exist, even if you cannot immediately recognize the force. For example, when a rolling ball hits a wall and bounces, its motion quickly changes. That change in motion is caused by the wall creating a force that changes the direction of the ball's motion.

Stopping a moving object Let's keep playing golf and assume you are on a perfectly level golf course with no friction at all. Once the golf ball is moving, how can you stop it? If you do nothing the ball will keep rolling at the same speed in the same direction *forever*. When the net force is zero, objects continue moving with the exact same motion they already have. This idea is known as Newton's first law of motion. The only way to stop the ball is to apply a force in a direction opposite its motion.

Newton's first law **Newton's first law** says that objects continue the motion they already have unless they are acted on by a net force. If the net force is zero, an object at rest will stay at rest. If an object is acted upon by **unbalanced forces**, its motion will change.

Inertia **Inertia** is the property of an object that resists changes in motion. To understand inertia, imagine trying to move a bowling ball and a golf ball (Figure 14.2). Which needs more force? Of course, the bowling ball needs more force to get it moving at the same speed as the golf ball (assuming the forces act for the same length of time). The bowling ball also needs more force to stop. A bowling ball has more inertia than a golf ball. The greater an object's inertia, the greater the force needed to change its motion. Because inertia is an important idea, Newton's first law is sometimes called the law of inertia.

Mass Inertia comes from mass. Objects with more mass have more inertia and are more resistant to changes in their motion. Mass is measured in kilograms (kg). A golf ball has a mass of 0.05 kilograms, and the average bowling ball has a mass of 5 kilograms. A bowling ball is 100 times as massive, so it has 100 times the inertia.

VOCABULARY

Newton's first law - an object at rest will stay at rest and an object in motion will stay in motion with the same velocity unless acted on by an unbalanced force.

inertia - the property of an object that resists changes in its motion.

unbalanced forces - result in a net force on an object that can cause changes in motion.

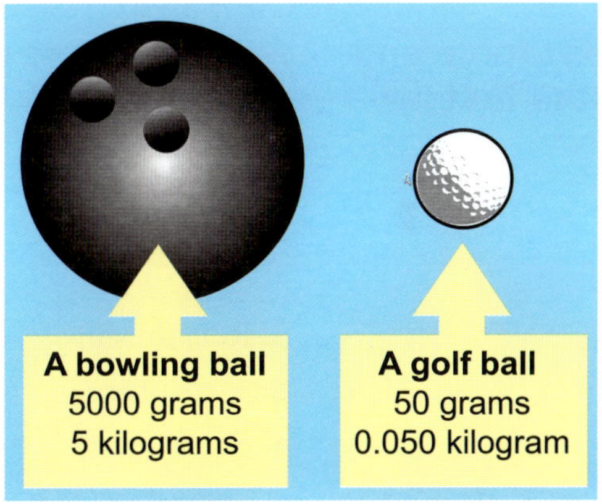

Figure 14.2: *A bowling ball has more mass than a golf ball. The bowling ball is harder to move because it has more inertia.*

The net force

Multiple forces When you hit a golf ball, the force from the club is not the only force that acts on the ball (Figure 14.3). The ball's weight, the normal force from the ground, and friction are also acting. Which force determines how the ball moves?

Net force causes motion You are right if you are thinking "all forces together." The motion of an object depends on the *net force* acting on it. There is almost always more than one force present because gravity acts on all objects.

Adding forces Recall that force is a vector. Adding up forces can be different from simply adding numbers because the *directions* of the forces matter. To find the net force, you must include positive and negative signs to account for the directions of the forces. Newton's first law is often written in terms of the net force: *an object at rest will stay at rest and an object in motion will continue in motion at constant velocity UNLESS there is a net force.*

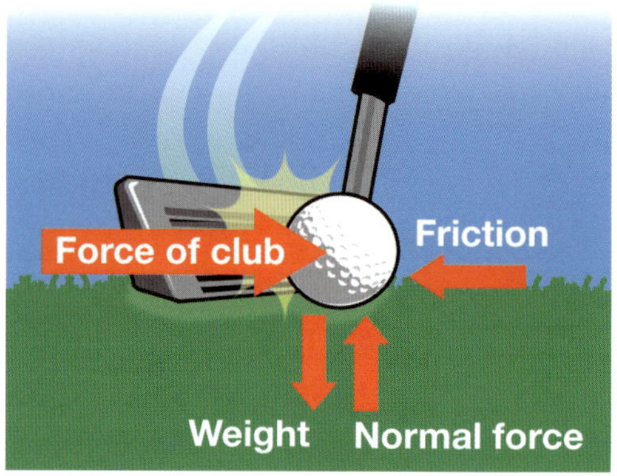

Figure 14.3: *Four forces act on a golf ball. The net force determines how it moves.*

Net force and the first law

A car drives along the highway at constant velocity. Find the car's weight and the friction force.

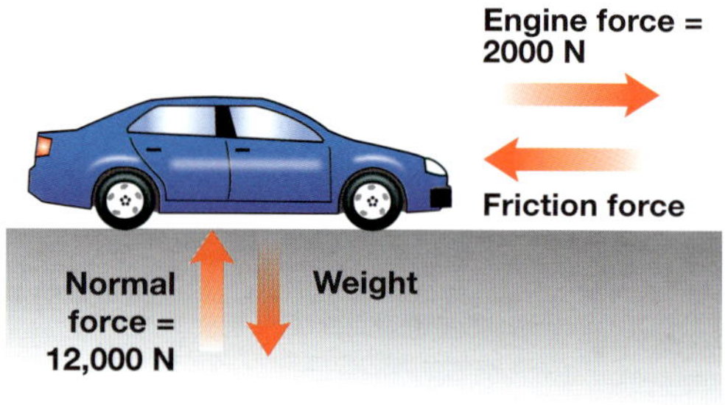

1. Looking for: You are asked for the weight and the friction force.
2. Given: You are given the normal force and engine force. The car is moving at a constant velocity.
3. Relationships: Newton's first law states that if the car is moving at a constant velocity, the net force must be zero.
4. Solution: The weight balances the normal force, so the weight is -12,000 N. The engine force balances the friction force, so the friction force is -2,000 N.

Your turn...

a. Identify the forces on the same car if it is stopped at a red light on level ground. **Answer**: The friction force and engine force are zero. The weight and normal force are still each 12,000 N

b. As you sit on a chair, the chair exerts a normal force of 550 N on you. If you weigh 600 N, what is the normal force of the ground on your feet? **Answer**: 50 N

Forces always come in matched pairs

Throwing a ball Newton's first law applies to the motion of an *individual* object. However forces must be applied by something! Think about throwing a basketball (Figure 14.4). You feel the ball push back against your hand as you throw it. You know you apply a force to the ball to make it move. Where does the force against your hand come from?

An imaginary skateboard contest Imagine a skateboard contest between Newton and an elephant. They can only push against each other, not against the ground. The fastest one wins. The elephant knows it is much stronger and pushes off Newton with a huge force thinking he will surely win. But will he?

The winner The result of the giant push from the elephant is that Newton flies away with a great speed and the puzzled elephant moves backwards with a much smaller speed. Newton wins — and will always win. No matter how hard the elephant pushes, Newton always moves away at a greater speed. In fact, Newton doesn't have to push at all and he still wins. Why?

Forces always come in pairs You already know it takes force to make both Newton and the elephant move. Newton wins because *forces always come in pairs*. The elephant pushes against Newton and that *action* force pushes Newton away. The elephant's force against Newton creates a *reaction* force against the elephant. The action and reaction forces are equal in strength. Newton has a lot less mass (inertia) so he gains more speed.

Figure 14.4: *You experience the third law (action-reaction) whenever you apply force to any object, such as a basketball.*

MY JOURNAL

Think of three examples of action reaction that you experienced before class today. Write each one down and identify the action and reaction forces. Also write down what object each force acts on. Hint: the action and reaction forces never act on the same object.

The third law: action and reaction

The first and second laws The first law of motion applies to single objects. The first law says an object will remain at rest or in motion at constant velocity unless acted upon by a net force.

The third law The third law of motion deals with pairs of objects. This is because *all forces come in pairs*. **Newton's third law** states that every action force creates a reaction force that is equal in strength and opposite in direction.

Every action force creates a reaction force that is equal in strength and opposite in direction.

Force pairs There can never be a single force, alone, without its action-reaction partner. Forces *only* come in action-reaction pairs. The force exerted by the elephant (action) moves Newton since it acts on Newton. The reaction force acting back on the elephant is what moves the elephant.

The labels "action" and "reaction" The words action and reaction are just labels. It does not matter which force is called action and which is reaction. You choose one to call the action and then call the other one the reaction (Figure 14.5).

Why action and reaction forces do not cancel each other out Why don't action and reaction forces cancel each other out? The reason is *action and reaction forces act on different objects*. For example, think again about throwing a ball. When you throw a ball, you apply the action force to the ball, creating the ball's acceleration. The reaction is the ball pushing back against your hand. The action acts on the ball and the reaction acts on your hand. The forces do not cancel because they act on different objects. You can only cancel forces acting on the same object (Figure 14.6).

VOCABULARY

Newton's third law - For every action force, there is a reaction force equal in strength and opposite in direction.

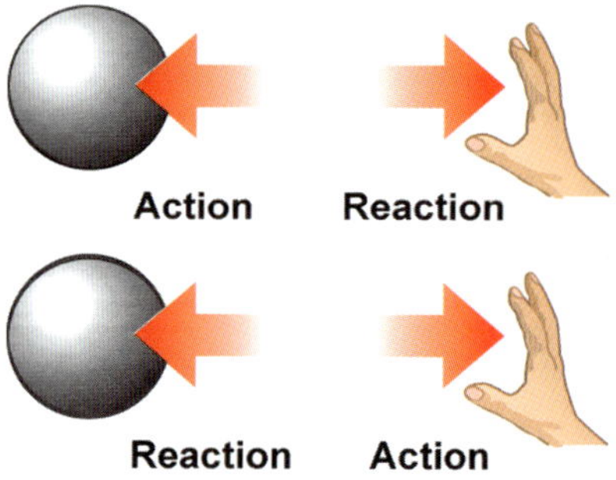

Figure 14.5: *It doesn't matter which force you call the action and which the reaction.*

Figure 14.6: *Action and reaction forces do not cancel.One force acts on the ball, and the other force acts on the hand.*

Action and reaction forces

A skateboard example Think carefully about moving the usual way on a skateboard. Your foot presses backward against the ground (Figure 14.7). The force acts *on* the ground. However, *you* move, so a force must act on you. Why do you move? What force acts on you? You move because the action force of your foot against the ground creates a reaction force of the ground against your foot. You "feel" the ground because you sense the reaction force pressing on your foot. The reaction force is what makes you move because it acts on *you*.

Figure 14.7: *You move forward because of the reaction force of the ground on your foot.*

Draw diagrams When sorting out action and reaction forces it is helpful to draw diagrams. Draw each object apart from the other. Represent each force as an arrow in the appropriate direction.Here are some guidelines to help you sort out action and reaction forces:

- Both are always there whenever any force appears.
- They always have the exact same strength.
- They always act in opposite directions.
- They always act on different objects.
- Both are real forces and can cause changes in motion.

Action and reaction

A woman with a weight of 500 N is sitting on a chair. Describe an action-reaction pair of forces.

1. Looking for: You are asked for a pair of action and reaction forces.
2. Given: You are given one force in newtons.
3. Relationships: Action-reaction forces are equal and opposite, and act on different objects.
4. Solution: The force of 500 N exerted by the woman on the chair seat is an action. The chair seat acting on the woman with an upward force of 500 N is a reaction.

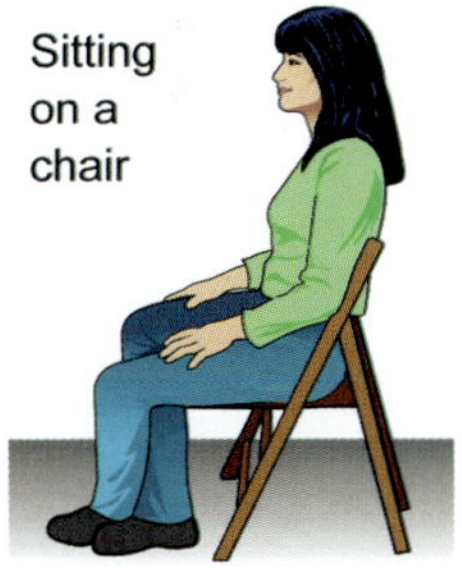

Your turn...

a. A baseball player hits a ball with a bat. Describe an action-reaction pair of forces. **Answer:** The force of the bat on the ball accelerates the ball. The force of the ball on the bat (reaction) slows down the swinging bat (action).

b. Earth and its moon are linked by an action-reaction pair. **Answer:** Earth attracts the moon (action) and the moon attracts Earth (reaction) in an action-reaction pair. Both action and reaction are due to gravity.

Collisions

The effect of forces

Newton's third law tells us that any time two objects hit each other, they exert equal and opposite forces on each other. However, the *effect* of the force is not always the same. Imagine two hockey players moving at the same speed toward each other, one with twice the mass of the other. The force on each during the collision is the same strength, but they do not have the same change in motion.

More mass results in less acceleration

The person with more mass has more inertia. More force is needed to change his motion. Because of his greater inertia, the more massive skater will have a smaller change in motion during the collision. The forces on each skater are always exactly equal and opposite. The two skaters have different changes in motion because they have different amounts of inertia, *not because the forces are different.*

Auto collisions

The same is true of vehicles in a collision. When a large truck hits a small car, the forces are equal (Figure 14.8). However, the small car experiences a much greater change in velocity much more rapidly than the big truck.

Safety features

Riding in a vehicle with a large mass does not guarantee passengers are safe in a collision. Large SUVs are more likely to roll over during accidents, and rollovers frequently cause injuries. Auto manufacturers conduct crash tests to help them improve the design of cars. Safety features such as seat belts, airbags, and antilock brakes help make cars safer (Figure 14.9).

Figure 14.8: *The car has less inertia, so accelerates more and becomes more damaged than the truck.*

Figure 14.9: *Safety features help passengers avoid injury during a collision.*

14.1 Section Review

1. A force is any action that changes motion. For each of the following situations, identify what creates the force.
   a. A flag flaps back and forth at the top of a flagpole.
   b. A soccer ball is passed from one player to another.
   c. A large piece of hail falls to the ground.
   d. The tide goes from high to low at the shore (you might have to do a little research to get this one if you don't know already).
2. Which has more inertia - a shopping cart full of groceries or an empty shopping cart?

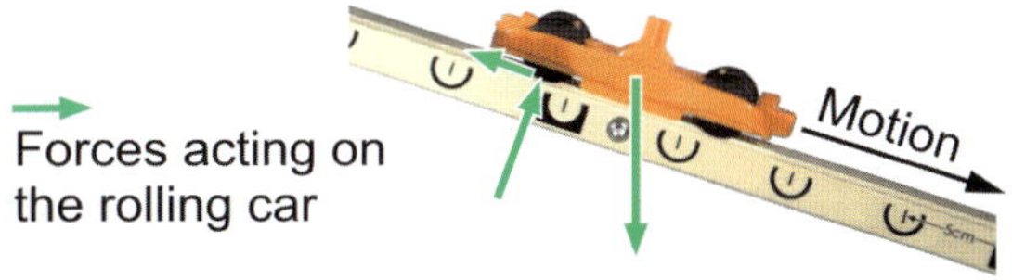

3. The net force acting on a car rolling down a ramp is the combination of three forces. One of the forces is the ramp pushing up to support the car.
   a. Name the other two forces acting on the car.
   b. Which of these two forces helps the motion of the car?
   c. Which of these two forces acts against the motion of the car?
4. Emilio tries to jump to a nearby dock from a canoe that is floating in the water. Instead of landing on the dock, he falls into the water beside the canoe. Use Newton's third law to explain why this happened.
5. Molly is sitting motionless on a playground swing. The downward force is 500 N. How much force is there on *each* of the two swing chains? Draw a diagram to go with your answer.
6. Two teams participate in a tug-of-war contest. Describe the action-reaction force pair that will determine who wins the contest. (Hint: action - reaction force pairs act on different objects!)

SOLVE IT!

Squid Science

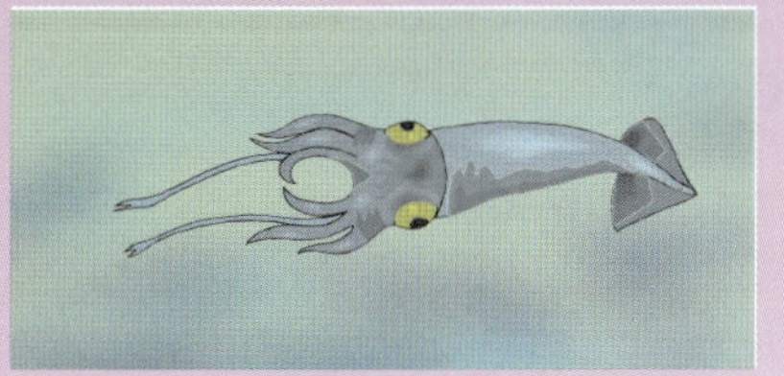

A squid takes water into its body chamber and rapidly pushes it out of a backward-facing tube. What are the action - reaction forces in this example? Draw a diagram to go with your answer.

Did you know that in September 2004, Japanese scientists took over 500 photos of a giant squid? The animal was nearly 25 feet long! This was the first ever record of a live giant squid in the wild. Check out National Geographic's website for more information and photos.

14.2 Newton's Second Law

Newton's first law says that a force is needed to change an object's motion. But what kind of change happens? The answer is *acceleration*.The amount of acceleration depends on both the force and the mass according to Newton's second law. The second law is probably the most well-used relationship in all of physics.

VOCABULARY

acceleration - the rate of change of velocity.

Acceleration

Definition of acceleration What happens if you coast down a long hill on a bicycle? At the top of the hill, you move slowly. As you go down the hill, you move faster and faster—you accelerate. **Acceleration** is the rate at which your velocity (speed with direction) changes. If your speed increases by 1 kilometer per hour (km/h) each second, then your acceleration is 1 km/h per second.

Time	Speed
0 (start)	0 (start)
1 second	1 km/h
2 seconds	2 km/h
3 seconds	3 km/h
4 seconds	4 km/h
5 seconds	5 km/h

Time	Speed
0 (start)	0 (start)
1 second	2 km/h
2 seconds	4 km/h
3 seconds	6 km/h
4 seconds	8 km/h
5 seconds	10 km/h

Steeper hills Your acceleration depends on the steepness of the hill. If the hill is a gradual incline, you have a small acceleration, such as 1 km/h per second. If the hill is steeper, your acceleration is greater, perhaps 2 km/h per second. On the gradual hill, your speed increases by 1 km/h every second. On the steeper hill, it increases by 2 km/h every second.

Acceleration and direction If an object's acceleration is *zero*, the object must be moving at a constant speed in a straight line (or stopped). Acceleration occurs whenever there is a change in speed, direction, or both (Figure 14.10). For example, a car driving around a curve at a constant speed is accelerating (in the "physics" sense) because its direction is changing.

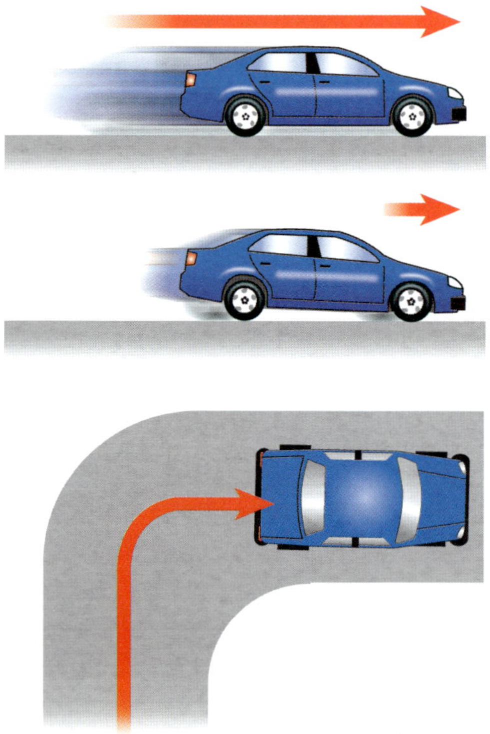

Figure 14.10: *A car can change its velocity by speeding up, slowing down, or turning.*

Units of acceleration

Speed units and time units

Acceleration is the rate of change of an object's speed. To calculate acceleration, you divide the change in speed by the amount of time it takes for the change to happen. In the example of the bicycle, acceleration was given in kilometers per hour per second. This unit can be abbreviated as km/h/s. Notice that two time units are included in the unit for acceleration. One unit of time is part of the speed unit, and the other is the time over which the speed changed.

Metric units

If the change in speed is in meters per second and the time is in seconds, then the unit for acceleration is m/s/s or *meters per second per second*. An acceleration of 10 m/s/s means that the speed increases by 10 m/s *every second*. If the acceleration lasts for three seconds, then the speed increases by a total of 30 m/s (3 seconds × 10 m/s/s).

What do units of seconds *squared* mean?

An acceleration in m/s/s is often written m/s^2 (meters per second squared). The steps below show you how to simplify the fraction m/s/s to get m/s^2. Saying *seconds squared* is just a math-shorthand way of talking. It is better to think about acceleration in units of speed change per second (that is, meters per second *per second*).

$$\text{Acceleration} = \frac{\text{Change in speed}}{\text{Change in time}}$$

How we get units of m/s^2

Plug in values		*Clear the compound fraction*		*Final units*
$\dfrac{50\,\frac{m}{s}}{s}$	$=$	$50\,\dfrac{m}{s} \times \dfrac{1}{s} = 50\,\dfrac{m}{s \times s}$	$=$	$50\,\dfrac{m}{s^2}$

Acceleration in m/s^2

Nearly all physics problems will use acceleration in m/s^2 because these units agree with the units of force (newtons). One newton is the force needed to accelerate a one-kilogram mass at a rate of 1 m/s^2 (Figure 14.11). If you measure speed in centimeters per second, you may have to convert to meters per second before calculating acceleration. This is especially true if you do any calculations using force in newtons.

Newton

One newton (N) is the force it takes to change the speed of a 1 kg mass by 1 m/s in 1 second.

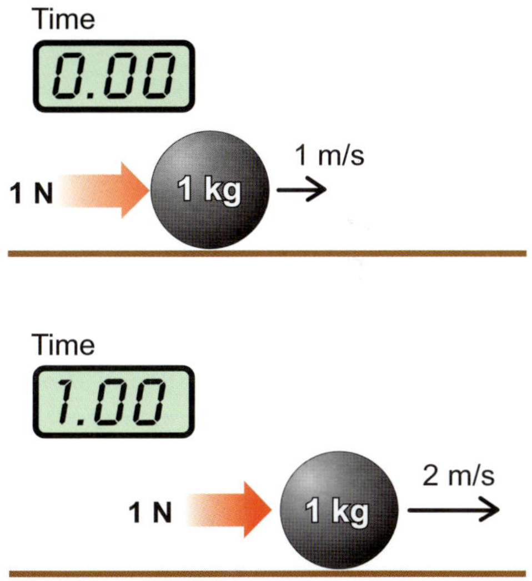

Figure 14.11: *The force unit the newton is defined in terms of the acceleration it can create.*

Calculating acceleration

The equation for acceleration

To calculate acceleration, you divide the change in speed by the time over which the speed changes. For example, if a bicycle's speed increases from 2 m/s to 6 m/s, its change in speed is 4 m/s. Because two speeds are involved, subscripts are used to show the difference. The starting speed is v_1, and the ending speed is v_2.

ACCELERATION

$$a = \frac{v_2 - v_1}{t}$$

Acceleration (m/s²) → a

Change in speed (m/s) → $v_2 - v_1$

Time (s) → t

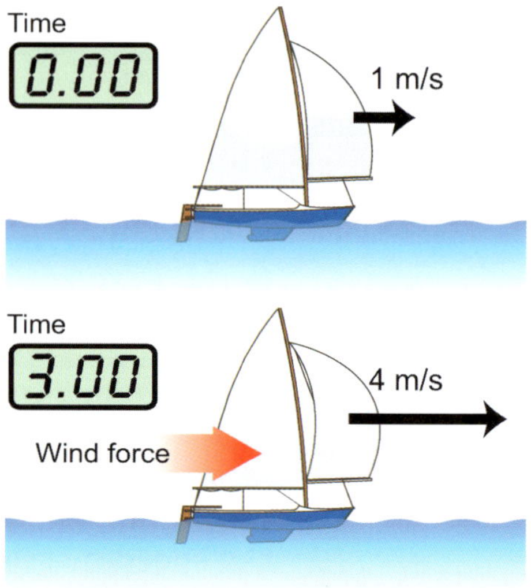

What is the acceleration?

Figure 14.12: *An acceleration example with a sailboat.*

Positive and negative acceleration

If an object *speeds up*, it has a *positive acceleration*. If it *slows down*, it has a *negative acceleration*. In physics, the word acceleration is used to describe to any change in speed, positive or negative. However, people sometimes use the word *deceleration* to describe the motion that is slowing down.

Calculating acceleration

A sailboat moves at 1 m/s. A strong wind increases its speed to 4 m/s in 3 seconds (Figure 14.12). Calculate the acceleration.

1. Looking for: You are asked for the acceleration in meters per second.
2. Given: You are given the starting speed in m/s (v_1), final speed in m/s (v_2), and the time in seconds.
3. Relationships: Use the formula for acceleration: $a = \frac{v_2 - v_1}{t}$
4. Solution:

$$a = \frac{4 \text{ m/s} - 1 \text{ m/s}}{3 \text{ s}} = \frac{3 \text{ m/s}}{3 \text{ s}} = 1 \text{ m/s}^2$$

Your turn...

a. Calculate the acceleration of an airplane that starts at rest and reaches a speed of 45 m/s in 9 seconds. **Answer:** 5 m/s^2

b. Calculate the acceleration of a car that slows from 50 m/s to 30 m/s in 10 seconds. **Answer:** -2 m/s^2

Force, mass, and acceleration

Newton's second law

Force causes acceleration, and mass resists acceleration. **Newton's second law** relates the force on an object, the mass of the object, and its acceleration.

Force causes acceleration, and mass resists acceleration.

Newton's second law

The relationships between force, mass, and acceleration are combined in Newton's second law.

NEWTON'S SECOND LAW

$$a = \frac{F}{m}$$

Acceleration (m/s^2) → a; F ← Force (N); m ← Mass (kg)

Force and acceleration

The stronger the force on an object, the greater its acceleration. In mathematical terms, the acceleration of an object is *directly proportional* to the applied force. This means that increasing force increases acceleration. If twice the force is applied, the acceleration is twice as great.

Mass and acceleration

The greater the mass, the smaller the acceleration for a given force (Figure 14.13). That means acceleration is *inversely proportional* to mass. Increasing mass decreases the acceleration caused by a given force. For example, an object with twice the mass will have half the acceleration if the same force is applied.

VOCABULARY

Newton's second law - acceleration is force ÷ mass.

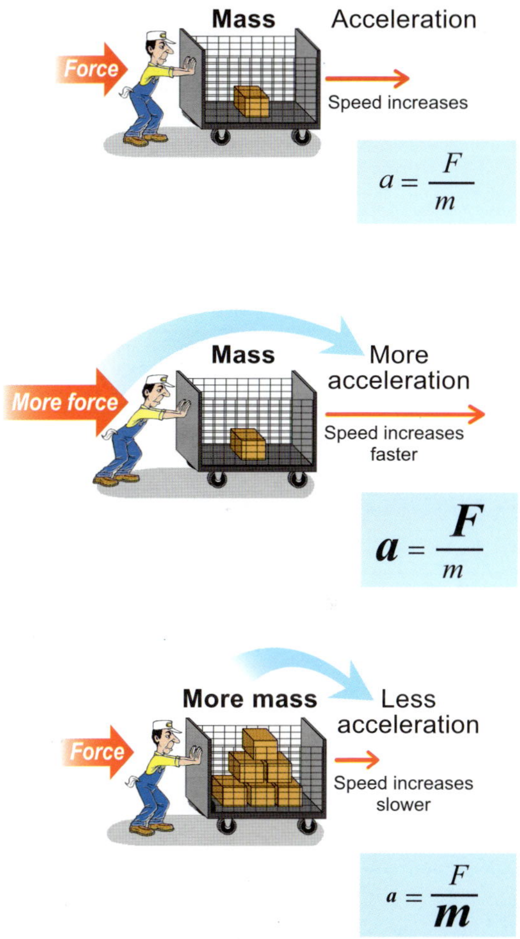

Figure 14.13: *More force causes more acceleration, and more mass causes less acceleration*

Applying the second law

Writing the second law You can use Newton's second law to calculate force, mass, or acceleration if two of the three are known. The way you write the formula depends on what you want to know. Three ways to write the law are summarized in Figure 14.14.

Net force and the second law Newton's second law explains the effect of a force, but what if there is more than one force on an object? In this case, the force that matters is the *net force*. You must consider all the forces that are acting and add them up to find the net force. Then you can use the net force to calculate acceleration. If you work in the other direction, calculating force from mass and acceleration, it is the net force that you get from the second law.

To use Newton's second law properly, keep the following important ideas in mind.

1. The *net* force is what causes acceleration.
2. If there is *no* acceleration, the net force *must* be zero.
3. If there *is* acceleration, there *must* also be a net force.
4. The force unit of newtons is based on kilograms, meters, and seconds

Use...	... if you want to find...	... and you know...
$a = \frac{F}{m}$	accel. (a)	force (F) and mass (m)
$F=ma$	force (F)	acceleration (a) and mass (m)
$m = \frac{F}{a}$	mass (m)	acceleration (a) and force (F)

Figure 14.14: *The three forms of the equation for Newton's second law.*

Newton's second law

A car has a mass of 1,000 kg. If a net force of 2,000 N is exerted on the car, what is its acceleration?

1. Looking for: You are asked for the car's acceleration.

2. Given: You are given its mass in kilograms and the net force in newtons.

3. Relationships: $a = \frac{F}{m}$

4. Solution: $a = \frac{2000\text{ N}}{1000\text{ kg}} = \frac{2\text{kg}\square\text{m/sec}^2}{\text{kg}} = 2\text{ m/sec}^2$

Your turn...

a. As you coast down the hill on your bicycle, you accelerate at 0.5 m/s^2. If the total mass of your body and the bicycle is 80 kg, with what force is gravity pulling you down the hill? **Answer:** 40 kg·m/s^2 or 40 N

b. You push a grocery car with a force of 30 N and it accelerates at 2 m/s^2. What is its mass? **Answer:** 15 kg

Newton's second law and gravity

The definition of free fall

An object is in **free fall** if it is accelerating due to the force of gravity and no other forces are acting on it. A ball dropped off a cliff is in free fall until it hits the ground (Figure 14.15). Objects in free fall on Earth accelerate downward at 9.8 m/s^2, the **acceleration due to gravity**. Because this acceleration is used so frequently in physics, the letter g is used to represent its value. When you see the letter *g* in a physics question, you can substitute the value 9.8 m/s^2.

Gravitational force, mass, and distance

Gravitational force exists between *all* objects that have mass. **Newton's law of universal gravitation** says the strength of the force depends on the mass of the objects and the distance between them. The greater the masses and the smaller the distance, the greater the force. You do not notice gravity between ordinary objects because it takes a huge amount of mass to create enough force to notice. You feel the force of gravity between you and Earth because the planet's mass is huge.

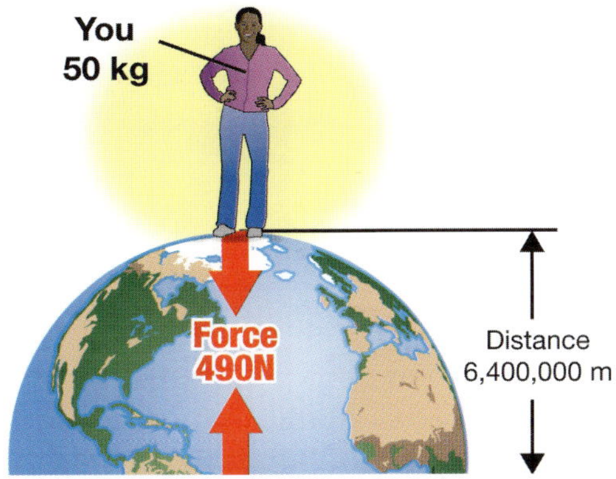

Weight

Your *weight* is the force of gravity between you and Earth. It depends on your mass, Earth's mass, and your distance from Earth's center. You calculated weight with the formula $W=mg$. The g in the weight formula is the same *g* that describes the acceleration due to gravity. The value of *g* depends on Earth's mass and the distance between its center and surface. If you travel to a planet or moon with a different mass and/or radius, the value of *g* and your weight would change.

ă VOCABULARY

free fall - the motion of an object acted on only by the force of gravity.

acceleration due to gravity - the acceleration of an object in free fall, 9.8 m/s^2 on Earth.

Newton's law of universal gravitation - the force of gravity between objects depends on their masses and the distance between them.

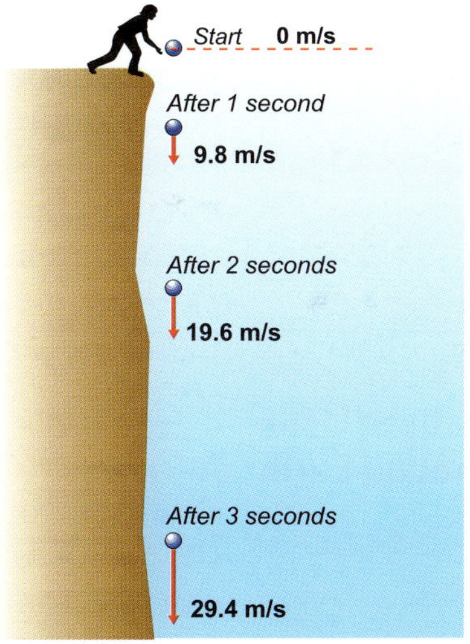

Figure 14.15: *An object in free fall accelerates at 9.8 m/s^2. Its speed increases by 9.8 m/s every second.*

Orbital motion

Gravitational force and acceleration

Newton's second law can be used to explain the motion of planets, moons, and satellites in orbit. For example, the moon moves in a 384,000 km orbit around Earth at nearly constant speed (Figure 14.16). While the moon's speed is constant, its direction is not. Therefore the moon is accelerating. The gravitational force between the Moon and Earth causes the acceleration.

Why the moon doesn't fall to Earth

The moon **orbits** or moves around Earth just as Earth and the other planets orbit the sun. But why doesn't the force of gravity just pull the moon into Earth? To answer that question, imagine kicking a ball off the ground at an angle. If you kick it at a slow speed, it curves and falls back to the ground. The faster you kick the ball, the farther it goes before hitting the ground. If you kick it fast enough, the curve of the ball's path matches the curvature of Earth. The ball goes into orbit instead of falling back to Earth.

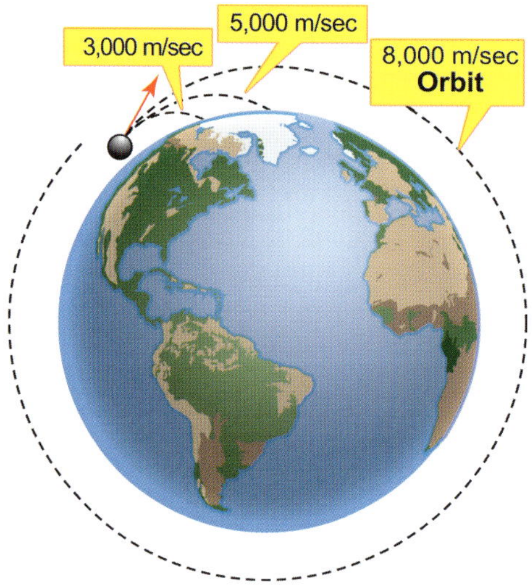

The moon falls around Earth

The same idea applies to the motion of the moon. The orbiting moon *falls around Earth*. But as it falls, Earth curves away beneath it. What do you think would happen if the gravitational force between the moon and Earth were to disappear? The moon's inertia would cause it to move in a straight line at a constant speed. The moon would fly off into space!

Satellites

Many human-made satellites orbit Earth providing weather data, communications services and scientific research. These satellites orbit much closer than the moon and must be launched at speeds high enough to reach orbit and not fall back to Earth.

VOCABULARY

orbit - the motion of one object around another caused by gravitational force.

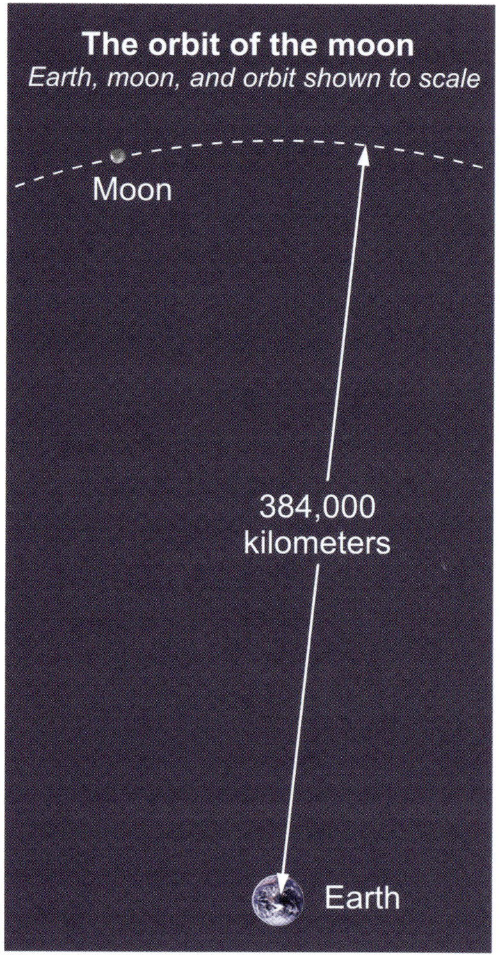

Figure 14.16: *The orbit of the moon shown to scale.*

14.2 Section Review

1. Does a car accelerate when it goes around a corner at a constant speed? Explain your answer.
2. Nearly all physics problems will use the unit m/s^2 for acceleration. Explain why the seconds are squared (in other words, why isn't the unit simply stated as m/s?)
3. A rabbit starts from rest and is moving at 6 m/sec after 3 seconds. What is the average acceleration of the rabbit? (Figure 14.17)
4. You are running a race and you speed up from 3 m/s to 5 m/s in 4 seconds.
   a. What is your change in speed?
   b. What is your acceleration?
5. Explain how changing force or mass affects the acceleration of an object, and provide examples to support your answer.
6. A tow truck pulls a 1,500-kilogram car with a net force of 4,000 newtons. What is the acceleration of the car?
7. A potato launcher uses a spring that can apply a force of 20 newtons to potatoes. A physics student launched a 100-gram potato, a 150-gram potato, and a 200-gram potato with the launcher. Which potato had the greatest acceleration?
8. What factors does *weight* depend on?

9. Earth exerts a gravitational force on the moon. Why doesn't the force of gravity pull the moon into the Earth?
10. An experiment measures the speed of a 250 kilogram motorcycle every two seconds (Figure 14.18). The motorcycle moves in a straight line. What is the net force acting on the motorcycle?

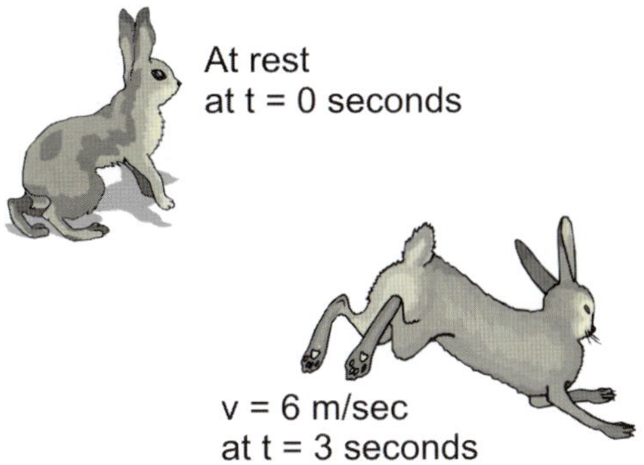

Figure 14.17: *What is the average acceleration of this rabbit? (question 3)*

Speed	Time
0 m/s	0 s
5 m/s	2 s
10 m/s	4 s
15 m/s	6 s
20 m/s	8 s

Figure 14.18: *Speed and time data for a 250 kg motorcycle and rider (question 10).*

Skateboarding with Sir Isaac Newton

Who knew that an apple falling from a tree would inspire flying acrobats? The popular story about Sir Isaac Newton sitting under the apple tree and the universal law of gravitation helps explain much of how skateboarding works. Skateboarders are defying gravity by leaping and skidding over and onto obstacles at top speeds. It also seems like they are defying the laws of physics. In reality, skateboarders are using the laws of physics to make their amazing moves. The sport demonstrates Newton's three laws.

The laws of motion

Sir Isaac Newton, a scientist from the 17th century, came up with the three laws of motion that explain why objects move or don't move.

- Newton's first law is the law of inertia states that an object that is at rest tends to stay at rest. An object that is in motion tends to stay in motion with the same speed and in the same direction unless an outside force acts on it. Once in motion, a skateboarder will continue to move forward (or stay in motion) unless some outside force affects them. If the outside force is a concrete wall, you can guess what happens. The skateboarder's motion changes fast. Since concrete so often provides the "outside force" that acts on their motion, skateboarders wear protective gear.
- Newton's second law of motion relates to the acceleration of an object. Acceleration depends upon the mass of the object and the force(s) acting on the object.
- Newton's third law is probably the best known of his laws of motion. It states that every action has an equal and opposite reaction.

How things have changed

When skateboarding was invented, a skateboard was simply a two-by-four on roller skate wheels. The goal was to start at the top of a hill and ride the skateboard to the bottom without crashing. It wasn't long before skateboarders wanted the sport to be more challenging.

Today, skateboarders use boards with wheels that reduce friction. They make incredible moves. Even the most basic skateboarding moves rely on the laws of physics. A good example is skateboarders in half-pipes. The half-pipe is a U-shaped ramp that usually has a flat section in the middle. In the half-pipe, controlled acceleration is absolutely essential.

On flat ground, gaining speed is easy. Push with one foot, and you accelerate. Half-pipes are a little trickier. They require "pumping." Pumping makes use of Newton's second law. When skateboarders pump, they drop down into a crouch as they roll through the flat bottom of the U-shaped half-pipe. As they enter the sloped part of the ramp, called the transition, they straighten their legs. By standing up in this way, skateboarders raise their center of mass. The sudden shift in their center of mass gives them more energy. More energy means acceleration. The shift in the center of mass is in the direction the skateboarder is moving. The skateboard accelerates in that direction.

This is like pumping on a swing. The swing goes higher if you lift your legs while passing through the bottom of the swing's arc. At the top of the arc, you drop your legs. The more times you repeat this movement, the more energy you gain. Gained energy translates into swinging higher. This same concept is what helps skateboarders gain speed in the half-pipe until they have the enough height and speed to perform all kinds of stunts.

Motion in action

How does Newton's third law affect skateboards? In essence, it is what allows them to move. For each of the skateboarder's actions, there is an equal and opposite reaction. A skateboarder pushes against the concrete at the top of a half-pipe. The concrete pushes back. The skateboarder is in motion. Friction between the wheels and the concrete is a force that acts on the motion. Gravity is another force that acts on the skateboarder's motion. Have you ever watched skateboarders in a half-pipe? If so, you know that they overcome the force of friction in some pretty dramatic ways.

In one impressive stunt, skateboarders soar above the half-pipe and perform a move called the "frontside 180." It's a trick that puts a little more spin on physics. Skateboarders appear to hang in the air for a moment after flying out of the half-pipe. Then they skate back down the ramp.

Physics tells us that if something is rotating, it will continue to rotate unless a twisting force stops it from rotating. The same also holds true if something is not rotating. It will need a twisting force to start it rotating. Skateboarders use their arms to create this twisting force. Also known as torque, this twisting force allows the skateboarder to turn around in mid-air. By pulling their arms in as they twist them, skateboarders create torque. That's how they seem to defy gravity in a successful frontside 180.

The evolution of skateboarding will continue to change with each new generation. Skating stunts and styles will change dramatically, but the science will remain the same.

Questions:

1. Do skateboarders actually defy the laws of physics? Why or why not?
2. How does "pumping" increase a skater's acceleration?
3. What is torque and how does a skateboarder use it?

CHAPTER ACTIVITY Making a Spool Car

Newton's three laws can be used to explain the motion of everyday objects: from a car driving down the highway to the moon orbiting around Earth. The first law says that objects at rest tend to stay at rest and objects in motion tend to stay in motion with constant speed and direction. The second law explains the relationship between force and acceleration ($a = F/m$). According to the third law, an action created by one object results in an equal but opposite reaction on another object. In this activity you will build a car and apply Newton's laws to explain how it works and why it moves as it does.

Materials:

Thread spool, assorted rubber bands, approximately 2-3 cm long, metal washer - approximately 2 cm in diameter, piece of masking tape, pencil, paper clip, carpeted floor, rug, or fabric on which to run the car

What you will do

1. Attach a rubber band to the paper clip.
2. Slide the paper clip partially through the center of the spool, leaving the rubber band exposed at one end.
3. Place the washer over the rubber band and slide a pencil through the loop.
4. Push the paper clip through to the other end of the spool.
5. Adjust the paper clip so it lies flat against the spool and holds the rubber band in place. Use a piece of tape to secure it.

6. Turn the pencil several times to twist the rubber band. Place the car on the carpeted floor and release it.
7. Experiment with the car until you can get it to move in a straight line. Adjusting the position of the pencil may be helpful.
8. Determine the number of turns of the pencil that gives the greatest distance of travel.
9. If time allows, experiment with other rubber bands until you have made a spool car that goes as far and as straight as possible. Race your car against your classmates to determine who has the best car.
10. Try to run your car on a smooth floor and observe what happens.

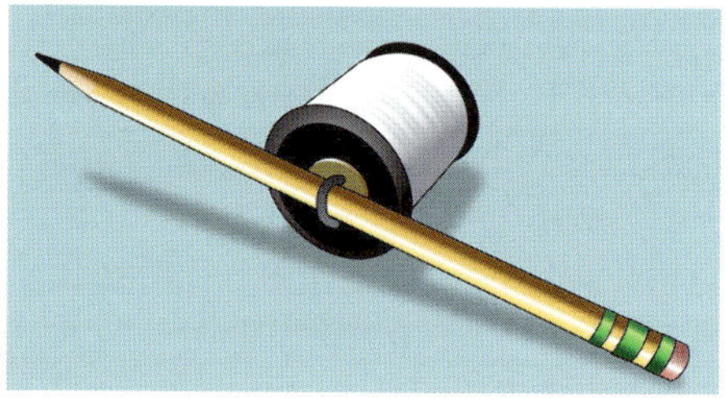

Applying your knowledge

a. What were you giving to the car when you turned the pencil?
b. Did winding the rubber band a greater number of turns always make the car go farther? Why do you think this is?
c. What was the force that caused the car to move forward?
d. Describe what happened when you ran the car on a smooth floor. Why was there a difference in the motion?
e. Explain how each of Newton's three laws relates to the motion of the car.

Chapter 14 Assessment

Vocabulary

Select the correct term to complete the sentences.

orbit	Newton's first law	acceleration
acceleration due to gravity	Newton's second law	inertia
force	Newton's third law	law of universal gravitation
	free fall	

Section 14.1

1. A push or a pull is called a(n) ____.
2. ____ states that objects continue the motion they already have unless they are acted upon by a force.
3. ____ states that every action creates a reaction that is equal in strength and opposite in direction.
4. The tendency to resist a change in motion is called ____.

Section 14.2

5. The rate at which an object's velocity changes is called ____.
6. The relationship between the force on an object, the mass of the object, and its acceleration is described by ____.
7. When an object is acted upon by only gravity, it is said to be in ____.
8. How the strength of the gravitational force depends on the distance between two objects is described by Newton's ____.
9. The motion of one object around another caused by gravitational force is called a(n) ____.
10. The acceleration of an object in free fall, 9.8 m/s^2 on Earth, is called "____".

Concepts

Section 14.1

1. Which of the following best describes an application of Newton's first law?
    a. When one object strikes another object, there is an equal and opposite force.
    b. As the force on an object increases, the acceleration of the object increases.
    c. An object rolling along a flat surface with no friction will keep moving at the same velocity
2. What quality of a bowling ball enables it to plow through a set of pins without slowing down significantly?
3. What happens to the inertia of an object if its mass is decreased?
4. If the sum of all the forces on a ball is zero, what is the ball doing?
5. What does it mean to say that the "net force" determines an object's acceleration?
6. What two quantities do you need to include when defining a force?
    a. Speed and direction
    b. Strength and direction
    c. Acceleration and time
    d. Distance and direction
7. When a bug traveling west hits the windshield of a car that is traveling east, what can be said about the collision?
    a. The bug feels a stronger force than the car.
    b. The car and the bug feel the same size force.
    c. The car accelerates more than the bug.
    d. The bug does not accelerate due to the force.

8. What is the reaction force to the force on a baseball from a bat?

9. A brick is sitting on a table. The force of gravity pushes down on the brick. What prevents the brick from accelerating downward and falling through the table?

Section 14.2

10. Which of the following is an acceleration?
    a. 12 m/s^2 down
    b. 5 m/s up
    c. 8 N West

11. Describe three different ways you could cause an acceleration on a moving car.

12. Write an **A** if the object is accelerating. Write an **X** if the object is not accelerating.
    a. ____ A car travels in circles at constant speed.
    b. ____ A ball rolls at a constant velocity.
    c. ____ A book sits motionless on a table.
    d. ____ A scooter skids to a stop.
    e. ____ A ball is thrown up in the air.
    f. ____ An apple falls to the ground.

13. If you are applying the brakes on your bicycle, are you accelerating? Explain.

14. A net force of 30 N is applied to an object with a mass of 15 kg. The object accelerates at 2 m/s^2. If half the force is applied, what happens to the acceleration?
    a. it doubles
    b. it stays the same
    c. it quadruples
    d. it halves

15. What does it mean when one quantity is "directly proportional" to another quantity?

16. Which of the following is the equivalent unit to a newton?
    a. m/s^2
    b. m/s
    c. $kg \cdot m/s^2$

17. "*g*" is called:
    a. gravity.
    b. free fall.
    c. the acceleration due to gravity.

18. Explain why a satellite doesn't fall into Earth even though it is being pulled toward Earth by gravity.

19. Name a unit for measuring:
    a. mass
    b. weight
    c. force

20. Why are the units for mass and weight different?

21. According to the law of universal gravitation, what two things do you need to know to determine the force of gravity between two objects?

22. What forces are acting on an object in free fall?

Problems

Section 14.1

1. What is the net force on the refrigerator shown to the right?

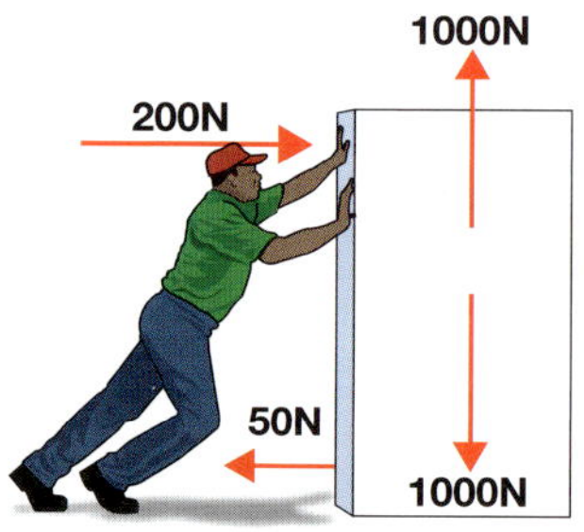

2. While an object is moving at 20 m/s, a 5 N force pushes the object to the left. At the same time, a 5 N force is pushing the object to the right. What will the object's velocity be after 10 seconds?
3. Which has more inertia, a 1-kg ball or a 10-kg ball?
4. A 3-kg hammer hits a 0.03-kg nail with a force of 1000 N. With how much force does the nail hit the hammer?
5. Identify at least three action-reaction pairs in the picture of the firefigher below:

6. A 3000-kg car collides with a 5000-kg truck. The acceleration of the car due to the force of the collision is 2 m/s^2. What is the acceleration of the truck due to the force of the collision?
7. Jane has a mass of 40 kg. She pushes on a 50-kg rock with a force of 100 N. What force does the rock exert on Jane?

Section 14.2

8. If a skateboarder starts from rest and starts accelerating at 3 m/s^2, how fast will she be traveling after 1 second? After two seconds? How fast would she be traveling if she continued accelerating for 15 seconds?
9. A car starts out with a velocity of 5 m/s and accelerates for 4 seconds, reaching a final velocity of 29 m/s.
   a. What is the car's acceleration?
   b. If the car started at 29 m/s and ended at 5 m/s after 4 seconds, what would its acceleration be? How is this different from the answer above?
10. What is the acceleration of a truck with a mass of 2,000 kg when its brakes apply a force of 10,000 N?
11. A 20 N force accelerates a baseball at 140 m/s^2 (briefly!). What is the mass of the baseball?
12. Gina is pushing a 10-kg box with 50 N of force toward the east. Dani is pushing the same box at the same time with 100 N of force toward the west. Assuming there is no friction, what is the acceleration and direction the box moves?
13. A cheetah can accelerate at 7 m/s^2, and the average cheetah has a mass of 40 kg. With what average force does the cheetah push against the ground?
14. If you have a mass of 75 kg, what is your weight in newtons on Earth?
15. When a ball is first dropped off a cliff in free fall, it has an acceleration of 9.8 m/s^2. What is its acceleration as it gets closer to the ground? Assume no air friction.

UNIT 6

Astronomy

TRY THIS AT HOME

How do satellites orbit Earth? Place a world globe on the floor, or use a large ball placed on top of an empty flowerpot. Choose a small ball, such as a tennis or racquetball. Carefully make a small slit in the ball, and shove the knotted end of a piece of string though the slit.

Imagine the large ball is Earth and the small ball is a satellite. Hold the satellite's string over the large ball, and figure out how to make the satellite move in a direction and at a speed that will allow it to orbit Earth. What did you have to do to make this work?

Chapter 15

The Solar System

Earth is a planet that is just right for living things — and among them are people who have long wondered if other planets have life. Mars and Europa (a moon of Jupiter) are good candidates for having extraterrestrial life, but are only just candidates. Space probes have explored only a tiny fraction of the surfaces of Mars and Venus looking for signs of life, and the small amount of evidence collected gives no definite answers. If you were asked to describe a creature that could live on each of the planets (or moons) in the solar system, what characteristics would it have? What would it eat? How would it move? A creature on Venus might have to live at a surface temperature of 500°C. Neptune's environment is frozen; what type of creature could live there? In this chapter, you will learn about the vast, unexplored territories that are the planets and moons of the solar system.

Key Questions

1. ***What is the solar system and how does it stay together?***

2. ***How do the other planets in the solar system compare with Earth? Could they support life?***

2. ***What else is there in the solar system besides the sun and planets?***

Footnote: On August 24, 2006, the International Astronomical Union (IAU) passed a new definition of a planet. The new definition excludes Pluto as a planet. According to the new definition, Pluto is classified as a "dwarf planet."

Image courtesy of NASA/JPL-Caltech

15.1 The Solar System

Ancient observers noticed that five bright objects seemed to wander among the stars at night. They called these objects *planets*, from the Greek word meaning "wandering star," and named them Mercury, Venus, Mars, Jupiter, and Saturn. In A.D. 140, the Greek astronomer Ptolemy "explained" that planets and the moon orbited Earth. For the next 1,400 years, people believed those ideas, until science proved Ptolemy wrong.

How the solar system was discovered

Planets shine by reflecting sunlight

Today we know that **planets** are not stars. Stars give off their own light. We see the planets *because they reflect light from the sun*. For example, Venus appears as a crescent like the moon, becoming dark at times. This is because Venus does not give off its own light. When Earth is on the same side of the sun as Venus, we see Venus's shadowed side (Figure 15.1 top). The phases of Venus were discovered by Galileo in the 1600s and were part of the evidence that eventually overturned Ptolemy's model of the solar system.

Changing ideas about the solar system

Almost 100 years before Galileo, Polish astronomer Nicolaus Copernicus had proposed that the planets orbited the sun, but few believed him. Then came Galileo, using a telescope he built himself to make two discoveries that strongly supported Copernicus's ideas. First, he argued that the phases of Venus could not be explained if Earth were at the center of the planets (Figure 15.1). Second, he saw that there were four moons orbiting Jupiter. This showed that not everything in the sky revolved around Earth.

Discovery of the outer planets

The distant planets Uranus and Neptune are far from the sun and don't reflect much light back to Earth. These planets were not discovered until telescopes became large enough to see very faint objects. The dwarf planet Pluto is so far away that even today we have only a blurry image of it. Astronomers believe that many objects like Pluto may orbit the sun beyond Neptune's orbit in the *Kuiper Belt*. These objects reflect so little sunlight that the two largest (Pluto-sized) ones have only just recently been discovered.

VOCABULARY

planet - a massive object orbiting a star, like the Sun. A true planet has cleared the neighborhood around its orbit and has enough mass so that its gravity forms it into a spherical shape.

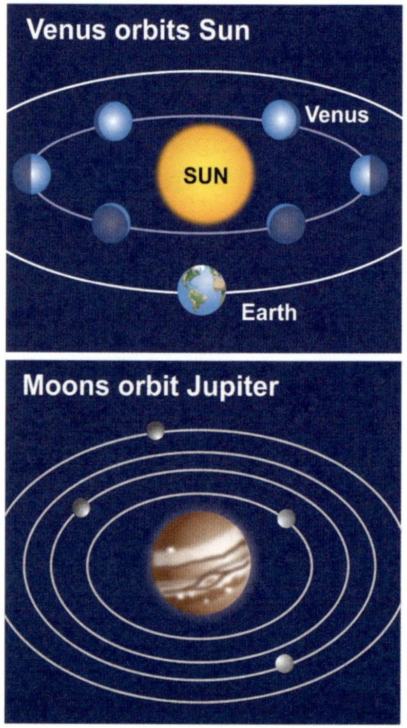

Figure 15.1: *Two of Galileo's discoveries that helped prove that Earth and the other planets orbit the sun. The top diagram shows how the phases of Venus are due to its orbit around the sun. The bottom diagram depicts moons orbiting Jupiter. This observation proved that not all objects revolve around Earth.*

Organization of the solar system

The sun, planets, and other objects

Today, we define the **solar system** as the sun and all objects that are gravitationally bound to the sun. The gravitational force of the sun keeps the solar system together just as gravity keeps the moon in orbit around Earth.

The solar system includes eight major planets and their moons (also called **planetary satellites**), and a large number of smaller objects (dwarf planets, asteroids, comets, and meteors).

VOCABULARY

solar system - the sun, planets, and their moons, and other objects that are gravitationally bound to the sun.

planetary satellite - small body of matter that orbits a planet.

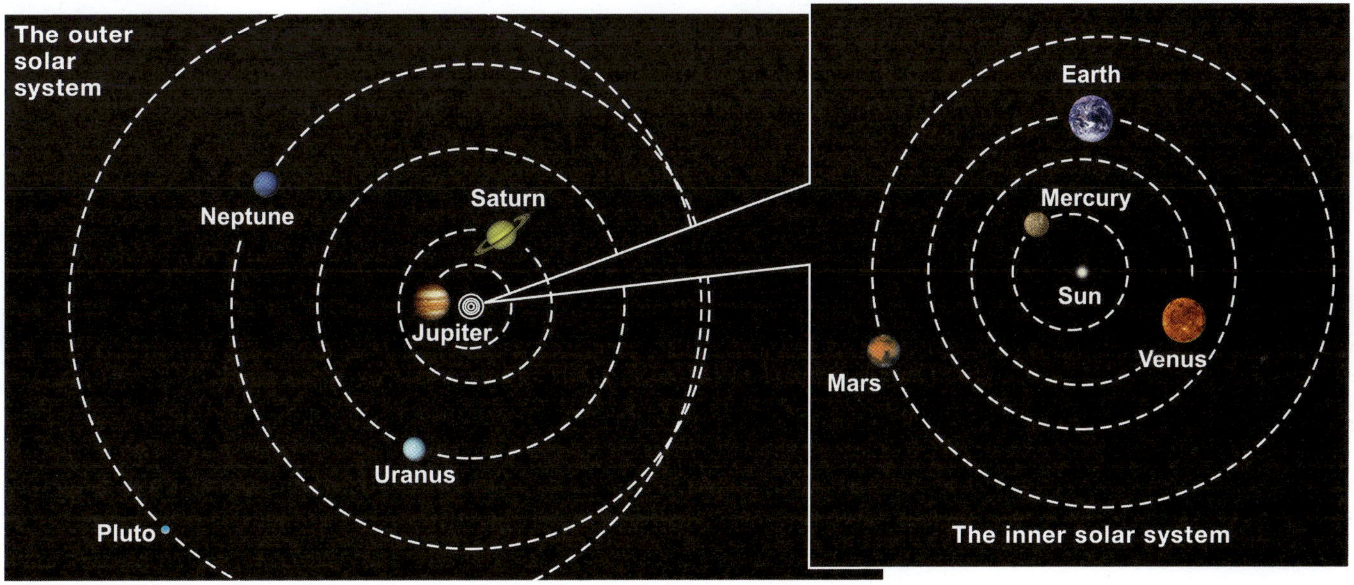

Inner and outer planets

The solar system is roughly divided into the inner planets (Mercury, Venus, Earth, Mars) and the outer planets (Jupiter, Saturn, Uranus, Neptune) The dwarf planet Pluto is the oldest known member of a smaller group of frozen worlds orbiting beyond Neptune. The diagram above shows the orbits of the planets to scale (the planets, however, are really MUCH smaller than shown). Notice that Neptune is farther from the sun than Pluto over part of its orbit.

The orbits of the planets are not true circles, but *ellipses*. While the actual paths are close to circles, the sun is not at the center, but is off to one side. For example, Mercury's orbit is shifted 21 percent to one side of the Sun.

Comparing size and distance in the solar system

Relative sizes The sun is by far the largest object in the solar system. The next largest objects are the planets Jupiter, Saturn, Uranus, and Neptune. As you can see from the scale diagram below, the planets Mercury, Venus, Earth, Mars, and the dwarf planet Pluto appear as small dots compared with the size of the sun.

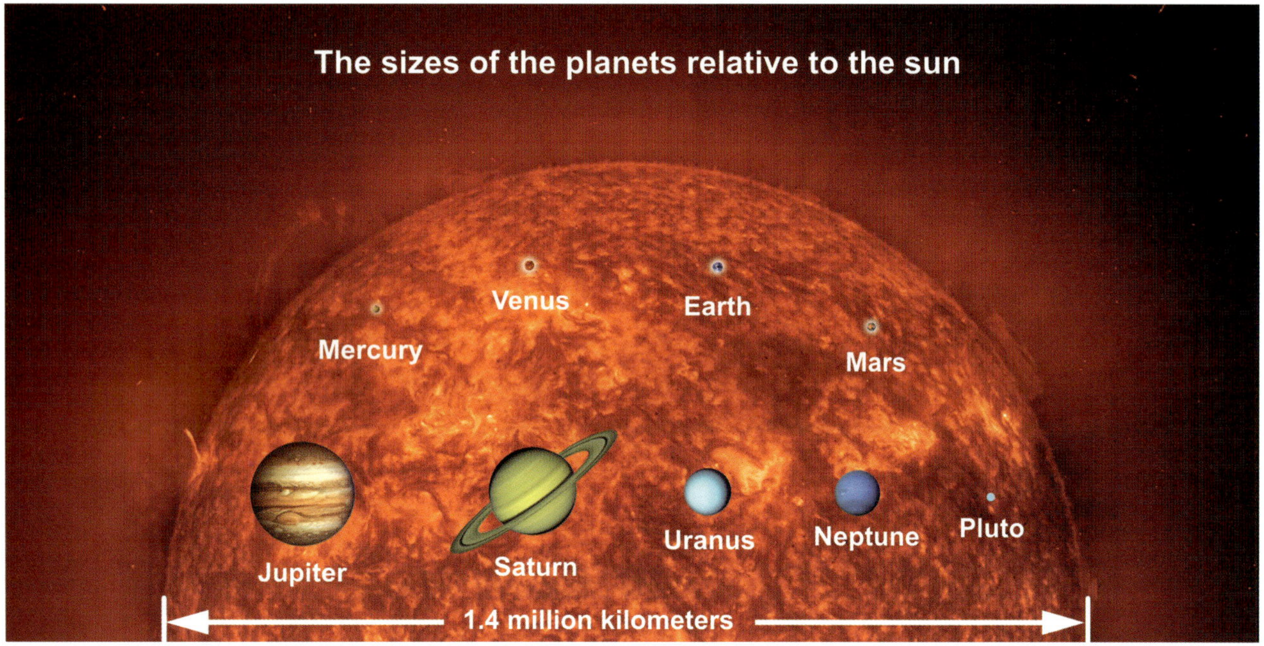

Distance Astronomers often use the distance of Earth from the sun as a measurement of distance in the solar system. One **astronomical unit** (AU) is equal to 150 million kilometers, or the distance from Earth to the sun. Mercury is 58 million kilometers from the sun. To convert this distance to astronomical units, divide it by 150 million kilometers (or 58 by 150). Mercury is 0.39 AU from the sun. Figure 15.3 lists the planets and the distance of each of them from the sun in astronomical units.

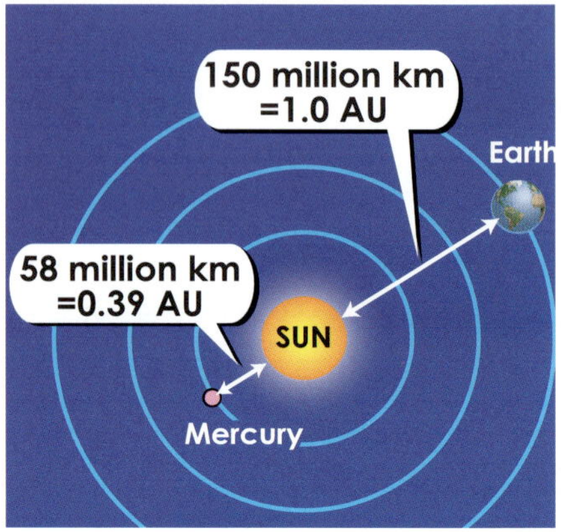

Figure 15.2: *One astronomical unit (AU) is equal to 150 million kilometers. If Earth is 1.0 AU from the sun, then Mercury, with a distance of 58 million kilometers, is 0.39 AU from the sun.*

Planet	Average distance from the sun (AU)
Mercury	0.39
Venus	0.72
Earth	1.0
Mars	1.5
Jupiter	5.2
Saturn	9.5
Uranus	19.2
Neptune	30.0
Pluto (dwarf)	39.4

Figure 15.3: *Distances of the planets from the sun in astronomical units (AU).*

Gravitational force

All objects attract The force of gravity that you are most familiar with is the one between you and Earth. We call this force your *weight*. But gravitational force is also acting between the sun, Earth, and the planets. *All objects* that have mass attract each other through gravitational forces. For example, a gravitational force exists between you and this book, but you cannot feel it because both masses are small (Figure 15.4). You don't notice the attractive force between ordinary objects because gravity is a relatively weak force.

Gravitational force is relatively weak It takes an extra-large mass to create gravitational forces that are strong enough to feel. You notice the gravity between you and Earth because Earth's mass is huge. We usually only notice gravitational forces when one of the objects has the mass of a star or planet.

Gravitational force and mass **Newton's law of universal gravitation** explains how the strength of the force depends on the mass of the objects and the distance between them. The force is *directly proportional* to each object's mass. This means the force goes up by the same factor as the mass. Doubling the mass of either of the objects doubles the force. Doubling both masses quadruples the force.

Gravitational force and distance The distance between objects also affects gravitational force. The closer objects are to each other, the stronger the force between them. The farther apart, the weaker the force. The decrease in gravitational force is proportional to the inverse square of the distance from the center of one object to the center of the other. Doubling the distance divides the force by four (2^2). If you are twice as far from an object, you feel one-fourth the gravitational force.

Gravity on Earth and the moon The strength of gravity on the surface of Earth is 9.8 N/kg. Earth and a one-kilogram object attract each other with 9.8 newtons of force. In comparison, the strength of gravity on the moon is only 1.6 N/kg. Your weight on the moon would be one-sixth what it is now. The moon's mass is much less than Earth's, so it creates less gravitational force.

VOCABULARY

Newton's law of universal gravitation - the force of gravity between objects depends on their masses and the distance between them.

Comparing gravitational forces between ordinary objects and between objects and planets

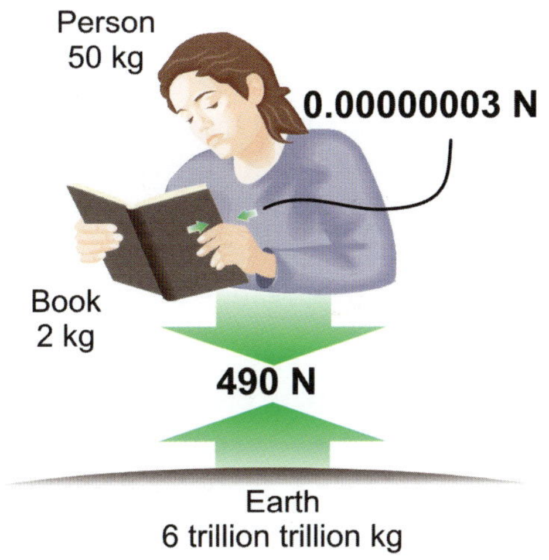

Figure 15.4: *The gravitational force between you and Earth is stronger than the force between you and your book because of Earth's large mass.*

Orbital motion

Why the moon does not fall to Earth

Earth and the other planets orbit the sun. Why doesn't the force of gravity pull the Earth into the sun (or the moon into Earth)? To answer the question, imagine kicking a ball off the ground at an angle (Figure 15.5). If you kick it at a slow speed, it curves and falls back to the ground. The faster you kick the ball, the farther it goes before hitting the ground. If you kick it fast enough, the curve of the ball's path matches the curvature of Earth. The ball goes into orbit instead of falling back to Earth.

Inertia and gravitational force

Orbital motion is caused by the interaction between inertia and gravitational force. According to Newton's first law, inertia causes objects to tend to keep moving in a straight line. Force is needed to change an object's speed or direction. Earth has a tendency to move in a straight line, but the gravitational force from the sun causes its direction of motion to curve toward the sun, into an orbit.

The size of an orbit depends on speed and mass

The radius of an orbit is a balance between gravity and inertia. Gravity gets stronger as a planet's orbit gets closer to the sun, forcing a tighter curve into the planet's motion. Increasing a planet's speed or mass has the opposite effect. Higher speed or mass increase the tendency of a planet to move in a straight line, resulting in larger, less curved orbits. Each planet orbits at the precise radius where its mass and speed are in balance with the gravity of the sun.

The shape of an orbit

The sun's gravity always pulls the planets toward it. This force would create a perfectly circular orbit IF a planet's velocity vector were *exactly* at right angles to its radius from the sun. As the solar system formed from swirling gases, interactions between planets caused slight variations in velocity vectors. As a result, the orbits of the planets are ellipses instead of perfect circles. Dwarf planet Pluto has the most elliptical orbit. However, the deviation from circular is quite small. Even Pluto's orbit is "squashed" only about 4 percent out of round. Much more significant, the sun is at a point called the *focus* that is offset from the center. This causes the distance from the sun to change as a planet orbits.

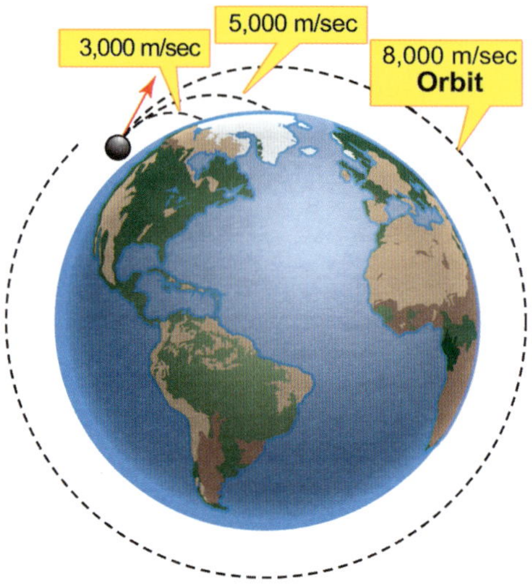

Figure 15.5: *An object launched at 8,000 meters per second will orbit Earth.*

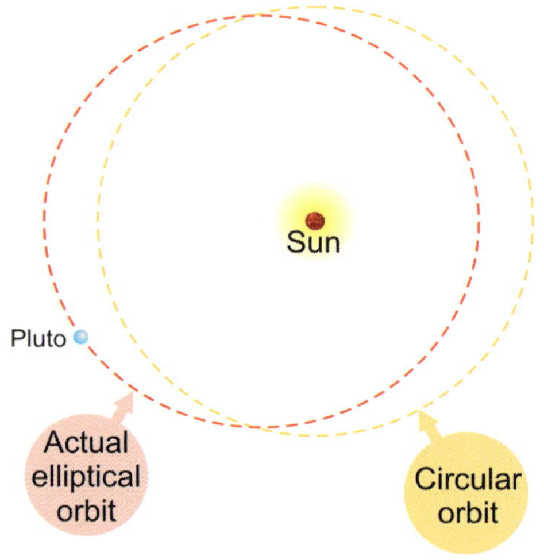

Figure 15.6: *Orbits are mathematically ellipses but are close to (shifted) circles.*

An overview of the planets

Classifying the planets

The planets are commonly classified in two groups. The **terrestrial planets** include Mercury, Venus, Earth, and Mars. The terrestrial (rocky) planets are mostly made of rock and metal. They have relatively high densities, slow rotations, solid surfaces, and few moons. The **gas planets** include Jupiter, Saturn, Uranus, and Neptune. They are made mostly of hydrogen and helium. These planets have relatively low densities, rapid rotations, thick atmospheres, and many moons. Pluto is neither terrestrial nor gas, but in a class of its own. Table 15.1 compares the planets.

VOCABULARY

terrestrial planets - Mercury, Venus, Earth, and Mars.

gas planets - Jupiter, Saturn, Uranus, and Neptune.

Table 15.1: Comparing properties of the planets

Property	Mercury	Venus	Earth	Mars	Jupiter	Saturn	Uranus	Neptune	Pluto (dwarf)
Diameter *(km)*	4,878	12,102	12,756	6,794	142,796	120,660	51,200	49,500	2,200
Mass *(kg)*	3.3×10^{23}	4.9×10^{24}	6.0×10^{24}	6.4×10^{23}	1.9×10^{27}	5.7×10^{26}	8.7×10^{25}	1.0×10^{26}	1.3×10^{22}
Density *(g/cm^3)*	5.44	5.25	5.52	3.91	1.31	0.69	1.21	1.67	1.75
Average distance from sun *(million km)*	58	108	150	228	778	1430	2870	4500	5910
Major moons *(#)*	0	0	1	2	39	30	21	8	1
Strength of gravity *(N/kg)*	3.7	8.9	9.8	3.7	23.1	9.0	8.7	11.0	0.6
Surface temperature *(°C)*	-170 to +400	+450 to +480	-88 to +48	-89 to -31	-108	-139	-197	-201	-223
Rotation period *(Earth days)*	59	243	1	1.03	0.41	0.43	0.72	0.67	6.4
Revolution period *(Earth years)*	0.24	0.62	1	1.9	12	29	84	165	249
Orbital speed *(km/sec)*	47.89	35.04	29.80	24.14	13.06	9.64	6.80	5.43	4.74

Asteroids and comets

Asteroids

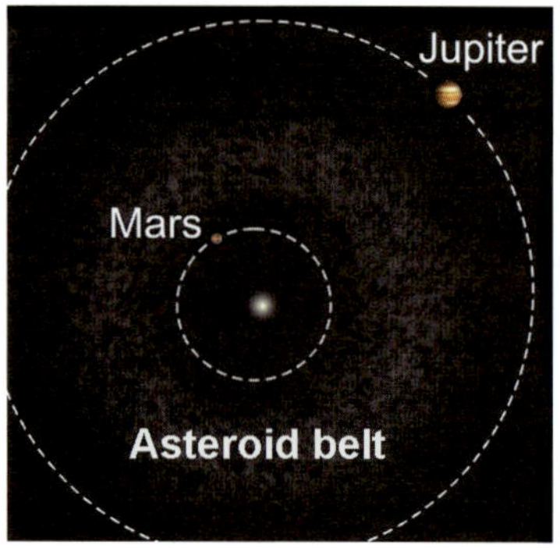

Between Mars and Jupiter, at a distance of 320 million to 495 million kilometers, there is a huge gap that cuts the solar system in two. This gap is called the *asteroid belt* because it is filled with thousands of small, rocky bodies called *asteroids*. An **asteroid** is an object that orbits the sun but is too small to be considered a planet. So far, more than 10,000 asteroids have been discovered and more are found each year.

The size of asteroids

Most asteroids are small — less than a kilometer in diameter — but many have been found that are over 250 kilometers in diameter. The largest asteroid, named Ceres, is 933 kilometers (580 miles) across. While the majority of asteroids are found in the asteroid belt, many have highly elliptical orbits that allow them to come close to Mercury, Venus, and even Earth. About 65 million years ago, a large asteroid hit Earth near Mexico, leaving a huge crater. Some scientists believe this event led to the extinction of the dinosaurs.

Comets

We believe **comets** are made mostly of ice and dust. The ones we can detect are about the size of an Earth mountain. Comets revolve around the sun in highly elliptical orbits. In 1997, the comet Hale-Bopp could be clearly seen in the night sky without a telescope. However, we still know little about the composition and structure of comets. Several recent spacecraft have made close approaches and each new piece of evidence they gather has lead to new insights about what comets are made of and how they formed.

Evolution of a comet

As a comet approaches the sun, some of its ice turns into gas and dust and forms an outer layer called a *coma*. The inner core of the comet is the *nucleus*. As a comet gets closer to the sun, it forms a *tail*. A comet's tail can stretch for millions of kilometers into space and faces away from the sun as the comet continues its orbit (Figure 15.7). Each time a comet passes the sun, it loses some mass.

VOCABULARY

asteroid - an object that orbits the Sun but is too small to be considered a planet.

comet - an object in space made mostly of ice and dust.

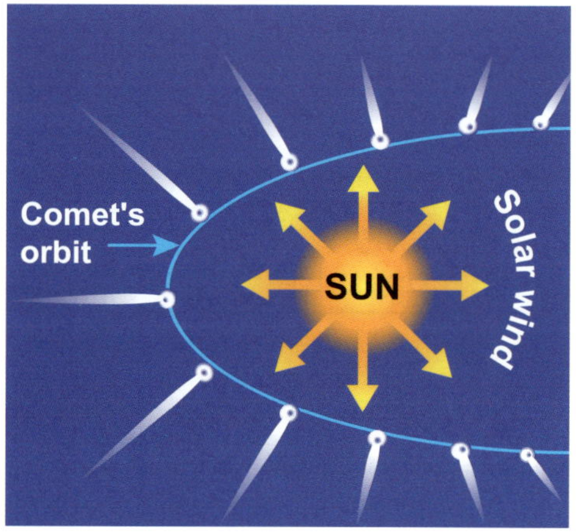

Figure 15.7: *A comet's tail faces away from the sun and can stretch for millions of kilometers in space.*

Meteors and meteorites

Meteors Occasionally, chunks of rock or dust break off from a comet or asteroid and form a **meteor**. Imagine a tennis ball traveling at about 30,000 miles per hour. That's about the size and speed of most meteors. These chunks of dust or rock travel through space and some of them end up hitting Earth's atmosphere. When this happens, meteors rub against air particles and create friction, heating them to more than 2,000°C. The intense heat vaporizes most meteors, creating a streak of light known as a "shooting star." Occasionally, larger meteors cause a brighter flash called a *fireball*. These sometimes cause an explosion that can be heard up to 30 miles away. If you live or find yourself away from any city lights, look at the sky on a clear night and chances are that, if you look long enough, you will see a meteor. On average, a meteor can be seen in the night sky about every 10 minutes.

Meteor showers When a comet nears the sun, a trail of dust and other debris burns off and remains in orbit around the sun. As Earth orbits the sun, it passes through this debris, creating a *meteor shower* as the small bits of dust burn up in the atmosphere. During a meteor shower, you can see tens and even hundreds of meteors per hour. Because Earth passes the same dust clouds from comets each year, meteor showers can be predicted with accuracy.

Meteorites

If a meteor is large enough to survive the passage through Earth's atmosphere and strike the ground, it becomes a **meteorite**. Meteorites are thought to be fragments from collisions involving asteroids. Most meteorites weigh only a few pounds or less and cause little damage when they hit. Most fall into the oceans that cover almost three-quarters of our planet's surface. Meteor Crater in Winslow, Ariz., is believed to have been caused by a giant, 50-meter diameter meteorite about 50,000 years ago. The Holsinger meteorite (Figure 15.8) is the largest known piece of this 300,000-ton meteorite, most of which vaporized on impact.

VOCABULARY

meteor - a chunk of burning rock traveling through Earth's atmosphere.

meteorite - a meteor that passes through Earth's atmosphere and strikes the ground.

Figure 15.8: *The Holsinger meteorite is a large piece of a much larger meteorite that blasted out Meteor Crater in Arizona about 50,000 years ago. This meteorite, while no taller than your thigh, weighs 1,400 lbs.*

15.1 Section Review

1. Do we see planets because, like the sun, they are sources of light?
2. Name the planets in order, starting nearest to the sun.
3. What is an astronomical unit?
4. Gravitational force gets weaker as _____ increases and gets stronger as the _____ of the objects increases.
5. Gravity exists between all objects with mass. So why is it that you don't you notice the force of gravity between you and all of the objects around you?
6. Is a satellite orbiting Earth free from Earth's gravity? Why or why not?
7. Which planet-like object is neither a gas planet nor a terrestrial planet?
8. Use Table 15.1 to answer the following questions:
    a. Which planet is the largest? The smallest?
    b. On which planet is gravity the strongest? The weakest?
    c. A day is the time it takes a planet to rotate once on its axis. Which planet has the longest day? The shortest day?
    d. A year is the time it takes a planet to revolve once around the sun. Which planet has the longest year? The shortest year?
    e. Which planet is the most dense? The least dense?
    f. Which planet is approximately 10 AU from the sun?
9. Why are we able to see a certain comet one year but not again until many years later?
10. What is the difference between a meteor and a meteorite?
11. What is the asteroid belt and where is it located?
12. Why are the orbits of the planets slightly elliptical instead of being perfect circles?
13. Compared with Earth's diameter, Saturn's diameter is roughly: a) the same b) 5 times larger c) 10 times larger d) 50 times larger

CHALLENGE

Use the data from Table 15.1 to make a graph of surface temperature vs. distance from the sun for the nine planets. Graph the distance on the *x*-axis and the temperature on the *y*-axis. Use these values for the surface temperature of the four inner planets:

Mercury 167 °C; Venus 465 °C, Earth 15 °C, Mars -65 °C.

What does your graph show you about the relationship between temperature and distance from the sun?

Do the planets perfectly follow this relationship?

What other factors might affect the surface temperature of the planets?

15.2 The Planets

The eight major planets of our solar system together contain 250 times the surface area of Earth. This vast territory includes environments baked by heat and radiation (Mercury) and frozen colder than ice (Neptune). Venus, the most Earth-like planet in size, has a surface atmosphere of hot dense sulfuric acid that would be instantly fatal to any form of life on Earth. Our own crystal blue world is unique in having the right balance of temperature and environment to sustain life — or is it? Might there be unusual kinds of life unknown to us on the other planets? Scientists have recently discovered living organisms that feed off hot sulfur emissions from volcanoes on the ocean floor. These organisms might be able to survive on Venus. With a combined surface area 1,700 times the size of North America, the planets are an unexplored frontier full of discoveries waiting to be made.

Figure 15.9: *Mercury was named for the messenger of the Roman gods because of its quick motion in the sky (image from radar maps, NASA)*

Mercury

Mercury

Mercury, the closest planet to the sun, is the smallest in both size and mass. Mercury appears to move quickly across the night sky because its period of revolution is the shortest of all of the planets. Only 40 percent larger than Earth's moon, Mercury is a rocky, cratered world, more like the moon than like Earth. Like the moon, Mercury has almost no atmosphere (except for traces of sodium). Mercury has no moons.

Surface environment

Of all the planets, Mercury has the most extreme variations in temperature. The side of Mercury that faces the sun is very hot, about 400°C, while the other side is very cold, about -170°C. This is partly because Mercury's rotation is locked in a 3:2 ratio with its orbit. The planet completes three "Mercury days" every two "Mercury years." This also translates into one day on Mercury being about 59 Earth days long, and a year on Mercury being not much longer, about 88 Earth days.

Mercury facts

Type: Rocky
Moons: none
Distance from sun: 0.39 AU
Diameter: 0.38 of Earth
Surface gravity: 38% of Earth
Surface temp.: -170 to 400°C
Atmosphere: none
Length of day: 59 Earth days
Length of year: 88 Earth days
Shortest flight to Earth: 2.3 AU
Travel time from Earth: 3 months

Venus

Venus is similar to Earth as a planet

Venus appears as the brightest planet in the evening sky and is the third brightest observable object (after the sun and moon). Venus was named after the Roman goddess of love because of its beautiful, shiny appearance. Of the planets, Venus is closest to Earth in terms of size, surface gravity, and rocky composition. Venus is slightly smaller than Earth and, like Earth, has volcanic activity indicating an active geology. But there the similarity ends. The dense, acidic, furnace-like surface conditions on Venus are not at all Earth-like.

Figure 15.10: *This radar map was colored to match Venus's surface colors, normally hidden by clouds. (NASA)*

Venus's surface is unpleasant

Venus has a thick atmosphere which is mostly (96 percent) carbon dioxide at a surface pressure 90 times that of Earth. Carbon dioxide traps heat; the greenhouse effect makes Venus the hottest planet in the solar system. The surface temperature is more than 500°C, hot enough to melt lead and zinc. Venusian clouds are not water, but corrosive sulfuric acid (H_2SO_4) formed from the sulfur emitted by many active volcanoes. The first successful landing on Venus was the Soviet probe Venera 7 in 1970. This tough lander broadcast the first images of the rocky surface in the brief 23 minutes it lasted before the corrosive atmosphere destroyed it. More recently, Venus was studied by the US Magellan (1989-94) and Messenger (2004) missions, and by the European Venus Express orbiter (2005).

Surface of Venus - Venera 13 - Soviet Academy of Sciences/U.S.S.R.

Venus day and year

Venus is one of three planets that rotate "backward," that is, east to west. Its rotation is the slowest of all of the planets; Venus makes a little less than one rotation for each revolution around the sun. This means that a day on Venus is 243 Earth days, while a year is shorter that that, just 225 Earth days. Like Mercury, Venus has no moons.

Venus

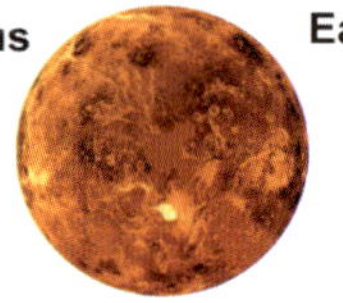

Earth

Venus facts

Type: Rocky

Moons: none

Distance from sun: 0.72 AU

Diameter: 0.95 of Earth

Surface gravity: 91% of Earth

Avg. surface temp.: 460°C

Atmosphere: dense, 96% CO_2

Length of day: 243 Earth days

Length of year: 225 Earth days

Shortest flight to Earth: 2.7 AU

Travel time from Earth: 3 1/2 mo.

Earth and moon

Earth Earth is a small, rocky planet with an atmosphere that is made of mostly nitrogen (78 percent N_2) and oxygen (21 percent O_2). Earth is one of only two bodies in the solar system known to have liquid water (the other is Europa, a moon of Jupiter). Earth has an active geology, including volcanoes and crustal movement. Earth's atmosphere, along with its vast oceans and moderate temperature range, supports an incredible variety of life. *As far as we know,* Earth is the only planet in the solar system to support life. Although space probes have begun searching, the ultimate answer to the question of life on other planets may have to wait until humans can look in person.

The moon Earth's single rocky moon is about one-quarter the diameter of Earth. At a distance of 385,000 kilometers, the moon is about 30 Earth-diameters away from the planet, completing one orbit every 29 days.

The seasons Earth's orbit is within 2 percent of a perfect circle. The seasons are caused by the 23-degree tilt of Earth's axis of rotation relative to its orbit. When Earth is on one side of the sun, the northern hemisphere receives a greater intensity of sunlight because the sun passes nearly straight overhead once per day, making it summer. Six months later, on the opposite side of Earth's orbit, the northern hemisphere tilts away from the sun. This spreads the sunlight over a larger surface area. The lower intensity of sunlight each day makes for winter.

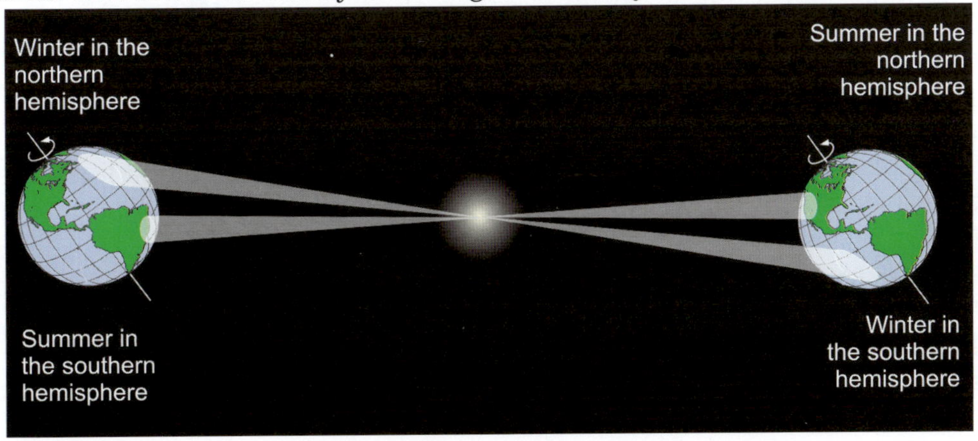

Figure 15.11: *Earth is the only planet not named after a Roman god. Its name comes from Old English "oerthe," meaning land or country (NASA photo).*

Earth at a glance

Type: Rocky

Moons: one

Distance from sun: 1 AU

Diameter: 12,800 km

Surface gravity: 9.8 N/kg

Avg. surface temp.:10°C

Atmosphere: dense, N_2, O_2

Length of day: 24 hours

Length of year: 365.25 days

Mars

Mars The fourth planet out from the sun, Mars appears as a reddish point of light in the night sky. Mars is a relatively small rocky planet with a mass only 11 percent the mass of Earth. Mars has two tiny, irregular-shaped moons named Deimos and Phobos. Both are much smaller than Earth's moon and are more like asteroids.

Figure 15.12: *Mars was named after the Roman god of war. (ESA photo)*

The surface of Mars The surface of Mars has deserts, huge valleys, craters, and volcanic mountains even larger than those on Earth. However, Mars's "air" is mostly carbon dioxide and less than 1 percent the density of Earth's atmosphere. Like Earth, Mars has polar ice caps, but they are made of a combination of water and frozen carbon dioxide. Because of the thin atmosphere and the planet's distance from the sun, temperatures are below 0°C most of the time. Because it is tilted like Earth, Mars also has seasons. A day on Mars (24.6 hours) is similar in length to an Earth day. But Mars's larger orbit makes a Martian year (687 days) almost twice as long as an Earth year.

Mars was different in the past Mars is cold and dry today, but there is strong evidence that Mars was much wetter and had a thicker atmosphere in the past. Aerial photos of the Martian surface show erosion and patterns of riverbeds similar to those formed by flowing water on Earth. Even today, there is evidence of water beneath the Martian surface. Several robot space probes have landed on Mars searching for life but the results have been inconclusive. As Earth's nearest match in climate, Mars will probably be the first planet in the solar system to be explored by humans.

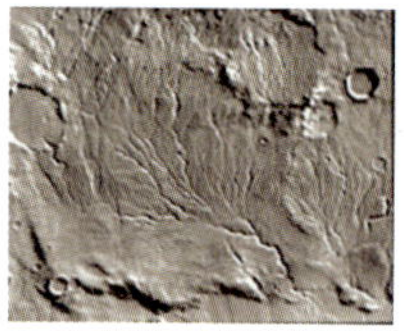

Mars Facts

Type: Rocky
Moons: 2
Distance from sun: 1.5 AU
Diameter: 0.53 of Earth
Surface gravity: 38% of Earth
Avg. surface temp.: -50°C
Atmosphere: thin, CO_2
Length of day: 24.6 hours
Length of year: 687 Earth days
Shortest flight to Earth: 2.7 AU
Travel time from. Earth: 3 1/2 mo.

Jupiter

Jupiter

The fifth planet out from the sun, Jupiter is by far the largest. Jupiter's mass is greater than the combined mass of all of the other planets. Jupiter also spins the fastest, rotating about once every 10 hours. In composition, Jupiter is much different from the rocky, inner planets like Earth. Jupiter's average density is only 1.3 g/cm^3 compared with Earth's density of 5.1 g/cm^3. Jupiter is a gas planet composed mostly of hydrogen and helium, similar to the sun. In fact, if Jupiter were larger it would be a star, like the sun.

Jupiter's environment

Jupiter does not have a solid surface. In fact, Jupiter is more liquid than gaseous or solid — more than half of its volume is an ocean of liquid hydrogen. Its atmosphere is about 88 percent hydrogen, 11 percent helium, and 1 percent methane, ammonia, and other gases. The atmospheric pressure below Jupiter's thick clouds is more than a million times that of Earth. A huge storm called the Great Red Spot has been observed in Jupiter's atmosphere for more than 300 years.

Jupiter's fascinating moons

With 63 known moons, Jupiter is like a mini solar system. In 1995, when the US Galileo probe took these photographs, the four largest moons became some of the most fascinating objects in the solar system. Io, Europa, Ganymede, and Callisto are like small planets. Because it is heated by gravitational forces from Jupiter itself, Io looks like a boiling pizza and is covered with smoking sulfur volcanoes. Europa has a surface layer of ice as much as 20 kilometers thick. Beneath the ice is a vast ocean of liquid water that may even be warm enough to support life. Ganymede, the largest moon in the solar system, has a magnetic field like Earth. No other moons have this feature. Even pock-marked Callisto has many mysteries.

Figure 15.13: *Jupiter was king of the Roman gods. The planet's brightness inspired its name. (NASA photo)*

Jupiter facts

Type: Gas giant
Moons: 63 plus faint rings
Distance from sun: 5.2 AU
Diameter: 11.2 × Earth
Surface gravity: 253% of Earth
Avg. atmos. temp.: -108°C
Atmosphere: 90% H, 10% He
Length of day: 10 Earth hours
Length of year: 11.9 Earth years
Shortest flight to Earth: 12 AU
Travel time from Earth: 15 months.

Saturn

Saturn Saturn, at almost 10 times the size of Earth, is the second largest planet. Similar to Jupiter's, Saturn's atmosphere is mostly hydrogen and helium. Saturn also spins quickly, with a day on Saturn lasting about 11 Earth hours. As with Jupiter, Saturn's rapid rotation is one contributor to huge planetary storms in its atmosphere. Because of its distance from the sun, a year on Saturn is about 29 Earth years.

Saturn's rings

The most striking feature of Saturn is its system of rings, which are visible from Earth with a telescope. Saturn's rings are made up of billions of particles of rock and ice ranging from microscopic to the size of a house. Although they are hundreds of thousands of kilometers wide, the rings are less than 100 meters thick *(NASA photo)*.

Saturn has many moons Saturn, again like Jupiter, has many natural satellites. There are eight bigger moons and 39 smaller ones known as of this writing. Some of the smaller moons act as "shepherds" keeping the particles in Saturn's rings confined through a complex waltz of gravity.

Titan is the largest moon

Photo courtesy of NASA

Titan is Saturn's largest moon, and like Jupiter's large moons, is like a small planet. It has an atmosphere of nitrogen and a surface pressure comparable to Earth's. Astronomers have found spectroscopic evidence of organic molecules in Titan's atmosphere, raising the possibility of life there. Titan is very cold, with an average temperature of -183°C. We know little about its surface because of its dense cloud cover.

Figure 15.14: *Because of its slow orbit around the sun, Saturn was named after the Roman god of agriculture and time.*

Saturn facts

Type: Gas giant
Moons: 47 plus rings
Distance from sun: 9.5 AU
Diameter: 9.4 × Earth
Surface gravity: 1.06% of Earth
Avg. atmos. temp.: -139°C
Atmosphere: 96% H, 3% He
Length of day: 10.7 Earth hours
Length of year: 29.5 Earth years
Shortest flight to Earth: 22 AU
Travel time from Earth: 2.2 years

Uranus and Neptune

Uranus and Neptune are similar

Both Uranus and Neptune are huge cold gas planets very much like Jupiter and Saturn. Both are about four times the diameter of Earth, considerably smaller than Jupiter or Saturn. Like the other gas giants, these planets' atmospheres are mostly hydrogen and helium, similar to the sun. Scientists believe all nine planets condensed out of the same cloud of interstellar material as the sun. The smaller inner planets could not hold onto their lighter gases (hydrogen and helium) and their exposed cores became the rocky planets. Under their deep atmospheres, the gas giants also have rocky cores.

Uranus

The seventh planet from the sun, Uranus can barely be seen without a good telescope and was not discovered until 1781. It rotates "backward" and has an axis that is tilted 98 degrees to the plane of its orbit. A day on Uranus is only 18 Earth hours, but a year takes 84 Earth years. Uranus has at least 21 moons, all of them relatively small. Titania, the largest, has only 4 percent the mass of Earth's moon.

Photo courtesy of NASA

Neptune

Neptune, the eighth planet from the sun, is the outermost of the gas planets. It was discovered in 1846 and its discovery almost doubled the diameter of the known solar system because of its great distance from the sun. Neptune's orbit is nearly a perfect circle; only Venus has a more circular orbit. Neptune has a series of faint rings invisible from Earth but that have been seen in photographs taken by space probes such as Voyager. Neptune has eight known moons, six of which were found in photographs taken by Voyager 2 in 1989. Of the eight moons, only Triton is bigger than a few hundred kilometers.

Photo courtesy of NASA

Uranus facts

Type: Gas giant
Moons: 27 plus rings
Distance from sun: 19.1 AU
Diameter: 4 × Earth
Surface gravity: 90% of Earth
Avg. atmos. temp.: -197°C
Atmosphere: 82% H, 15% He
Length of day: 17 Earth hours
Length of year: 84 Earth years
Shortest flight to Earth: 43 AU
Travel time from Earth: 4.4 years

Neptune facts

Type: Gas giant
Moons: 13 plus rings
Distance from sun: 30 AU
Diameter: 3.9 × Earth
Surface gravity: 114% of Earth
Avg. atmos. temp.: -201°C
Atmosphere: 96% H, 3% He
Length of day: 16 Earth hours
Length of year: 165 Earth years
Shortest flight to Earth: 67 AU
Travel time from Earth: 8.1 years

Triton, Pluto, and the far outer system

Triton and Pluto are similar

Triton is Neptune's largest moon (Figure 15.15). Pluto is a dwarf planet, and most of the time the farthest from the sun. Triton and Pluto are similar objects in both composition and size. In fact, Pluto is slightly smaller than Triton and only a fraction larger than Earth's moon. Some astronomers believe Pluto may actually be an "escaped" moon of Neptune.

Figure 15.15: *Very little is known about Pluto since it is so far from the sun. No space probe has yet visited this cold icy dwarf planet. This image from the Hubble Space Telescope shows Pluto and its single "moon", Charon (NASA photo).*

Triton

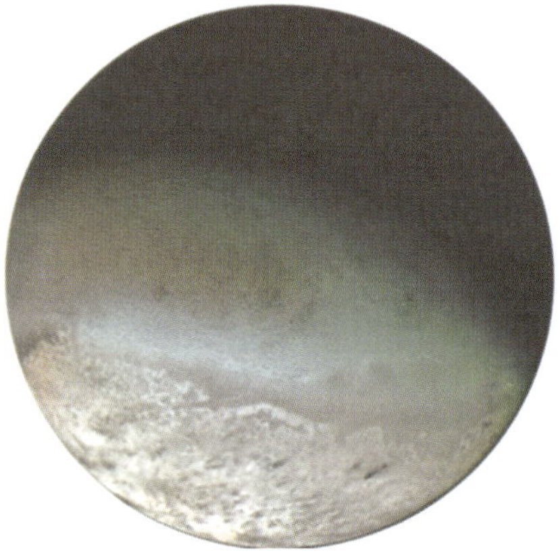

Triton was not discovered until 1846 and not seriously investigated until the US probe Voyager 2 in 1989. Triton is about three-quarters the diameter of Earth's moon, but its mass is much lower. Triton's low density of 2.2 g/cm^3 points to a mix of rock and ice. Alone of the moons in the solar system, Triton revolves around Neptune opposite from Neptune's direction of rotation. *(NASA photo)*

Pluto

Discovered in 1930, Pluto was named for the Roman god of the underworld. The first dwarf planet discovered, Pluto rotates slowly — one turn every six days — and backward. Its orbit is strongly elliptical and Pluto crosses the path of Neptune for about 20 years out of the 249 years it takes to revolve around the sun. Because their orbits are not in the same plane, Neptune and Pluto will never collide. Because it is so far away, little is known about Pluto.

Are there 8, 9, or 11+ planets?

Outside the orbit of Pluto is a region called the Kuiper Belt. The Kuiper Belt stretches to 1,000 AU and is believed to contain many asteroid-size and a few Pluto-size objects. As of this writing, two Pluto-size bodies have been found, nicknamed Sedna and Xena. To avoid confusion, astronomers no longer count Pluto as a planet. Instead, Pluto is grouped along with Sedna, Xena, and similar distant bodies in the Kuiper Belt Objects (or KBOs).

Pluto Facts

Type: Rock and ice
Moons: 1
Distance from sun: 39.2 AU
Diameter: 0.31 × Earth
Surface gravity: 2% of Earth
Avg. atmos. temp.: -223°C
Atmosphere: almost none
Length of day: 153 Earth hours
Length of year: 248 Earth years
Shortest flight to Earth: 88 AU
Travel time from Earth: 9 years

15.2 Section Review

1. Which planet has the most extreme temperature variations?
2. Which planet looks brightest in the sky?
3. Mercury is most similar to:
    a. Earth's moon.
    b. Pluto.
    c. Venus.
    d. Mars.
4. Which planet is closest to Earth in size, gravitational strength, and composition?
5. What happened to the space probe that first landed on Venus?
6. What is the cause of Earth's seasons?
7. Why do scientists believe the surface of Mars may have contained liquid water in the past?
8. What important feature do Europa and Earth have in common?
9. What makes up Saturn's rings?
10. Is Saturn the only planet with rings?
11. The gas giant planets have atmospheres made of hydrogen and helium. What evidence does this give scientists about the formation of the planets?
12. Why is Neptune sometimes farther from the sun than Pluto?
13. Which three planets rotate backward?

MY JOURNAL

Suppose you were given the opportunity to travel to another planet or a moon of another planet. Would you go? Why or why not? Would you go to Pluto, knowing the trip would last 20 years? What if you could bring along anything and anyone you wanted? Write an essay exploring your answers to these questions.

FOOTNOTE:

On August 24, 2006, the International Astronomical Union (IAU) passed a new definition of a planet. The new definition excludes Pluto as a planet. According to the new definition, Pluto is classified as a "dwarf planet."

ASTRONOMY CONNECTION

Jupiter's Volcanic Moon

We know that volcanoes have shaped the Earth's surface. A volcanic eruption can quickly change the landscape of an area right before our eyes. In fact there are several active and potentially active volcanoes in the state of California, Oregon and Washington. These volcanoes have shaped and changed the environment of these areas.

Did you ever wonder if volcanoes exist on other planets? There are scientists who study volcanoes throughout our solar system. They are called planetary volcanologists. One such scientist is Dr. Rosaly Lopes, who studies volcanoes on Earth as well as other planets at NASA's Jet Propulsion Laboratory in Pasadena, California.

Dr. Rosaly Lopes standing on the Pu'u O'o eruption of the Kilauea volcano on Hawaii's Big Island

The study of Volcanism

Volcanology is the study of volcanoes that combines geology, physics, chemistry, and mathematics. Here on Earth scientists study and monitor volcanoes in hopes of being able to successfully predict volcanic eruptions. There are several key techniques used to monitor volcanoes. On the Earth's surface monitoring can be done locally, observed remotely by aircraft or by satellites orbiting Earth. Scientista measure ground movement, emission of gases and changes in temperature. Scientists from all over the world report their findings of volcanic patterns and behavior monthly. The Smithsonian Institution in the U.S. publishes them in the *Bulletin of the Global Volcanism Network.*

NASA's Galileo mission to Jupiter

Today scientists can explore volcanic activities far beyond the boundaries of Earth. In 1989 the Galileo mission was launched aboard the Space Shuttle Atlantis. The spacecraft was equipped with cameras and scientific instruments needed to collect data. The mission was named after Galileo Galilei an Italian scientist. In 1610, Galileo Galilei discovered the four major moons around the planet Jupiter. The four moons named Io, Europa, Ganymede and Callsito are also called the Galilean Moons in honor of Galileo.

The Galileo spacecraft was equipment with instruments such as the Near-Infrared Mapping Spectrometer (NIMS), Solid State Imaging System (SSI) and Photopolarimeter Radiometer. These remote-sensing instruments provided amazing facts and evidence about the fiery volcanoes on Jupiter's moon, Io.

Volcanic intensity on Io

In December of 1995 the Galileo spacecraft entered into orbit around Jupiter. A series of "flybys" and "close encounters" were the primary objective of the mission. The purpose of these "flybys" and "close encounters" were to collect pictures and scientific information. The spacecraft beamed the pictures and data back to Earth for scientist to study. Dr. Rosaly Lopes was a member of the Galileo Flight project. As an expert on planetary volcanism she worked from 1996 to 2001 on the NIMS team. She helped to plan and analyze data on the fascinating moon Io. She and her team were responsible for the discovery of 71 volcanoes on Jupiter's volcanic moon, Io.

Image - courtesy of NASA

Today, Io is considered to be the most volcanically active place in the solar system. Some of the first images of Io were beamed back to Earth from the spacecraft Voyager in 1979. The surface is an array of colors that include red, yellow, white, black and green. The moon's coloring lead to the nickname "Pizza Moon."

Flyby observations

Dr. Lopes realized that flying too close to Io could be dangerous because of Jupiter's magnetic field and intense radiation. The first planned "flyby" took place in 1995, but no images were recorded due to equipment failure. Additional "flybys" were planned and each time volcanic eruptions were observed. Dr. Lopes realized that the plumes of smoke and deposits were responsible for the changing appearance of Io's surface. The brilliant red and yellow coloring of the surface is evidence of sulfur and sulfur dioxide. As the temperature of sulfur changes so do the colors of the moon. Red is an indication of an active or recently volcano on Io.

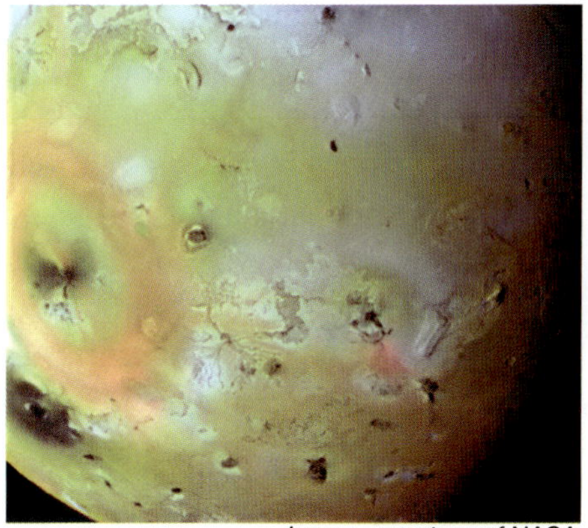

Image - courtesy of NASA

The Galileo spacecraft was so successful that NASA extended the mission three times. After 14 years and 4,631,778,000 kilometers the spacecraft was intentionally destroyed on September 21, 2003. It will forever change our view of the solar system as scientists continue to explore alien territory.

Questions:

1. How do scientist study volcanic activity on Earth?
2. What was the purpose of the Galileo mission?
3. Why was the appearance of Io's surface constantly changing?
4. Why was it risky for the Galileo spacecraft to fly close to Io?

Chapter 15 Connection

CHAPTER ACTIVITY Alien Design

The plants and animals that live on Earth are uniquely suited to Earth's environment. As humans, our bodies are able to withstand the Earth's temperate climate. We survive by breathing the air in Earth's atmosphere, and by drinking the fresh water in Earth's rivers and lakes. Although no life has yet been found on other planets, it may be possible for some form of life to live on another planet in our solar system, and scientists are continually brainstorming what life might be like on a different planet than Earth. In this activity, you will create an organism that could live on another planet. Follow the guidelines to develop your organism, and be creative!

What you will do

1. Choose any of the planets in our solar system, except Earth.
2. Develop an organism or animal that can survive on this planet. Make sure you explain how your organism overcomes the harsh climate present on its planet. For example, if an animal lived on Mercury, it would need special protection to survive extreme hot and cold temperatures.
3. Explain how your organism exchanges elements with its atmosphere or soil, and how it moves around the planet. For instance, how does it "breathe" if there is no atmosphere? How would an organism move around a planet such as Jupiter, which does not have a solid surface? How would your organism deal with the extra gravity on a planet such as Neptune?

Planet	Temperature range	Weight of a 100 lb Earthling	Length of Day	Length of Year	Interesting Fact
Mercury	-300°F to 870°F	38 lbs	59 days	88 days	No atmosphere; many craters
Venus	850°F	91 lbs	243 days	225 days	Dense atmosphere mostly CO_2 and N_2
Mars	-190°F to 98°F	38 lbs	24 hours	687 days	Water trapped in frozen poles
Jupiter	-244°F	254 lbs	10 hours	11.8 years	No solid surface; H_2O and H_2 oceans
Saturn	-300°F	108 lbs	10 hours	29.5 years	No solid surface, icy rings
Uranus	-300°F	91 lbs	17.2 hours	84 years	Atmosphere mostly H_2, He, methane; possible water
Neptune	-370°F	119 lbs	16 hours	165 years	Atmosphere mostly H_2, He, methane
Pluto (dwarf)	-390°F	8 lbs	7 days	248 years	Cold, remote; Sun looks like a bright star in the sky

Chapter 15 Assessment

Vocabulary

Select the correct term to complete the sentences.

Newton's law of universal gravitation	terrestrial planets	solar system
	gas planets	meteorite
meteor	light years	astronomical unit
comet	asteroid	

Section 15.1

1. An explanation of the force that exists between all objects with mass is given by ____.
2. A rocky body orbiting the sun but too small to be called a planet is called a(n) ____.
3. The distance from Earth to the sun, often used as a unit of measure for large distances, is named the ____.
4. Jupiter, Saturn, Uranus, and Neptune, made mostly of hydrogen and helium, are called the ____.
5. A small piece of an ateroid or comet that breaks off and is vaporized in Earth's atomosphere is called a(n) ____.
6. The sun and the eight planets and their moons orbiting the sun are referred to as the ____.
7. While traveling close to the sun in its highly elliptical orbit, a(n) ____ develops a tail that can stretch for milions of kilometers into space.
8. The planets including Mercury, Venus, Earth, and Mars are commonly called the ____.
9. A meteor that does not burn up as it passes through Earth's atmosphere is known as a(n) ____.

Section 15.2

No vocabulary words in this section

Concepts

Section 15.1

1. Copernicus suggested that the planets orbit the sun. What discoveries did Galileo make to support Copernicus' ideas?
2. Why is the sun at the center of our solar system?
3. Which is the best unit for comparing relative distances within the solar system?
   a. Astronomical units (AU)
   b. Light years
   c. Kilometers
4. Name the factors that determine the strength of the force between two masses.
5. For each of the following, tell whether it reflects or emits light:
   a. Earth
   b. Mars
   c. The sun
   d. The moon
   e. Stars
6. The moon appears bright in the sky, but it does not produce its own light. Why can we see it shining so brightly?
7. Why is it so difficult to see if there are planets around other stars than the sun?
8. What is the difference between a meteor and a meteorite?
9. Compare asteroids to comets by filling in the blanks of the table below:

Object	Size	Material	Orbit Shape	Location
Asteroid				
Comet				

10. Why does a comet form a visible tail as it approaches the sun?
11. Are the gas planets made up only of gas?

Section 15.2

12. Which planet has the most moons?
13. Earth has a day that is 24 hours long. Which planet has a day of about the same length as Earth's?
14. Mercury closer to the sun than Venus, but Venus has higher surface temperatures. Explain why.
15. Seasons are mainly caused by:
    a. the distance between Earth and the sun.
    b. the tilt of Earth's axis.
    c. the orbit of Earth.
16. Which planet is the closest to Earth?
17. How do Saturn's rings stay in place?
18. Which planet has a day that is longer than its year?
19. Which planets, beside Earth, have an atmosphere?
20. What is the Great Red Spot observed on Jupiter?
21. What is important about Jupiter's moon, Io?
22. Name the three brightest observable objects from Earth.
23. Which planet has a climate most like Earth's? What sort of opportunity does this represent?

Problems

Section 15.1 and 15.2

1. A moon rock weighs 8.5 pounds on the moon. How much would this rock weigh on Earth? (strength of gravity on the moon = 1.6 N)
2. An astronaut has a mass of 60 kilograms.
    a. What is the astronaut's weight on the surface of Earth?
    b. What is the astronaut's weight on the surface of the moon?
3. The moon is approximately 385,000 km from Earth. What is this distance in astronomical units?
4. Newton's law of universal gravitation explains the strength of the gravitational attraction between Earth and the moon.
    a. If the mass of Earth suddenly doubled, what would happen to the gravitational force between Earth and the moon?
    b. If the mass of Earth and the mass of the moon were **both** doubled, what would happen to the gravitational attraction between them?
    c. If the distance from Earth to the moon were doubled, what would happen to the gravitational attraction between them?
5. Neptune's mass is about 17 times greater than Earth's mass. Would your weight be 17 times greater if you visited Neptune?
6. What is the relationship between a planet's distance from the sun and its orbital speed?
7. The average distance from Earth to the sun is:
    a. 1 light year.
    b. 1 astronomical unit.
    c. 385,000 km
8. Why does the sun feel warmer during summer and colder during winter in the northern hemisphere?

Chapter 16

The Sun and Stars

Stargazing is an awe-inspiring way to enjoy the night sky, but humans can learn only so much about stars from our position on Earth. The Hubble Space Telescope is a school-bus-size telescope that orbits Earth every 97 minutes at an altitude of 353 miles and a speed of about 17,500 miles per hour. The Hubble Space Telescope (HST) transmits images and data from space to computers on Earth. In fact, HST sends enough data back to Earth each week to fill 3,600 feet of books on a shelf. Scientists store the data on special disks. In January 2006, HST captured images of the Orion Nebula, a huge area where stars are being formed. HST's detailed images revealed over 3,000 stars that were never seen before. Information from the Hubble will help scientists understand more about how stars form. In this chapter, you will learn all about the star of our solar system, the sun, and about the characteristics of other stars.

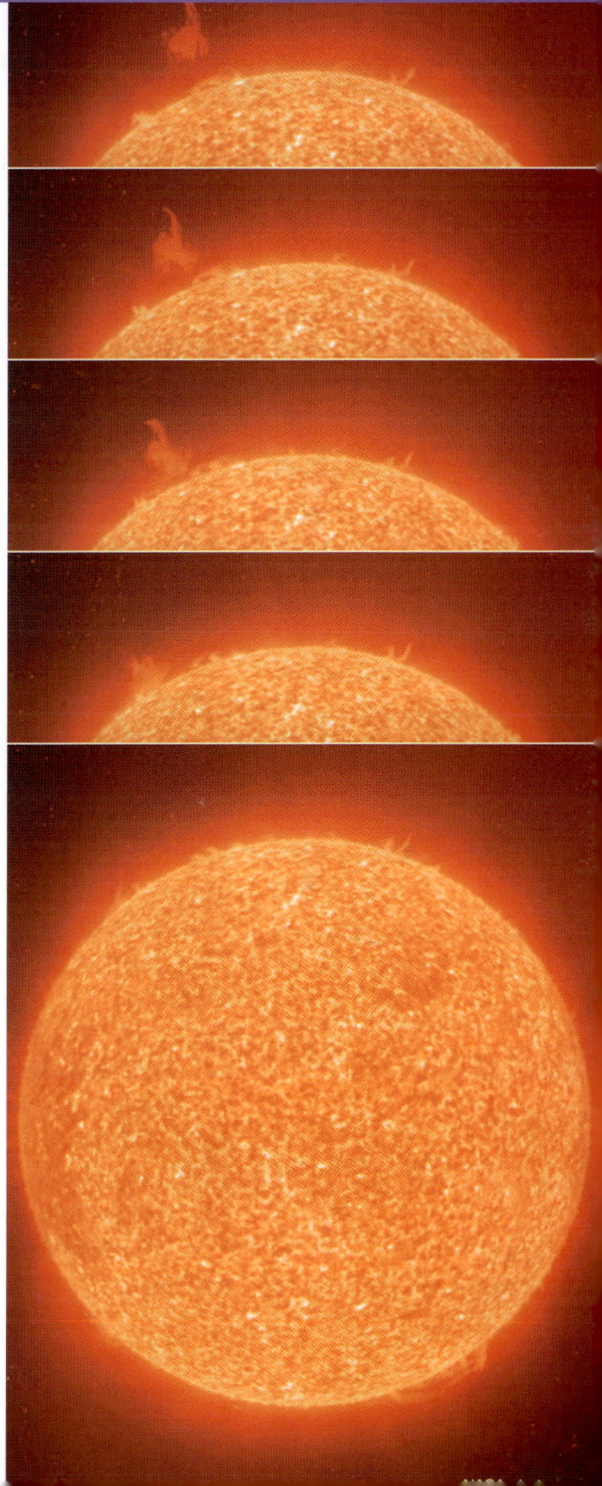

Key Questions

1. *Why do stars shine?*
2. *What kinds of stars are there?*
3. *How are stars formed, and do any other stars have planets?*

16.1 The Sun and the Stars

What are stars? Where did they come from? How long do they last? During most of the day, we see only one star, the sun, which is 150 million kilometers away. On a clear night, about 6,000 stars can be seen without a telescope. Ancient astronomers believed that the sun and the stars were different from each other. Today we know that the sun is just one star like all the others in the night sky. The others appear to be so dim because they are incredibly far away. The closest star to Earth is Alpha Centauri: 4.3 light years (41 trillion kilometers). That is 7,000 times farther away than Pluto.

Why stars shine

Nuclear fusion

A **star** is essentially an enormous, hot ball of gas held together by gravity. Gravity squeezes the density of stars so tightly in the core that the electrons are stripped away and the bare nuclei of atoms almost touch each other. At this high density, **nuclear fusion** occurs, releasing tremendous amounts of energy. The nuclear fusion that powers the sun combines four hydrogen atoms to make helium, converting two protons to neutrons in the process (Figure 16.1). The minimum temperature required for fusion to occur is 7 million°C. The sun's core reaches a temperature of 15 million°C.

The dense core of a star

The high density and temperature needed for fusion occurs in the center of a star. The density at the sun's core is about 158.0 g/cm^3. This is about 18 times the density of solid copper. In order to reach this high density, a star must have a mass much larger than a planet. For example, the sun has a mass about 330,000 times larger than the mass of Earth.

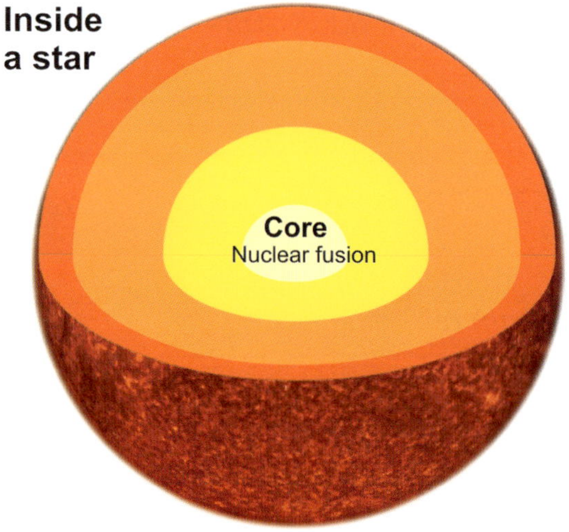

VOCABULARY

star - an enormous hot ball of gas held together by gravity which produces energy through nuclear fusion reactions in its core.

nuclear fusion - reactions which combine light elements such as hydrogen into heavier elements such as helium, releasing energy.

The fusion reactions in the sun combine hydrogen to make helium

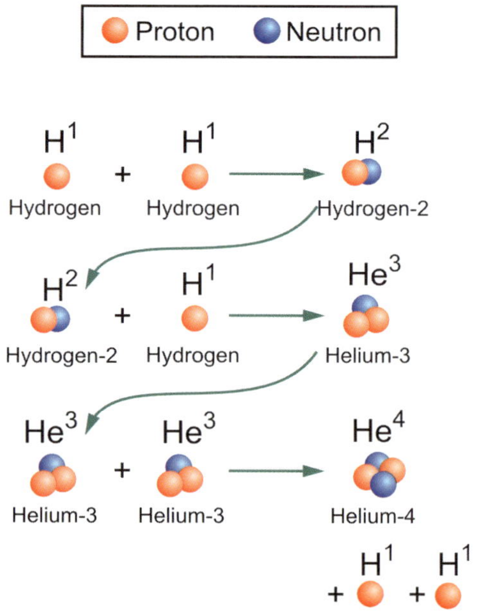

Figure 16.1: *One of several nuclear fusion reactions that release energy in the sun by combining hydrogen into helium.*

The anatomy of the sun

The sun has three regions

Because the sun is made of gas, its surface is hard to define. The apparent surface that we can see from a distance is called the *photosphere,* which means "sphere of light." Just above it is the *chromosphere*. This is a very hot layer of plasma, a high-energy state of matter. The *corona* is the outermost layer of the sun's atmosphere, extending millions of kilometers beyond the sun. Both the corona and chromosphere can be seen during a total eclipse of the sun, as shown in Figure 16.2.

Sunspots

A safe method for viewing the sun is to use a telescope to project its image onto a white surface. (You should NEVER look directly at the sun.) When the sun is observed in this way, small, dark areas can be seen on its surface. These areas, called *sunspots*, may look small, but they can be as large as Earth. **Sunspots** are areas of gas that are cooler than the gases around them. Because they don't give off as much light as the hotter areas, they appear as dark spots on the photosphere (Figure 16.3).

Anatomy of the sun

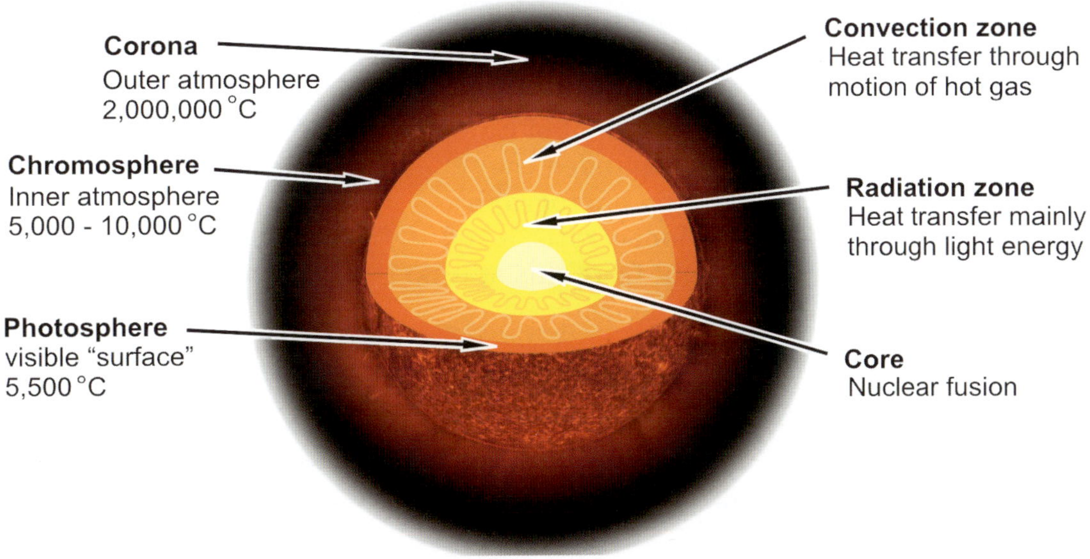

Figure 16.2: *The sun's corona and chromosphere can be seen during a total eclipse.*

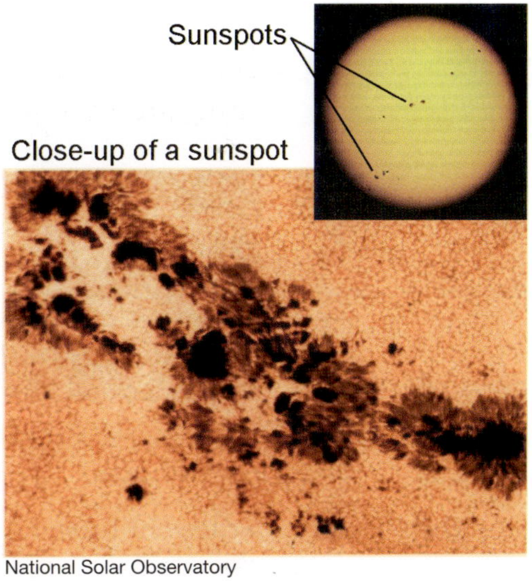

Figure 16.3: *Sunspots appear as dark spots on the photosphere.*

Types of stars

How are stars classified? Stars come in a range of sizes and temperatures. Our sun is almost an average star in many ways. Some stars are larger and hotter. Other stars are smaller and cooler. Astronomers classify stars according to *size*, *temperature*, *color,* and *brightness*.

Sizes of stars Stars come in a range of masses. The largest stars have a mass of about 60 times the mass of the sun. The smallest stars are about one-twelfth the mass of the sun. This is about the minimum required to create enough gravitational pressure to ignite fusion reactions in the core. The sun is a medium-sized star (Figure 16.4), as is Alpha Centauri, the nearest star to the sun.

Giant stars

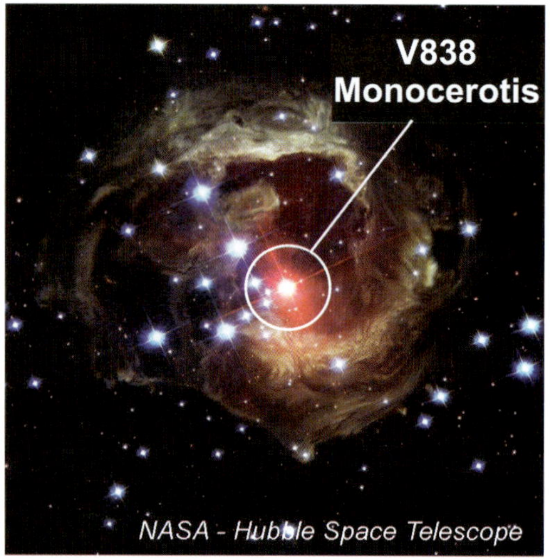

Stars vary in size as well as mass. There are two types of giant stars. Blue giant stars are hot and much more massive than the sun. Rigel in the constellation of Orion is a blue giant star. Red giants are of similar mass to the sun and much cooler. The red giants are huge because they began as sunlike stars but have expanded out past the equivalent of Earth's orbit. As they expanded they cooled down. The photograph shows V838 Monocerotis, a red giant star. Light from this star is illuminating the nebula around it.

Dwarf stars Stars that are smaller than the sun come in two main categories, *dwarfs* and *neutron stars*. Dwarf stars are about the size of the smaller planets. Sirius B, the largest known dwarf star, is slightly smaller than Earth. Neutron stars are even smaller. Their diameter is only 20 to 30 kilometers, about the size of a big city.

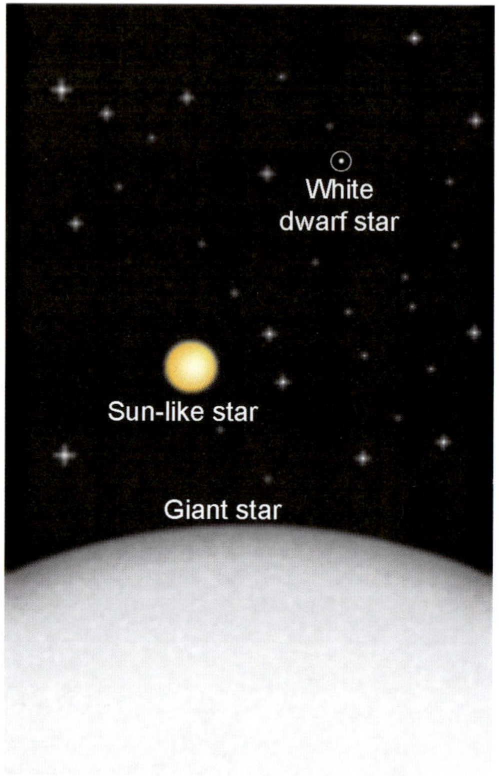

Figure 16.4: *Comparing different sizes of stars.*

Distances to the nearest stars

What is a light year? Because distances in space are huge, scientists have to use units much larger than kilometers or meters. You may have heard of *light years* (ly). Light years are a common way to measure distance outside the solar system. One **light year** is the *distance* that light travels through space in one year. A light year is a unit of distance, *not time*.

Calculating a light year In space, light travels at the amazing speed of exactly 299,792 kilometers per second (approximately 300,000 kilometers per second). How far will it travel in one year? Recall that *speed* = distance ÷ time. This means we can calculate the distance light travels in one year by multiplying the speed of light by time (by rearranging the variables). However, to get the correct value, we must also convert seconds into years since the value for the speed of light contains seconds. Here's how to solve the problem:

$$\begin{aligned}\text{1 light year (ly)} &= \text{speed of light} \times \text{time}\\ &= (300{,}000 \text{ km/sec}) \times \left(1 \text{ year} \times \frac{365 \text{ days}}{1 \text{ year}} \times \frac{24 \text{ hours}}{1 \text{ day}} \times \frac{3600 \text{ sec}}{1 \text{ hour}}\right)\\ &= (300{,}000 \text{ km/sec}) \times (31{,}536{,}000 \text{ sec})\\ &= 9{,}460{,}000{,}000{,}000 \text{ km} \quad \text{or} \quad 9.46 \times 10^{12} \text{km}\end{aligned}$$

A light year is the distance light travels in one year through space (9.46×10^{12} kilometers).

Why we need units as large as light years The stars are light years distant from Earth. By comparison, even Pluto is only 5.4 *light hours* away from the sun. The nearest stars are 4.3 light years away, 7,000 times farther away than Pluto. Our best rockets travel at 30 kilometers per second. That speed would take you from Los Angeles to San Francisco in *25 seconds*. Even at this enormous speed it would take *40,000 years* to reach the nearest star. If humans are ever to venture beyond the solar system, we clearly need to develop faster ways to travel.

VOCABULARY

light year - the distance light travels through space in one year, 9.46×10^{12} km.

Object	Distance from Earth (light years)
Sirius (brightest star in the sky)	8.8
Betelgeuse (appears as a red star in the sky)	700
Crab Nebula (remnant of an exploded star)	4,000

Figure 16.5: *Distance from Earth (in light years) of some well-known objects in the universe.*

Temperature and color

Temperatures of stars

If you look closely at the stars on a clear night, you might see a slight reddish or bluish tint to some stars. This is because stars' surface temperatures are different. Red stars are cooler than white stars, and blue stars are the hottest. The table below lists some stars, their colors, and their surface temperatures.

Table 16.1: Stars, their colors, and their surface temperatures

Star	Color	Temperature range (°C)
Betelgeuse	red	2,000 to 3,500
Arcturus	orange	3,500 to 5,000
Sun	yellow	5,000 to 6,000
Polaris	yellow-white	6,000 to 7,500
Sirius	white	7,500 to 11,000
Rigel	blue-white	11,000 to 25,000
Zeta Orionis	blue	25,000 to 50,000

Figure 16.6: *Sirius, the Dog Star in the constellation of Canis Majoris, is a good example of a white star.*

The color of light is related to its energy. Red light has the lowest energy of the colors we can see. Blue and violet light have the most energy. Yellow, green, and orange are in between. White light is a mixture of all colors at equal brightness.

Color and temperature

When matter is heated, it first glows red at about 600°C. As the temperature increases, the color changes to orange, yellow, and finally white. The graph in Figure 16.7 shows the colors of light given off at different temperatures. The curve for 2,000°C crosses red and yellow, but drops off before getting to blue. That means a surface at 2,000°C gives off mostly red and some yellow. At 10,000°C a star gives off an even mix from red to blue so it appears white. At 20,000°C the emitted light is white with a bluish color.

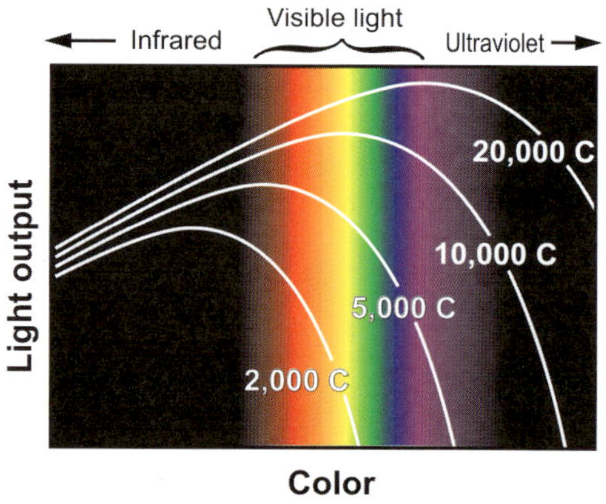

Figure 16.7: *The range of light given off by a star depends on its temperature. Stars at 2,000°C give off mostly red and some yellow light. At 10,000°C a star gives off an even mix from red to blue, so the light appears white.*

Brightness and luminosity

Light radiates in all directions You can see a bare light bulb from anywhere in a room because the bulb emits light in all directions. When the rays of light are represented by arrows, the light coming from a bulb looks like Figure 16.8. A star also radiates light equally in all directions.

Light intensity From experience, you know that as you move away from a source of light, the **brightness** decreases. *Brightness*, also called *intensity*, describes the amount of light energy per second falling on a surface, such as the ground, your eye, or a telescope (Figure 16.9). The brightness of a star described as the *light reaching Earth*.

Light intensity follows an inverse square law For a distant source of light like a star, the brightness decreases as the inverse square of the distance. For example, a star that is twice as far away will appear only 1/4 as bright because 1/4 is $1/2^2$. A star that is three times as far away will appear 1/9 as bright ($1/9 = 1/3^2$).

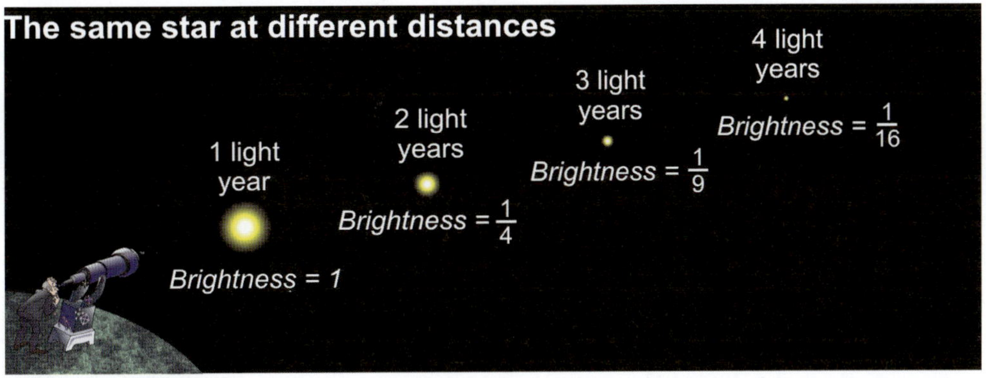

Luminosity The brightness of a star also depends on how much light the star gives off. This is called a star's **luminosity**. Luminosity is the total amount of light given off by a star in all directions. Luminosity is a fundamental property of a star whereas brightness depends on both luminosity and distance. *To understand stars, we wish to know their luminosity*. All we can measure is their brightness. To find the luminosity of a star we need to know both its brightness and its distance from Earth. We can then apply the inverse square law to calculate the luminosity from the brightness and distance.

VOCABULARY

brightness - measures the amount of light reaching Earth.

luminosity - the total amount of light given off by a star.

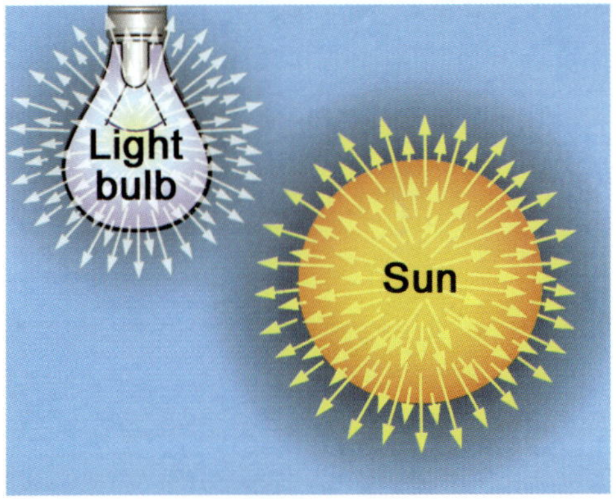

Figure 16.8: *Light emitted from the sun or from a light bulb.*

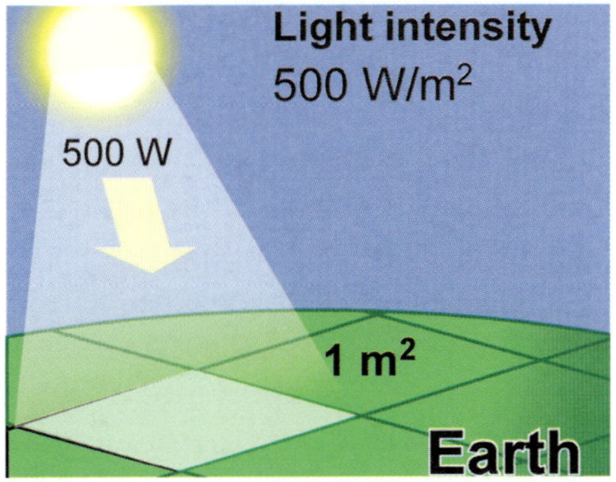

Figure 16.9: *Brightness is the intensity or amount of light per second falling on a surface. In summer, the brightness of sunlight reaches 500 W/m^2.*

The temperature and luminosity of stars

H-R diagrams

In the early 1900s, the Danish astronomer Ejnar Hertzsprung and American astronomer Henry Russell developed an important tool for studying stars. They made a graph that showed the temperature of the stars on the x-axis and the luminosity on the y-axis. The result is called the *Hertzsprung-Russell*, or H-R diagram. Each dot on the diagram below represents a star with a particular luminosity and temperature.

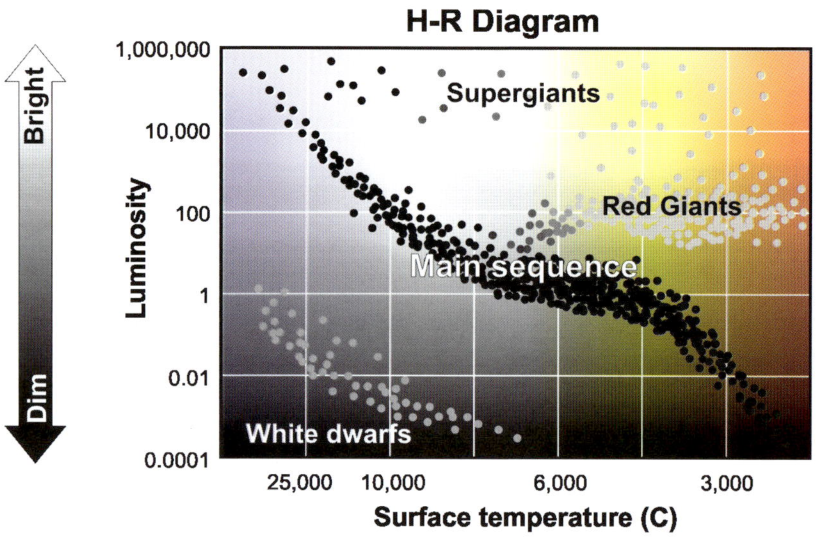

Reading H-R diagrams

H-R diagrams are useful because they help astronomers categorize stars into groups. Stars that fall in the band that stretches diagonally from cool, dim stars to hot, bright stars are called **main sequence stars**. Main sequence stars, like the sun, are in a very stable part of their life cycle. **White dwarfs** are in the lower left corner of the diagram. These stars are hot and dim and cannot be seen without a telescope. **Red giants** appear in the upper right side of the diagram. These stars are cool and bright and some can be seen without the aid of a telescope. **Supergiants**, both red and blue, are found in the extreme upper portion of the diagram. You can observe red and blue supergiants in the constellation Orion (Figure 16.10).

VOCABULARY

main sequence star - a stable star in the main category in the H-R diagram.

white dwarf - a small star with a high temperature and low brightness.

red giant - a large star with low temperature and high brightness.

supergiant - very large, bright star that may be blue or red, depending on its temperature.

Figure 16.10: *If you locate Orion in the night sky, you can see Betelgeuse, a red supergiant, and Rigel, a blue supergiant. It is easy to find this constellation because of the three stars that form its belt.*

16.1 Section Review

1. What is the basic process through which the sun releases energy?
2. Describe the three regions of the sun.
3. What are sunspots?
4. List the four variables astronomers use to classify stars.
5. How does the size of the sun compare with the size of other stars?
6. What is a light year?
7. Regulus, the brightest star in the constellation Leo, is approximately 77 light years from Earth. Which year did Regulus give off the light you see when looking at the star today?
8. What can you tell about a star by looking at its color?
9. What happens to the brightness of a star as you observe it from farther away?
10. Explain the difference between a star's brightness and its luminosity.
11. Suppose one star is three times farther away than another. If both stars have the same luminosity, how will their brightness compare?
12. Describe the four types of stars categorized in a Hertzsprung-Russell diagram.
13. True or false: A white dwarf star is about the same size as the sun.
14. The star in Figure 16.11 is a:
    a. Red giant star.
    b. Blue giant star.
    c. Main sequence star.
    d. White dwarf star.

CHALLENGE

A light year is the distance light travels in one year. Other units can be defined according to the distance light travels in a certain amount of time. For example, a light second is the distance light travels in one second.

Calculate the number of meters each of the following units represents:

1. Light second
2. Light minute
3. Light nanosecond (a nanosecond is one-billionth of a second)

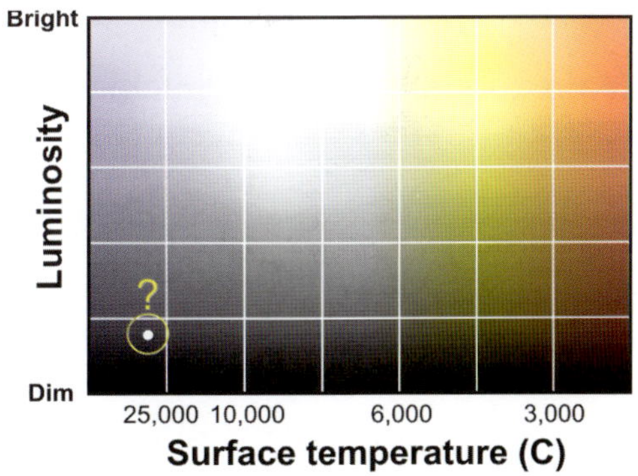

Figure 16.11: *The H-R diagram for question 15.*

16.2 The Life Cycles of Stars

Like living organisms, stars have life cycles. Of course, stars are not truly “alive” but astronomers sometimes use the terms “born,” “live,” and “die” to represent parts of that cycle. Our sun, a medium-size star, was “born” about 5 billion years ago. Because most medium-size stars have a life ‘1span of around 10 billion years, it will live for another 5 billion years before it dies. Stars that are larger than the sun have shorter life spans.

Nebulae, birth, and life span of stars

How are stars born? A star, regardless of its size, begins its life inside a huge cloud of gas (mostly hydrogen) and dust called a **nebula** (Latin for “mist”). Gravitational forces cause denser regions of the nebula to collapse, forming a *protostar*. A **protostar** is the earliest stage in the life cycle of a star. The gases at the center of the protostar continue to collapse, causing pressure and temperature to rise. A protostar becomes a star when the temperature and pressure at its center become great enough to start nuclear fusion. This is the nuclear reaction in which hydrogen atoms are converted into helium atoms and energy is released. Figure 16.12 shows a portion of the Eagle Nebula, the birthplace of many stars.

Main sequence stars Once nuclear fusion begins, a star is in the *main sequence* stage of its life cycle (Figure 16.13). This is the longest and most stable part of a star’s life. The time a star stays on the main sequence depends on the star’s mass. The sun will stay on the main sequence for about 10 billion years. You might think that high-mass stars live longer than low-mass stars because they contain more hydrogen fuel for nuclear fusion. However, the opposite is true. High-mass stars burn brighter, and hotter, using up their hydrogen faster than low-mass stars. Consequently, high-mass stars have much shorter life spans. For example, a star with 60 times the mass of the sun only stays on the main sequence for a few million years, a lifetime a thousand times shorter than the sun’s.

VOCABULARY

nebula - a huge cloud of dust and gas from which stars form.

protostar - the first stage in the life cycle of a star.

Figure 16.12: *A NASA/HST photo of a portion of the Eagle Nebula. The bright area is lit by young stars forming from clouds of molecular hydrogen (H_2).*

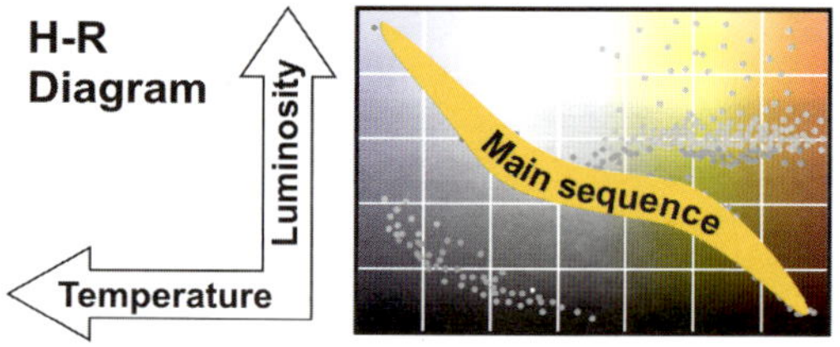

Figure 16.13: *The main sequence on an H-R diagram is where most stars spend most of their lives.*

Planetary systems and binary stars

Other stars have planetary systems

A star with orbiting planets is called a **planetary system**. Until the last decade, no one knew whether planets were commonly formed with stars or whether solar systems like our own were rare. However, as of this writing, more than 150 planets have been discovered around nearby stars. Because they give off no light of their own, planets are very hard to see against the brightness of a star. Astronomers had to devise very clever techniques to find them. Scientists now believe that planets are a natural by-product of the formation of stars. Therefore, planets of some type should exist around many (if not most) stars in the universe.

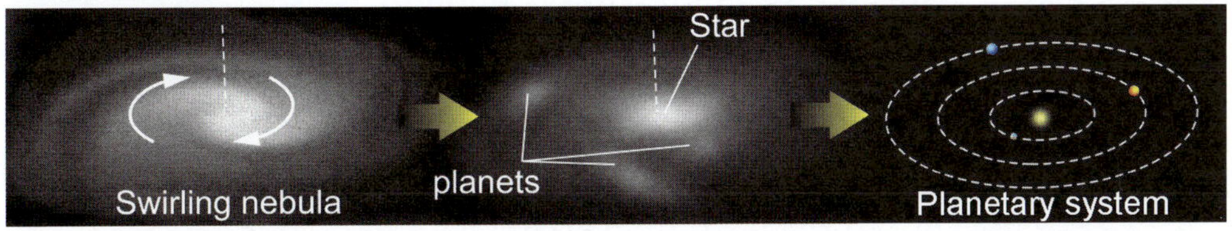

How our solar system was formed

A planetary system (like the solar system) forms out of the same nebula that creates the star. The protostar that became the sun also contained small amounts of other elements such as carbon, nickel, iron, aluminum, and silicon. As the protostar swirled inward on itself, it flattened into a disk perpendicular to its axis of rotation. Matter that was rotating too fast to fall inward and become the sun eventually amassed into planets. This explains why all of the planets formed in the same plane around the sun, and why they all orbit in the same direction.

Binary stars

Many gas clouds have enough swirling matter that multiple stars form, possibly with planets as well. A **binary star** is a system with two stars that are gravitationally tied and orbit each other. Binary stars are common. Mizar, the middle star in the handle of the Big Dipper, was the first binary star discovered, in 1650 by the Italian astronomer Giovanni Riccioli. Modern telescopes show that those two stars are actually four. Both Mizar A and Mizar B are themselves binary stars, making this a four-star system (Figure 16.14). About half of the 60 nearest stars are in binary (or multiple) star systems.

VOCABULARY

planetary system - a star and its planets.

binary star - a system of two stars orbiting each other that are gravitationally tied together.

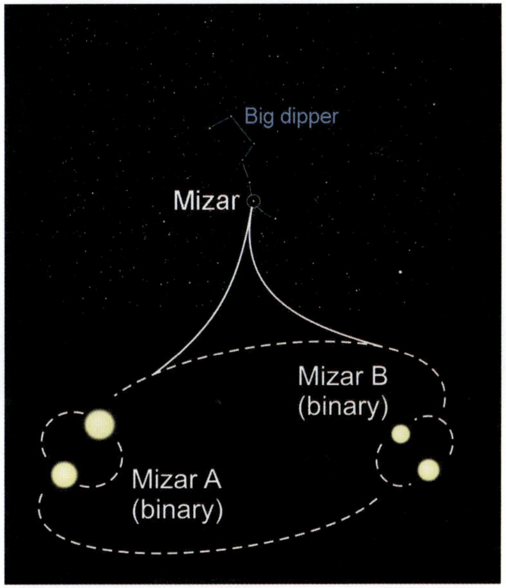

Figure 16.14: *Mizar, a star in the Big Dipper, is actually a double binary star system containing two pairs of two stars.*

The old age of sunlike stars

The formation of a red giant

Eventually, the core of a star runs out of hydrogen. Gravity then causes the core to contract, raising the temperature. At higher temperatures, other nuclear fusion reactions occur that combine helium to make carbon and oxygen. The hotter core radiates more energy, pushing the outer layers of the star away. The star expands into a red giant as the outer layers cool. In its red giant phase, our sun will expand to beyond the orbit of Mars, and the inner planets, including Earth, will be incinerated. Fortunately, this event is still 4 billion years in the far future.

White dwarf stars

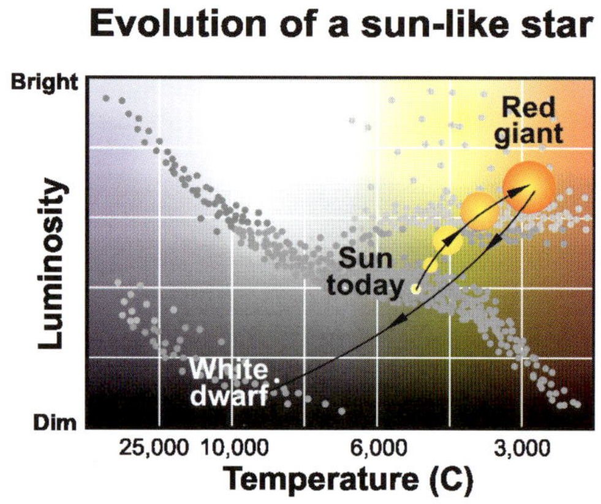

Sunlike stars don't have enough mass (gravity) to squeeze their cores any hotter than what is needed to fuse helium into carbon and oxygen. Once the helium is used up, the nuclear reactions essentially stop. With no more energy flowing outward, nothing prevents gravity from crushing the matter in the core together as close as possible. At this stage, the core glows brightly and is called a *white dwarf*. It is about the size of Earth, yet has the same mass as the sun. Because of its high density, a spoonful of matter from a white dwarf would weigh about the same as an elephant on Earth.

Planetary nebulae

During the white-dwarf stage, the outer layers of the star expand and drift away from the core. In the most extravagant stars this creates a **planetary nebula** (Figure 16.15). The planetary nebula contains mostly hydrogen and helium, but also some heavier elements that were formed in the core. Over time, the matter in a planetary nebula expands out into the rest of the universe and becomes available for forming new stars. Planetary nebulae are one of nature's ways of recycling the matter in old stars and distributing new elements.

VOCABULARY

planetary nebula - the expanding outer shell of a sunlike star. This matter is blown away as the core shrinks to become a white dwarf.

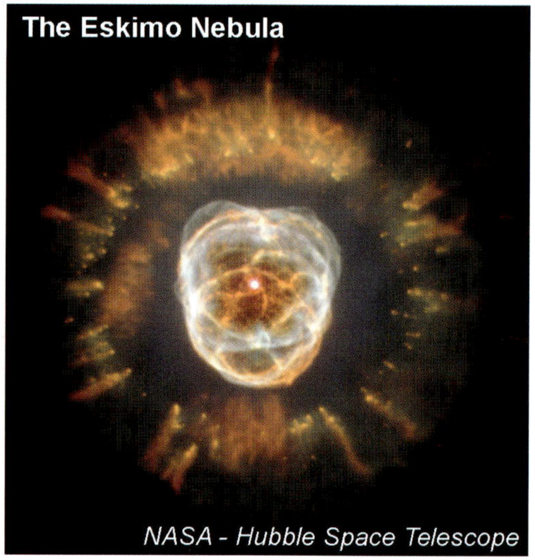

Figure 16.15: *A planetary nebula forms when a star blows off its outer layers leaving its bare core exposed as white dwarf. This may occur with stars that have a mass between 1.5 and 5 times the mass of the sun.*

Supernovae and synthesis of the elements

The origin of the elements Scientists believe the early universe was mostly hydrogen, with a small amount of helium and a trace of lithium. Heavier elements such as carbon and oxygen did not exist. So, where did they come from? All the heavier elements are created by nuclear fusion inside the cores of stars — including the elements in your body. Every carbon atom in your body, which is 53 percent of the solid matter of your body, was once inside a star.

The creation of elements Stars of more than 12 times the mass of the sun have a violent end. As the core runs out of helium, gravity compresses and heats the core hot enough for other types of nuclear fusion to start. The new fusion reactions combine carbon and oxygen into neon, sodium, magnesium, sulfur, silicon, and even heavier elements up to iron. Nuclear fusion reactions are *exothermic*, releasing energy only up to iron (Fe, atomic number 26). After that, the reactions become *endothermic*, using energy rather than releasing it. When the core of the star contains mostly iron, nuclear fusion stops.

Supernovas If a star's iron core reaches 1.4 times the mass of the sun, gravity becomes strong enough to combine electrons and protons into neutrons. The core of the star collapses in moments to form a single "nucleus" a tiny fraction of its former size. The rest of the star rushes in to fill the empty space left by the core then bounces back off the nucleus with incredible force. The result is a spectacular explosion called a **supernova**. A supernova is brighter than 10 billion stars and can outshine an entire galaxy for a few seconds. More than 90 percent of the mass of the star is blown away (Figure 16.16). During this brief period, heavier elements such as gold and uranium are created, as atomic nuclei are smashed together.

Neutron stars The light and heat produced by a supernova fades over time, and the remnants become a nebula that can be used to make more stars. All that remains of the original star is a core made entirely of neutrons called a *neutron star*. This super-dense object is no more than a few kilometers in diameter!

VOCABULARY

supernova - the explosion of a very large star.

Figure 16.16: *The Crab nebula is the remains of a supernova that occurred in AD 1054 and was recorded by Chinese astronomers.*

Examining light from stars

What is spectroscopy?

We know what elements are in distant stars from the light the stars produce. Astronomers analyze the light given off by stars, and other "hot" objects in space in order to figure out what they are made of and their temperatures. **Spectroscopy** is a tool of astronomy in which the electromagnetic waves (including visible light) produced by a star or other object (called its spectrum) is analyzed.

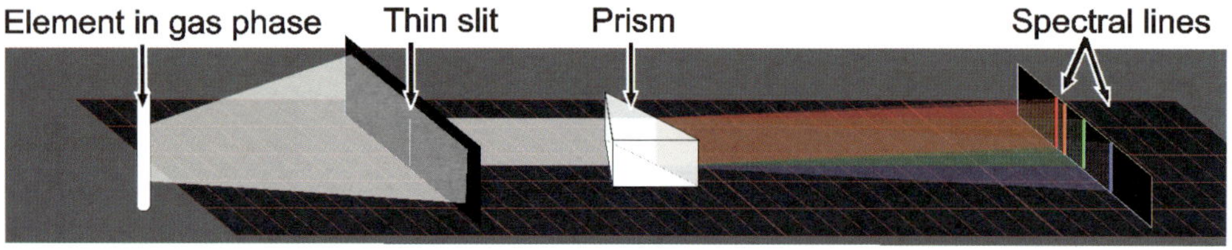

Chemical composition of stars

Astronomers use a tool called a *spectrometer* to split light into a *spectrum* of colors. A spectrometer displays lines of each color along a scale that measures the wavelength of light in nanometers (nm). Light waves are extremely small: A nanometer is one-billionth of a meter. Each element has its own unique pattern of lines—like a fingerprint. For example, when light from hydrogen is examined with a spectrometer, four lines are seen: red, blue-green, blue-violet and violet (Figure 16.17). Astronomers use spectroscopy to determine what elements are present in stars. A star's speed, temperature, rotation rate, and magnetic field can also be determined from its spectrum.

The composition of the sun

In 1861, Sir William Huggins, an English amateur astronomer, used spectroscopy to discover that the sun and the stars are made mostly of hydrogen. A few years later, Sir Joseph Norman Lockyer observed a line at the exact wavelength of 587.6 nanometers. Since no known element had a line at this wavelength, he concluded that this must be an undiscovered element and named it helium, after the Greek name for the sun, *Helios*. Today, we know that hydrogen is the most common element in the universe, with helium second (Figure 16.18). Helium is abundant in the sun but rare on Earth.

VOCABULARY

spectroscopy - a method of studying an object by examining the visible light and other electromagnetic waves it creates.

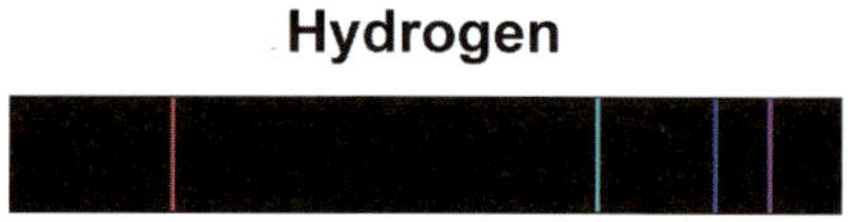

Figure 16.17: *When hydrogen gives off light, four visible lines are seen at 656 nm (red), 486 nm (blue-green), 434 nm (blue-violet), and 410 nm (violet) on the scale of a spectrometer.*

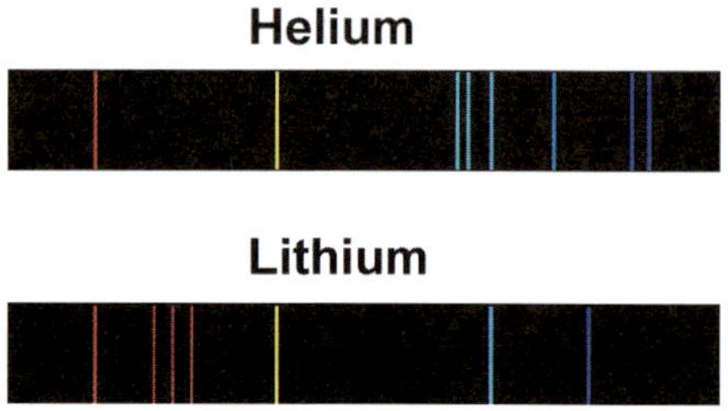

Figure 16.18: *Spectral lines for helium and lithium. Each element has its own unique pattern of spectral lines (spectrum).*

16.2 Section Review

1. How many years do scientists believe are left before the sun runs out of hydrogen in its core and leaves the main sequence?
2. All stars begin inside a huge cloud of dust and gas called a _____.
3. What force causes a nebula to form a protostar?
4. What happens to the temperature and pressure in a protostar as it collapses?
5. How long does the main sequence stage of a sunlike star normally last?
6. Why do stars with smaller masses burn longer than stars with larger masses?
7. What happens when a sunlike star runs out of hydrogen fuel in its core?
8. How do scientists believe heavier elements such as carbon and oxygen were created?
9. What is a supernova?
10. Is it probable that the sun will become a supernova? Why or why not?
11. What can scientists learn about a star by studying the light it gives off?
12. How is the spectrum for an element similar to a person's fingerprint?
13. Compared with the age of the sun, a blue giant star is likely to be:
    a. Younger.
    b. About the same age.
    c. Older.
14. Fusion reactions that combine light elements release energy only until what element is created?

CHALLENGE

The last supernova to be observed in our galaxy occurred in 1604. It was named Kepler's supernova after the German astronomer Johannes Kepler (1571-1630). The supernova was visible to the naked eye as the brightest object in the night sky.

Use the internet to research Kepler's supernova to find the answers to the following questions:

1. Was the supernova named after Kepler because he was the first person to see it?
2. How many light years from Earth was the supernova?
3. How did the occurrence of the supernova help support Galileo's view of the universe?

RESEARCH CONNECTION

Big Bear Solar Observatory

Bring your sunglasses - you are about to visit a city with more than 300 sunny days a year. Big Bear Lake, Calif., two hours east of Los Angeles and high (starting about 6,700 feet) in the San Bernardino Mountains and San Bernardino National Forest, has something that not all of Southern California can boast: four seasons. Because of all these features, it has proven an ideal location for a solar observatory. Changeable climate provides a great opportunity for studying the sun under all sorts of conditions.

Big Bear Solar Observatory/ New Jersey institute of Technology

A premier site, an improbable location

Big Bear Solar Observatory (BBSO) was built in 1969 by the California Institute of Technology. Since 1997, it has been operated by the New Jersey Institute of Technology. The National Aeronautics and Space Administration (NASA), the National Science Foundation (NSF), the Air Force and the Navy, and other agencies finance the observatory.

Big Bear Solar Observatory
New Jersey Institute of Technology

The observatory is located on an island in the center of Big Bear Lake. It might seem obvious that high altitude and clear skies would benefit an observatory - but in the middle of a lake? In water, however, scientists have clearer images of the sun than on land.

The science behind this starts, of course, with the sun. When the sun heats the ground, convection occurs, the warm air rising and the cold air sinking in vertical circulation. Water absorbs more heat than does the ground. With less heat rising, the convection currents over the water are smaller than those on land. A natural light wind blowing across the lake also helps to keep images clear - which keeps the observatory one of the best in the world.

The importance of our nearest star

For Earth, the sun is not merely important, it is essential. It is the planet's source of heat, energy, and light, weather and climate. The sun's natural furnace is what makes life possible for each and every creature on Earth.

The sun is the star closest to Earth. Because of that "nearness," scientists are able to see and study the surface of the sun. Other stars are just too far away to view their surface features. From the sun, scientists learn about stars in general.

Big Bear Solar Observatory staff study the sun, solar phenomena, and the solar atmosphere, made up of:

- Photosphere: the visible surface of the sun.
- Chromosphere: the irregular layer above the photosphere.
- Corona: the outer atmosphere.

Scientists observe solar flares, explosions from the sun that importantly can disrupt communications, satellites, and other systems on Earth. They also observe sunspots, which are dark, cool areas on the photosphere, and prominences, which are arched clouds of gas extending from the sun.

Mighty telescopes

BBSO houses several powerful telescopes. As you know, viewing the sun directly is dangerous and safety measures are always required. But these solar telescopes have cameras and filters specifically crafted for viewing the sun. They magnify the sun and provide detailed images of surface events.

Big Bear Solar Observatory/ New Jersey institute of Technology

At the top of the observatory are four main telescopes. Two are used to observe sunspots, flares, and prominences. The observatory monitors and predicts solar flares and provides reports to interested groups. A third, smaller telescope examines the entire sun. On clear days, this telescope works from dawn to dusk, taking images every 30 seconds. The fourth telescope measures earthshine, which is sunlight that reflects onto the darkened portion of the moon and then back again to Earth. Earthshine can provide scientists with information about our planet's temperature, atmosphere, and global warming.

In November 2005, BBSO replaced its dome with a larger one to accommodate a new solar telescope that will be the largest in the world. It will help scientists to better study solar flares, space weather, and sunspots.

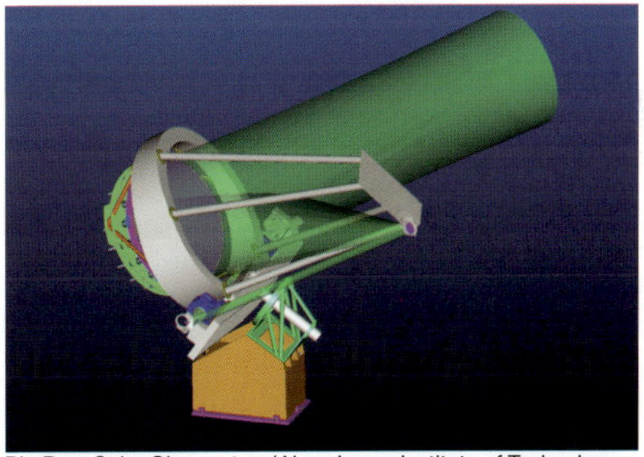

Big Bear Solar Observatory/ New Jersey Institute of Technology

One thing solar observers know for sure is that the sun is always changing. It is important for us to understand that solar activity so that we can predict its impact on our lives and Earth's future.

Questions:

1. Why is the sun important to people on Earth?
2. What exactly does Big Bear Solar Observatory observe and how does it do it?
3. What is ideal about the observatory's location?

CHAPTER ACTIVITY Sunspots

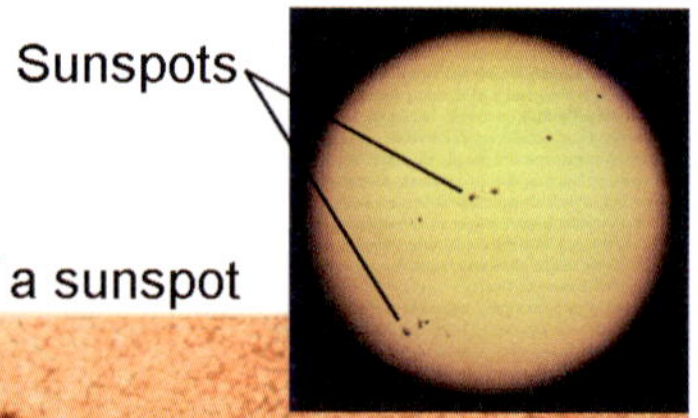

Close-up of a sunspot

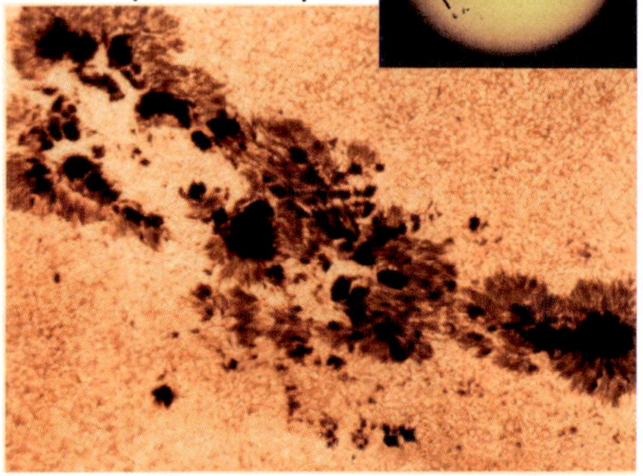

National Solar Observatory

Sunspots are large, dark regions that appear on the surface of the Sun. They appear dark because they are cooler than the areas around them. Sunspots are caused by the Sun's magnetic field. They slowly grow over a few days and then decay over a few days. The largest sunspots may last a few weeks. In the early 1600's, astronomers used the movement of sunspots across the surface of the Sun to figure out that the Sun's rotational period was about 27 days. In this activity, you will determine the diameter of the Sun and the number of sunspots on the Sun.

Materials:

cardboard, piece of aluminum foil (3 cm × 3 cm), tape, a pin or sharp point to puncture a hole in the aluminum foil, ruler, white paper, solar telescope (sunspotter)

Measure the diameter of the sun

1. First, make a pinhole viewer: Cut a square (2 cm × 2 cm) in the center of the cardboard. Take the aluminum foil and tape it over the opening. Using the pin, puncture a small hole in the center of the aluminum foil.
2. Hold the pinhole viewer so the light of the Sun passes through the hole onto a piece of paper held behind the viewer. Hold the viewer and paper as far apart as you can.
3. Measure the diameter of the Sun's image on the paper.
4. Measure the distance between the pinhole viewer and the image on the paper. Now, calculate the sun's diameter!

Distance from Earth to the Sun: 149,600,000 km
Measured diameter of the Sun: ________________ cm
Measured distance from viewer to paper: __________ cm

$$\text{Diameter of Sun} = \left(\frac{\text{Diameter of image of Sun}}{\text{Distance from viewer to paper}}\right) \times \text{Distance from Earth to Sun}$$

Measure the size and number of sunspots

5. Using a solar telescope (sunspotter), project the image of the Sun onto a flat surface, then trace the sun's perimeter and any sunspots on a white piece of paper.

Applying your knowledge

a. Using the diameter of the Sun that you calculated, figure out the size of one of the sunspots viewed through the sunspotter. Tell which sunspot you used and how you found your answer.

b. Do you see any patterns associated with sunspot activity on the graph below?

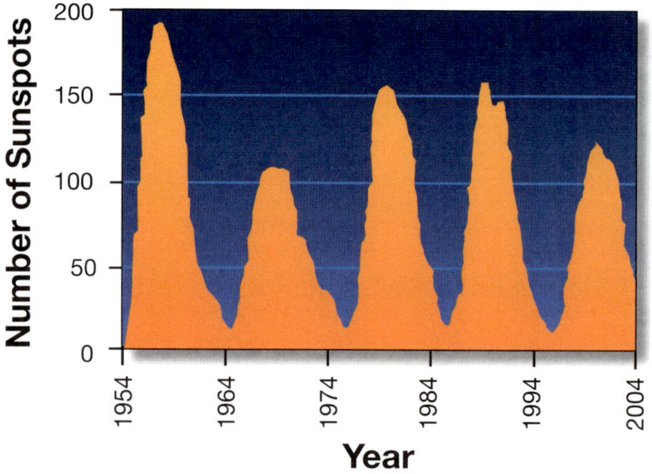

c. The real diameter of the sun is 1.35 million km. By what percent did your estimated diameter of the sun differ from the actual diameter?

Chapter 16 Assessment

Vocabulary

Select the correct term to complete the sentences.

red giant	sunspots	spectroscopy
nebula	white dwarf	main sequence stars
supergiant	protostar	protostar
supernova	nuclear fusion	binary star
light years	brightness	stars

Section 16.1

1. Small, dark areas on the surface of the sun representing relatively cooler areas are named ____.
2. A reaction in which hydrogen atoms combine to form helium, releasing large amounts of energy is called ____.
3. ____ are enormous, hot balls of gas held together by gravity.
4. A ____ is the distance light travels in one year.
5. A very large, bright star that may be blue or red, depending upon its temperature, is called a ____.
6. A stable star in the HR diagram's main category is a ____ star.
7. A large star with low temperature and high brightness is a(n) ____.

Section 16.2

8. A system of two stars tied together by gravity and orbiting each other is known as a(n) ____.
9. A huge cloud of dust from which stars are formed is known as a ____.
10. A method of studying an object by examining the visible light and other electromagnetic waves it creates is known as ____.
11. The name for a small star with high temperature and low brightness is ____.
12. The first stage in the life cycle of a star is known as a ____.
13. Astronomers call the explosion of a very large star a ____.

Concepts

Section 16.1

1. Why are solar flares important to the average person in the United States?
2. How are solar flares formed?
3. Arrange the layers of the sun listed below in order from innermost core to the outermost layer:
   a. chromosphere
   b. core
   c. corona
   d. photosphere
4. Describe two changes due to the fusion of hydrogen.
5. List three conditions that must exist for the continuous fusion of hydrogen to occur.
6. Describe the color changes that occur in matter as it is heated.
7. Which has the highest surface temperature?
   a. A red star
   b. A white star
   c. A blue star
8. Where in a star does fusion occur
9. Name the two most common elements in the universe. Are they also the most common elements on Earth?

10. Describe our sun in terms of color, brightness, size and temperature compared to other stars in the galaxy.

11. Why is an H-R diagram useful to astronomers?

Section 16.2

12. A spectrometer is a tool astronomers find very useful.
    a. How does it work?
    b. What information can an astronomer find using a spectrometer?

13. What does the lifespan of a star depend on?

14. List the stages in the life cycle of a typical medium-sized star like the sun.

15. Describe what astronomers believe will happen to our sun after it leaves the main sequence.

16. Do stars shine forever?

17. Explain why all the planets in our solar system orbit in the same direction.

Problems

Section 16.1

1. The Andromeda galaxy is 2.3 million light years away from Earth. What does this equal in kilometers?

2. Which star would you expect to have a longer lifespan, the sun or Sirius, a star whose mass is about twice as great as the sun?

3. Three identical blue giant stars are located 100 light years, 200 light years, and 300 light years away from Earth. How do their brightnesses compare from Earth?

4. A supernova was observed in 1987. If the star that collapsed into this supernova was 169,000 light years away, how long ago did this explosion occur?

5. Use the H-R diagram below to answer the following questions.

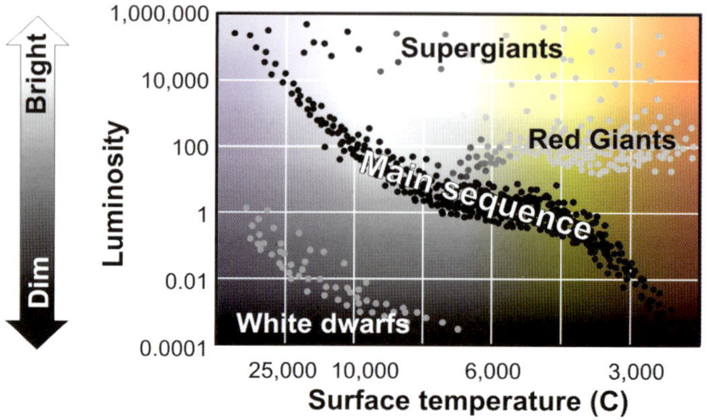

    a. What is the main difference between a typical white dwarf star and a typical supergiant?
    b. Which category of star is Vega, with a surface temperature of 10,000°C and a luminosity of 100?
    c. Predict which category Vega will enter next in its evolution.

Section 16.2

6. Write whether each of the following elements are formed by the early universe, nuclear fusion in a star, or supernova:
    a. iron (Fe)
    b. gold (Au)
    c. helium (He)
    d. hydrogen (H)
    e. carbon (C)

Chapter 17
Galaxies and the Universe

How big is *everything*? How long is *forever*? In science, *everything* means the universe including all matter and energy. In science, forever means the amount of time the universe exists or will exist. As a start to answering these deep questions, think about the night sky. It's dark. But why is the night sky dark? Imagine the universe was infinite, stretching off to forever in all directions. That means in any direction you looked, you would eventually see a star. If there were stars in every possible direction, the night sky should be light, not dark. This puzzled people for a long time and became known as Olber's Paradox.

The solution to Olber's Paradox is that the universe is not infinite in all directions, but has a finite size. Nor is the universe forever. Time had a beginning between 10 billion and 20 billion years ago and time may or may not have an end. It is amazing where you find yourself as you try to understand something so simple as why the night sky is dark. Of course, we are not sure all our answers are correct. Read ahead about galaxies and the universe. What do *you* think?

Key Questions

1. ***How is the universe organized?***
2. ***Did the universe have a beginning? Does it have an end?***
3. ***How can science answer questions like these?***

17.1 Galaxies

Early civilizations believed that Earth was the center of the universe. In the 16th century, we became aware that Earth is a small planet orbiting a medium-size star. It was only in the 20th century that we became aware that the sun is one of billions of stars in the Milky Way galaxy, and that there are billions of other galaxies in the universe. In the past 50 years, astronomers have found evidence that the universe is expanding and that it originated 10 billion to 20 billion years ago. In this section you will learn about galaxies and theories about how the universe began.

What is a galaxy?

The discovery of other galaxies

A **galaxy** is a huge group of stars, dust, gas, and other objects bound together by gravitational forces. The sun, along with an estimated 200 billion other stars, belongs to the **Milky Way galaxy**. The Milky Way is a typical *spiral galaxy* (Figure 17.1). From above, it would look like a giant pinwheel, with arms radiating out from the center. Although some stars are in *globular clusters* above and below the main disk, the majority are arranged in a disk that is more than 100,000 light years across and only 3,000 light years thick.

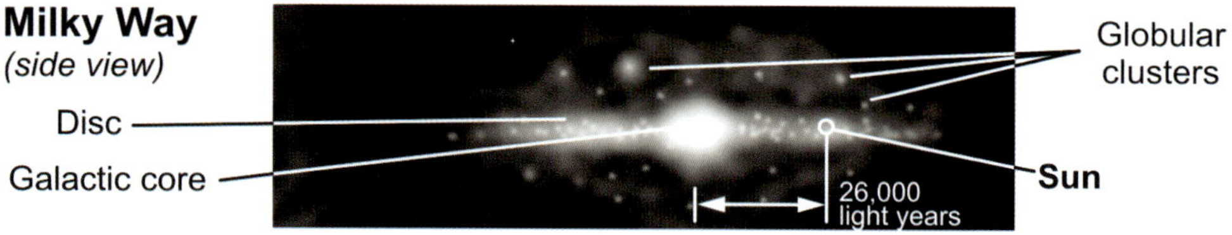

Our sun is 26,000 light years from the center

The disk of the Milky Way is a flattened, rotating system that contains young to middle-aged stars, along with gas and dust (Figure 17.2). The sun sits about 26,000 light years from the center of the disk and revolves around the center of the galaxy about once every 250 million years. When you look up at the night sky, you are looking through that disk of the galaxy. On a crystal clear night, you can see a faint band of light stretching across the sky. This is the combined light of billions of stars in the galaxy, so numerous that their light merges.

VOCABULARY

galaxy - a group of stars, dust, gas, and other objects held together by gravitational forces.

Milky Way galaxy - the spiral galaxy to which our solar system belongs.

Figure 17.1: *The Whirlpool galaxy is a typical spiral, like the Milky Way.*

Figure 17.2: *How the Milky Way appears in the night sky. We are looking in from the edge of the disk.*

Types of galaxies

The discovery of other galaxies

At the turn of the 20th century astronomers believed the Milky Way galaxy *was* the entire universe. As telescopes got better, though, some "smudges" that were thought to be nebulae in the Milky Way were recognized to be whole galaxies far outside our own. The discovery was made in the 1920s by Edwin Hubble, an American astronomer. When he focused a huge telescope on an object thought to be a nebula in the constellation Andromeda, Hubble could see that the "nebula" actually consisted of faint, distant stars. He named the object the Andromeda galaxy. Just since Hubble's time (1889-1953), astronomers have discovered a large number of galaxies. In fact, many galaxies are detected each year using the famous telescope launched into orbit in 1990: the Hubble Space Telescope, or HST.

Galaxy shapes

Astronomers classify galaxies according to their shape. *Spiral galaxies* like the Milky Way consist of a central, dense area surrounded by spiraling arms. *Barred spiral* galaxies have a bar-shaped structure in the center. *Elliptical galaxies* look like the central portion of a spiral galaxy without the arms. *Lenticular galaxies* are lens-shaped with a smooth, even distribution of stars and no central, denser area. *Irregular galaxies* exhibit peculiar shapes and do not appear to rotate like those galaxies of other shapes. Figure 17.3 shows examples of some galaxy shapes.

Galaxies change shape over time

The shapes of galaxies change over time. It is impossible to actually see the changes in a single galaxy, since it takes hundreds of millions of years. However, by looking at many galaxies, astronomers can see similar galaxies at different times in their histories. This observational data has allowed astronomers to develop computer based models which calculate how a galaxy changes over hundreds of millions of years. It is now believed that the barred spiral form is just one phase of a regular spiral. Computer simulations show how the "bar" forms and disappears repeatedly as a spiral galaxy rotates.

Figure 17.3: *Some representative galaxy shapes.*

Distances between galaxies

Galaxies are a million times farther away than stars

The distances between stars are 10,000 times greater than the distances between planets. *The distances between galaxies are a million times greater than the distances between stars.* For example, the distance from Earth to the nearest star is 4.3 light years, but from Earth to the Whirlpool galaxy is over 30 million light years.

The local group of galaxies

The Milky Way belongs to a group of about 30 galaxies called the local group. This group includes the Large Magellanic Cloud (179,000 light years) and the Small Magellanic Cloud (210,000 ly). These Magellanic Clouds are small, irregular galaxies of less than 100,000 stars. The local group also includes Andromeda, an elliptical galaxy 2.9 million light years away (Figure 17.4).

Galactic collisions

Galaxies move through space singly and in groups. Galaxies even collide with each other in slow dances of stars that take millions of years to complete (Figure 17.5)

Determining the distance to nearby galaxies

Figuring out the distance between galaxies is one of the more difficult tasks in astronomy. A faint (low brightness) object in the night sky could be a dim object that is relatively nearby or a bright object that is far, far away. The most reliable method for estimating the distance to a galaxy is to find a star whose luminosity is known. If the luminosity is known, the inverse square law can be used to find the distance from the observed brightness.

Distant galaxies

This method works for the closest galaxies. However, the vast majority of galaxies are too far away to see single stars even with the best telescopes. Beyond 150 million light years, astronomers compare size and type with closer galaxies to estimate the luminosity of the farther ones. This method is not as accurate and, consequently, the distances to far galaxies are known only to within a factor of two.

Figure 17.4: *The Andromeda galaxy is an elliptical galaxy in our local group.*

Figure 17.5: *Two galaxies that are near to colliding.*

The central black hole

The center of the galaxy

Since we are located in the outer part of the galaxy, dust between the stars blocks out much of the visible light coming from objects in the disk. Because of this, astronomers use infrared and radio telescopes to study our galaxy. They have learned that the center of the galaxy is crowded with older stars and hot dust. Recent studies have suggested that a black hole, with a mass of more than a million suns, exists at the very center of the galaxy.

Evidence for the black hole theory

The evidence for a huge black hole comes from measurements of the orbital speeds of stars and gas at the center of the galaxy. In one study, an infrared telescope was used to measure the orbital speeds of 20 stars over a three-year period. It was determined that these stars were orbiting at speeds of up to 1,000 kilometers per second (3 million miles per hour). This extremely high orbital speed requires an object with a mass that is over 2 million times that of the sun.

Relativity predicts black holes

One of the strangest predictions of Einstein's theory of relativity is the existence of *black holes*. To understand a black hole, consider throwing a ball fast enough to leave the Earth completely. If the ball does not go fast enough, the Earth's gravity eventually pulls it back. The minimum speed an unpowered projectile must have to escape the planet's gravity is called the *escape velocity*. The stronger gravity becomes, the higher the escape velocity.

The escape velocity of a black hole

If gravity becomes strong enough, the escape velocity can reach the speed of light. A *black hole* is an object with such strong gravity that its escape velocity equals or exceeds the speed of light. When the escape velocity equals the speed of light, nothing can get out because nothing can go faster than light. In fact, even light cannot get out, because in general relativity, light is affected by gravity. The name *black hole* comes from the fact that anything that falls in never comes out. Since no light can get out, the object is "black" (Figure 17.7). To make a black hole, a very large mass must be squeezed into a relatively tiny space. For example, to make the Earth into a black hole, you would have to squeeze the mass of the entire planet down to the size of your thumb.

Figure 17.6: *The core of the Milky Way is in the direction of the constellation Sagittarius. Astronomers believe a huge black hole lives at the core of the Milky Way and most other large galaxies.*

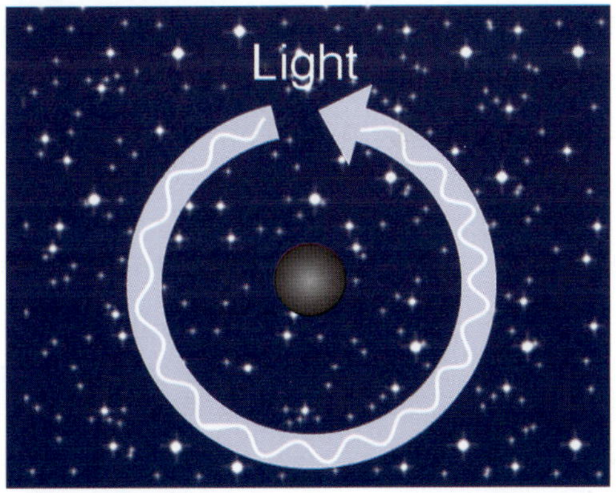

Figure 17.7: *Light from a black hole cannot escape because the escape velocity is higher than the speed of light.*

The expanding universe

Sirius is moving away from Earth

In the 1890s, astronomers began to use spectroscopy to study the stars and other objects in space. One of the first stars they studied, Sirius, had spectral lines in the same pattern as the spectrum for hydrogen. However, these lines did not have the exact same measurements as those for hydrogen. Instead, they were shifted toward the red end of the visible spectrum. This was a puzzle at first, until scientists realized that a red-shifted spectrum meant Sirius was moving away from Earth. They could even determine how fast Sirius was moving away by measuring the amount that the lines had shifted toward red (Figure 17.8).

Redshift

Redshift is caused by relative motion that increases the distance from the source to the observer. The faster the source of light is moving away from the observer, the greater the redshift. The opposite (blueshift) happens when an object is moving toward the observer. A star moving toward Earth would show a spectrum for hydrogen that was shifted toward the blue end of the scale.

Discovery of the expanding universe

In the late 1920s, Hubble began to measure the distance and redshift of galaxies. Much to his surprise, he discovered that the farther away a galaxy was, the faster it was moving away from Earth. By the early 1930s, he had enough evidence to prove that galaxies were moving away from each other with a speed proportional to the distance between them. This concept came to be known as the *expanding universe*.

The Big Bang theory

The expanding universe was a great surprise to scientists. Before Hubble's discovery, people believed the universe had existed in its same form for all time. The fact that the universe was expanding implies the universe must have been smaller in the past than it is today. In fact, it implies that the universe must have had a *beginning*. Astronomers today believe the universe exploded outward from a single point smaller than an atom into the vast expanse of galaxies and space we see today. This idea is known as the Big Bang theory.

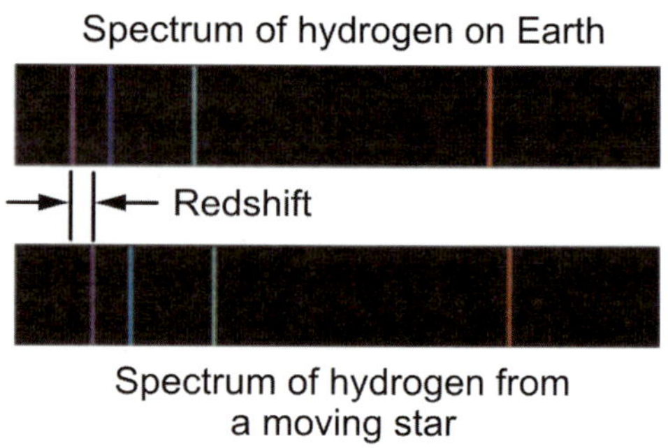

Figure 17.8: *The top diagram shows the hydrogen spectrum for an object that is not moving. The bottom diagram shows the hydrogen spectral lines for a moving star. While the lines are in the exact same pattern, they have shifted toward the red end of the spectrum.*

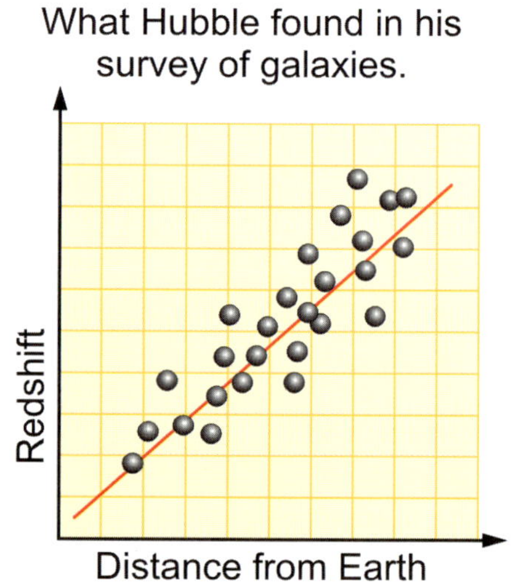

Figure 17.9: *Hubble discovered that the redshift (speed) of galaxies increased proportionally to their distance.*

The Big Bang theory

The Big Bang theory

The **Big Bang** theory says the universe began as a huge explosion between 10 billion and 20 billion years ago. According to this theory, all matter and energy started in a space smaller than the nucleus of an atom. Suddenly, (no one knows why) a huge explosion occurred that sent everything that makes up the universe spraying out in all directions. In its first moments of existence, the universe was an extremely hot fireball of pure energy that began to expand rapidly.

Protons and neutrons form at 4 minutes

As the universe expanded, it cooled down as its energy spread out over a larger volume. About four minutes after the Big Bang, the universe had cooled enough that protons and neutrons could stick together to form the nuclei of atoms. Because atoms were still flying around with high energy, heavy nuclei were smashed apart. Only one helium atom survived for every 12 hydrogen atoms. Almost no elements heavier than helium survived. When we look at the matter in the universe today, we see this ratio of hydrogen to helium left by the Big Bang, with the exception of elements formed later in stars.

Matter and light decouple in 700,000 years

For the next 700,000 years the universe was like the inside of a star: hot ionized hydrogen and helium. At the age of 700,000 years, the universe had expanded enough to become transparent to light. At this point, the light from the fireball was freed from constant interaction with hot matter. The light continued to expand separately from matter and became the cosmic background radiation we see today.

Stars and galaxies form

When the universe was about 1 billion years old, it had expanded and cooled enough that galaxies and stars could form. At this point the universe probably began to look similar to how it looks today. The sun and solar system formed about 4 billion years ago, by which time the universe was 12 billion years old.

Unresolved questions

While scientists feel relatively confident about the overall picture, they are not confident about the details. For example, recent observations suggest the expansion of the universe is accelerating. This is a puzzle because, if anything, gravity should be slowing the expansion down.

VOCABULARY

Big Bang - theory that the universe began as a huge explosion 10 billion to 20 billion years ago.

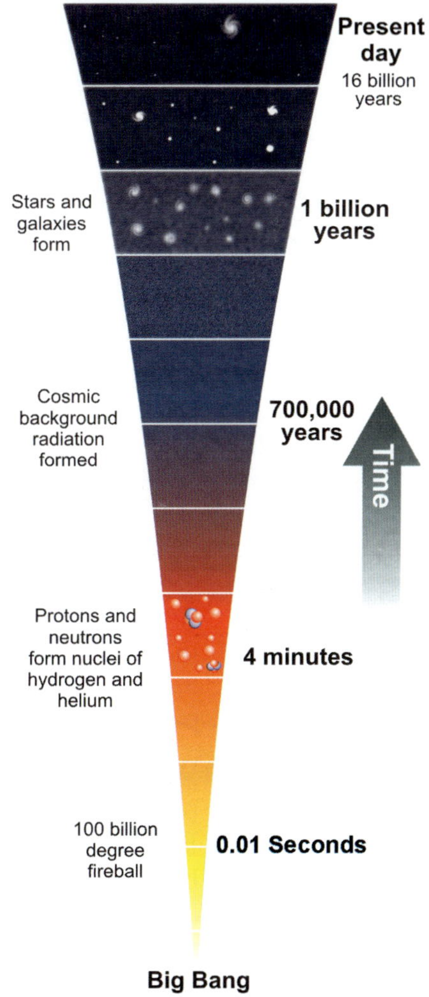

Figure 17.10: *Big Bang timeline.*

Evidence for the Big Bang theory

Evidence for the Big Bang

When it was first introduced, not everyone believed the Big Bang. In fact, the name "Big Bang" was made up by scientists to mock the theory. Unfortunately for them, the name stuck. As with any new theory, the Big Bang became more accepted as new scientific tools and discoveries established more evidence.

The fact that galaxies are expanding away from each other is a strong argument for the Big Bang. As far as we can look into the universe, we find galaxies are expanding away from each other. On average, we do not see galaxies coming toward each other (Figure 17.11).

Figure 17.11: *The observed expansion of the universe is strong evidence for the Big Bang theory.*

Microwave background radiation

In the 1960s, Arno Penzias and Robert Wilson, two American astrophysicists, were trying to measure electromagnetic waves given off by the Milky Way. No matter how they refined their technique, they kept detecting a background noise that interfered with their observations. This noise seemed to be coming from all directions and had little variation in frequency.

When you light a match, the flame bursts rapidly from the first spark and then cools as it expands. When the Big Bang exploded, it also created hot radiation. This radiation has been expanding and cooling for 16 billion years. The radiation is now at a temperature only 2.7°C above absolute zero and it fills the universe. The *cosmic background radiation* is the "smoke" from the Big Bang that fills the room (that is, the universe), even 16 billion years later. The "noise" that Penzias and Wilson found was the cosmic microwave background radiation predicted by the Big Bang theory (Figure 17.12).

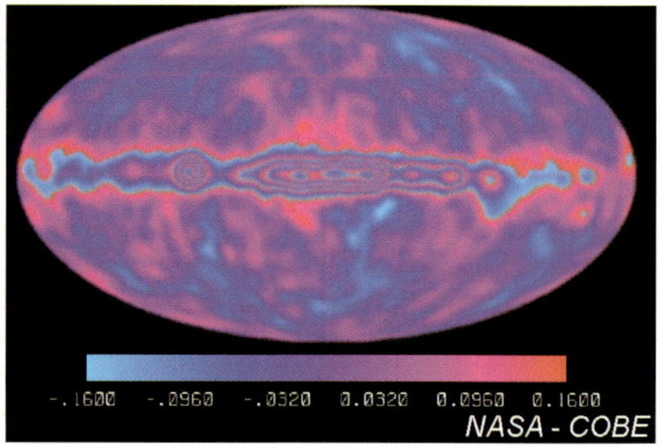

Figure 17.12: *The COBE satellite measured this image of the cosmic background radiation. (NASA)*

Ratios of the elements

We have other evidence that supports the Big Bang theory. The proportion of hydrogen to helium is consistent with the physics of the Big Bang (Figure 17.13). Elements heavier than hydrogen and helium are formed in stars. When stars reach the end of their life cycles, they spread heavy elements such as carbon, oxygen, and iron out into the universe. If the universe were significantly older, there would be more heavy elements present compared with hydrogen and helium.

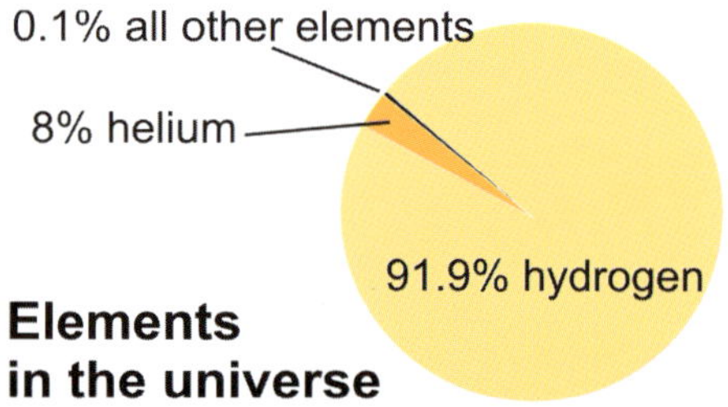

Figure 17.13: *The universe is mostly hydrogen with a small amount of helium and tiny amounts of other elements.*

17.1 Section Review

1. Which is bigger, a supergiant star or a galaxy?
2. In which galaxy do we live?
3. The number of stars in our home galaxy is closest to:
   a. 200.
   b. 200,000.
   c. 200,000,000.
   d. 200 billion.
4. What two important discoveries are credited to astronomer Edwin Hubble?
5. List four galaxy shapes.
6. The distances between galaxies are in the range of:
   a. 100 kilometers.
   b. 100 light years.
   c. 1 million light years.
   d. 1 billion light years.
7. How do astronomers estimate the distance between galaxies?
8. What do scientists believe is at the center of the Milky Way galaxy?
9. Why does a black hole appear black?
10. How did astronomers discover that the star Sirius was moving away from Earth?
11. What did Hubble discover about the relationship between a galaxy's location and speed?
12. According to the Big Bang theory, how large was the universe before it exploded and expanded in all directions?
13. How many years did it take for stars to begin to form after the Big Bang?
14. What evidence for the Big Bang did the scientists Arno Penzias and Robert Wilson discover?

CHALLENGE

When reading about black holes, you learned that Earth would have to be squeezed into the size of a marble one centimeter in diameter to make its gravity strong enough to be a black hole.

Use what you learned about density in Chapter 4 to calculate how dense Earth would be if it were compressed to this size. Compare your calculated density to that of common materials such as lead and steel..

A Spectacular Show of Lights

Do you think of weather as an Earth-only phenomenon? Then it will broaden your horizons to know there is weather in outer space, and, just like on Earth, conditions there change constantly.

Weather on our planet is affected by factors such as temperature, wind, precipitation, and by the sun's heat and light. In space, the sun drives the weather. Explosions on the sun's surface can cause radiation storms, changes in magnetic fields, and movement of energetic particles. The fast-moving solar wind carries charged particles away from the sun. Some of these particles and energy are able to penetrate Earth's upper atmosphere, and when this happens, an aurora is created - a spectacular show of lights.

Photo - collection of Dr. Herbert Kroehl, NGDC, NOAA

Charged particles and auroras

Our planet is a magnetic object. Magnetic field lines travel between the north and south poles. Charged electron particles trapped in these field lines create what is known as Earth's magnetic field (also called the magnetosphere).

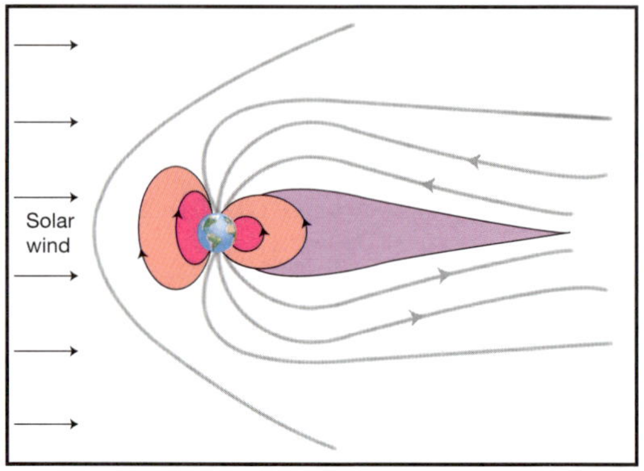

This magnetosphere protects Earth from solar wind that contains energetic particles and radiation. An aurora is caused by energy from the sun. Charged particles from the sun travel down Earth's magnetic field lines forming oval shapes. These so-called auroral ovals cover each magnetic pole. When electrons hit Earth's upper atmosphere, light particles are released. These colliding solar particles and atmospheric gases create aurora lights.

Generally, aurorae occur at higher latitudes, closer to the magnetic poles. Auroral ovals are usually located at 60° and 70° latitude, north and south. At 45°, aurorae are seen about fives times per year. Above 55°, aurorae are visible almost every night if viewing conditions are right. In November 2004, auroral ovals could be seen as far south as Arizona, whose latitude is 31° to 37°.

Dazzling details

An aurora looks like streaks of colored light in the dark night sky. Commonly seen aurora colors include green, yellow, and red. The gases present in the atmosphere determine the color. The electromagnetic radiation in an aurora also includes infrared and ultraviolet, neither of which the human eye can see.

The aurora in the Northern Hemisphere is called aurora borealis or the northern lights, and is popularly viewed from Alaska, Canada, Iceland, Scandinavia, and Russia. In the Southern Hemisphere, the southern lights, or aurora australis, tend to occur in remote, barely populated areas.

This photo of aurora australis was taken at the Antarctica South Pole Weather Station in 1979.

Clear, dark nights are best for seeing an aurora. Cloud cover, moonlight, and city lights can interfere with viewing.

Space Environment Center

The National Oceanic & Atmospheric Administration (NOAA) and the Air Force operate the Space Environment Center (SEC) in Boulder, Colo. The center provides real-time information about solar events and space weather that may affect people and equipment. It issues alerts and warnings to any organization that needs space weather information.

Space weather storms can damage satellites, interrupt cell phone systems, and disable electric power grids, which are networks of power lines and equipment that provide electrical power to large areas. In 1989, Montreal, Quebec, lost power for nine hours because of a large geomagnetic storm.

Space weather can also affect short wave radio systems and Global Positioning Systems (GPS). GPS is a satellite system that tracks the location of vehicles, planes, and boats. GPS relies on accurate and timely data that can be hampered severely by geomagnetic storms. NASA also requires timely information to make decisions about launching satellites and manned space vehicles.

NOAA's Polar Orbiting Environmental Satellites (POES) contain monitors and detectors that gather data as they travel over the poles. POES provides information about aurora size, intensity, and location. The SEC website supplies regularly updated auroral oval images and data.

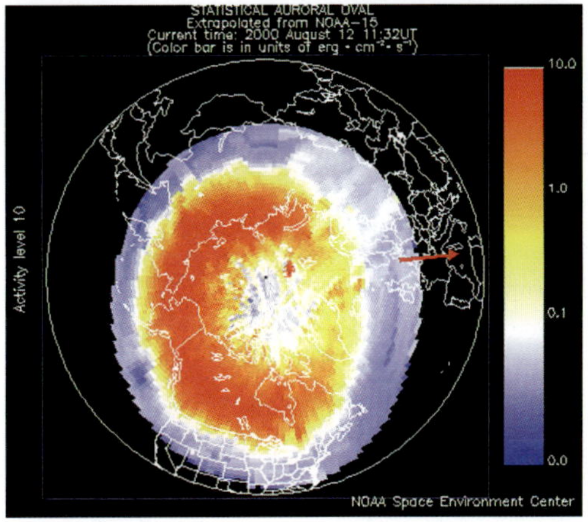

The SEC uses magnetometers to measure geomagnetic activity. Data is provided to create indices or guides to determine the likelihood of an aurora. The K index provides data on a scale from 0 to 9. The higher the value the more likely an aurora will occur. The Kp index (also on a 0-9 scale) is updated every three hours and collects data from 13 geomagnetic observatories. These values tell how far south an aurora can be seen past the magnetic north pole.

Maybe some night your location, viewing conditions, and the geomagnetic activity all will be right for you to witness a spectacular natural show of lights.

Questions:

1. What causes an aurora?
2. What is the SEC and what is its role?
3. What conditions are needed for viewing an aurora?

Chapter 17 Assessment

Vocabulary

Select the correct term to complete the sentences.

galaxy	Big Bang theory

1. Our solar system is located in the Milky Way ____.
2. The ____ explains how the universe began.

Concepts

1. Name and describe the four shapes used to classify galaxies.
2. To which galaxy does our sun belong and what shape does our galaxy take?
3. Describe the Milky Way galaxy. Refer to:
    a. the position of our sun,
    b. its dimensions in light years, and
    c. the location of a black hole.
4. What is the Big Bang theory?
5. Describe three pieces of evidence that support the Big Bang theory.
6. According to the Big Bang theory, how old is the universe?
7. As observed from Earth, the light from a distant star is shifted toward the red end of the visible spectrum. What would this indicate to an astronomer?
8. Everything that exists, including all matter and energy, is known as the:
    a. solar system.
    b. galaxy.
    c. universe.
9. In the 1920's, Edwin Hubble measured that the further away a galaxy was, the more redshift it had. What was the significance of this increased redshift?
10. It is believed that the universe is:
    a. expanding.
    b. shrinking.
    c. not changing.
11. Where and when was the helium found in the sun formed?
12. Where were the elements carbon, oxygen, and iron formed?
13. A black hole is:
    a. a hole in the universe that is black in color.
    b. the black regions of space you can see between stars.
    c. an object with such strong gravity that nothing can escape it.
14. Why can't light escape from a black hole?

Problems

1. Rank the following from smallest to largest in mass:
    a. Milky Way galaxy
    b. our sun
    c. the universe
    d. Earth
    e. the black hole at the center of the Milky Way Galaxy.
2. Rank the following from closest to farthest away from Earth:
    a. Andromeda galaxy
    b. our sun
    c. the moon
    d. the star, Sirius

Glossary

A glossary is an alphabetical list of important words found in the sections of this book. Use this glossary just as you would use a dictionary: to find out the meaning of unfamiliar words. This glossary gives the meaning that applies to the words as they are used in the sections of this book. As with any subject, science has its own vocabulary. The study of science is more meaningful if you know the language of science.

A

absolute zero – the lowest temperature there can be, equal to -273°C or -460°F.

acceleration – the rate of change of velocity.

acceleration due to gravity – the acceleration of an object in free fall, 9.8 m/s^2 on Earth.

accuracy – describes how close a measurement is to the true value.

acid – a substance that produces hydronium ions when dissolved in water. Acids have pH less than 7.

activation energy – energy needed to break chemical bonds in the reactants to start a reaction.

addition reaction – two or more substances chemically combine to form a new compound.

alkali metals – elements in the first group of the periodic table.

alloy – a solution of two or more solids.

amino acids – organic molecules that are the building blocks of proteins.

amorphous – solids that do not have a repeating pattern of molecules or atoms.

analysis – the process of evaluating data. Analysis may include thinking, creating graphs, doing calculations, and discussing ideas with others.

Archimedes' principle – states that the buoyant force is equal to the weight of the fluid displaced by an object.

asteroid – an object that orbits the Sun but is too small to be considered a planet.

astronomical unit (AU) – the distance between Earth and the Sun (1 AU).

atom – the smallest particle of matter that retains the identity of its element.

atomic mass – the average mass of all the known isotopes of an element, expressed in amu.

atomic mass unit (amu) – a unit of mass equal to 1.66×10^{-24} grams, which is one twelfth the mass of the isotope carbon-12.

atomic number – the number of protons in the nucleus. The atomic number determines what element the atom represents.

average – a mathematical process in which you add up all the values, then divide the result by the number of values.

average speed – the total distance divided by the total time for a trip.

B

balanced forces – result in a zero net force on an object.

base – a substance that produces hydroxide ions when dissolved in water. Bases have a pH greater than 7.

Big Bang – theory that the universe began as a huge explosion 10 billion to 20 billion years ago.

binary star – a system of two stars orbiting each other that are gravitationally tied together.

boiling point – the temperature at which a substance changes from a liquid to a gas.

brightness – measures the amount of light reaching Earth.

brittleness – the tendency to crack or break; the opposite of elasticity.

buoyancy – the measure of the upward force a fluid exerts on an object that is submerged.

C

calorie – a unit of energy equal to 4.184 joules or the energy needed to heat 1 gram of water by 1°C.

carbohydrates – energy-rich sugars and starches.

catalyst – a chemical that allows a reaction to have a much lower activation energy than it normally would have.

cellular respiration – the reactions in cells that release energy from glucose.

Celsius scale – temperature scale in which water freezes at 0 degrees and boils at 100 degrees.

charged – a condition where there is an excess of positive or negative charge.

chemical bond – a bond formed between atoms through the sharing or transferring of electrons.

chemical change – transforms one kind of matter into another kind, which may have different properties.

chemical equation – an equation of chemical formulas that shows the exact numbers of atoms and compounds in a chemical reaction.

chemical formula – identifies the number and element of each type of atom in a compound.

chemical properties – characteristics of matter that can only be observed when one substance changes into a different substance, such as iron into rust.

chemical reaction – a process that rearranges chemical bonds to create new substances.

comet – an object in space made mostly of ice and dust.

compound – a substance whose smallest particles include more than one element chemically bonded together.

compression –a squeezing force that can act on a spring.

concentration – the ratio of solute to solvent in a solution.

conclusion – a statement of what was learned in an experiment or observation.

condensation – change from gas to liquid at a temperature below the boiling point.

constant speed – speed of an object that travels the same distance each second.

control variable – a variable that is kept constant in an experiment.

convection – the transfer of heat through the motion of fluids such as air and water.

covalent bond – a type of chemical bond formed by shared electrons.

crystalline – solids that have an orderly, repeating pattern of molecules or atoms.

D

data – information collected during an experiment or other scientific inquiry. Data are often values of variables measured in an experiment.

deduce – to figure something out from known facts using logical thinking.

density – the mass of matter per unit volume.

dependent variable – in an experiment, a variable that responds to changes in the independent variable.

dissolve – to separate and disperse a solid into individual molecules or ions in the presence of a solvent.

distance – the amount of space between two points.

ductility – the ability to bend without breaking.

E

elasticity – the ability to be stretched or compressed and then return to original size.

electric charge – a fundamental property of matter that comes in two types, called positive and negative.

electrical conductor – a material that allows electricity to flow through easily.

electron – a particle with an electrical charge (-e) found inside of atoms but outside the nucleus.

element – a pure substance that cannot be broken down into other elements.

elementary charge – the smallest unit of electric charge that is possible in ordinary matter; represented by the lowercase letter *e*.

endothermic – a reaction is endothermic if it uses more energy than it releases.

energy – a quantity that measures the ability to cause change in the physical system. Energy is measured in joules.

enzymes – special proteins that are catalysts for chemical reactions in living things.

equilibrium – occurs when a solution has the maximum concentration of dissolved solute; the dissolving rate equals the rate at which molecules come out of solution (un-dissolve).

equilibrium – state in which the net force on an object is zero.

error – the difference between a measurement and the true value.

evaporation – change from liquid to gas at a temperature below the boiling point.

exothermic – a reaction is exothermic if it releases more energy than it uses.

experiment – a situation specifically set up to investigate relationships between variables.

experimental variable – a variable that changes in an experiment.

F

Fahrenheit scale – temperature scale in which water freezes at 32 degrees and boils at 212 degrees.

fats – energy-rich hydrocarbon chain molecules.

fluid – a form of matter that flows when any force is applied, no matter how small. Liquids and gases are fluids.

force – a push, pull, or any action which has the ability to change motion. Force is measured in newtons.

free fall – the motion of an object acted on only by the force of gravity.

free-body diagram – a diagram showing all the forces acting on an object.

friction – a force that resists the motion of objects or surfaces.

G

galaxy – a group of stars, dust, gas, and other objects held together by gravitational forces.

gas planets – Jupiter, Saturn, Uranus, and Neptune.

gram (g) – a unit of mass smaller than a kilogram. One kg equals 1,000 g.

graph – a mathematical diagram showing one variable on the vertical (y) axis and the second variable on the horizontal (x) axis.

group – a column of the periodic table is called a group.

H

halogens – elements in the group containing fluorine, chlorine, and bromine, among others.

heat – thermal energy that is moving.

heat conduction – the transfer of heat by the direct contact of particles of matter.

hypothesis – an unproven or preliminary explanation that can be tested by comparison with scientific evidence. Early hypotheses are rarely correct and are often modified as new evidence becomes available.

I

independent variable – in an experiment, a variable that is changed by the experimenter and/or causes changes in the dependent variable.

inertia – the property of an object that resists changes in its motion.

inquiry – a process of learning that starts with questions and proceeds by seeking the answers to the questions.

insoluble – a substance is insoluble in a particular solvent if it does not dissolve in that solvent.

insulator – a material that slows down or stops the flow of either heat or electricity.

intermolecular forces – forces between separate atoms and molecules that are attractive at a distance but repulsive at close range.

ion – an atom that has an electric charge different from zero. Ions are created when atoms gain or lose electrons.

ionic bond – a bond that transfers an electron from one atom to another resulting in attraction between oppositely charged ions.

isotopes – atoms of the same element that have different numbers of neutrons in the nucleus.

J

joule (J) – the unit of energy in the metric (SI) system. One joule is enough energy to push with a force of one newton for a distance of one meter.

K

kilogram (kg) – the basic metric (SI) unit of mass.

L

light year – the distance light travels in one year (9.5 trillion km).

luminosity – the total amount of light given off by a star.

M

magnitude – describes the size component of a vector.

main sequence star – a stable star in the main category in the H-R diagram.

mass – the amount of matter an object contains.

mass number – the number of protons plus the number of neutrons in the nucleus.

matter – everything that has mass and takes up space.

melting point – the temperature at which a substance changes from a solid to a liquid.

metal – elements that are typically shiny and good conductors of heat and electricity.

meteor – a chunk of burning rock traveling through Earth's atmosphere.

meteorite – a meteor that passes through Earth's atmosphere and strikes the ground.

mixture – a substance that includes more than one type of element and/or compound.

molecule – a group of atoms held together by covalent bonds in a specific ratio and shape.

mutation – change in the sequence of base pairs in DNA that may be passed on to successive generations.

N

natural law – the set of rules that are obeyed by every detail of everything that occurs in the universe, including living creatures and human technology.

nebula – a huge cloud of dust and gas from which stars form.

net force – the sum of two or more forces on an object.

neutral – a condition where the total positive charge is canceled by the total negative charge. Matter is neutral most of the time.

neutron – a particle with zero charge found in the nucleus of atoms.

newton (N) – the unit of force in the metric (SI) system.

Newton's first law – an object at rest will stay at rest and an object in motion will stay in motion with the same velocity unless acted on by an unbalanced force.

Newton's law of universal gravitation – the force of gravity between objects depends on their masses and the distance between them.

Newton's second law – acceleration is force ÷ mass.

Newton's third law – for every action force, there is a reaction force equal in strength and opposite in direction.

noble gases – elements in the group containing helium, neon, and argon, among others.

nonmetal – elements that are poor conductors of heat and electricity.

normal force – the force a surface exerts on an object that is pressing on it.

nuclear reaction – a process that changes the nucleus of an atom and may turn one element into a completely different element.

nucleic acids – biological molecules such as DNA that have the ability to store the genetic code.

nucleus – the tiny core at the center of an atom containing most of the atom's mass and all of its positive charge.

O

objective – describes evidence that documents only what actually happened as exactly as possible.

orbit – the motion of one object around another caused by gravitational force.

organic chemistry – the chemistry of carbon and carbon compounds.

origin – a fixed reference point.

oxidation number – indicates the charge of an atom when an electron is lost, gained, or shared in a chemical bond. An oxidation number of +1 means an electron is lost, -1 means an electron is gained.

P

period – a row of the periodic table is called a period.

periodic table – a chart that organizes the elements by their chemical properties and increasing atomic number.

periodicity – the repeating pattern of chemical and physical properties of the elements.

pH – pH measures the acidity of a solution.

pH scale – the pH scale goes from 1 to 14 with 1 being very acidic and 14 being very basic. Pure water is neutral with a pH of 7.

phases of matter – the different forms matter can take; commonly occur as solid, liquid, or gas.

physical properties – characteristics of matter that can be seen through direct observation such as density, melting point, and boiling point.

planet – a massive object orbiting a star, like the Sun. A true planet has cleared the neighborhood around its orbit and has enough mass so that its gravity forms it into a spherical shape.

planetary nebula – the expanding outer shell of a sunlike star. This matter is blown away as the core shrinks to become a white dwarf.

planetary satellite – small body of matter that orbits a planet.

planetary system – a star and its planets.

polar – describes a molecule that has charge separation, like water.

polymer – material in which individual molecules are made of long chains of repeating units.

polymerization – a series of addition reactions that join small molecules into large chain molecules.

position – the location of an object compared to a reference point.

precipitate – a solid product that comes out of solution in a chemical reaction.

pressure – a distributed force per unit area that acts within a fluid.

procedure – a description of an experiment that details the equipment used, the techniques used, and the data collected.

products – the new substances which result from a chemical reaction.

protein synthesis – using the information in DNA to assemble proteins from amino acids.

proteins – large molecules found in animal and plant tissue.

proton – a particle with an electric charge (+e) found in the nucleus of atoms.

protostar – the first stage in the life cycle of a star.

R

radioactive – a nucleus is radioactive if it spontaneously breaks up, emitting particles or energy in the process.

reactants – the substances which are combined and changed in the chemical reaction.

red giant – a large star with low temperature and high brightness.

repeatable – describes evidence that can be seen independently by others if they repeat the same experiment or observation in the same way.

S

saturated – a solution is saturated if it contains as much solute as the solvent can dissolve.

saturated fat – a fat molecule in which each carbon is bonded with two hydrogen atoms.

scientific method – a process of learning that begins with a hypothesis and proceeds to prove or change the hypothesis by comparing it with scientific evidence.

significant – a difference between two measured results is significant if the difference is greater than the error in measurement.

sliding friction – the friction force that resists the motion of an object moving across a surface.

slope – the ratio of the rise (vertical change) to the run (horizontal change) of a line on a graph.

solar system – the sun, planets, and their moons, and other objects that are gravitationally bound to the sun.

solid – a phase of matter with a definite shape and constant volume.

solubility – the amount of solute that can be dissolved under certain conditions.

solute – any component of a solution other than the solvent.

solution – a mixture of two or more substances that is uniform at the molecular level.

solvent – the component of a solution that is present n the greatest amount.

specific heat – a material property that tells how much energy is needed to change the temperature by one degree.

spectral line – a bright colored line in a spectroscope.

spectroscope – an instrument that separates light into a spectrum.

spectroscopy – a method of studying an object by examining the visible light and other electromagnetic waves it creates.

spectrum – the characteristic colors of light given off or absorbed by an element.

speed – the distance an object travels divided by the time it takes.

stable – a nucleus is stable if it stays together.

static electricity – the buildup of either positive or negative charge; made up of isolated, motionless charges.

static friction – the friction force that resists the motion between two surfaces that are not moving.

steel – an alloy of iron and carbon.

strength – the ability to maintain shape under the application of forces.

supergiant – very large, bright star that may be blue or red, depending on its temperature.

supernova – the explosion of a very large star.

supersaturated – a concentration greater than the maximum solubility.

system – a small group of related things that work together.

T

temperature – a measurement of hot or cold that depends on the thermal energy in a material.

tension – a pulling force that acts in a rope, string, or other object.

terrestrial planets – Mercury, Venus, Earth, and Mars.

theory – a scientific explanation supported by much evidence collected over a long period of time.

thermal conductor – a material that allows heat to flow easily.

thermal energy – energy that is due to difference in temperature. Thermal energy comes from kinetic energy of individual atoms.

thermal equilibrium – a condition where temperatures are the same and no heat flows.

thermometer – an instrument used to measure temperature.

U

unbalanced forces – result in a net force on an object that can cause changes in motion.

unsaturated fat – a fat molecule that has less hydrogen atoms than a saturated fat.

V

valence electrons – electrons in the highest unfilled energy level of an atom. These electrons participate in chemical bonds.

value – the particular number (with units) or choice that a variable may have.

variable – a quantity that can be precisely specified, often with a numerical value.

vector – a quantity that includes both a magnitude (size) and a direction.

velocity – speed with direction.

W

weight – the downward force of gravity acting on mass.

white dwarf – a small star with a high temperature and low brightness.

Index

The index gives the page numbers where you can find a word, definition, or information about a topic. You can use the index when you are studying and need to find information quickly. The index is a good place to look up a vocabulary word to get more information about the meaning of a word.

C

D

G

H

I

J

K

L

M

N

O

P

Q

R

S

T

These pages provide you with the standards that are taught in this book and are required learning for the state of California.

STANDARD SET 1. Motion		Completed
1. The velocity of an object is the rate of change of its position. As a basis for understanding this concept:	1.a. Students know position is defined in relation to some choice of a standard reference point and a set of reference directions.	☐
	1.b. Students know that average speed is the total distance traveled divided by the total time elapsed and that the speed of an object along the path traveled can vary.	☐
	1.c. Students know how to solve problems involving distance, time, and average speed.	☐
	1.d. Students know the velocity of an object must be described by specifying both the direction and the speed of the object.	☐
	1.e. Students know changes in velocity may be due to changes in speed, direction, or both.	☐
	1.f. Students know how to interpret graphs of position versus time and graphs of speed versus time for motion in a single direction.	☐

STANDARD SET 2. Forces		
2. Unbalanced forces cause changes in velocity. As a basis for understanding this concept:	2.a. Students know a force has both direction and magnitude.	☐
	2.b. Students know when an object is subject to two or more forces at once, the result is the cumulative effect of all the forces.	☐
	2.c. Students know when the forces on an object are balanced, the motion of the object does not change.	☐
	2.d. Students know how to identify separately the two or more forces that are acting on a single static object, including gravity, elastic forces due to tension or compression in matter, and friction.	☐
	2.e. Students know that when the forces on an object are unbalanced, the object will change its velocity (that is, it will speed up, slow down, or change direction).	☐
	2.f. Students know the greater the mass of an object, the more force is needed to achieve the same rate of change in motion.	☐
	2.g. Students know the role of gravity in forming and maintaining the shapes of planets, stars, and the solar system.	☐

STANDARD SET 3. Structure of Matter		Completed
3. Each of the more than 100 elements of matter has distinct properties and a distinct atomic structure. All forms of matter are composed of one or more of the elements. As a basis for understanding this concept:	3.a. Students know the structure of the atom and know it is composed of protons, neutrons, and electrons.	☐
	3.b. Students know that compounds are formed by combining two or more different elements and that compounds have properties that are different from their constituent elements.	☐
	3.c. Students know atoms and molecules form solids by building up repeating patterns, such as the crystal structure of NaCl or long-chain polymers.	☐
	3.d. Students know the states of matter (solid, liquid, gas) depend on molecular motion.	☐
	3.e. Students know that in solids the atoms are closely locked in position and can only vibrate; in liquids the atoms and molecules are more loosely connected and can collide with and move past one another; and in gases the atoms and molecules are free to move independently, colliding frequently.	☐
	3.f. Students know how to use the periodic table to identify elements in simple compounds.	☐

STANDARD SET 4. Earth in the Solar System (Earth Sciences)		
4. The structure and composition of the universe can be learned from studying stars and galaxies and their evolution. As a basis for understanding this concept:	4.a. Students know galaxies are clusters of billions of stars and may have different shapes.	☐
	4.b. Students know that the Sun is one of many stars in the Milky Way galaxy and that stars may differ in size, temperature, and color.	☐
	4.c. Students know how to use astronomical units and light years as measures of distance between the Sun, stars, and Earth.	☐
	4.d. Students know that stars are the source of light for all bright objects in outer space and that the Moon and planets shine by reflected sunlight, not by their own light.	☐
	4.e. Students know the appearance, general composition, relative position and size, and motion of objects in the solar system, including planets, planetary satellites, comets, and asteroids.	☐

STANDARD SET 5. Reactions		Completed
5. Chemical reactions are processes in which atoms are rearranged into different combinations of molecules. As a basis for understanding this concept:	5.a. Students know reactant atoms and molecules interact to form products with different chemical properties.	☐
	5.b. Students know the idea of atoms explains the conservation of matter: In chemical reactions the number of atoms stays the same no matter how they are arranged, so their total mass stays the same.	☐
	5.c. Students know chemical reactions usually liberate heat or absorb heat.	☐
	5.d. Students know physical processes include freezing and boiling, in which a material changes form with no chemical reaction.	☐
	5.e. Students know how to determine whether a solution is acidic, basic, or neutral.	☐

STANDARD SET 6. Chemistry of Living Systems (Life Sciences)		
6. Principles of chemistry underlie the functioning of biological systems. As a basis for understanding this concept:	6.a. Students know that carbon, because of its ability to combine in many ways with itself and other elements, has a central role in the chemistry of living organisms.	☐
	6.b. Students know that living organisms are made of molecules consisting largely of carbon, hydrogen, nitrogen, oxygen, phosphorus, and sulfur.	☐
	6.c. Students know that living organisms have many different kinds of molecules, including small ones, such as water and salt, and very large ones, such as carbohydrates, fats, proteins, and DNA.	☐

STANDARD SET 7. Periodic Table		
7. The organization of the periodic table is based on the properties of the elements and reflects the structure of atoms. As a basis for understanding this concept:	7.a. Students know how to identify regions corresponding to metals, non-metals, and inert gases.	☐
	7.b. Students know each element has a specific number of protons in the nucleus (the atomic number) and each isotope of the element has a different but specific number of neutrons in the nucleus.	☐
	7.c. Students know substances can be classified by their properties, including their melting temperature, density, hardness, and thermal and electrical conductivity.	☐

STANDARD SET 8. Density and Buoyancy		Completed
8. All objects experience a buoyant force when immersed in a fluid. As a basis for understanding this concept:	8.a. Students know density is mass per unit volume.	☐
	8.b. Students know how to calculate the density of substances (regular and irregular solids and liquids) from measurements of mass and volume.	☐
	8.c. Students know the buoyant force on an object in a fluid is an upward force equal to the weight of the fluid the object has displaced.	☐
	8.d. Students know how to predict whether an object will float or sink.	☐

STANDARD SET 9. Investigation and Experimentation		
9. Scientific progress is made by asking meaningful questions and conducting careful investigations. As a basis for understanding this concept and addressing the content in the other three strands, students should develop their own questions and perform investigations. Students will:	9.a. Plan and conduct a scientific investigation to test a hypothesis.	☐
	9.b. Evaluate the accuracy and reproducibility of data.	☐
	9.c. Distinguish between variable and controlled parameters in a test.	☐
	9.d. Recognize the slope of the linear graph as the constant in the relationship $y = kx$ and apply this principle in interpreting graphs constructed from data.	☐
	9.e. Construct appropriate graphs from data and develop quantitative statements about the relationships between variables.	☐
	9.f. Apply simple mathematical relationships to determine a missing quantity in a mathematic expression, given the two remaining terms (including speed = distance/time, density = mass/volume, force = pressure × area, volume = area × height).	☐
	9.g. Distinguish between linear and nonlinear relationships on a graph of data.	☐